MICROSCOPY AND ANALYSIS

Edited by **Stefan G. Stanciu**

Microscopy and Analysis

http://dx.doi.org/10.5772/61531
Edited by Stefan G. Stanciu

Contributors

Frank Winterroth, Katsutoshi Miura, Andrea Falqui, Alberto Casu, Cristiano Di Benedetto, Alessandro Genovese, Sergio Lentijo Mozo, Elisa Sogne, Efisio Zuddas, Francesca Pagliari, John Miller, Paola Pesantez Cabrera, Marianne Sowa, Wei Song, Naoki Fukutake, Hisashi Miyafuji, Toru Kanbayashi, Juan M. Bueno, Francisco J. Avila, Pablo Artal, Kenji Mizoguchi, Hirokazu Sakamoto, Eiichi Mori, Hiroyuki Tahara, Toshio Naito, Taka-Aki Hiramatsu, Hideki Yamochi, Hideyuki Arimoto, Keiichiro Namai, Javier Adur, Hernandes Carvalho, Carlos Cesar, Víctor Casco, Reema Adel Khorshed, Cristina Lo Celso, Marco Wiltgen, Marcus Bloice, Leonid Bolotov, Toshihiko Kanayama, Gitanjal Deka, Fu-Jen Kao, Shi-Wei Chu, Fan Wu, Nan Yao, Alexander Chaika

CBS, Edition **2017**

Published by InTech

Janeza Trdine 9, 51000 Rijeka, Croatia .

Notice

Statements and opinions expressed in the chapters are these of the individual contributors and not necessarily those of the editors or publisher. No responsibility is accepted for the accuracy of information contained in the published chapters. The publisher assumes no responsibility for any damage or injury to persons or property arising out of the use of any materials, instructions, methods or ideas contained in the book.

Publishing Process Manager Dajana Pemac

Technical Editor SPi Global

Cover Designer InTech Design Team

Additional hard copies can be obtained from orders@intechopen.com

Microscopy and Analysis , Edited by Stefan G. Stanciu
p. cm.
Print ISBN 978-953-51-2578-5
Online ISBN 978-953-51-2579-2

Contents

Preface

The word "microscopy" has its origins in the Greek language where the term "mikro" means "small" or "little", while "skopein" means "to look at,", so literally "microscope" means *an instrument for looking at small things*. So why the interest in such instruments? That's because *humans* have a deeply *curious* nature, and an infinity of questions that cross their minds cannot be answered without looking at small objects and scenes that are impossible to be seen with the naked eye. Although the underlying mechanisms of curiosity are not yet well understood, no one can argue that curiosity is equivalent to the desire to learn and to gain knowledge. This desire made Zacharias Jansen and his father, Hans, two Dutch spectacle makers, to be thrilled to see in the 1590s that by placing several lenses in a tube, an enlarged view of an object near its end could be observed, much larger than it was possible with any simple magnifying glass. Not much later, in the seventeenth century, the Englishman Robert Hooke was constructing the first microscope and using it to see and describe biological organisms. His book *Micrographia* , published in 1665, was the first illustrated volume on microscopy containing detailed accounts of 60 "observations" of objects examined microscopically. *Micrographia's* illustrations of a world were not accessible to the human eyes before created sensation, taking its readers by storm. For example, Samuel Pepys, a Fellow of the Royal Society, was writing in his diary: "Before I went to bed, I sat up till 2'o clock in my chamber, reading of Mr. Hookes Microscopical Observations, the most ingenious book that I ever read in my life." It is in *Micrographia* where the term "cell" was first introduced, a term that is currently used to describe the basic structural and functional biological unit of living organisms, the smallest unit of life that can replicate independently. Looking at thin cork slices, Hooke remarked structures resembling to pores, which to him looked similar to the small room in which a monk inhabited, the cellula, thus deriving the name: ". . . I could exceedingly plainly perceive it to be all perforated and porous, much like a Honey-comb, but that the pores of it were not regular. . . . these pores, or cells, . . . were indeed the first microscopical pores I ever saw, and perhaps, that were ever seen, for I had not met with any Writer or Person, that had made any mention of them before this. . .." These structures that puzzled Hooke were the dead cells of cork; although Hooke himself was going to observe as well living cells later, it was someone else who was to witness for the first time a live cell under a microscope, Antonie van Leeuwenhoek. The Royal Society in London was releasing a first letter from this self-educated Dutch scientific explorer 8 years later after *Micrographia* had been published. This letter, entitled "A specimen of some Observations made by a Microscope", contrived by M. Leewenhoeck in Holland, lately communicated by Dr. Regnerus de Graaf", presented microscopic observations on mold, bees, and lice. Further letters followed in which he provided his findings on different subjects using microscopes developed by him, the details of most of which he refused to reveal, preferring instead to provide his interpretations of the imaged scenes. Many of his letters dealt with the description of specific forms of microorganisms, which he referred to as "animalcules." These included protozoa and other unicellular organisms, like bacteria. Some of Leeuwenhoek's initial findings were met with both skepticism and open ridicule, but this was until Hooke was to return to his microscopes, which he had given up because of eye strain, and verify Leeuwenhoek's observations and confirm his findings. Leeuwenhoek was also the first to find and describe in his letters the sperm cells of animals and humans and to see that the fertilization process requires the sperm cell to enter the egg cell, which put an end to previous theories of spontaneous generation that revolved around the idea that certain forms of life such as fleas could arise from inanimate matter. Antonie van Leeuwenhoek was elected to the Royal Society in February 1680, and although he considered this to be a high honor, he did not attend the induction ceremony in London and never attended the meetings of the Royal Society. By the time of his death in 1723, Leeuwenhoek had written more than 550 letters to different scientific institutions, of which around 200 letters had been published by the Royal Society. Ever since those times, microscopes represent tools of the utmost importance for a wide range of disciplines. Without them, it would have been impossible to stand where we stand today in terms of understanding the structure and functions of organelles and cells, tissue composition and metabolism, or the causes behind various pathologies and their progression. Our knowledge on basic and advanced materials is also intimately intertwined to the realm of microscopy, and progress in key fields of micro- and nanotechnologies critically depends on high-resolution imaging systems. While Hooke and Leeuwenhoek were placing efforts on looking at small things with microscopes that relied on optical magnification, at this time a wide variety of imaging systems are available, relying on various contrast mechanisms. Light and optical magnification remain fundamental for the microscopy realm, but "looking" at small things is now possible also by using nanostructured probes that are scanned across a sample's surface to assess its topography or sense various other properties, by using beams of accelerated electrons to interact with a sample of interest and provide information on its structure, by exploiting sound interaction with matter, and by many other approaches. This volume includes 16 chapters that address highly significant scientific subjects from diverse areas of microscopy and analysis. Nine of these chapters deal with optical microscopy topics, while the remaining seven refer to nonoptical microscopy subjects. The authors present in this volume their work or review recent trends, concepts, and applications, in a manner that is accessible to quite a broad readership audience from both within and outside their specialist area. I am confident that this volume will be of great value not only to those actively involved in the addressed fields but also to those with passive but constant interest in these scientific areas and to those who will have their first encounter with microscopy and analysis when reading the contained chapters. In the end, I would like to express my deepest thanks to each of the authors for his or her fine contributions to this project.

Stefan G. Stanciu, PhD
Center for Microscopy-Microanalysis and Information Processing
University Politehnica of Bucharest
Romania

Optical Microscopy

Quantum Image-Forming Theory for Calculation of Resolution Limit in Laser Microscopy

Naoki Fukutake

Additional information is available at the end of the chapter

http://dx.doi.org/ 10.5772/63494

Abstract

Here we show what determines the optical resolution in laser microscopy. We define the expanded resolution limit (spatial frequency cutoff) that includes the classic Abbe definition as $2\,NA/\lambda$, where λ is the wavelength. The resolution limit can approximately be redefined as the frequency cutoff $\alpha NA/\lambda$, where α is the constant that depends on the optical process occurring in the sample. In the case of the optical process originating from the linear susceptibility $\chi^{(1)}$, the resolution limit is well known as the Abbe definition, namely, $\alpha = 2$. However, when other optical processes are harnessed to form the image through laser microscopy, the resolution limit can differ. We formulate a theoretical framework that can calculate the expanded resolution limits of all kinds of laser microscopy utilizing coherent, incoherent, linear, and nonlinear optical processes.

Keywords: image-forming theory, nonlinear optical microscopy, optical transfer function, optical resolution limit, light-matter interaction

1. Introduction

The resolution limit (spatial frequency cutoff) of optical microscopy is usually described as $2\,NA/\lambda$, where λ is the wavelength [1]. For example, bright field microscopy indicates the resolution limit of $2\,NA/\lambda$ at a maximum. However, this resolution limit is restricted to optical microscopy that utilizes the optical process derived from the linear susceptibility $\chi^{(1)}$, such as bright field microscopy. Since fluorescence is a $\chi^{(3)}$-derived optical process, the resolution limit of optical microscopy with fluorescence can differ from that of bright field microscopy. Although conventional fluorescence microscopy exhibits the resolution limit of $2\,NA/\lambda$, microscopy that achieves the full potential of fluorescence, such as structured illumination

microscopy (SIM), can reach 4 NA/λ [2]. In general, the resolution limit in optical microscopy becomes different according to the kind of optical process.

Laser microscopy is composed of an excitation optical system and a signal-collection optical system, and the signals are acquired point by point to reconstruct an image. One of the typical examples of laser microscopy is confocal fluorescence microscopy, which has been widely used as an optical imaging technique. Confocal fluorescence microscopy can acquire high-resolution optical images with depth sectioning by means of focused laser excitation and a pinhole in front of a detector, which eliminates out-of-focus signals. Confocal fluorescence microscopy harnesses fluorescence as an optical process to increase the optical resolution, compared with microscopy with a $\chi^{(1)}$-derived optical process. Confocal microscopy that utilizes a $\chi^{(1)}$-derived optical process has a resolution limit (frequency cutoff) of $NA_{ex}/\lambda_{ex} + NA_{col}/\lambda_{ex}$, where λ_{ex} is the wavelength of the excitation beam, NA_{ex} is the numerical aperture of the objective in the excitation system, and NA_{col} is the numerical aperture of the objective in the signal-collection system [3, 4], while confocal fluorescence microscopy theoretically indicates the resolution limit of $2NA_{ex}/\lambda_{ex} + 2\,NA_{col}/\lambda_{fl}$, where λ_{fl} is the wavelength of the fluorescence. Note that in conventional (wide field) fluorescence microscopy, since the entire specimen is excited evenly, which corresponds to the condition $NA_{ex} = 0$, the resolution limit becomes $2\,NA_{col}/\lambda_{fl}$.

In addition to confocal fluorescence microscopy, various laser microscopy techniques have recently been used to visualize biological specimens in three dimensions by harnessing many kinds of optical processes, such as two-photon excited fluorescence (TPEF), second-order harmonic generation (SHG), third-order harmonic generation (THG), coherent anti-Stokes Raman scattering (CARS), and stimulated Raman scattering (SRS) [5–11]. Depending on the optical process, each microscopy exhibits its own feature of image formation. In incoherent optical processes, such as fluorescence and TPEF, since the vacuum field is involved in the phenomena along with the excitation beam, the signals emitted from different molecules in the specimen do not interfere. In contrast, in coherent optical processes, such as $\chi^{(1)}$-derived phenomenon, SHG, THG, CARS, and SRS, because the processes are caused only by coherent excitation laser beams, the signals emitted from different molecules interfere. Although the coherence of the optical process influences the image-forming properties of laser microscopy, the basic concept is that the image of the linear or nonlinear susceptibility distribution $\chi^{(i)}$ (x, y, z) in the specimen is formed by microscopy regardless of coherence. From a perspective other than coherence, laser microscopy can be categorized into two types. In the first type, as the wavelength of the signal is different from that of the excitation beam, the signal can be separated from the excitation beam, resulting in the image being formed only by the signal. In the second type, since the signal has the same wavelength as the excitation beam, interference between the signal and the excitation beam is observed. It will be shown that the image-forming properties and resolution limits of both types can be dealt with in the identical framework.

Although the image-forming properties of each microscopy technique are well known, the unified theory does not exist that can deal with the image-forming properties and the resolution limits of all kinds of laser microscopy in the identical framework. If the unified image-forming theory is developed, it enables one to overview all microscopy techniques with any optical processes, such as linear, nonlinear, coherent, and incoherent processes, which can lead to the invention of a new microscopy technique. In this chapter, we formulate a unified framework that utilizes double-sided Feynman diagrams to discuss all microscopy applications by use of a unique technique. With our framework, the resolution limits of laser microscopy techniques will be able to be redefined with respect to each optical process. Moreover, we will lead to some important conclusions about laser microscopy. Although only laser microscopy is discussed here, our theory can be applied to any type of optical microscopy.

2. Theoretical framework

2.1. Model description

We begin by defining the imaging system (laser microscopy) in our model. Laser microscopy is composed of an excitation system to focus the laser beam onto a sample and a signal-collection system to gather the signal generated from the sample. A schematic of laser microscopy is shown in **Figure 1**, in which the coordinate systems are given. We assume in what follows that three-dimensional (3-D) sample-stage scanning is conducted instead of laser scanning, but it does not influence the optical resolution. In laser microscopy, usually one or two excitation beams are employed to generate the signal. The electric field of the signal is emitted from the molecule excited by the electric fields of excitation beams, and the signal field propagates through the signal-collection system. The signals are acquired point by point with a photodetector to reconstruct the 3-D image.

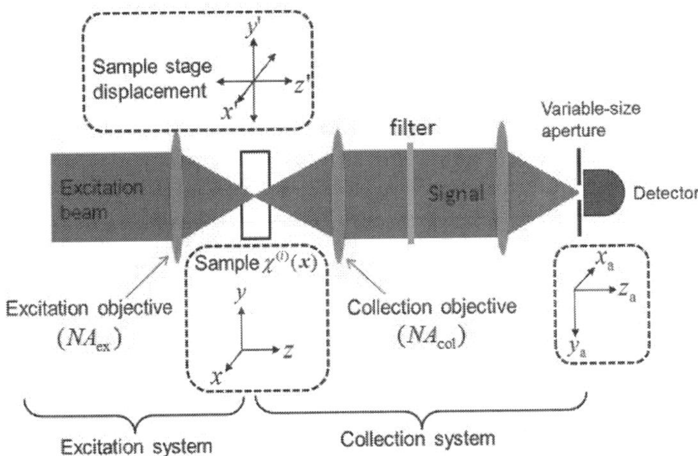

Figure 1. Schematic of laser microscopy (transmission type).

For simplicity, the first Born approximation is applied to understand the true nature of the optical resolution. In this approximation, the multiple scattering and depletion of the beam are neglected, which usually holds true for nearly transparent samples, such as a biological specimen. If the multiple scattering and depletion were intense, the image acquired would become deformed to some degree. We assume that both the excitation and signal-collection systems are 1× magnification systems with no aberration, which does not change the essence of their image-forming properties. In our model, the scalar diffraction theory is employed. The linear or nonlinear susceptibility distribution $\chi^{(i)}(x, y, z)$ in the sample plays a role as the object in the imaging system. The polarization $P(x, y, z)$ is induced by the excitation electric field, and the induced polarization emits the signal electric field. Hereafter, we express the electric field as a complex function.

2.2. Transmission linear confocal microscopy

We start with transmission linear confocal microscopy, in which a $\chi^{(1)}$-derived optical process occurs in the sample. **Figure 2** shows the double-sided Feynman diagram and the energy-level diagram describing the $\chi^{(1)}$-derived optical process. The polarization is induced by the excitation beam focused onto the sample, and the signal emitted by the polarization is gathered and delivered into the photodetector through the signal-collection system. We express the electric field distribution in the sample formed by the focused excitation beam as $E_{ex}(x, y, z)$. The Fourier transform of $E_{ex}(x, y, z)$ is shaped like a portion of a spherical shell, as shown in **Figure 3**, which represents the distribution of the wavenumber vector. When the sample-stage displacement (x', y', z') is zero, the polarization distribution becomes

$$P(x,y,z) = \chi^{(1)}(x,y,z)E_{ex}(x,y,z), \tag{1}$$

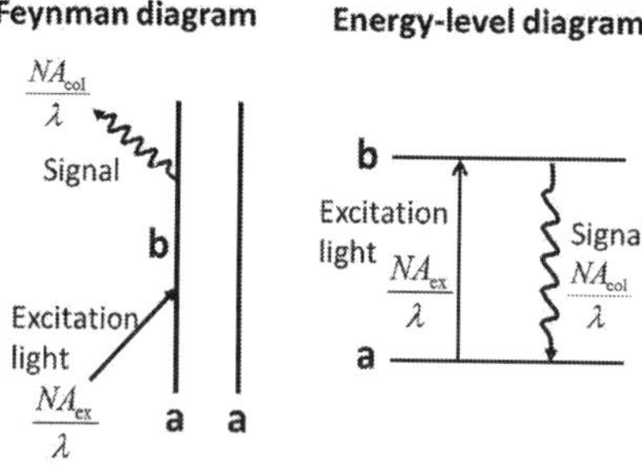

Figure 2. Double-sided Feynman diagram and energy-level diagram for the $\chi^{(1)}$-derived optical process. λ is the wavelength. The frequency cutoff (x-y direction) can be calculated by using the diagrams, which will be discussed in more detail in a later section.

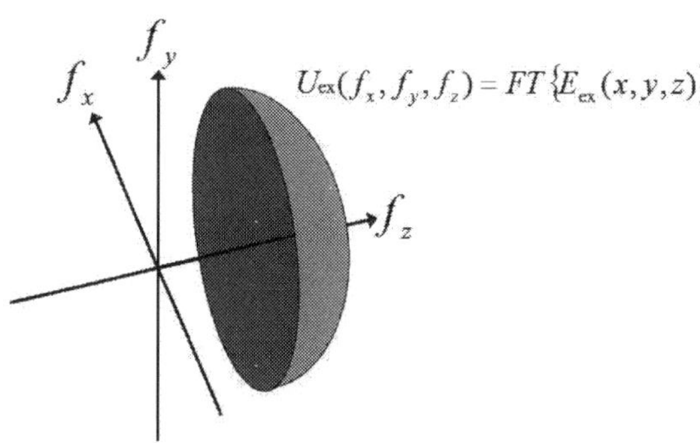

Figure 3. Distribution of the wavenumber vector for the beam focused by the objective. This is referred to as the "3-D pupil function".

where $\chi^{(1)}(x, y, z)$ is the linear susceptibility distribution in the sample and we presume that the electric permittivity ϵ_0 of free space is unity: $\epsilon_0 = 1$. We assume that the signal emitted from a single point in the sample located at the origin $((x, y, z) = (0, 0, 0))$ of the object coordinate forms the electric field distribution $E_{col}(x_a, y_a, z_a)$ in the detection space through the signal-collection system. The Fourier transform of $E_{col}(x_a, y_a, z_a)$ is also shaped like a portion of a spherical shell located on the $+ k_z$ side. Using $E_{col}(x_a, y_a, z_a)$ and $P(x, y, z)$, the electric field distribution $E_T(x_a, y_a, z_a)$ in the detection space in the case of arbitrary $\chi^{(1)}(x, y, z)$ is given by

$$E_T(x_a, y_a, z_a) = \iiint P(x, y, z) E_{col}(x_a - x, y_a - y, z_a - z)\, dx\, dy\, dz, \tag{2}$$

when $(x', y', z') = (0, 0, 0)$. Taking into account the sample-stage displacement (x', y', z'), we recast Eq. (2) as

$$E_T(x', y', z'; x_a, y_a, z_a)$$
$$= \iiint \chi^{(1)}(x + x', y + y', z + z') E_{ex}(x, y, z) E_{col}(x_a - x, y_a - y, z_a - z)\, dx\, dy\, dz. \tag{3}$$

In addition to the signal, we need to consider the electric field distribution formed in the detection space by the excitation beam itself through the excitation system and signal-collection system, which functions as the local oscillator.

For simplicity, we consider the case of the condition $NA_{ex} = NA_{col}$. The image intensity $I^t(x', y', z')$ acquired by our imaging system can be written as

$$I^t(x', y', z') = \iiint | \{-iE_{ex}(x_a, y_a, z_a) + E_T(x', y', z'; x_a, y_a, z_a)\}|^2 a(x_a, y_a)\delta(z_a)\, dx_a\, dy_a\, dz_a, \tag{4}$$

where $a(x_a, y_a)$ is the two-dimensional function representing the detector size, $\delta(z_a)$ stands for the Dirac delta function, the excitation beam $E_{ex}(x_a, y_a, z_a)$ acts as the local oscillator, and $-i$ before $E_{ex}(x_a, y_a, z_a)$ stems from the Gouy phase shift. To obtain the image intensity $I_c^t(x', y', z')$ for confocal microscopy, we can substitute $\delta(x_a)\,\delta(y_a)$ into $a(x_a, y_a)$ in Eq. (4). From Eqs. (3) and (4), we obtain

$$
\begin{aligned}
I_c^t&\left(x',y',z'\right)\\
&=\left|-iE_{ex}\left(0,0,0\right)+\iiint \chi^{(1)}\left(x+x',y+y',z+z'\right)E_{ex}\left(x,y,z\right)E_{col}\left(-x,-y,-z\right)dxdydz\right|^2\\
&\approx\left|E_{ex}\left(0,0,0\right)\right|^2\\
&\quad+i\left\{E_{ex}\left(0,0,0\right)\right\}^*\iiint \chi^{(1)}\left(x+x',y+y',z+z'\right)E_{ex}\left(x,y,z\right)E_{col}\left(-x,-y,-z\right)dxdydz\\
&\quad-iE_{ex}\left(0,0,0\right)\iiint\left\{\chi^{(1)}\left(x+x',y+y',z+z'\right)\right\}^*\left\{E_{ex}\left(x,y,z\right)\right\}^*\left\{E_{col}\left(-x,-y,-z\right)\right\}^* dxdydz.
\end{aligned}
\tag{5}
$$

Although $E_{ex}(x, y, z)$ and $E_{col}(x, y, z)$ are complex functions, $E_{ex}(x, y, z)\,E_{col}(-x, -y, -z) \equiv \mathrm{ASF}_t(x, y, z)$ approaches a real function under the condition $\mathrm{NA}_{ex} = \mathrm{NA}_{col}$. Regarding $\mathrm{ASF}_t(x, y, z)$ as the real function, Eq. (5) reduces to

$$
I_c^t(x',y',z') \approx \left|E_{ex}(0,0,0)\right|^2 - 2E_{ex}(0,0,0)\iiint \mathrm{Im}\{\chi^{(1)}(x+x',y+y',z+z')\}\mathrm{ASF}_t(x,y,z)dxdydz.
\tag{6}
$$

This equation shows that only absorbing objects can be observed and phase objects cannot be visualized. The first term in Eq. (6) leads to low-contrast images. The function $\mathrm{ASF}_t(x, y, z)$ is referred to as the amplitude spread function (ASF), and the coherent transfer function (CTF) is calculated by Fourier transforming the ASF.

2.3. Reflection linear confocal microscopy

Next, we deal with reflection linear confocal microscopy (see **Figure 4**), in which a $\chi^{(1)}$-derived optical process is harnessed. The excitation beam focused onto the sample by the excitation objective induces the polarization, and the signal generated from the polarization is gathered and delivered into a photodetector with the same objective. The excitation and signal-collection systems share a common objective. Unlike in transmission linear confocal microscopy, the excitation beam does not interfere with the signal. For reflection microscopy, the electric field distribution of the signal emitted from a single-point object in the sample $E_{col}(x_a, y_a, z_a)$ formed in the detection space through the signal-collection system is replaced by $E'_{col}(x_a, y_a, z_a)$, where the Fourier transform of $E'_{col}(x_a, y_a, z_a)$ is located on the $-k_z$ side (see **Figure 5**). With the arbitrary sample-stage displacement $(x', y'\, z')$, the electric field distribution $E_R(x', y'\, z'; x_a, y_a\, z_a)$ in the detection space is given by

$$E_R\left(x',y',z';x_a,y_a,z_a\right)=\iiint\chi^{(1)}\left(x+x',y+y',z+z'\right)E_{ex}\left(x,y,z\right)E'_{col}(x_a-x,y_a-y,z_a-z)dxdydz. \qquad (7)$$

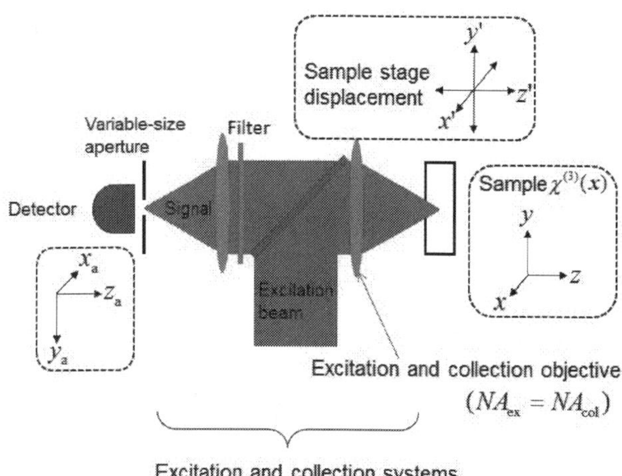

Figure 4. Schematic of reflection microscopy.

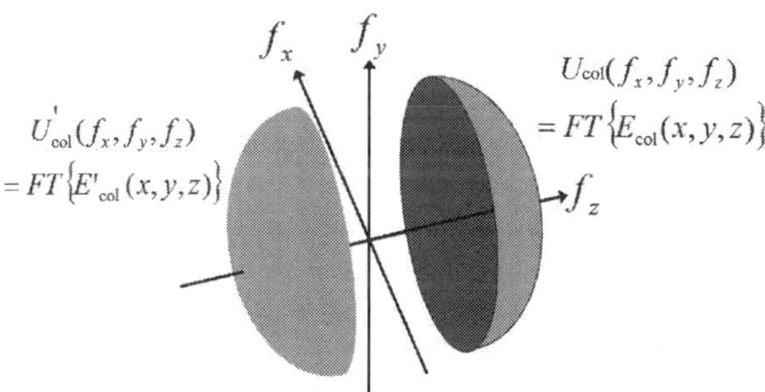

Figure 5. Distribution of the wavenumber vector (3-D pupil function) for the signal focused by the signal-collection objective.

As only the signal with no local oscillator forms the image, the image intensity $I_c^r(x',\,y',\,z')$ acquired by reflection linear confocal microscopy is written as

$$\begin{aligned}
I_c^r\left(x',y',z'\right) &= \iiint\left|\left\{E_R\left(x',y',z';x_a,y_a,z_a\right)\right\}\right|^2\delta(x_a)\delta(y_a)\delta(z_a)dx_ady_adz_a \\
&= \left|\iiint\chi^{(1)}\left(x+x',y+y',z+z'\right)E_{ex}\left(x,y,z\right)E'_{col}\left(-x,-y,-z\right)dxdydz\right|^2.
\end{aligned} \qquad (8)$$

The function $E_{ex}(x, y, z)E'_{col}(-x, -y, -z) \equiv ASF_r(x, y, z)$ is defined as the ASF for reflection linear confocal microscopy. Since we have the relation $E_{col}(x, y, z) = E'_{col}(-x, -y, -z)$, the ASF in reflection linear confocal microscopy $ASF_r(x, y, z)$ becomes equal to $E_{ex}(x, y, z)E_{col}(x, y, z)$. While the ASF in transmission linear confocal microscopy $ASF_t(x, y, z)$ approaches a real function, that in reflection linear confocal microscopy $ASF_r(x, y, z)$ is inevitably a complex function. In reflection linear confocal microscopy, the mixture image of the real and imaginary parts of the linear susceptibility is visualized.

2.4. Coherent nonlinear microscopy

We expand the image-forming formulas for $\chi^{(1)}$-derived optical processes to the general formulas for $\chi^{(i)}$-derived optical processes. In this subsection, we deal with coherent microscopy, which utilizes coherent optical processes. As an example, we first consider coherent anti-Stokes Raman scattering (CARS) microscopy, in which the two excitation beams (pump and Stokes) are used to generate the CARS signal (see **Figure 6**). In CARS microscopy, the pump and Stokes beams are temporally and spatially overlapped to generate the CARS signal such that the frequency difference between the pump and Stokes is tuned to match a particular Raman-active vibration frequency. The resonant CARS emission is several orders of magnitude greater than that from spontaneous Raman scattering. CARS microscopy provides chemically selective image contrast based on the intrinsic vibrational modes of molecular species, avoiding the need for labels. In addition, CARS imaging systems also employ near-infrared lasers to maximize imaging depth and minimize photodamage to cells. When the sample-stage displacement is zero, $(x', y', z') = (0, 0, 0)$, the polarization distribution becomes

$$P_{CARS}(x,y,z) = \chi^{(3)}_{CARS}(x,y,z)E_p(x,y,z)\{E_S(x,y,z)\}^* E_p(x,y,z), \tag{9}$$

where $\chi^{(3)}_{CARS}(x, y, z)$ denotes the nonlinear susceptibility for CARS and $E_p(x, y, z)$ and $E_s(x, y, z)$ stand for the electric field distributions in the sample formed by the pump and Stokes beams focused through the excitation system, respectively. In the same manner as in the previous subsection, taking into account the sample-stage displacement (x', y', z'), in transmission microscopy, the electric field distribution $E_{CARS}(x', y', z'; x_a, y_a, z_a)$ in the detection space can be written as

$$E_{CARS}(x',y',z';x_a,y_a,z_a) = \iiint \chi^{(3)}_{CARS}(x+x',y+y',z+z')E_p(x,y,z)\{E_S(x,y,z)\}^* E_p(x,y,z)$$
$$\times E_{col}(x_a-x,y_a-y,z_a-z)dxdydz, \tag{10}$$

where $E_{col}(x, y, z)$ is calculated by using the wavelength of the CARS signal. Note that for reflection microscopy, $E_{col}(x, y, z)$ is replaced by $E'_{col}(x, y, z)$. As the wavelength of the CARS signal is different from those of the pump and Stokes beams, only the CARS signal can be detected by using a filter. The image intensity $I_{CARS}(x', y', z')$ acquired by CARS microscopy is given by

$$I_{CARS}(x',y',z') = \iiint |\{E_{CARS}(x',y',z';x_a,y_a,z_a)\}|^2 \, a(x_a,y_a)\delta(z_a)dx_a dy_a dz_a. \tag{11}$$

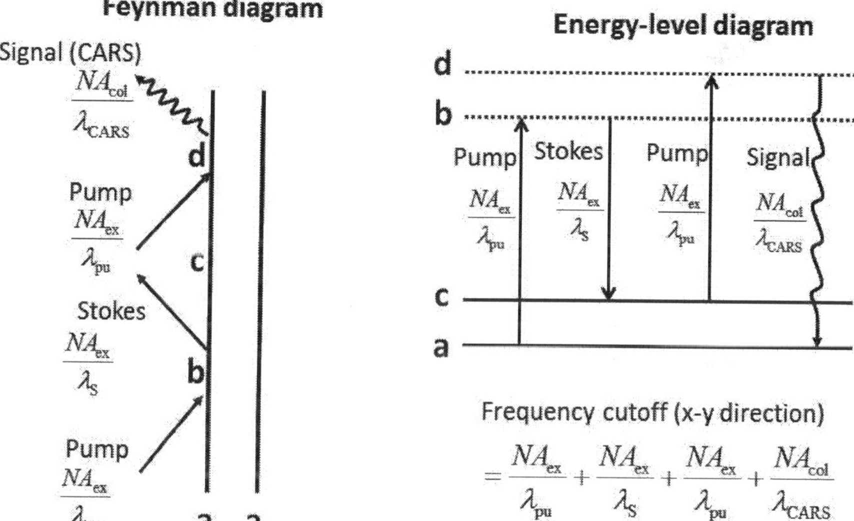

Figure 6. Double-sided Feynman diagram and energy-level diagram for CARS. λ_{pu} is the wavelength of the pump beam, λ_S is the wavelength of the Stokes beam, and λ_{CARS} is the wavelength of the CARS signal. The relation $2/\lambda_{pu}-1/\lambda_S = 1/\lambda_{CARS}$ is satisfied.

In confocal CARS microscopy, the image intensity $I^c_{CARS}(x', y', z')$ reduces to

$$\begin{aligned} I^c_{CARS}(x',y',z') &= |\{E_{CARS}(x',y',z';0,0,0)\}|^2 \\ &= |\iiint \chi^{(3)}_{CARS}(x+x',y+y',z+z')E_p(x,y,z)\{E_S(x,y,z)\}^* E_p(x,y,z)E_{col}(-x,-y,-z)dxdydz|^2 \\ &= |\iiint \chi^{(3)}_{CARS}(x+x',y+y',z+z')\text{ASF}_{CARS}(x,y,z)dxdydz|^2 . \end{aligned} \tag{12}$$

where $\text{ASF}_{CARS}(x, y, z) \equiv E_p(x, y, z)\{E_S(x, y, z)\}^* E_p(x, y, z)E_{col}(-x, -y, -z)$. The CTF of CARS microscopy is calculated by Fourier transforming $\text{ASF}_{CARS}(x, y, z)$. The maximum value of the frequency cutoff, which means the grating with the largest grating vector that can be resolved, is determined by the CTF. It is proven that the frequency cutoff of nonconfocal CARS microscopy does not change compared with that of confocal CARS microscopy [4]. In general, in coherent microscopy, the theoretical resolution limits (frequency cutoffs) are identical between confocal and nonconfocal systems.

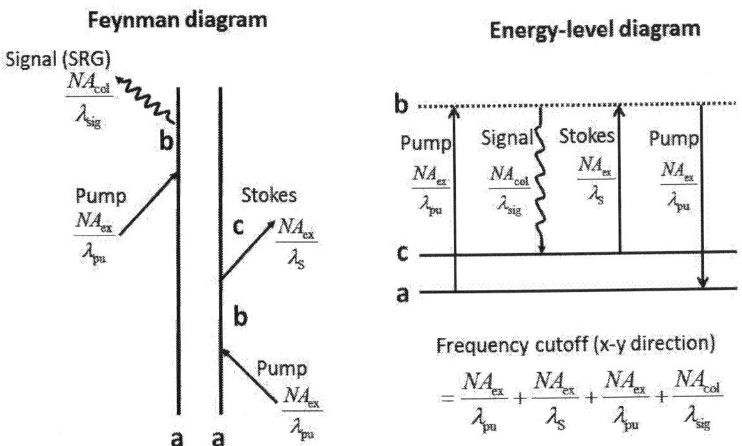

Figure 7. Double-sided Feynman diagram and energy-level diagram for SRG. λ_{sig} is the wavelength of the SRG signal. Note that $\lambda_S = \lambda_{\text{sig}}$.

As the next example, we consider stimulated Raman gain (SRG) microscopy, in which the pump and Stokes beams are employed as the excitation beams in common with CARS microscopy. As the wavelength of the SRG signal is identical with that of the Stokes beam (see **Figure 7**), the SRG signal interferes with the Stokes beam, which acts as the local oscillator in transmission microscopy. The pump beam is modulated and the SRG signal with the same wavelength as the Stokes beam can be extracted by demodulating the Stokes beam with the lock-in amplifier. In the signal-collection system, the pump beam is blocked by the filter. When $(x', y', z') = (0, 0, 0)$, the polarization distribution is given by

$$P_{\text{SRG}}(x,y,z) = \chi^{(3)}_{\text{SRG}}(x,y,z)\{E_p(x,y,z)\}^* E_S(x,y,z)E_p(x,y,z), \tag{13}$$

where $\chi^{(3)}_{\text{SRG}}(x, y, z)$ represents the nonlinear susceptibility for SRG. With the sample-stage displacement (x', y', z'), in transmission microscopy, the electric field distribution of the SRG signal $E_{\text{SRG}}(x', y', z'; x_a, y_a, z_a)$ in the detection space is written as

$$E_{\mathrm{SRG}}\left(x',y',z';x_a,y_a,z_a\right)=\iiint \chi^{(3)}_{\mathrm{SRG}}\left(x+x',y+y',z+z'\right)\left\{E_{\mathrm{p}}\left(x,y,z\right)\right\}^{*}E_{\mathrm{S}}\left(x,y,z\right)E_{\mathrm{p}}\left(x,y,z\right)$$
$$\times E_{\mathrm{col}}\left(x_a-x,y_a-y,z_a-z\right)dxdydz, \tag{14}$$

where $E_{\mathrm{col}}(x, y, z)$ in this case must be calculated by using the wavelength of the SRG signal. With the filter that blocks the pump beam, the total intensity $I^t_{\mathrm{SRG}}(x', y', z')$ identified by the detector of transmission SRG microscopy is given by

$$I^t_{\mathrm{SRG}}\left(x',y',z'\right)=\iiint\left|-iE_S\left(x_a,y_a,z_a\right)+E_{\mathrm{SRG}}\left(x',y',z';x_a,y_a,z_a\right)\right|^2 a(x_a,y_a)\delta(z_a)dx_a dy_a dz_a, \tag{15}$$

where the Stoke beam $-iE_S(x_a, y_a, z_a)$ with the Gouy phase shift $(-i)$ functions as the local oscillator. In confocal transmission SRG microscopy, substituting $\delta(x_a)\,\delta(y_a)$ into $a(x_a, y_a)$, $I^t_{\mathrm{SRG}}(x', y', z')$ reduces to

$$I^{ct}_{\mathrm{SRG}}\left(x',y',z'\right)$$
$$=\left|-iE_S\left(0,0,0\right)+\iiint \chi^{(3)}_{\mathrm{SRG}}\left(x+x',y+y',z+z'\right)\right.$$
$$\left.\times\left\{E_{\mathrm{p}}\left(x,y,z\right)\right\}^{*}E_{\mathrm{S}}\left(x,y,z\right)E_{\mathrm{p}}\left(x,y,z\right)E_{\mathrm{col}}\left(-x,-y,-z\right)dxdydz\right|^2$$
$$\approx\left|E_S\left(0,0,0\right)\right|^2$$
$$+i\left\{E_S\left(0,0,0\right)\right\}^{*}\iiint \chi^{(3)}_{\mathrm{SRG}}\left(x+x',y+y',z+z'\right)\left|E_{\mathrm{p}}\left(x,y,z\right)\right|^2 E_{\mathrm{S}}\left(x,y,z\right)E_{\mathrm{col}}\left(-x,-y,-z\right)dxdydz$$
$$-iE_S\left(0,0,0\right)\iiint\left\{\chi^{(3)}_{\mathrm{SRG}}\left(x+x',y+y',z+z'\right)\right\}^{*}\left|E_{\mathrm{p}}\left(x,y,z\right)\right|^2\left\{E_{\mathrm{S}}\left(x,y,z\right)\right\}^{*}$$
$$\times\left\{E_{\mathrm{col}}\left(-x,-y,-z\right)\right\}^{*}dxdydz$$
$$=\left|E_S\left(0,0,0\right)\right|^2-2E_S\left(0,0,0\right)\iiint \mathrm{Im}\left\{\chi^{(3)}_{\mathrm{SRG}}\left(x+x',y+y',z+z'\right)\right\}ASF_{\mathrm{SRG}}\left(x,y,z\right)dxdydz, \tag{16}$$

where we used the fact that $ASF_{\mathrm{SRG}}(x, y, z)\equiv |E_{\mathrm{p}}(x, y, z)|^2 E_S(x, y, z)E_{\mathrm{col}}(-x, -y, -z)$ approaches a real function under the condition $NA_{\mathrm{ex}} = NA_{\mathrm{col}}$. Note that the sign of $\mathrm{Im}\{\chi^{(3)}_{\mathrm{SRG}}\}$ is negative in SRG. The first term in Eq. (16) can be eliminated with lock-in detection. The CTF is calculated by Fourier transforming $ASF_{\mathrm{SRG}}(x, y, z)$.

We also consider "nonconfocal" transmission SRG microscopy, which is normally used to achieve a high signal intensity. Although the detector is normally placed at the plane conjugate to the pupil of the collection objective in nonconfocal microscopy, we deal with microscopy in which the detector is placed at the image plane conjugate to the sample plane to discuss confocal microscopy and nonconfocal microscopy in the same theoretical framework. Note that in nonconfocal microscopy, the image does not change regardless of detector position. Therefore, to simplify the equation, we calculate the intensity value at a certain sample-stage displacement (x', y', z') by three-dimensionally integrating the signal intensity in the detection space. The image intensity is proportional to the above-mentioned calculation result. The image intensity $I_{\mathrm{SRG}}^{\mathrm{nct}}(x',y',z')$ acquired by the detector of nonconfocal transmission SRG microscopy is given by

$$I_{\mathrm{SRG}}^{\mathrm{nct}}\left(x',y',z'\right)$$
$$\propto \iiint \left|\left\{-iE_{\mathrm{S}}\left(x_{\mathrm{a}},y_{\mathrm{a}},z_{\mathrm{a}}\right)+E_{\mathrm{SRG}}\left(x',y',z';x_{\mathrm{a}},y_{\mathrm{a}},z_{\mathrm{a}}\right)\right\}\right|^{2}dx_{\mathrm{a}}dy_{\mathrm{a}}dz_{\mathrm{a}}$$
$$=\iiint [|E_{\mathrm{S}}\left(x_{\mathrm{a}},y_{\mathrm{a}},z_{\mathrm{a}}\right)|^{2}+i\left\{E_{\mathrm{S}}\left(x_{\mathrm{a}},y_{\mathrm{a}},z_{\mathrm{a}}\right)\right\}^{*}$$
$$\times \iiint \chi_{\mathrm{SRG}}^{(3)}\left(x+x',y+y',z+z'\right)\left|E_{\mathrm{p}}\left(x,y,z\right)\right|^{2}E_{\mathrm{S}}\left(x,y,z\right)E_{\mathrm{col}}\left(x_{\mathrm{a}}-x,y_{\mathrm{a}}-y,z_{\mathrm{a}}-z\right)dxdydz$$
$$-iE_{\mathrm{S}}\left(x_{\mathrm{a}},y_{\mathrm{a}},z_{\mathrm{a}}\right)\iiint \chi_{\mathrm{SRG}}^{(3)}\left(x+x',y+y',z+z'\right)\left|E_{\mathrm{p}}\left(x,y,z\right)\right|^{2}\left\{E_{\mathrm{S}}\left(x,y,z\right)\right\}^{*}$$
$$\times \left\{E_{\mathrm{col}}\left(x_{\mathrm{a}}-x,y_{\mathrm{a}}-y,z_{\mathrm{a}}-z\right)\right\}^{*}dxdydz]dx_{\mathrm{a}}dy_{\mathrm{a}}dz_{\mathrm{a}}$$
$$=\iiint |E_{\mathrm{S}}\left(x_{\mathrm{a}},y_{\mathrm{a}},z_{\mathrm{a}}\right)|^{2}dx_{\mathrm{a}}dy_{\mathrm{a}}dz_{\mathrm{a}}$$
$$+i\iiint \chi_{\mathrm{SRG}}^{(3)}\left(x+x',y+y',z+z'\right)\left|E_{\mathrm{p}}\left(x,y,z\right)\right|^{2}\left|E_{\mathrm{S}}\left(x,y,z\right)\right|^{2}dxdydz$$
$$-i\iiint \left\{\chi_{\mathrm{SRG}}^{(3)}\left(x+x',y+y',z+z'\right)\right\}^{*}\left|E_{\mathrm{p}}\left(x,y,z\right)\right|^{2}\left|E_{\mathrm{S}}\left(x,y,z\right)\right|^{2}dxdydz$$
$$=\iiint |E_{\mathrm{S}}\left(x_{\mathrm{a}},y_{\mathrm{a}},z_{\mathrm{a}}\right)|^{2}dx_{\mathrm{a}}dy_{\mathrm{a}}dz_{\mathrm{a}}$$
$$-2\iiint \mathrm{Im}\left\{\chi_{\mathrm{SRG}}^{(3)}\left(x+x',y+y',z+z'\right)\right\}^{*}\left|E_{\mathrm{p}}\left(x,y,z\right)\right|^{2}\left|E_{\mathrm{S}}\left(x,y,z\right)\right|^{2}dxdydz. \tag{17}$$

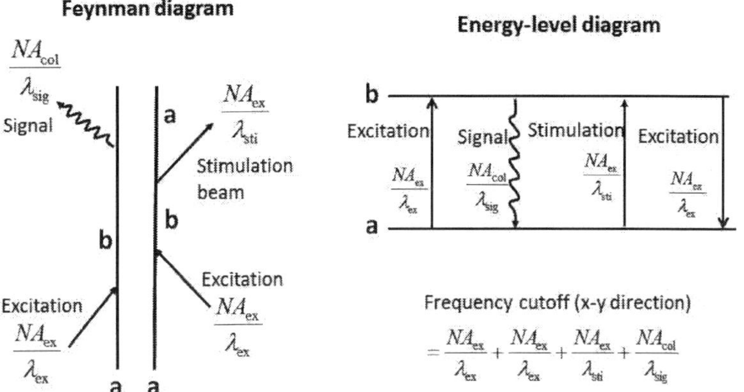

Figure 8. Double-sided Feynman diagram and energy-level diagram for stimulated emission. λ_{ex} is the wavelength of the excitation beam, λ_{sti} is the wavelength of the stimulation beam, and λ_{sig} is the wavelength of the SE signal. A two-level system is assumed.

Here we used the relations $E_{col}(-x, -y, -z) = \{E_{col}(x, y, z)\}^*$ and $E_S(x_a, y_a, z_a) \otimes E_{col}(x_a, y_a, z_a) = E_S(x_a, y_a, z_a)$ under the condition $NA_{ex} = NA_{col}$, where $\otimes$ represents the convolution. The first term in Eq. (17) can be eliminated with lock-in detection. The ASF for nonconfocal SRG microscopy is $|E_p(x, y, z)|^2 |E_S(x, y, z)|^2$, which is nearly equal to the ASF for confocal microscopy.

We then consider nonconfocal reflection SRG microscopy. Interestingly, in nonconfocal reflection SRG microscopy, the reflection light $E_R(x', y', z'; x_a, y_a z_a)$ generated by the $\chi^{(1)}$-derived optical process plays a role as the local oscillator. The image intensity $I_{SRG}^{ncr}(x', y', z')$ acquired by the detector of nonconfocal reflection SRG microscopy is given by

$$
\begin{aligned}
I_{SRG}^{ncr} &\left(x', y', z'\right) \\
&\propto \iiint \left| \left\{ E_R\left(x', y', z'; x_a, y_a, z_a\right) + E_{SRG}\left(x', y', z'; x_a, y_a, z_a\right) \right\} \right|^2 dx_a dy_a dz_a \\
&= \iiint \left| \iiint \chi^{(1)}\left(x+x', y+y', z+z'\right) E_S\left(x,y,z\right) E'_{col}\left(x_a-x, y_a-y, z_a-z\right) dx dy dz \right. \\
&\qquad + \iiint \chi^{(3)}_{SRG}\left(x+x', y+y', z+z'\right) \left\{E_p\left(x,y,z\right)\right\}^* E_S\left(x,y,z\right) E_p\left(x,y,z\right) \\
&\qquad \left. \times E'_{col}\left(x_a-x, y_a-y, z_a-z\right) dx dy dz \right|^2 dx_a dy_a dz_a .
\end{aligned}
\tag{18}
$$

As the local oscillator in this case does not have a Gouy phase shift, the real part of $\chi^{(3)}_{SRG}(x, y, z)$ is mainly observed. To see this, we consider a single-point object: $\chi^{(1)}(x, y, z) = \delta(x, y, z)$ and $\chi^{(3)}_{SRG}(x, y, z) = \varepsilon_r \delta(x, y, z) + i\varepsilon_i \delta(x, y, z)$, where $\varepsilon_r, \varepsilon_i \ll 1$. Eq. (18) reduces to

$$
\begin{aligned}
I_{SRG}^{ncr} &\\
&\propto \iiint \left| E_S\left(-x', -y', -z'\right) E'_{col}\left(x_a+x', y_a+y', z_a+z'\right) \right. \\
&\qquad + \varepsilon_r \left| E_p\left(-x', -y', -z'\right) \right|^2 E_S\left(-x', -y', -z'\right) E'_{col}\left(x_a+x', y_a+y', z_a+z'\right) \\
&\qquad + \left. i\varepsilon_i \left| E_p\left(-x', -y', -z'\right) \right|^2 E_S\left(-x', -y', -z'\right) E'_{col}\left(x_a+x', y_a+y', z_a+z'\right) \right|^2 dx_a dy_a dz_a \\
&\approx \left| E_S\left(-x', -y', -z'\right) \right|^2 \iiint \left| E'_{col}\left(x_a, y_a, z_a\right) \right|^2 dx_a dy_a dz_a \\
&\qquad + 2\varepsilon_r \iiint \left| E'_{col}\left(x_a, y_a, z_a\right) \right|^2 dx_a dy_a dz_a \, \mathrm{Re}\left\{ \left| E_p\left(-x', -y', -z'\right) \right|^2 \left| E_S\left(-x', -y', -z'\right) \right|^2 \right\} \\
&\qquad - 2\varepsilon_i \iiint \left| E'_{col}\left(x_a, y_a, z_a\right) \right|^2 dx_a dy_a dz_a \, \mathrm{Im}\left\{ \left| E_p\left(-x', -y', -z'\right) \right|^2 \left| E_S\left(-x', -y', -z'\right) \right|^2 \right\} .
\end{aligned}
\tag{19}
$$

The third term in Eq. (19) becomes zero, and the first term can be eliminated with lock-in detection. Eventually, only the real part ε_r remains.

The last example for coherent nonlinear microscopy is stimulated emission (SE) microscopy, in which the pump beam (electric field distribution: $E_p(x, y, z)$) and the stimulation beam (electric field distribution: $E_{sti}(x, y, z)$) are employed to generate the signal (see **Figure 8**). In SE microscopy, the SE signal has the same wavelength as that of the stimulation beam, and the SE signal interferes with the stimulation beam, which functions as the local oscillator in transmission microscopy. The pump beam is modulated and the SE signal can be extracted by demodulating the stimulation beam interfering with the SE signal. In analogy with SRG microscopy, replacing $E_S(x_a, y_a, z_a)$ with $E_{sti}(x_a, y_a, z_a)$ and $E_{SRG}(x', y', z'; x_a, y_a, z_a)$ by $E_{SE}(x', y', z'; x_a, y_a, z_a) \equiv \iiint \chi_{SE}^{(3)}(x + x', y + y', z + z') \, | \, E_p(x, y, z) \, |^2 \, E_{sti}(x, y, z) \, E_{col}(x_a\text{-}x, y_a\text{-}y, z_a\text{-}z) \, d_x d_y d_z$, the image intensity $I_{SE}^{nct}(x', y', z')$ acquired by the detector in nonconfocal transmission SE microscopy is written as

$$
\begin{aligned}
I_{SE}^{nct} \\
\propto & \iiint \left| \left\{ -iE_{sti}\left(x_a, y_a, z_a\right) + E_{SE}\left(x', y', z'; x_a, y_a, z_a\right) \right\} \right|^2 dx_a \, dy_a \, dz_a \\
= & \iiint \Bigg[\left| E_{sti}\left(x_a, y_a, z_a\right) \right|^2 + i\left\{ E_{sti}\left(x_a, y_a, z_a\right) \right\}^* \\
& \times \iiint \chi_{SE}^{(3)}\left(x + x', y + y', z + z'\right) \left| E_p\left(x, y, z\right) \right|^2 E_{sti}\left(x, y, z\right) E_{col}\left(x_a - x, y_a - y, z_a - z\right) dx\,dy\,dz \\
& - iE_{sti}\left(x_a, y_a, z_a\right) \iiint \left\{ \chi_{SE}^{(3)}\left(x + x', y + y', z + z'\right) \right\}^* \left| E_p\left(x, y, z\right) \right|^2 \left\{ E_{sti}\left(x, y, z\right) \right\}^* \\
& \times \left\{ E_{col}\left(x_a - x, y_a - y, z_a - z\right) \right\}^* dx\,dy\,dz \Bigg] dx_a \, dy_a \, dz_a \\
= & \iiint \left| E_{sti}\left(x_a, y_a, z_a\right) \right|^2 dx_a \, dy_a \, dz_a \\
& + i \iiint \chi_{SE}^{(3)}\left(x + x', y + y', z + z'\right) \left| E_p\left(x, y, z\right) \right|^2 \left| E_{sti}\left(x, y, z\right) \right|^2 dx\,dy\,dz \\
& - i \iiint \left\{ \chi_{SE}^{(3)}\left(x + x', y + y', z + z'\right) \right\}^* \left| E_p\left(x, y, z\right) \right|^2 \left| E_{sti}\left(x, y, z\right) \right|^2 dx\,dy\,dz \\
= & \iiint \left| E_{sti}\left(x_a, y_a, z_a\right) \right|^2 dx_a \, dy_a \, dz_a \\
& - 2 \iiint \mathrm{Im}\left\{ \chi_{SE}^{(3)}\left(x + x', y + y', z + z'\right) \right\} \left| E_p\left(x, y, z\right) \right|^2 \left| E_{sti}\left(x, y, z\right) \right|^2 dx\,dy\,dz,
\end{aligned}
\tag{20}
$$

where $\chi_{SE}^{(3)}(x, y, z)$ represents the nonlinear susceptibility for SE. In Eq. (20), we used the relations $E_{col}(-x, -y, -z) = \{E_{col}(x, y, z)\}^*$ and $E_{sti}(x_a, y_a, z_a) \otimes E_{col}(x_a, y_a, z_a) = E_{sti}(x_a, y_a, z_a)$ under the condition $NA_{ex} = NA_{col}$. The first term can be eliminated with lock-in detection. The ASF for nonconfocal SE microscopy is $| E_p(x, y, z) |^2 | E_{sti}(x, y, z) |^2$, which is nearly equal to the

ASF for confocal microscopy. Although this equation for SE microscopy is described by the same notation as SRG microscopy, the Feynman diagram differs.

2.5. Incoherent microscopy

To deal with incoherent optical processes, the vacuum field around the sample needs to be reckoned in our calculation. We assume that $|0\rangle_{(x,y,z)}$ denotes the amplitude of the quantum vacuum zero-point effect at the (x,y,z) position in the sample. For spontaneous Raman scattering, the Stokes beam $E_S(x_a, y_a, z_a)$ is replaced by the vacuum field $|0\rangle_{(x_a,y_a,z_a)}$:

$$
\begin{aligned}
I^t_{Ra}&(x',y',z')\\
&= \iiint \left| -i\,|0\rangle_{(x_a,y_a,z_a)} + E_{SRG}(x',y',z';x_a,y_a,z_a) \right|^2 a(x_a,y_a)\delta(z_a)\,dx_a dy_a dz_a\\
&= \iiint \left| -i\,|0\rangle_{(x_a,y_a,z_a)} + \iiint \chi^{(3)}_{SRG}(x+x',y+y',z+z')\{E_p(x,y,z)\}^* E_p(x,y,z)|0\rangle_{(x,y,z)} \right.\\
&\qquad \left. \times F_{col}(x_a-x,y_a-y,z_a-z)\,dxdydz \right|^2 a(x_a,y_a)\delta(z_a)\,dx_a dy_a dz_a.
\end{aligned}
\tag{21}
$$

We then consider the Fourier expansion of $|0\rangle_{(x,y,z)}$ into the plane wave basis $|0\rangle_{(k_x,k_y,k_z)}$:

$$
|0\rangle_{(x,y,z)} = \frac{1}{(2\pi)^3}\iiint |0\rangle_{(k_x,k_y,k_z)} e^{-i(k_x x + k_y y + k_z z)}\,dk_x dk_y dk_z,
\tag{22}
$$

$$
|0\rangle_{(x_a,y_a,z_a)} = \frac{1}{(2\pi)^3}\iiint C(k_x,k_y,k_z)|0\rangle_{(k_x,k_y,k_z)} e^{-i(k_x x_a + k_y y_a + k_z z_a)}\,dk_x dk_y dk_z,
\tag{23}
$$

where (k_x, k_y, k_z) is the wavenumber vector, $|0\rangle_{(k_x,k_y,k_z)}$ stands for the Fourier component of the vacuum field, and $C(k_x, k_y, k_z)$ is the cone-shaped function representing the wave vectors of the Fourier components of the vacuum field that can pass through the signal-collection objective with NA_{col} and reach the detector. In the case of the fixed angular frequency of the pump, the nonlinear susceptibility of SRG is a function of the angular frequency of the Stokes beam: $\omega_S = c\sqrt{k_x^2 + k_y^2 + k_z^2}$, where c is the light speed in vacuum. Therefore, we can replace $\chi_{SRG}^{(3)}(x, y, z)$ in Eq. (21) with $\chi_{SRG}^{(3)}(x, y, z)L(k_x, k_y, k_z)$, where $L(k_x, k_y, k_z)$ is a spherically symmetrical function (typically a complex Lorentzian function of ω_S). We then obtain

$$I_{Ra}^t(x', y', z') = \iiint |A_{Ra}(x', y', z'; x_a, y_a, z_a)|^2 \, a(x_a, y_a)\delta(z_a)dx_a dy_a dz_a, \tag{24}$$

with

$$
\begin{aligned}
&\left| A_{Ra}\left(x', y', z'; x_a, y_a, z_a\right)\right|^2 \\
&= \left| -i\,|0\rangle_{(x_a, y_a, z_a)} + \iiint \chi_{SRG}^{(3)}\left(x+x', y+y', z+z'\right)L\left(k_x, k_y, k_z\right)\left\{E_p\left(x, y, z\right)\right\}^* E_p\left(x, y, z\right) \right. \\
&\qquad \left. \times |0\rangle_{(x,y,z)} E_{col}\left(x_a - x, y_a - y, z_a - z\right)dx\,dy\,dz \right|^2 \\
&= \left| -i\frac{1}{(2\pi)^3}\iiint C\left(k_x, k_y, k_z\right)|0\rangle_{(k_x, k_y, k_z)}e^{-i\left(k_x x_a + k_y y_a + k_z z_a\right)}dk_x dk_y dk_z \right. \\
&\qquad + \iiint \chi_{SRG}^{(3)}\left(x+x', y+y', z+z'\right)L\left(k_x, k_y, k_z\right)\left\{E_p\left(x, y, z\right)\right\}^* E_p\left(x, y, z\right) \\
&\qquad \left. \times \frac{1}{(2\pi)^3}\iiint |0\rangle_{(k_x, k_y, k_z)}e^{-i\left(k_x x + k_y y + k_z z\right)}dk_x dk_y dk_z\, E_{col}\left(x_a - x, y_a - y, z_a - z\right)dx\,dy\,dz \right|^2 \\
&\approx \frac{1}{(2\pi)^6}\iiint\iiint C\left(k_x, k_y, k_z\right)C^*\left(k_x', k_y', k_z'\right)_{(k_x', k_y', k_z')}\langle 0|0\rangle_{(k_x, k_y, k_z)}e^{-i\left\{(k_x - k_x')x_a + (k_y - k_y')y_a + (k_z - k_z')z_a\right\}} \\
&\qquad \times dk_x dk_y dk_z dk_x' dk_y' dk_z' \\
&\quad + i\frac{1}{(2\pi)^6}\iiint\iiint\iiint L\left(k_x, k_y, k_z\right)C^*\left(k_x', k_y', k_z'\right)_{(k_x', k_y', k_z')}\langle 0|0\rangle_{(k_x, k_y, k_z)}e^{i\left(k_x x_a + k_y y_a + k_z z_a\right)} \\
&\qquad \times e^{-i\left(k_x' x + k_y' y + k_z' z\right)}dk_x dk_y dk_z dk_x' dk_y' dk_z'\, \chi_{SRG}^{(3)}\left(x+x', y+y', z+z'\right)|E_p\left(x, y, z\right)|^2 \\
&\qquad \times E_{col}\left(x_a - x, y_a - y, z_a - z\right)dx\,dy\,dz \qquad + \text{c.c.}
\end{aligned}
\tag{25}
$$

In common with SRG, the fourth term in the above equation is negligible also in spontaneous Raman scattering. Thus, we omitted the fourth term. Carrying on the calculation, we obtain

$$\left|A_{\mathrm{Ra}}\left(x',y',z';x_{\mathrm{a}},y_{\mathrm{a}},z_{\mathrm{a}}\right)\right|^{2}$$

$$=\frac{1}{\left(2\pi\right)^{3}}\iiint\iiint C\left(k_{x},k_{y},k_{z}\right)P^{*}\left(k_{x}',k_{y}',k_{z}'\right)\delta\left(k_{x}-k_{x}'\right)\delta\left(k_{y}-k_{y}'\right)\delta\left(k_{z}-k_{z}'\right)$$

$$\times e^{-i\left\{\left(k_{x}-k_{x}'\right)x_{\mathrm{a}}+\left(k_{y}-k_{y}'\right)y_{\mathrm{a}}+\left(k_{z}-k_{z}'\right)z_{\mathrm{a}}\right\}}dk_{x}dk_{y}dk_{z}dk_{x}'dk_{y}'dk'$$

$$+i\frac{1}{\left(2\pi\right)^{3}}\iiint\iiint\iiint L\left(k_{x},k_{y},k_{z}\right)C^{*}\left(k_{x}',k_{y}',k_{z}'\right)\delta\left(k_{x}-k_{x}'\right)\delta\left(k_{y}-k_{y}'\right)\delta\left(k_{z}-k_{z}'\right)$$

$$\times e^{i\left(k_{x}x_{\mathrm{a}}+k_{y}y_{\mathrm{a}}+k_{z}z_{\mathrm{a}}\right)}e^{-i\left(k_{x}'x+k_{y}'y+k_{z}'z\right)}dk_{x}dk_{y}dk_{z}dk_{x}'dk_{y}'dk_{z}'\ \chi_{\mathrm{SRG}}^{(3)}\left(x+x',y+y',z+z'\right)$$

$$\times\left|E_{\mathrm{p}}\left(x,y,z\right)\right|^{2}E_{\mathrm{col}}\left(x_{\mathrm{a}}-x,y_{\mathrm{a}}-y,z_{\mathrm{a}}-z\right)dxdydz\qquad+\mathrm{c.c.}$$

$$=\frac{1}{\left(2\pi\right)^{3}}\iiint\left|C\left(k_{x},k_{y},k_{z}\right)\right|^{2}dk_{x}dk_{y}dk_{z}$$

$$+i\frac{1}{\left(2\pi\right)^{3}}\iiint\iiint L\left(k_{x},k_{y},k_{z}\right)C^{*}\left(k_{x}',k_{y}',k_{z}'\right)e^{-i\left\{\left(x-x_{\mathrm{a}}\right)k_{x}+\left(y-y_{\mathrm{a}}\right)k_{y}+\left(z-z_{\mathrm{a}}\right)k_{z}\right\}}dk_{x}dk_{y}dk_{z}$$

$$\times\chi_{\mathrm{SRG}}^{(3)}\left(x+x',y+y',z+z'\right)\left|E_{\mathrm{p}}\left(x,y,z\right)\right|^{2}E_{\mathrm{col}}\left(x_{\mathrm{a}}-x,y_{\mathrm{a}}-y,z_{\mathrm{a}}-z\right)dxdydz\qquad+\mathrm{c.c.}$$

$$=\frac{1}{\left(2\pi\right)^{3}}\iiint\left|C\left(k_{x},k_{y},k_{z}\right)\right|^{2}dk_{x}dk_{y}dk_{z}$$

$$+i\iiint\chi_{\mathrm{SRG}}^{(3)}\left(x+x',y+y',z+z'\right)\left|E_{\mathrm{p}}\left(x,y,z\right)\right|^{2}\left|E_{\mathrm{col}}\left(x_{\mathrm{a}}-x,y_{\mathrm{a}}-y,z_{\mathrm{a}}-z\right)\right|^{2}dxdydz\qquad+\mathrm{c.c.}$$

$$=\frac{1}{\left(2\pi\right)^{3}}\iiint\left|C\left(k_{x},k_{y},k_{z}\right)\right|^{2}dk_{x}dk_{y}dk_{z}$$

$$-2\iiint\mathrm{Im}\left\{\chi_{\mathrm{SRG}}^{(3)}\left(x+x',y+y',z+z'\right)\right\}\left|E_{\mathrm{p}}\left(x,y,z\right)\right|^{2}\left|E_{\mathrm{col}}\left(x_{\mathrm{a}}-x,y_{\mathrm{a}}-y,z_{\mathrm{a}}-z\right)\right|^{2}dxdydz.$$

$$(26)$$

Here we used the following equations:

$$\left(k_{x}',k_{y}',k_{z}'\right)^{\langle 0|0\rangle\left(k_{x},k_{y},k_{z}\right)}=\left(2\pi\right)^{3}\delta\left(k_{x}-k_{x}'\right)\delta\left(k_{y}-k_{y}'\right)\delta\left(k_{z}-k_{z}'\right),\qquad(27)$$

$$\frac{1}{\left(2\pi\right)^{3}}\iiint L\left(k_{x},k_{y},k_{z}\right)C^{*}\left(k_{x},k_{y},k_{z}\right)e^{-i\left\{\left(x-x_{\mathrm{a}}\right)k_{x}+\left(y-y_{\mathrm{a}}\right)k_{y}+\left(z-z_{\mathrm{a}}\right)k_{z}\right\}}dk_{x}dk_{y}dk_{z}$$

$$=E_{\mathrm{col}}^{*}\left(x_{\mathrm{a}}-x,y_{\mathrm{a}}-y,z_{\mathrm{a}}-z\right).\qquad(28)$$

The first term in Eq. (26) corresponds to the vacuum field that cannot be observed. As only the difference from the vacuum state can be measured, the detected signal becomes

$$I_{\mathrm{Ra}}^{\mathrm{de}}\left(x',y',z'\right)\approx$$

$$\iiint\left\{\iiint\mathrm{Im}\left\{-2\chi_{\mathrm{SRG}}^{(3)}\left(x+x',y+y',z+z'\right)\right\}\left|E_{\mathrm{p}}\left(x,y,z\right)\right|^{2}\left|E_{\mathrm{col}}\left(x_{\mathrm{a}}-x,y_{\mathrm{a}}-y,z_{\mathrm{a}}-z\right)\right|^{2}dxdydz\right\}\qquad(29)$$

$$\times a\left(x_{\mathrm{a}},y_{\mathrm{a}}\right)\delta\left(z_{\mathrm{a}}\right)dx_{\mathrm{a}}dy_{\mathrm{a}}dz_{\mathrm{a}}.$$

Note that $\mathrm{Im}\{\chi_{\mathrm{SRG}}^{(3)}(x, y, z)\}$ is a negative value. This equation is well known as the image-forming formula of confocal microscopy with a finite detector size for incoherent optical processes [12]. In incoherent microscopy, the size of the detector influences the optical resolution, according to Eq. (29). In nonconfocal incoherent microscopy, only the pump beam $|E_{\mathrm{p}}(x, y, z)|^2$ determines the optical resolution, while in confocal incoherent microscopy, both the pump beam $|E_{\mathrm{p}}(x, y, z)|^2$ and the signal $|E_{\mathrm{col}}(-x, -y, -z)|^2$ affect the resolution limit, resulting in better resolution.

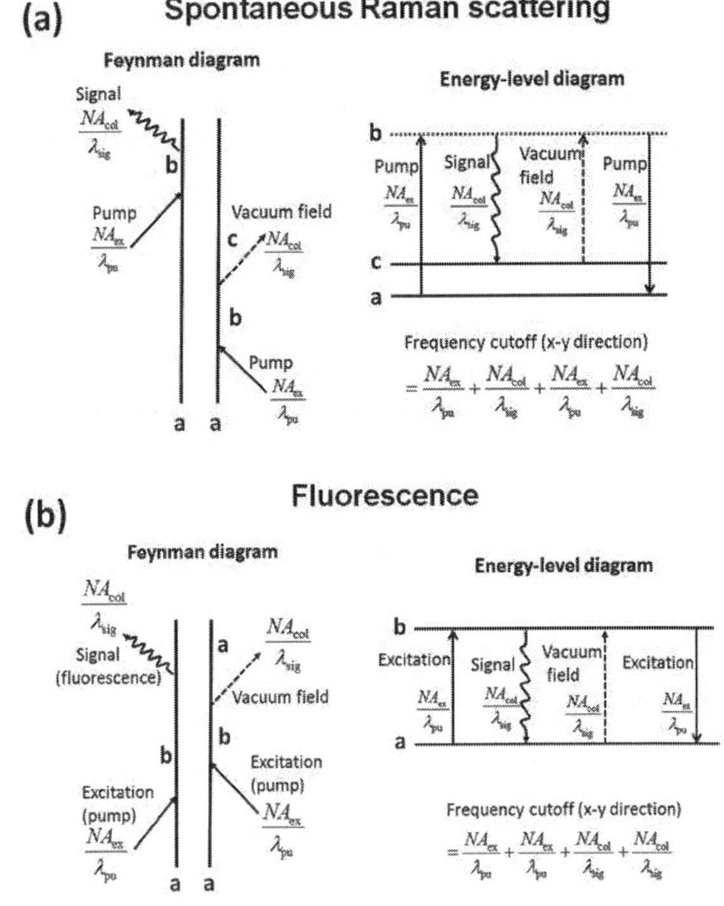

Figure 9. Double-sided Feynman diagram and energy-level diagram for (a) spontaneous Raman scattering and (b) fluorescence.

To discuss the maximum value of the frequency cutoff, we concentrate on confocal microscopy. Substituting $\delta(x_a)\,\delta(y_a)$ into $a(x_a, y_a)$, we obtain the image-forming formula of confocal Raman microscopy as follows:

$$I_{\mathrm{Ra}}^{c}(x',y',z') = \iiint \mathrm{Im}\{-2\chi_{\mathrm{SRG}}^{(3)}(x+x',y+y',z+z')\}\,|\,E_{p}(x,y,z)\,|^{2}|\,E_{\mathrm{col}}(-x,-y,-z)\,|^{2}\,dxdydz. \tag{30}$$

Another example of an incoherent optical process is fluorescence. For the image-forming formula of confocal fluorescence microscopy, we only replace $\chi_{\mathrm{SRG}}^{(3)}(x,\,y,\,z)$ with $\chi_{\mathrm{SE}}^{(3)}(x,\,y,\,z)$, namely,

$$I_{\mathrm{flu}}^{c}(x',y',z') = \iiint \mathrm{Im}\{-2\chi_{\mathrm{SE}}^{(3)}(x+x',y+y',z+z')\}\,|\,E_{p}(x,y,z)\,|^{2}|\,E_{\mathrm{col}}(-x,-y,-z)\,|^{2}\,dxdydz. \tag{31}$$

Although the ASF of spontaneous Raman scattering and that of fluorescence microscopy are represented by the same notation, it is notable that the double-sided Feynman diagrams or energy-level diagrams for these two processes are different, as described in **Figure 9**.

3. Optical resolution limit

In Section 2, we covered all types of laser microscopy, which include (i) coherent microscopy with the signal wavelength different from the excitation wavelength, (ii) coherent microscopy with the signal wavelength identical with the excitation wavelength, and (iii) incoherent microscopy.

In type (i), the signal can be measured by blocking the excitation beam with the filter. In type (ii), the signal can be extracted by lock-in detection, except linear microscopy that utilizes a $\chi^{(1)}$-derived optical process. Type (ii) has the local oscillator interfering with the signal, while type (i) does not have the local oscillator. Type (iii) is described in the same fashion as type (ii). Interestingly, in type (iii), the vacuum field plays a role as the local oscillator.

In this section, we form the framework that can discuss the resolution limit for all types of laser microscopy by using double-sided Feynman diagrams and energy-level diagrams. It is well known that the image of the object is formed in three dimensions by laser microscopy with a finite optical resolution, determined by the NA of the excitation and signal-collection systems and the wavelengths of the excitation beams and signal. In addition, we show that the type of optical process occurring in the sample also influences the optical resolution. In our model, the distribution of the linear or nonlinear susceptibility $\chi^{(i)}(x,\,y,\,z)$ corresponds to the object we would like to visualize. While the double-sided Feynman diagram was developed to calculate the quantity of susceptibility, which depends on the type of optical process involved, we provide the calculation method for the optical resolution of laser microscopy, which is also linked to the double-sided Feynman diagram.

Our model includes coherent microscopy and incoherent microscopy. Regardless of the coherence, we deal with all types of laser microscopy in the identical framework. In coherent microscopy, the coherent interaction between the excitation laser beam and the molecule occurs, and the corresponding susceptibility distribution is visualized in the image. In incoherent microscopy, an incoherent optical process, which is caused by the excitation beam and the vacuum field lying in the object space, takes place, and the image of the corresponding susceptibility distribution is created. Incoherent microscopy exhibits the incoherent property inherited from the vacuum field. For incoherent microscopy, the equation is formulated by partially using quantum optical notation.

In incoherent microscopy, the definition of the optical resolution becomes straightforward by utilizing the optical transfer function (OTF), while in coherent microscopy, because the OTF does not exist, some definitions are conceivable. In this section, for both coherent microscopy and incoherent microscopy, we define the resolution limit as the largest grating vector that can be resolved, when the three-dimensional grating of the susceptibility is observed as the object. By using our theoretical framework described below, the frequency cutoffs (resolution limits) of coherent microscopy and incoherent microscopy can be compared.

3.1. Diagram technique

We introduce the double-sided Feynman diagram to discuss the frequency cutoff. Originally, this diagram was developed to count and categorize the optical processes and calculate the nonlinear susceptibility of each one. Here we connect the diagram to the frequency cutoffs of linear, nonlinear, coherent, and incoherent microscopy. The diagram can deal with all optical processes, including incoherent processes, such as fluorescence and spontaneous Raman scattering. In coherent microscopy, the resolution limits of confocal and nonconfocal systems are identical, while in incoherent microscopy, the confocal system exhibits the better optical resolution than the nonconfocal system. Note that in coherent microscopy, the images of the confocal and nonconfocal systems indicate the different contrasts. To discuss the theoretical maximum value of the frequency cutoff, we deal with the confocal system for both coherent microscopy and incoherent microscopy. We consider the ASF and its Fourier transform: CTF. For incoherent microscopy, although the point spread function (PSF) is ordinarily used instead of ASF, in this section we refer to PSF for incoherent microscopy as ASF, to integrate coherent microscopy and incoherent microscopy into the identical framework.

The essential part of the image-forming formula for all types of microscopy can be written as

$$I(x',y',z') = \iiint O^{(i)}(x+x',y+y',z+z')\mathrm{ASF}(x,y,z)dxdydz, \tag{32}$$

or the square of its modulus. Here $O^{(i)}(x,y,z)$ corresponds to the object originating from $\chi^{(i)}(x,y,z)$. To discuss the largest grating vector that can be resolved, we consider 3-D grating as the object. In this case, we can just concentrate on Eq. (32), because the resolution limit does not change regardless of whether Eq. (32) is squared or not. The Fourier transform of Eq. (32) is given by

$$\tilde{I}\left(f_x, f_y, f_z\right) = \tilde{O}^{(i)}\left(f_x, f_y, f_z\right) CTF\left(f_x, f_y, f_z\right), \tag{33}$$

where $\tilde{A}$ means the Fourier transform of A and $(f_x, f_y, f_z) = (k_x/2\pi, k_y/2\pi, k_z/2\pi)$, which corresponds to the grating vector. For convenience, we use the definition that the Fourier transform of $\{ASF(x, y, z)\}^*$ is $\{CTF(f_x, f_y, f_z)\}^*$. The CTF expresses the existence range of the grating vector that can be resolved. In transmission linear confocal microscopy, for example, $ASF(x, y, z)$ is equal to $E_{ex}(x, y, z)E_{col}(-x, -y, -z)$, and Fourier transforming it leads to $CTF(f_x, f_y, f_z) = U_{ex}(-f_x, -f_y, -f_z) \otimes U_{col}(f_x, f_y, f_z)$, where $U_{col}(f_x, f_y, f_z)$ and $U_{ex}(f_x, f_y, f_z)$ stand for the Fourier transforms of $E_{col}(x, y, z)$ and $E_{ex}(x, y, z)$, respectively. Considering the Ewald sphere helps in understanding the CTF. The Ewald sphere in this case has the same radii as $U_{col}(f_x, f_y, f_z)$ and $U_{ex}(f_x, f_y, f_z)$, which are partial spheres (3-D pupil functions) as mentioned above. The phase-matching condition (momentum conservation law), $k_{sig} = k_{ex} + K$, is satisfied with the Ewald sphere, where k_{ex} and k_{sig} are the wavenumber vectors of the excitation light and the signal, respectively, and K is the grating vector in the sample. Unless the phase-matching condition is satisfied as shown in **Figure 10**, the signal cannot be generated. Consequently, the resolvable grating vector is restricted to the range determined by the CTF.

In analogy with the above formulation, also for any laser microscopy, the phase-matching condition is taken into account. Since the focused excitation beam is composed of numerous plane waves, all combinations of the excitation plane wave need to be considered. In coherent microscopy, only if the sum of the wavenumber vector of each excitation plane wave and the grating vector of the susceptibility is equal to the wavenumber vector of the signal that can be collected by the signal-collection system, the signal can be generated and detected by the signal-collection system. The phase-matching condition (e.g., $k_{sig} = k_{ex1} - k_{ex2} + \cdots + k_{exn} + K$) can be connected to the double-sided Feynman diagram and energy-level diagram as follows:

1. For the right-pointing arrow in the Feynman diagram or the up-pointing arrow in the energy-level diagram, the wavenumber vector of the excitation light corresponds to $+k_{ex}$.

2. For the left-pointing arrow in the Feynman diagram or the down-pointing arrow in the energy-level diagram, the wavenumber vector of the excitation light corresponds to $-k_{ex}$.

   The focused excitation beams and the signal contain many plane waves whose wavenumber vectors lie on the 3-D pupil functions. The ASF (e.g., $E_{ex1}(x)E_{ex2}^*(x)\cdots E_{exn}(x)E_{col}(-x)$) obeys the following rule.

3. For the right-pointing arrow in the Feynman diagram or the up-pointing arrow in the energy-level diagram, the electric field distribution formed by the excitation beam corresponds to $E_{ex}(x)$.

4. For the left-pointing arrow in the Feynman diagram or the down-pointing arrow in the energy-level diagram, the electric field distribution formed by the excitation beam corresponds to $E_{ex}^*(x)$.

5. For the wavy-line arrow, the electric field distribution $E_{col}(-x)$ formed by the signal through the signal-collection system is applied.

The CTF (e.g., $U_{ex1}(-f) \otimes U_{ex2}^{*}(f) \otimes \cdots \otimes U_{exn}(-f) \otimes U_{col}(f)$) satisfies the following rule.

6. For the right-pointing arrow in the Feynman diagram or the up-pointing arrow in the energy-level diagram, the 3-D pupil function for the excitation beam corresponds to $U_{ex}(-f)$.

7. For the left-pointing arrow in the Feynman diagram or the down-pointing arrow in the energy-level diagram, the 3-D pupil function for the excitation beam corresponds to $U_{ex2}^{*}(f)$.

8. For the wavy-line arrow, the 3-D pupil function for the signal $U_{col}(f)$ is applied.

As an example of coherent microscopy, **Figure 10** describes the relation between the CTF and the phase-matching condition represented by the 3-D pupil function. The figure shows the case of linear confocal microscopy, but in the case of nonlinear coherent microscopy, the nonzero region of the CTF becomes larger and the missing cone in the z direction disappears. While the CTF can be calculated by the above rule, the frequency cutoff in the x-y direction can be evaluated more easily with the following rule.

9. Each arrow for the excitation is connected to NA_{ex}/λ', where λ' is the wavelength of the corresponding beam, such as pump or Stokes.

10. The arrow for the signal is connected to NA_{col}/λ_{sig}.

11. The maximum possible value of the frequency cutoff in the x-y direction is given by the sum of all the above-mentioned values: $\Sigma\{NA/\lambda\}$.

In incoherent microscopy, the vacuum field, which contains the virtual photons with the wavenumber vectors in all directions, plays a role as one of the excitation light. The vacuum field is described by the right-pointing dashed arrow in the double-sided Feynman diagram and up-pointing dashed arrow in the energy-level diagram. The vacuum field has its own rule as follows.

12. The wavenumber vector for the vacuum field corresponds to $+k_{vac}$.

13. For the ASF, the electric field distribution for the vacuum field corresponds to $E_{col}^{*}(-x)$.

14. For the CTF, the 3-D pupil function for the vacuum field corresponds to $U_{col}^{*}(-f)$.

15. For the frequency cutoff in the x-y direction, the corresponding value for the vacuum field is NA_{col}/λ_{sig}, which is the same value as that of the signal.

Note that in incoherent microscopy, the CTF is referred to as the OTF and the ASF becomes the PSF. The vacuum field around the sample includes the Fourier components that have the wavenumber vectors also on the side opposite to the excitation beam. As a result, for reflection microscopy, $|E_{col}(x_a-x, y_a-y, z_a-z)|^2$ in Eq. (29) can be replaced by $|E_{col}'(x_a-x, y_a-y, z_a-z)|^2$, but both become the same function if the NAs are identical. Thus,

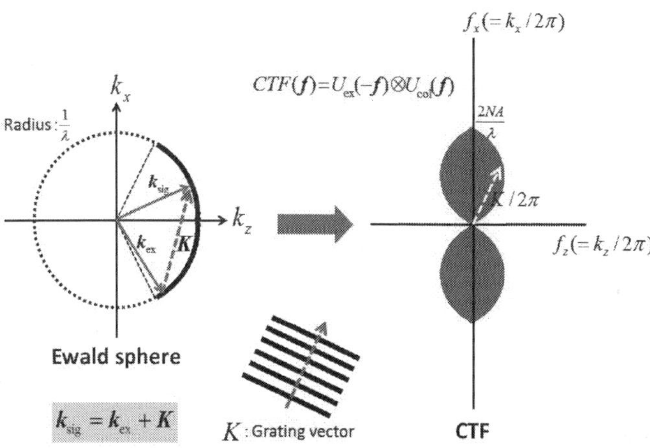

Figure 10. The relation between the phase-matching condition and the CTF for transmission linear confocal microscopy.

unlike in coherent microscopy, it turns out that the OTF of transmission microscopy is identical with that of reflection microscopy in incoherent microscopy. As an example of incoherent microscopy, we take the transmission fluorescence confocal microscopy shown in **Figure 11**, where the relation between the CTF and the phase-matching condition is described with the 3-D pupil function.

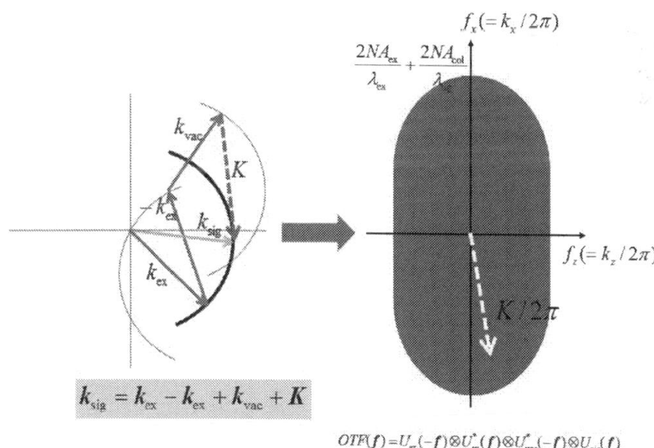

Figure 11. The relation between the phase-matching condition and the CTF for the transmission fluorescence confocal microscopy. Note that $U_{col}^*(-f)=U_{vac}^*(-f)$ and $U_{col}(f)=U_{sig}(f)$

4. Discussion

As stated above, the maximum possible resolution limit is determined by the kind of optical process in force. For illustration, **Figure 12** shows the calculation results of the CTF for CARS, stimulated Raman loss (SRL), SRG, and third-order harmonic generation (THG) microscopy [4]. For reference, the double-sided Feynman diagram and the energy-level diagram describing SRL and THG are shown in **Figure 13**. SRL microscopy and SRG microscopy have the same resolution limit, while CARS microscopy shows better optical resolution than the two former techniques. The CTF of THG microscopy exhibits peculiar properties in which the value of the origin in the spatial frequency domain is zero.

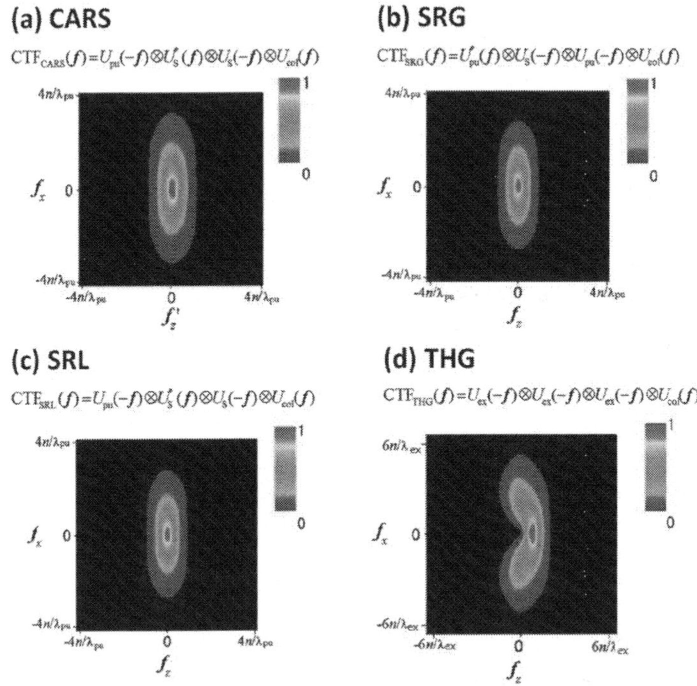

Figure 12. Calculation samples of the CTF for (a) CARS, (b) SRG, (c) SRL, and (d) THG microscopy.

From our theory, some interesting results are also obtained. In reflection coherent microscopy, the uniform region of the susceptibility disappears in the image, as does the interface whose normal is perpendicular to the optical axis. As an example to see the difference between reflection and transmission microscopy, the CTF of transmission and reflection CARS micro-scopy are shown in **Figure 14** [13]. In transmission THG microscopy, the dot and interface of the susceptibility are emphasized in the image and the uniform region vanishes. In reflection

CARS microscopy and transmission THG microscopy, the grating of susceptibility cannot be resolved, but by assembling the interference microscopy where the signal interferes with the local oscillator generated separately, the grating becomes resolved and then the optical resolution can be defined.

In incoherent microscopy such as fluorescence and spontaneous Raman scattering microscopy, the vacuum field as well as the excitation beam are involved in the optical process and contribute to the increase in the frequency cutoff. It is noteworthy about incoherent microscopy that the OTFs of the transmission and reflection microscopy becomes equal. On the other hand, in coherent microscopy such as SHG, THG, CARS, SRG, and SRL microscopy, the CTFs of transmission and reflection microscopy differ from each other.

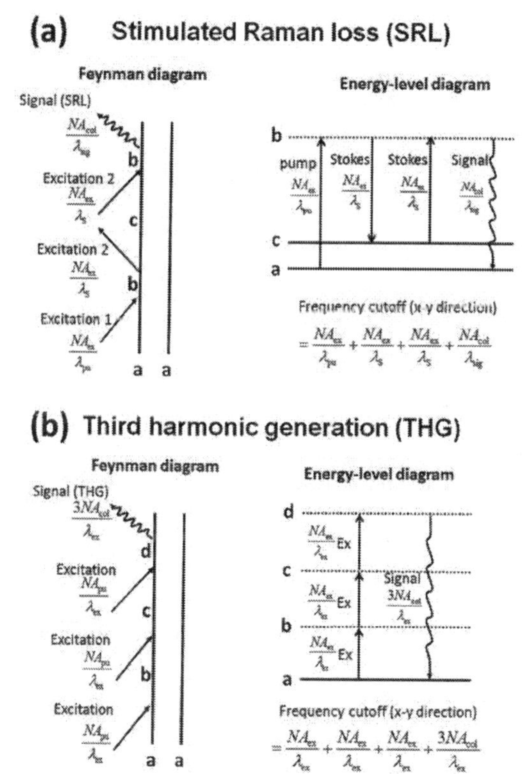

Figure 13. Double-sided Feynman diagram and energy-level diagram for (a) SRL and (b) THG.

Without restricting laser scanning (stage-scanning) microscopy, we can conjecture the following theorem of the resolution limit for all types of microscopy, which visualize $\chi^{(i)}(x, y, z)$ through a variety of optical processes.

4.1. Theorem

- If there is no a priori information on the object (sample), the resolution limit (the maximum value of frequency cutoff) is determined by the diagram describing the optical process. As long as the optical process described by a certain diagram is used to visualize $\chi^{(i)}(x, y, z)$, the resolution limit calculated from the diagram cannot be surpassed regardless of how well the optical apparatus is devised.

The typical exception to the above theorem is localization microscopy, such as photo-activated localization microscopy (PALM) [14] and stochastic optical reconstruction microscopy (STORM) [15], which have a priori information on the object (isolated single-point object). Any microscopy application, including SIM and stimulated emission depletion (STED) microscopy [16], that does not have a priori information on the object should follow this theorem.

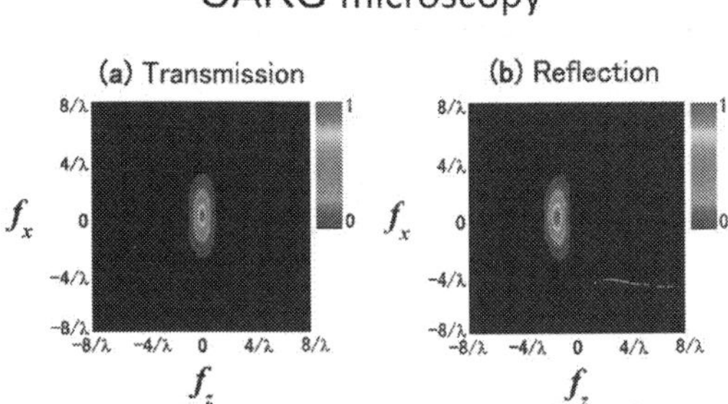

Figure 14. The CTF of transmission and reflection CARS microscopy.

5. Conclusions

We have constructed a theoretical framework to deal with the image formation of all kinds of microscopy by using the double-sided Feynman diagrams and energy-level diagrams describing optical processes. We discovered some rules to evaluate the resolution limit by using these diagrams. Our diagram technique can overview laser microscopy with any optical processes regardless of coherence or linearity. In our framework, the susceptibility distribution is visualized in the image, which blurs based on the optical resolution of each type of microscopy calculated from the diagram technique. Interestingly, in microscopy with an incoherent process, the vacuum field plays a role as part of the excitation light and contributes to the improvement of the optical resolution. In nonconfocal systems, which is commonly used to acquire a high-intensity signal particularly in nonlinear microscopy, the resolution limit of incoherent microscopy is determined by the excitation system only, whereas that of coherent microscopy is determined by both the excitation and signal-collection systems. In SRS microscopy, the transmission type mainly observes the imaginary part of the nonlinear susceptibility, while the reflection type can detect the real part.

Author details

Naoki Fukutake

Address all correspondence to: Naoki.Fukutake@nikon.com

Nikon Corporation, Yokohama, Japan

References

[1] M. Born and E. Wolf. Principles of Optics. 5th ed. Pergamon Press (Oxford); 1974.

[2] M. G. Gustafsson. Surpassing the lateral resolution limit by a factor of two using structured illumination microscopy. J. Microsc. 2000;198:82–87.

[3] M. Gu. Principles of three dimensional imaging in confocal microscopes. World Scientific (Singapore); 1996.

[4] N. Fukutake. Comparison of image-formation properties of coherent nonlinear microscopy by means of double-sided Feynman diagrams. J. Opt. Soc. Am. B. 2013;30:2665–2675.

[5] W. Denk, J. H. Strickler, and W. W. Webb. Two-photon laser scanning fluorescence microscopy. Science. 1990;248·73–76.

[6] W. R. Zipfel, R. M. Williams, and W. W. Webb. Nonlinear magic: multiphoton microscopy in the biosciences. Nat. Biotechnol. 2003;21:1369–1377.

[7] I. Freund and M. Deutsch. Second-harmonic microscopy of biological tissue. Opt. Lett. 1986;11:94–96.

[8] Y. Barad, H. Eisenberg, M. Horowitz, and Y. Silberberg. Nonlinear scanning laser microscopy by third-harmonic generation. Appl. Phys. Lett. 1997;70:922–924.

[9] M. D. Duncan, J. Reintjes, and T. J. Manuccia. Scanning coherent anti-Stokes Raman microscope. Opt. Lett. 1982;7:350–352.

[10] A. Zumbusch, G. R. Holtom, and X. S. Xie. Vibrational microscopy using coherent anti-Stokes Raman scattering. Phys. Rev. Lett. 1999;82:4142–4145.

[11] W. Freudiger, W. Min, B. G. Saar, S. Lu, G. R. Holtom, C. He, J. C. Tsai, J. X. Kang, and X. S. Xie. Label-free biomedical imaging with high sensitivity by stimulated Raman scattering microscopy. Science. 2008;322:1857–1861.

[12] S. Kawata, R. Arimoto, and O. Nakamura. Three-dimensional optical-transfer-function analysis for a laser-scan fluorescence microscope with an extended detector. J. Opt. Soc. Am. A. 1991;8:171–175.

[13] N. Fukutake. Coherent transfer function of Fourier transform spectral interferometric coherent anti-Stokes Raman scattering microscopy. J. Opt. Soc. Am. A. 2011;28:1689–1694.

[14] E. Betzig, G. H. Patterson, R. Sougrat, O. W. Lindwasser, S. Olenych, J. S. Bonifacino, M. W. Davidson, J. Lippincott-Schwartz, and H. F. Hess. Imaging intracellular fluorescent proteins at nanometer resolution. Science. 2006;313:1642–1645.

[15] M. J. Rust, M. Bates, and X. Zhuang. Sub-diffraction-limit imaging by stochastic optical reconstruction microscopy (STORM). Nature Methods. 2006;3:793–796.

[16] S. W. Hell and J. Wichmann. Breaking the diffraction resolution limit by stimulated emission: stimulated-emission-depletion fluorescence microscopy. Opt. Lett. 1994;19:780–782.

Chapter 2

Kinetic Model of Development and Aging of Artificial Skin Based on Analysis of Microscopy Data

Paola Pesantez-Cabrera, Cläre von Neubeck,
Marianne B. Sowa and John H. Miller

Additional information is available at the end of the chapter

http://dx.doi.org/10.5772/63402

Abstract

Artificial human skin is available commercially or can be grown in the laboratory from established cell lines. Standard microscopy techniques show that artificial human skin has a fully developed basement membrane that separates an epidermis with the corneal, granular, spinosal, and basal layers from a dermis consisting of fibroblasts in an extracellular matrix. In this chapter, we show how modeling can integrate microscopy data to obtain a better understanding of the development and aging of artificial human skin. We use the time-dependent structural information predicted by our model to show how irradiation with an electron beam at different times in the life of artificial human skin affects the amount of energy deposited in different layers of the tissue. Experimental studies of this type will enable a better understanding of how different cell types in human skin contribute to overall tissue response to ionizing radiation.

Keywords: artificial human skin, kinetic model, radiation exposure, protection by corneal layer, selective irradiation of epidermis

1. Introduction

Engineered human tissues provide a bridge between in vivo and in vitro studies by enabling the investigation of fundamental cellular mechanisms at a level of detail that is not possible with whole-animal models, while providing a tissue-like context specific to the organ under investigation (reviewed in reference [1]). Such models are routinely used for toxicology and radiation studies [2–6].

Artificial human skin, a well-known example of engineered human tissue, is available commercially but can also be developed from human cell lines in the laboratory [7, 8]. EpiDermFT™ (MatTek, Ashland, VA) is a widely used commercial product that exhibits a fully developed basement membrane that separates the epidermis, with keratinocytes in the corneal, granular, spinosal, and basal layers, from the dermis consisting of fibroblasts in an extracellular matrix.

Microbeam irradiation of EpiDermFT™ and similar skin-tissue models showed that skin exhibits a "radiation-induced bystander effect" (reviewed in reference [9]); cells that are not directly damaged by ionizing particles in the beam nevertheless exhibit biological responses similar to cell that receive direct damage. Hence, it appears that Interactionslayers and cell types plays a role among different layers and cell types play a role in the response of skin to ionizing radiation. Experiments that vary the dose delivered to different layers of artificial skin will contribute to our understand of these interactions.

Cole and coworker [10] showed that, due to its limited penetration in biological materials, the sensitivity of different components of a biological system can be investigated by low-voltage electron-beam irradiation. Recognizing that this technique could be useful in a layered system such as the skin, we calculated the penetration of electron beams of various energies into artificial skin [11] as a basis for the design of experimental studies.

We immediately realized that a detailed analysis of microscopic images of artificial skin was required to ensure accurate calculation of penetration depths. In this chapter, we trace the evolution of our analysis of microscopy data for the purpose of modeling the interaction of artificial skin with electron-beam irradiation. Initially, our focus was on irradiation at a fixed time dictated by experimental procedures coupled with the purchase of EpiDermFT™ from MatTek. Later, we realized that earlier delivery by MatTek and in-house production of artificial skin enabled irradiation studies at different stages in the development of the tissue model. This realization encouraged us to develop a kinetic model of artificial-skin growth and aging that would support simulations of electron-beam exposure at any time during its life cycle.

2. Morphology of the fully developed epidermis of artificial skin

Radiation biology experiments on EpiDermFT™ [6, 12, 14] are usually preformed approximately 3 weeks postseeding of the keratinocytes onto the dermal substrate. At this time, images such as those shown in **Figure 1** reveal the morphology of the mature skin model.

Image A in **Figure 1** (kindly provided by MatTek [13]) uses hematoxylin and eosin (H&E) staining to show the structure of EpiDermFT™ at the time of shipping, about 17 days postseeding of keratinocytes onto the dermal substrate. Horizontal lines added by us approximately delineate layers of the epidermis and suggest that, on average, the spinosal layer is about twice as thick as the basal layer, and granular layers have about the same thickness as the basal layer. Using the scale mark shown in **Figure 1A**, we estimate that the basal, spinosal, and granular layers are about 17, 37, and 17 μm thick, respectively. In this chapter, we refer to

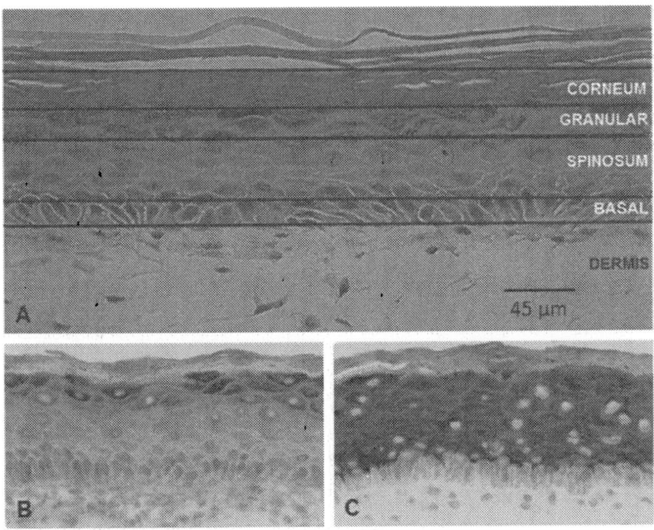

Figure 1. Representative histological sections showing morphology and differentiation in the EpiDermFT™ skin model 17–20 days after the seeding of keratinocytes onto the dermal substrate. Samples A, B, and C were stained for eosin, filaggrin, and keratin10, respectively. For filaggrin and keratin10, positive DAB staining appears dark in the image. All tissues were counterstained with hematoxylin. The scale bar in A applies to B and C as well.

the combined basal, spinosal, and granular layers as the "viable" epidermis to contrast it from the corneal layer of dead cells.

Cells in the stratum corneum have lost most of their intercellular adhesion so that even careful sample handling is likely to induce small air gaps such as those seen in the lower part of the corneal layer in **Figure 1A**. Experimental procedures involved in microscope slide preparation can destabilize the stratum corneum and are, most likely, responsible for large air gaps, such as those seen in the upper part of the corneal layer in **Figure 1A**. Consequently, these images are not a reliable source of data on corneal thickness.

MatTek also provided valuable information about the production of their human skin model. A nutrient layer is seeded with neonatal foreskin basal cells. It is unlikely that this population contains stem cells; hence, the basal layer has finite replication capacity due to transiently amplifying (TA) cell, about 80% of which are in a resting state at any one time. Ca+ ions in the nutrient layer diffuse into the epidermis but are trapped beneath the corneal layer, which creates a Ca+ gradient since water can diffuse into the liquid-air interface. As in normal skin, this Ca+ gradient is most likely responsible for the differentiation of keratinocytes as they are pushed toward the stratum corneum by cell replication in the basal layer.

Images B and C in **Figure 1** are prepared in Dr. Sowa's laboratory approximately 3 weeks postseeding of keratinocytes onto the dermal substrate [14]. When EpiDermFT™ skin samples were received from MatTek, they were placed in 2 ml of maintenance media and incubated at

37°C and 5% CO_2 as the manufacturer instructed. After 2 days, tissues were fixed in 4% paraformaldehyde, dehydrated, and embedded in paraffin wax using standard methods [15].

Five-micrometer sections were prepared using a Leica microtome and mounted onto coated slides (IMEB, Inc., San Marcos, CA). Sections were deparaffinized in xylene and rehydrated via a series of alcohol rinses. Sections were processed for antigen retrieval by immersion in a citrate acid solution (pH 6.0, 99°C) for 30 min, followed by immersion in 3% H_2O_2 for 10 min to block endogenous peroxidase activity. Slides were then washed three times in phosphate-buffered saline (PBS) for 10 min, blocked with 0.3% bovine serum albumin (BSA) for 1 h, and incubated in the primary antibody overnight at 4°C.

Immunofluorescence staining was performed in triplicate to ensure consistent results. Individual tissue sections were imaged using a Nikon Eclipse TE300 microscope with a Nikon Plan APO VC 60x/1.20 water immersion objective. A Retiga 1300 cooled charge-coupled device (CCD) camera (Qimaging) controlled by QCapture software was used to acquire the image. Image processing was performed in Image J (NIH; Bethesda, MD).

Images B and C in **Figure 1** show differentiation profiles for nonirradiated EpiDermFT™ skin-model samples using filaggrin and Keratin10, respectively. Filaggrin-positive staining defines the granular layer only. Keratin10-positive staining defines the combined spinosum and granular layers but is excluded from the basal layer. Image B seems to indicate that the granular layer contains cells both with and without a nucleus. Furthermore, granular cells with nuclei appear to be adjacent to the stratum corneum. Image C shows that cells in the spinosal layer vary in size with smaller cells nearer to the basal layer. We interpret this as evidence for cell growth as keratinocytes traverse the spinosal layer.

Thicknesses of the basal, spinosal, and granular layers revealed by images B and C in **Figure 1** are roughly in the same proportions as that suggested by the horizontal lines in **Figure 1A**. However, absolute thicknesses cannot be compared due to both sample variability and differences in ages of the sample when the images were acquired. Due to the time spent in shipping and equilibration, images B and C are for samples approximately 4 days older than the sample shown in **Figure 1A**. A significant shrinkage of the viable epidermis (basal, spinosal, and granular layers) during this time was reported in reference [14].

The thickness of the viable epidermis at any given point is stochastic, depending on the particular arrangement of cells in the basal, spinosal, and granular layers. One way to average over this intrinsic variability is to measure area over a fixed width, also called "field of view." This technique was used in reference [14] to measure the thickness of the viable epidermis in samples stained using a standard H&E protocol [15]. As stated above, MatTek normally ships EpiDermFT™ 17 days postseeding of keratinocytes onto the dermal substrate; however, they will ship samples at earlier time points in sample production if requested. Early shipment was desirable for some of the radiation biology studies conducted by Sowa and coworkers [9, 14], which allowed them to measure the thickness viable layers of the epidermis starting on day 17 postseeding. Results of these measurements are shown in **Table 1**.

Day	Relative thickness
17	1.00 ± 0.04
18	1.14 ± 0.06
19	0.93 ± 0.05
20	0.93 ± 0.06
21	0.76 ± 0.04
22	0.90 ± 0.16
23	0.64 ± 0.08
24	0.42 ± 0.06

Table 1. Relative thickness of the viable epidermis as a function of sample age.

In **Table 1**, the data have been normalized by the mean of areas observed on day 17 and uncertainties are ±1 standard deviation. Shrinkage starts between days 18 and 19 after an increase in thickness between days 17 and 18. With the exception of day 21, shrinkage of the viable epidermis is relatively minor between days 19 and 22. After day 22, shrinkage is linear with the thickness decreasing about 25% each day.

The biological reason for these changes in the thickness of the combined basal, spinosal, and granular layers is not clear. This is particularly true between days 17 and 22 when the variation is not systemic. The severe shrinkage after day 22 suggests that a dramatic change has occurred in the viable epidermis, possibly due to the exhaustion of replication capacity in the basal layer. If we assume that shrinkage is uniform across the basal, spinosal, and granular layers, we can easily include this shrinkage when we convert cell count from our kinetic model into layer thickness.

3. Properties of artificial skin epidermis revealed by confocal microscopy

The microscope images described in the previous section provide the quantitative information needed to model the viable layers of EpiDermFT™ skin-tissue samples at about 20 days after keratinocytes were seeded onto the dermal substrate. These images of stained vertical slices through the epidermis frequently display large air gaps in the corneal layer, such as those shown in **Figure 1A**, which we attribute to sample preparation. Hence, these images cannot provide reliable quantitative information about the thickness of the corneal layer. To obtain this type of information, we analyzed three-dimensional (3D) confocal microscopy of live samples [6].

Samples were stained overnight with SYTO13 and SYTO59 fluorescent nucleic acid stains (Invitrogen). Two colors (red and green) were chosen because of their high contrast and strong overlap with the excitation lasers of the confocal microscope. Both dyes were used at a final concentration of 10 µM in 3 ml of media. Stained samples were washed with PBS and placed in a 35-mm culture plate containing sufficient PBS to cover the tissue. To minimize sample movement during imaging, samples were placed on a thin coating of autoclaved petroleum jelly prior to the addition of the PBS. Images were acquired in 1-μm z-steps on a Zeiss laser scanning microscope (LSM) 710 scanning head confocal microscope with a Zeiss plan apo 40×/ 1.1 objective. Excitation lasers were 488 and 633 nm for the green and red emission channels, respectively. Laser dwell times were 1.27 μs for both channels. Two-dimensional (2D) and 3D image analyses were carried out using Volocity (Perkin Elmer, Waltham, MA).

Confocal microscopy of EpiDermFT™ skin-model samples treated with fluorescent nucleic acid stains, as described above, revealed the location of nuclei in the samples. **Figure 2** illustrates a detailed analysis of a confocal microscopy image designed to measure the thicknesses of the corneal layer and the viable epidermis at 20 days after seeding of keratinocytes onto the dermal substrate.

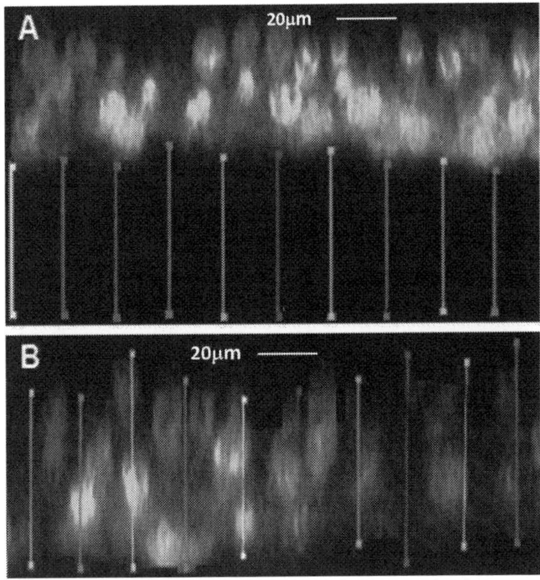

Figure 2. Side view of a 3D image of EpiDermFT™ epidermis obtained by confocal microscopy following treatment with fluorescent nucleic acid stains at 20 days after seeding of keratinocytes onto the dermal substrate. Vertical lines illustrate repeated measurements of the thickness of regions that did (B) and did not (A) take up the stain.

Vertical lines in panel B of **Figure 2**, which span the thickness of the epidermis where stained nuclei were observed by confocal microscopy, have an average length of 45.2 ± 0.7 μm. This distance agrees with the thickness of the combined basal and spinosal layers from **Figure 1A** after we allow for shrinkage between days 17 and 20, shown in **Table 1** (0.93 ± 0.06 × 54 μm = 50.2 ± 3.2 μm). For this association between information derived by two different types of microscopy to be valid, we must assume that granular cells do not take up the DNA stains, even though, as mentioned above, H&E-stained tissue sections seem to show nuclei in some granular cells.

Vertical lines in panel A of **Figure 2**, which span the thickness of the epidermis where DNA staining is not observed, have an average length of 55.0 ± 0.41 μm. Based on the assumption that granular cells do not take up the DNA stain, the combined granular and corneal layers have a thickness of 55.0 ± 0.41 μm on day 20. Assuming further that the shrinkage observed on day 20 is uniform across the viable epidermis, we calculate the thickness of 16 ± 1 μm ($17 \times 0.93 \pm 0.06$) for the granular layer on day 20, which allows us to estimate the corneal-layer thickness as 39 ± 1.41 μm ($55 \pm 0.41 - 16 \pm 1$) on day 20.

By itself, this estimate of the corneal-layer thickness on day 20 is not extremely useful; however, in conjunction with a kinetic model for the number of cells in the stratum corneum, it can be used to estimate the thickness of the corneal layer at all times after its first appearance 7–10 days postseeding of keratinocytes onto the dermal substrate. If we assume that the thickness of corneal cells is constant, then the ratio of cell number to layer thickness on day 20 is the same at all times. Our kinetic model, described in Section 5, was developed to predict corneal cell populations, which can be converted into corneal thickness.

Figure 3 shows the distribution of nuclei in the region of the epidermis that took up the DNA stains. To determine the number of nuclei in a specific volume at a specific depth, Z-stacks of optical sections were reconstructed in 10-μm sections starting at the bottom of the basal layer. Individual nuclei were identified by a threshold on the florescence intensity and the number in each 10-μm section was counted. The 2D image area was approximately 60,000 μm².

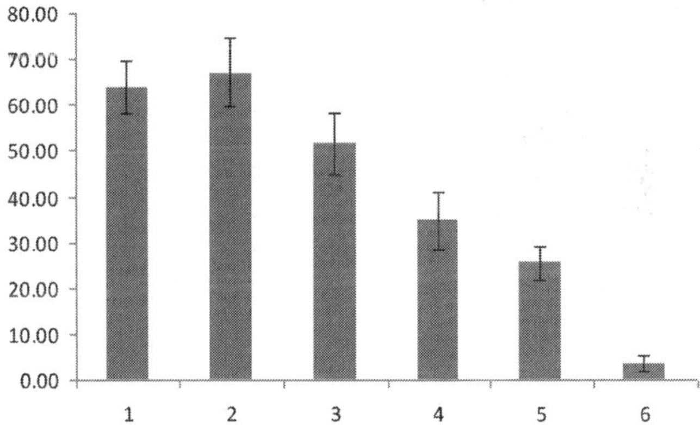

Figure 3. Number of nuclei (vertical axis) in successive 10-mm thick slices numbered 1-6 on the horizontal axis of a 3D confocal microscope image of EpiDermFT™ skin model starting on the basement membrane and ending in the granular layer.

The first two bars of the histogram in **Figure 3** are counts of nuclei in the basal layer and the start of the spinosal layer. Between 20 and 50 μm above the basal membrane, the number of nuclei counted in 10-μm sections decreases linearly. We interpret this decrease in nuclei per unit area as evidence for increasing cell volume as keratinocytes pass through the spinosal layer. A significant departure from this linear behavior occurs in the count of nuclei between

50 and 60 µm above the basal membrane. We associate the small number of nuclei in this 10-µm section as a small number of cells at the base of the granular layer that took up the DNA stain. The deviation from the linearity of nuclei counts between 50 and 60 µm is interpreted as a decrease in the number of cell susceptible to DNA staining, not to an increase in cell volume. Our kinetic model assumes that keratinocytes in the granular layer have the same volume as those cells at the top of the spinosal layer.

4. Simulation of electron-beam penetration into skin

PITS [16] is a radiation transport code that performs an event-by-event simulation of charged particles transferring their kinetic energy to electronic excitations of a medium. Simulated tracks are data objects containing, among other things, the Cartesian coordinates of energy deposition events. Properties of elastic collisions are not retained in the track object except as they influence the position of inelastic interactions in the stopping medium. Condensed phase effects [17] are included as described by Wilson et al. [18]. Primary electrons and all generations of secondary electrons are followed until their energy falls below 10 eV, the lowest ionization threshold in Dingfelder's model [17], after which residual kinetic energy is assigned to a final transfer point cast in an isotopically random direction and at an exponentially random distance.

For each electron-beam energy considered, 10^5 independent primary electrons were simulated as they transferred their energy to a liquid-water medium. For layers of the epidermis containing live cells, liquid-water provides a reasonable approximation due to their high water content; however, this approximation may not be valid for the corneal layer due to the low water content and small air gaps.

The density of the cellular material in the stratum corneum is slightly greater than water [19] but the presence of air gaps makes the average corneal density less than that of its cellular material. To allow for these competing factors in determining a water thickness with equivalent mass per unit area as the corneal layer, we used microscope images such as in **Figure 4**, to quantify the relative area of small air gaps. We consider these small air gaps as intrinsic to the corneal layer under normal sample handling during experiments.

Microscope slides of Vertical slices through EpiDermFT™ were prepared by methods descri-bed in Section 2 and stained with H&E (IMEB, Inc.). A Nikon Eclipse TE300 inverted micro-scope with a Nikon Plan APO 20/0.75 objective was used to image individual tissue slices. A Retiga 1300 cooled CCD camera (Qimaging) controlled by Volocity Acquisition (Improvision) software was used to acquire images and to take area measurements. A 12-bit gray scale of intensity determined the relative proportions of high- and low-density materials in the image. High-density material was associated with pixels with normalized gray scale intensity between 0.5 and 15%. The air pockets in the stratum corneum layer were associated with pixels in the 23–100% gray level intensity range.

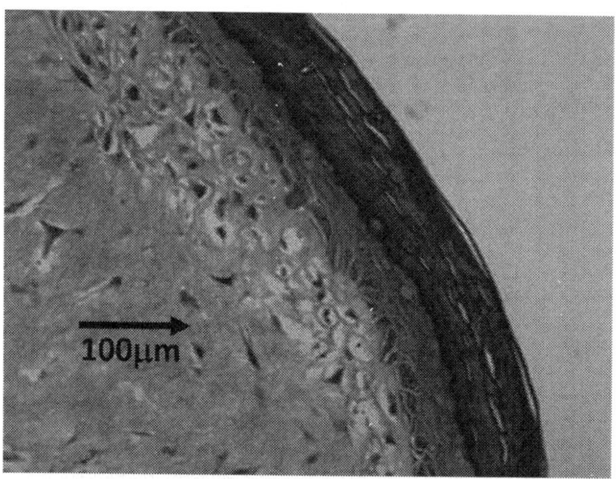

Figure 4. Microscope image of a vertical slice through the EpiDermFT™ skin-tissue model showing small air gaps in the corneal layer.

Area measurements of high and low intensity performed on 14 distinct regions of the stratum corneum gave a mean relative area of low-density material of 9.53% with a standard deviation of the mean 0.81%. Given the large number of cross-sections sampled, it is reasonable to assume that the relative volume of air in the stratum corneum is about 10%

Weigand et al. [19] measured the buoyant density of the stratum corneum cellular material from Caucasian and Black subjects by several techniques. They concluded that the sucrose density gradient method gave values closest to that of the natural state. Averages of repeated experiments with Caucasian samples, which should apply to the EpiDermFT™ skin model, ranged from 1.075 to 1.145 g/ml. Consequently, the increase in equivalent water thickness to account for the higher density of cellular material in the stratum corneum is approximately equal to the decrease in the thickness of water to account for air pockets. We conclude from these results that calculations of electron-beam penetration in a uniform water medium provide a reasonable approximation to the penetration of the EpiDermFT™ skin model, including the corneal layer.

Figure 5 shows the cumulative probability distribution of samples of the largest z-coordinate of energy transfer points in simulated tracks of 90-keV electrons stopping in a uniform liquid-water medium. Percentile points of this distribution give the thickness of water required to stop a specified fraction of electrons injected into the medium. For example, about 40 μm of water is required to stop about half of the electrons in a 90-keV beam. The dashed curve in **Figure 5** shows the fraction of beam energy deposited by events not exceeding a specified depth. The point a zero depth shows that about 12% of the beam energy is backscattered. By interpolation between calculations at 60 and 70 μm, we estimate that 4% of the beam energy is deposited at depths that exceed the 90th percentile of penetration.

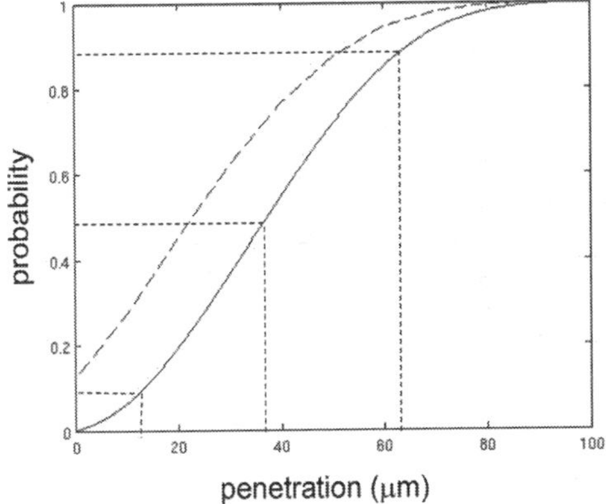

Figure 5. Penetration and energy deposited by a 90-keV electron beam in water. Solid curve is the cumulative probability of maximum z-coordinate of energy transfer points. Intersecting horizontal and vertical dashed lines show 10th, 50th, and 90th percentiles of penetration. Dashed curve shows the fraction of beam energy deposited by events not exceeding a specified depth, with the point a zero depth showing the fraction of beam energy backscattered. Reprinted with permission from Radiation Research.

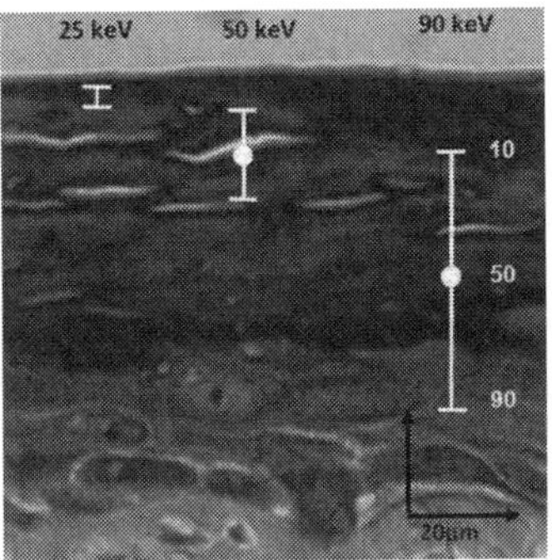

Figure 6. Microscope image of a vertical slice through the EpiDermFT™ skin-tissue model overlay showing calculated penetration of 25-, 50-, and 90-keV electron beams. Bars cover the 10–90th percentile of penetration with the 50th percentile at the center. Reprinted with permission from Radiation Research.

The overlay in **Figure 6** shows the penetration of 25-, 50-, and 90-keV electron beams superimposed on a microscope image of a vertical slice through the EpiDermFT™ skin model. The uncertainty bar is centered on Z_{p50} and the upper and lower extremes are Z_{p10} and Z_{p90}, respectively, where Z_{pX} is the depth to which X% of the electron beam is expected to penetrate. It is clear from the overlay in **Figure 6** that beam energies near 90 keV are required to irradiate keratinocytes in the epidermis that are undergoing active cell division, which is usually the population of greatest interest in radiation biology.

5. Kinetic model of epidermis formation and aging

In this section, we present a kinetic model for the development of the basal, spinosal, granular, and corneal layers of the EpiDermFT™ skin-tissue model. The main purpose for developing this model is to estimate the corneal layer thickness from the start of its formation 7–10 days after seeding of keratinocytes onto the dermal substrate to the end of the useful life of the sample for radiation biology studies due to shrinkage. Our model predicts the kinetics of the number cells in the corneal layer. If we assume that the thickness of a corneal cell is constant, then the measure of corneal layer thickness at 20 days (see Section 3) can be used to convert cell count into layer thickness at any time.

A kinetic model of artificial-skin development was reported in reference [9] that focused on the viable epidermis for comparison with experimental studies of skin homeostasis after exposure to ionizing radiation [9, 14]. The model in reference [9] starts with a rapid expansion of TA basal cells, with the population increasing sixfold in about 12 days. By contrast, the kinetic model described here starts with a confluent monolayer of the basal cells and the total basal-cell population, consisting of cycling TAs, noncycling TAs with replication potential, and the basal cells that have exhausted their replication capacity, is constant throughout the simulation.

As in reference [9], we assume that TA cells exhaust their replication capacity after five to three cycles, but the current model includes the information from MatTek that only about 20% of TA cells are cycling at any given time. This means that as individual TA cells exhaust their replication capacity, they are replaced by a TA cell with full replication capacity that has been held in reserve. The biological mechanism for delayed cycling of TA cells with replication capacity is unclear; however, as **Figure 7** shows, this modeling assumption allows the cycling TA population to be constant throughout most of our kinetic simulation. Replication in the

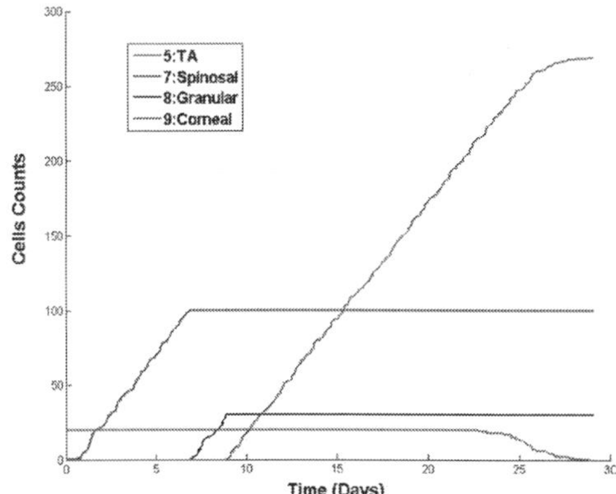

Figure 7. Predicted kinetics of cell count in the basal (orange), spinosal (blue), granular (black), and corneal (red) layers of the epidermis of an artificial-skin tissue, starting from the time with the basal layer is confluent.

basal layer goes to zero after about day 23 when reserve replication capacity is no longer available.

In the current model, the differentiation of keratinocytes is determined by their height above the basement membrane. We think this more correctly models differentiation driven by a Ca + gradient than the purely time-dependent transitions of the kinetic model in reference [9]. Height-dependent differentiation means that a spinosal cell is the consequence of each TA-cell replication. TA cycling produces two daughter TA cells, but some basal cells must move to the spinosal layer because the basal-cell layer is confluent

As can be seen in **Figure 7**, the spinosal-cell population is non-zero after the first TA-cell replication, which occurs at TA cycle time chosen randomly from a lognormal distribution with a mean TA cycle time of 31.2 h. The lognormal distribution of TA cycle times and its variance are the same as in reference [9] but the mean TA cycle time in the current model is significantly shorter. A shorter mean cycling time for TA cells in the current model is a direct consequence of our assumption, based on information from MatTek [15] that only about 20% of TA cell with replication potential are cycling at any given time. Consequently, a shorter mean cycling time is required to generate complete spinosal and granular layers 7–10 days postseeding of keratinocytes than is the case when all TAs with replication potential are cycling.

In our kinetic model, the corneal layer begins to form as soon as the granular layer reaches its full thickness. Exposure to air is crucial to the formation of a corneal layer but this requirement is not explicitly included in our model. **Figure 7** also shows that the rates of increase of spinosal, granular and corneal cell populations are all nearly equal, aside from random fluctuations, to a constant value determined by the rate of TA-cell replication in the basal layer. The rates of increase in the thickness of the spinosal, granular, and corneal layers are not the same because the volumes of individual cells in those layers are different. In addition, the volume of a spinosal cells is increasing with time since it entered the stratum spinosum.

Sharp transitions between increasing and constant cell populations in the spinosal and granular layers are an artifact of our model in which cell type is determined by a specified height of the cell above the basement membrane. If a TA cycle, which adds a cell to the stratum spinosum, makes the upper boundary of the spinosal layer exceed the height above the basement membrane allowed for spinosal cells, the oldest cell in the current stratum spinosum becomes a granular cell. A similar modeling assumption governs the transition of granular to corneal cells.

The transition between linearly increasing to constant corneal cell population is not sharp. This transition mirrors the loss of replication capacity in the basal layer, which occurs over about a 4-day time period. In **Figure 7**, this loss of replication capacity occurs between days 23 and 29. Various model parameters, including the mean TA cycle time and the fraction of cycling TA cells, determine the duration of replication capacity in the basal layer. parameters affect the time when replication capacity is exhausted relative seeding of keratinocytes onto the dermal substrate. We think that the decay of replication seen in **Figure 7** is reasonable because it correlates roughly with the onset of significant shrinkage of the viable epidermis observed experimentally [14] (see **Table 1**). After replication capacity is exhausted, cell count in all layers is constant. Cell count in the basal layer is always constant but transitions between cycling, resting with replication potential, and sterile TA cells occur throughout the kinetic simulation.

Even though cell count in the viable epidermis remains constant after complete formation of spinosal and granular layers, its thickness is not constant due to shrinkage. On converting cell count to layer thickness after shrinkage begins, we assume that its effect is uniform across all three layers, basal, spinosal, and granular, of the viable epidermis. After day 18, we assume that the volume of all cells in the viable epidermis decreases in accordance with the shrinkage fractions in **Table 1**. Although no data are available, it seems reasonable that corneal cells do not shrink since they are dead and have lost most of their water content. Hence, we can use the corneal thickness of 39 ± 1.41 μm deduced from confocal microscopy on day 20 to convert corneal cell count to corneal thickness at any time after its appearance, about day 8.5 shown in **Figure 7**.

A typical simulation begins by assigning a maximum of 3, 4, or 5 replications randomly to 100 basal cells, 20 of which are chosen randomly to be expressing this replication potential at the beginning of the simulation. The 20 active TA cells are assigned cycle times randomly drawn from a lognormal distribution with a variance of 0.2 and a mean cycling time chosen to generate complete spinosal and granular layers 7–10 days postseeding of keratinocytes, as reported by MatTek [15].

The active TA cell with the shortest cycling time is selected for replication, which forces a randomly selected TA cell off the basement membrane to become a spinosal cell and leaves a TA cell on the basement membrane that is capable of one less replication. The current simulation time is upgraded and a new random cycle time is assigned to the TA cell that just replicated, if it still has replication capability. If not, the TA cell that just replicated becomes a permanently resting TA cell and is replaced by a TA cell that can replicate, if the pool of TA cells with replication capacity is not empty.

If we knew the growth rate of spinosal cells, we would assign the spinosal cell generated by TA-cell replication a birth time equal to the current simulation time and volume equal to that of a basal cell. The age of any preexisting spinosal cells would increase by the TA cycle time and their volume would increase to reflect growth during that cycle time. The growth rate of spinosal cells cannot be determined from the available data; however, as discussed below, the difference in the volume of spinosal cells at different heights in the spinosal layer can be estimated from the data in **Figure 3**. This allows us to calculate the average volume of a spinosal cell, which we assign to all cells in the spinosal population.

After each TA-cell replication, we find the active TA cells with the shortest cycling time, which determine the size of the next time step in the simulation. However, before that time step can be taken, we must determine if a spinosal cell needs to transition to a granular cell, based on the current thickness of the spinosal layer. From the microscopy images in **Figure 1**, it is clear that the spinosal and granular layers are confluent; hence, the volume of the spinosal and granular layers at any time in the simulation is the sum of the volumes of the cells in those layers. The current thickness of the spinosal and granular layers is their current total volume divided by the area of skin being simulated, which is 100 times the cross-sectional area of a basal cell. If the current thickness of the spinosal layer exceeds the observed thickness of a fully formed spinosal layer, a spinosal cell transitions to a granular cell with an appropriate average volume change.

We assume that all granular cells have the same volume that does not change with time. As with the spinosal layer, we estimate the current thickness of the granular layer as its population volume divided by the simulation area. If this total thickness exceeds that observed for a fully developed granular layer, a granular cell transitions to a corneal cell. Updating the count of corneal cells, if needed, completes a simulation time step.

To conduct the series of steps in the simulation described above requires parameters derived from the analysis of microcopy data. A vertical slice through the skin-tissue model imaged by H&E staining 17 days postseeding of keratinocytes allowed us to estimate values of 17, 37, and 17 μm for the thickness of the basal, spinosal, and granular layers, respectively, after they are fully developed and before shrinkage begins. We compare these thicknesses to layer thicknesses after each time step of the simulation to determine if changes in cellular populations are needed. To conduct this test, we need the area of artificial skin being modeled. In our simulation, the basal-cell layer is always a confluent monolayer; hence, the mean separation of basal-cell nuclei, observed to be 14.9 μm by confocal microscopy after DNA staining [6], is an estimate of the mean lateral thickness of the basal cells. By this method, we estimate the cross-sectional area of the basal cells to be 14.9 μm^2, so a typical area of skin in our simulations is $100 \times (14.9\ \mu m)^2 = 2.22 \times 10^4\ \mu m^2$.

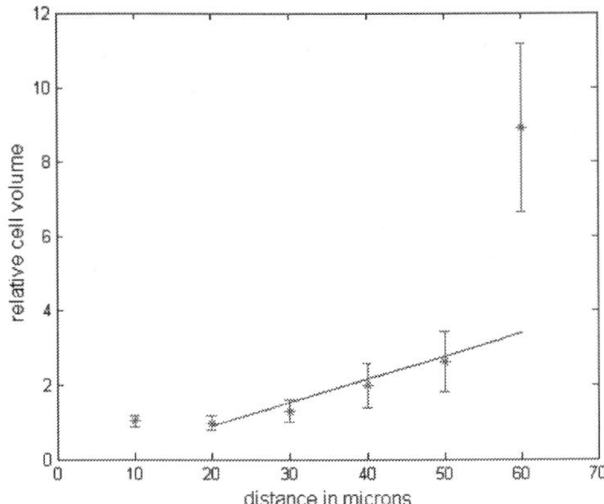

Figure 8. Change in the volume per cell as a function of distance above the basement membrane. Data were normalized to the volume of the basal cells by averaging the results for slices of 3D confocal microscopy data 0–10 and 10–20 μm above the basement membrane. The line fit to data at heights greater than 20 μm mainly reflects the changing volume of spinosal cells.

The analysis of microcopy data discussed in this chapter does not yield a direct measure of spinosal-cell growth rate. However, the analysis of confocal microcopy data for the distribution of nuclei as a function of height above the basement membrane, shown in **Figure 3**, can be used to estimate the change in spinosal-cell volume as a function of height above the basement membrane. As discussed in Section 3, the number of nuclei at various heights above the basement membrane was estimated by counting the number of stained object in a 10-μm slice with an image area of 60,000 μm². Dividing the volume of the slice, 600,000 μm³, by the observed count, we get an estimate of the volume per cell, assuming that each cell has a single nucleus. The counts of nuclei in slices 10 and 20 μm above the basement membrane mainly reflect the volume of the basal cells. The average of calculated volume per cell at 10 and 20 μm was used to normalize the data; hence, the data shown in **Figure 8** are the relative change in cell volume versus height above the basement membrane.

The linear trend in the relative volume per cell between 20 and 50 μm above the basement membrane is not reflected in the data at 60 μm. Based on **Figure 1A** and allowing for shrinkage on day 20 (see **Table 1**), we estimate that the combined thickness of the basal and spinosal layers on day 20, when the confocal data were obtained, is 50.2 μm. Hence, the data shown in **Figure 8** at 60 μm are mainly due to nuclear staining in the granular layer and reflect the uptake of the DNA stains more than the size of cells in the granular layer.

As the results in **Figure 8** show, a linear model is reasonable for the change of spinosal-cell volume as a function of their height in that layer. We express this model as

$$v_s(h) = v_b(1 + (g - 1)(h - \Delta_b) / \Delta_s \tag{1}$$

where v_b is the volume of a basal cell, $v_s(h)$ is the volume of a spinosal cell at height h above the basement membrane, Δ_b is the thickness of the basal layer, and Δ_s is the total thickness of the spinosal layer. The parameter g is the ratio of the volume of a granular cell to a basal cell, which can be verified by applying Eq. (1) at the maximum height of the spinosal layer. This parameter can be estimated by setting $(g-1)/\Delta_s$ to the slope of the linear fit shown in **Figure 8**, which gives a value of $g = 3.32$.

Equation (1) is not directly useful in our kinetic simulation because we do not know the height above the basement membrane of individual cells in the spinosal layer; however, since $v_s(h)$ is linear, we can easily calculate the average volume of a spinosal cell:

$$< v_s > = (v_b + v_g) / 2 = v_b(1 + g) / 2 = 2.16 v_b \tag{2}$$

We assign this average volume to every spinosal cell, which allows us to calculate the thickness of the spinosal layer at any time from the number of spinosal cells at that time. As explained above, knowing the thickness of the spinosal layer at any time is sufficient to determine when a spinosal cell makes a transition to a granular cell.

Using an average volume of spinosal cells also simplifies the relationship between the mean TA-cell cycling time and the mean time t_0 required in our kinetic simulation to produce a fully developed via epidermis, which marks the beginning of a corneal layer by exposed to the liquid-air interface. If the confluent basal-cell population is N and a fraction p of basal cells is cycling with a mean time T_c, then

$$t_0 = (T_c \, / \, p)(N_s \, / \, N + N_g \, / \, N) \tag{3}$$

where N_s and N_g are the number of spinosal and granular cells, respectively, in the fully developed spinosal and granular layers. Since $N = A\Delta_b/v_b$, where A is the area of epidermis being simulated, $N_s = A\Delta_s/{<}v_s{>}$, and $N_g = A\Delta_g/v_g$, where Δ_g is the thickness of the granular layer, Eq. (3) becomes

$$t_0 = (T_c \, / \, p)[(\Delta_s \, / \, \Delta_b)(v_b/ {<}v_s{>} +(\Delta_g \, / \, \Delta_b)(v_b \, / \, v_g)] \tag{4}$$

Taking the average of 7 and 10 days as the typical time when the corneal layer begins [15], Eq. (4) predicts the mean TA cycling time of 31.2 h. This completes the parameter estimates for our kinetic simulation.

6. Effect of corneal screening on irradiation of the viable epidermis

The results of our kinetic simulation of epidermis development shown in **Figure 7** together with the estimate of 39 ± 1.41 μm for corneal layer thickness at 20 days postseeding of keratinocytes onto the dermal substrate enable the calculation of the protection that the corneal layer provides to live cells in the viable epidermis. We only model radiation exposure delivered after day 8.5, model radiation exposures delivered after day 8.5, the day selected in our model for first appearance of the corneal layer based on information from MatTek [15] that viable epidermis (basal, spinosal, and granular layers) is completed 7–10 days postseeding of keratinocytes onto the dermal substrate.

The thickness of the corneal layer on the day of irradiation was obtained from the results for cell count in the corneal layer, shown in **Figure 7**, after combining several simulations to reduce random fluctuations and normalizing to a thickness of 39 μm on day 20. From day 8.5 to about day 25, the increase in corneal thickness is linear. Beyond day 30, its thickness is constant since the replication capacity of the basal layer has been exhausted.

Based on results from simulations of a 90-keV electron beam stopping in a liquid-water medium [11], we developed an interpolation procedure for the amount of energy deposited in a layer of a given thickness at a given depth in the medium. As shown in Section 4, a liquid-water medium is a reasonable approximation to the epidermis of skin, including the corneal layer. Hence, our interpolation procedure allows the accurate prediction of radiation exposure to specified regions of the epidermis without the computationally intense, event-by-event simulation of 90-keV electrons penetrating an aqueous medium. The dermal substrate, which is composed of human fibroblast in a collagen matrix, is thick enough to stop a 90-keV electron beam; hence, we estimate the energy deposited in the dermal substrate as the residual energy of electrons if they are not stopped in the epidermis.

The curve in **Figure 9** referred to the left-hand vertical axis shows our prediction for the energy deposited in the combined spinosal and basal layers depending on the day when the tissue was treated relative to the seeding of keratinocytes onto the dermal substrate. Spinosal and basal layers contain most of the cells with nuclei and that are undergoing growth and replication; consequently, they are the most sensitive to radiation exposure.

The results shown in **Figure 9** allow for the shrinkage of the viable epidermis observed to start on day 19 but with little significance until day 23, with the exception of day 21 where shrinkage is greater than on days 20 or 22 (see **Table 1**). The energy deposited in layers that contain nucleated cells decreases from a value about 50 keV per incident electron for irradiation on day 8.5, when the corneum is just beginning to form, to a value about 1 keV for irradiation on day 25. The lack of smoothness in these calculations after day 18 is due to observed shrinkage of the viable epidermis that begins on day 19.

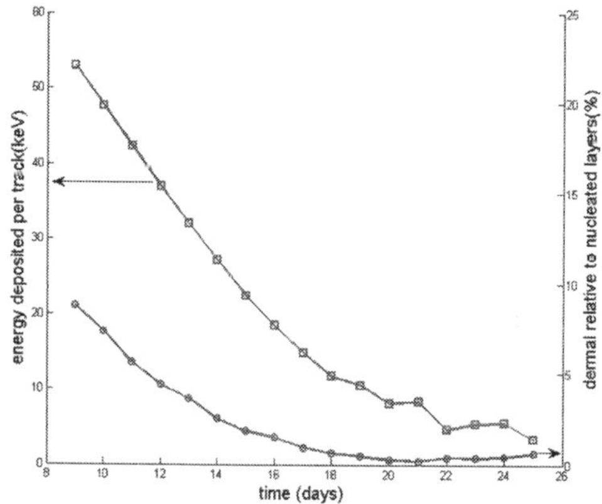

Figure 9. Effects of increasing corneum thickness on energy deposition in the combined basal and spinosum layers where keratinocytes are undergoing cell division and growth (left-hand vertical axis). The curve referred to the right-hand vertical axis is the energy deposited in the dermal substrate relative to that deposited in the combined basal and spinosal layers.

The curve in **Figure 9** referred to the right-hand vertical axis shows the energy deposited in the dermal substrate relative to that deposited in nucleated cells of the epidermis. When the corneal layer is thin, a significant fraction of the energy deposited reaches the fibroblast in the dermal substrate but this exposure is near zero for irradiation on day 21 when the corneal thickness has increased to about 40 μm. Comparing tissue responses at these different exposure times will not only show how screening by the corneal layer protects viable skin cells but may also reveal interactions between dermis and epidermis that are present when both components are exposed and absent when essentially all the radiation exposure is to epidermis alone.

Figure 9 suggests that irradiations performed around day 20 would deposit significant energy in the basal and spinosal layers with minimal exposure to the dermal substrate.

7. Conclusions

As artificial organotypic cell cultures become more widely used in research, their characterization becomes increasingly important. Morphology is a fundamental part of this characterization that can be accomplished through quantifiable microscopy techniques with sample preparation that preserves organ structure. The research described in this chapter concerned a part of artificial skin, the corneal layer, which is particularly prone to distortion in standard microtome-based slide preparation. To circumvent this difficulty, we correlated data from histologic staining methods [14] with data from 3D images of artificial-skin samples acquired by confocal microscopy [6]. However, this correlation of different types of microscope data required assumptions about the uptake of DNA stains by granular cells and a kinetic model to amplify the usefulness of confocal microscope data obtained at only one time point. In future work, direct measurements of corneal thickness can be obtained by new non-destructive techniques [20]. Despite these limitations on the available data, we achieved our objective of modeling the decrease in radiation exposure to live cells in the epidermis with increasing corneal thickness as the age of artificial skin samples increases.

Author details

Paola Pesantez-Cabrera[1], Cläre von Neubeck[2,3], Marianne B. Sowa[4] and John H. Miller[5*]

*Address all correspondence to: jhmiller@tricity.wsu.edu

1 School of Electrical Engineering and Computer Science, Washington State University, Pullman, WA, USA

2 German Cancer Consortium (DKTK), Dresden and German Cancer Research Center (DKFZ), Heidelberg, Germany

3 OncoRay, National Center for Radiation Research in Oncology, Faculty of Medicine and University Hospital Carl Gustav Carus, Technische Universität Dresden, Dresden, Germany

4 Space Biosciences Research Branch, NASA Ames Research Center, Moffett Field, CA, USA

5 School of Electrical Engineering and Computer Science, Washington State University Tri-Cities, Richland, WA, USA

References

[1] Griffith LG, Swartz MA. Capturing complex 3D tissue physiology in vitro. Nat Rev Mol Cell Biol 2006; 7:211–24.

[2] Belyakov OV, Mitchell SA, Parikh D, Randers-Pehrson G, Marino SA, Amundson SA, et al. Biological effects in unirradiated human tissue induced by radiation damage up to 1 mm away. Proc Natl Acad Sci U S A 2005; 102:14203–8.

[3] Sedelnikova OA, Nakamura A, Kovalchuk O, Koturbash I, Mitchell SA, Marion SA, et al. DNA double-strand breaks form in bystander cells after microbeam irradiation of three-dimensional human tissue models. Cancer Res 2007; 67:4295–302.

[4] Prise KM, Schettino G, Vojnovic B, Belyakov O, Shao C. Microbeam studies of the bystander response. J Radiat Res (Tokyo) 2009; 50 Suppl A:A1–6.

[5] Sowa MB, Chrisler WB, Zens KD, Ashjian EJ, Opresko LK. Three-dimensional culture conditions lead to decreased radiation induced cytotoxicity in human mammary epithelial cells. Mutat Res 2010; 687:78–83.

[6] von Neubeck C, Shankaran H, Geniza M, Kauer P, Robinson J, Chrisler W, et al., Integrated experimental and computational approach to understand the effects of heavy ion radiation on skin homeostasis. Integr Biol 2013; 5:1229–45.

[7] Gangatirkar P, Paquet-Fifield S, Li A, Rossi R, Kaur P. Establishment of 3D organotypic cultures using human neonatal epidermal cells. Nat Protoc 2007; 2:178–86.

[8] Vaughan MB, Ramirez RD, Brown SA, Yang JC, Wright WE, Shay JW. A reproducible laser-wounded skin equivalent model to study the effects of aging in vitro. Rejuvenation Res 2004; 7:99–110.

[9] Hei TK, Ballas LK, Brenner DJ, Geard CR. Advances in radiobiological studies using a microbeam. J Radiat Res (Tokyo) 2009; 50 Suppl A: A7–12.

[10] Zermeno A, Cole, A. Radiosensitive structures of metaphase and interphase hamster cells as studied by low-voltage electron beam irradiation. Radiat Res 1969; 39:669–84.

[11] Miller JH, Suleiman A, Chrisler WB, Sowa MB. Simulation of electron beam irradiation of skin tissue model. Radiat Res 2011; 175:113–8.

[12] Yang F, Waters KW, Miller JH, Gristenko MA, Zhao R, Du X, et al Phosphoproteomic profiling of human skin fibroblast cells reveals pathways and proteins affected by low doses of ionizing radiation. PLoS One 2010; 5(11) e14152 DOI: 10.1371/journal.pone. 0014152

[13] Mat Tek Corporation. The EpiDermFT™ skin model.http://www.mattek.com/pages/ products/epidermft/October 2, 2012

[14]   von Neubeck C, Geniza M, Kauer P, Robinson J, Chrisler W, Sowa MB. The effect of low dose ionizing radiation on homeostasis and functional integrity in an organotypic human skin model. Mut Res 2015; 775:10–18.

[15]   Luna LG (ed). Manual of histologic staining methods of the armed forces institute of pathology. 3rd edition. McGraw-Hill, New York; 1968.

[16]   Wilson WE, Nikjoo H. A Monte-Carlo code for positive ion track simulation. Radiat Environ Biophys 1999; 38:97–104.

[17]   Dingfelder M, Hantke D, Inokuti M, Paretzke HG. Electron inelastic-scattering cross sections in liquid water. Radiat Phys Chem 1998; 53:1–18.

[18]   Wilson WE, Miller JH, Lynch DJ, Lewis RR, Batdorf M. Analysis of low-energy electron track structure in liquid water. Radiat Res 2004; 161:591–6.

[19]   Weigand DA, Haygood C, Gaylor JR. Cell layers and density of Negro and Caucasian stratum coreum. J Invest Dermatol 1974; 62:563–8.

[20]   Mateus R, Abdalghafor H, Oliverira G, Hadgraft J, Lane ME. A new paradigm in dermatopharmacokinetics – confocal Raman spectroscopy. Int J Pharm 2013; 444:106–8.

Chapter 3

Automatic Interpretation of Melanocytic Images in Confocal Laser Scanning Microscopy

Marco Wiltgen and Marcus Bloice

Additional information is available at the end of the chapter

http://dx.doi.org/10.5772/63404

Abstract

The frequency of melanoma doubles every 20 years. The early detection of malignant changes augments the therapy success. Confocal laser scanning microscopy (CLSM) enables the noninvasive examination of skin tissue. To diminish the need for training and to improve diagnostic accuracy, computer-aided diagnostic systems are required. Two approaches are presented: a multiresolution analysis and an approach based on deep layer convolutional neural networks. For the diagnosis of the CLSM views, architectural structures such as micro-anatomic structures and cell nests are used as guidelines by the dermatologists. Features based on the wavelet transform enable an exploration of architectural structures at different spatial scales. The subjective diagnostic criteria are objectively reproduced. A tree-based machine-learning algorithm captures the decision structure explicitly and the decision steps are used as diagnostic rules. Deep layer neural networks require no a priori domain knowledge. They are capable of learning their own discriminatory features through the direct analysis of image data. However, deep layer neural networks require large amounts of processing power to learn. Therefore, modern neural network training is performed using graphics cards, which typically possess many hundreds of small, modestly powerful cores that calculate massively in parallel. Readers will learn how to apply multiresolution analysis and modern deep learning neural network techniques to medical image analysis problems.

Keywords: confocal laser scanning microscopy, skin lesions, multiresolution image analysis, convolutional neural networks, machine learning, computer-aided diagnosis

1. Introduction

The skin is the largest organ of the body. Its surface comprises up to two square meters. It is the organ that is in direct contact to the environment and is therefore exposed to several environmental influences such as sun radiation, temperature, infections. The skin consists of three main layers: the epidermis, the dermis and the hypodermis (subcutis), whereby each layer is subdivided into several sublayers (strata) [1]. As the outermost layer, the epidermis provides a protective barrier of the body's surface which keeps water in the body, protects against heat and ultraviolet radiation and prevents infections (caused by bacteria, fungi, parasites, etc.) [2, 3]. The horny layer (stratum corneum), which is the top layer of the epidermis, undergoes a continuous process of renovation (every 4 weeks). Keratinocytes, which represents 90% of the cell types in the epidermis, protect the body against ultraviolet radiation. Keratinocytes are derived from epidermal stem cells residing in the lower part of the epidermis (stratum basalis). During their lifetime, they migrate through the different strata of the epidermis. Via this process, they are pressed to the epidermis surface by the continuously succeeding cells. During the migration through the different strata, the keratinocytes cells undergo multiple stages of differentiation, whereby they change shape and composition and are filled with keratin. Different stages and corresponding strata are represented in **Figure 1**. Keratin, a structural protein, is the key structural material making up the outer layer of the epidermis and protects the cells from damage or stress. On their way to the outermost strata, the keratinocytes lose liquid and become hornier. Corneocytes are keratinocytes that have completed their differentiation program. They are dead cells in the stratum corneum and are shed off (by desquamation) as new ones come in. Keratinocytes protect against ultraviolet radiation by taking up melanosomes from epidermal melanocytes. The melanosomes are vesicles which contain the endogenous photo protectant molecule melanin. Melanocytes are melanin producing cells which comprise between 5 and 10% of the cells in the basal layer (stratum basalis) of the epidermis. The production of the skin pigment melanin is stimulated by ultraviolet radiation (melanogenesis). Melanocytes have several arm-like structures (dendrites) that stretch out to connect them with many keratinocytes. Once synthesized, melanin is contained in the melanosomes and moved along the dendrites to reach the keratinocytes. The melanin molecules are stored within keratinocytes (and melanocytes) in the perinuclear area, around the nucleus, where they protect the DNA against ultraviolet radiation. Thereby, a melanin molecule transforms nearly all the radiation energy in to heat. This is done by ultrafast internal conversation of the energy from the excited electronic states into vibrational modes. The ultrafast conversion shortens the lifetime of the excitation states and therefore prevents the formation of harmful free radicals.

The dermis is connected to the epidermis through a basement membrane (a thin sheet of fibres) and provides anchoring and nourishment for the epidermis. The dermis contains collagen (stability), elastic fibres (elasticity) and an extrafibrillar matrix as structural components. The papillary region (stratum papillae) in the dermis is composed of connective tissue which extends towards the epidermis. These finger-like projections are called papillae and strengthen the connection between the dermis and the epidermis. In addition to the structural components, blood vessels are present in the dermis providing nourishment for the dermal and epidermal cells. Furthermore, the dermis contains hair follicles, sweat glands and lymphatic vessels. (In addition to the presented components, the dermis also contains mechanoreceptors that enable the sense of touch and thermoreceptors that provide the sense of heat). The hypodermis is beneath the dermis. Its tasks comprise energy storage, heat insulation and the connection of the skin with inner structures like muscles and bones. The hypodermis consists primarily of loose connective tissue and adipocytes (fat cells), which are grouped together in lobules (subcutaneous fat). Furthermore, the hypodermis contains larger blood vessels and nerves than those found in the dermis.

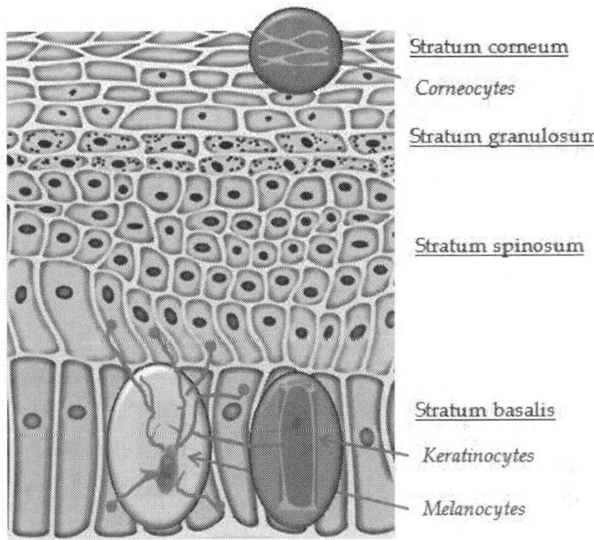

Figure 1. The layer architecture of the epidermis.

2. Malignant melanoma and benign nevi

The primary cause for the increasing number of melanomas is the extreme sun exposure during sun-bathing (especially for people with low levels of skin pigment). The malignant melanoma is a type of cancer that develops from the pigment containing melanocytes [4]. Melanomas are mainly caused by DNA damage resulting from the ultraviolet radiation [5]. It is observed that strongly pigmented people are less susceptible to (sun induced) melanomas, which demonstrates the protection function of melanin. At the early stage, melanocytes begin an out-of-control growth [5]. In a posterior stage (invasive melanoma), the melanoma may grow into the surrounding tissue and can spread out around the body through lymph or blood vessels deeper in the skin. People with melanomas at the early stage are treated by surgical removal of the skin lesion. In cases where the melanoma has spread out, patients are treated by immunotherapy or chemotherapy. Most people are cured if spreading has not occurred. Therefore, the early and reliable recognition of melanomas at the early stage is of special importance [6]. The difference between a benign or malignant tumour is its invasive potential. If a tumour lacks the ability to invade adjacent tissues and to metastasize then it is benign, whereas a malignant tumour is invasive or metastatic. A nevus (birthmark) is a sharply circumscribed and benign chronic lesion of the skin. The melanocytic nevus results from benign proliferation of the dendritic melanocytes. Due to the pigment melanin, they are mostly brown. Nevus cells are related to the melanocytes, but they show a lack of the dendrites and are oval in shape. They are typically arranged in cell nests. The majority of acquired nevi appear during the childhood up to young adults (the first two decades of life). A melanocytic nevus present at birth is called a congenital nevus. They are rarely about one in every 100 newborns. Nevi are harmless. However, 25% of malignant melanomas arise from pre-existing nevi.

3. Confocal laser scanning microscopy

In conventional microscopy, the entire field of a tissue sample is simultaneously illuminated by light and displayed. Although the brightest light intensity results from the focal point of the objective lens, other parts of the tissue are still illuminated, resulting in a large unfocused background section. This background noise diminishes the image quality. Both conventional and confocal laser scanning microscopy (CLSM) can use reflected light to image a tissue sample. The reflected light from the illuminated spot is then re-collected by the objective lens. In addition to the reflected light from the focal point, the scattered light from sample points outside the focus light (coming from places above or below the focus) is projected by the optical system of the microscope and therefore contributes to the image assembly. This causes a blurring and obscuring of the resulting image. Confocal microscopy overcomes this problem by placing a pinhole in the conjugate focal plane (hence the designation confocal) that allows only the light emitting from the desired focal spot to pass through [7]. Any light outside of the focal plane (the scattered light) is blocked. **Figure 2** shows the principle: the out of focus light (red), coming from places above the selected focal plane, is blocked by the pinhole in the conjugate focal plane. The (in focus) light from focal plane (blue) can pass through the pinhole and is detected. Therefore, a blurring is avoided and sharp and detailed images are produced (in other words: the image information from multiple depths in the sample is not superimposed). In confocal microscopy, a light beam is directed by a dichroic mirror to the objective lens where it is focused into a small focal volume at a layer within the tissue sample (**Figure 3**). A laser, with a near-infrared wavelength, is used as a coherent monochromatic light source. The same microscope objective gathers the reflected light from the illuminated spot in the sample. The dichroic mirror separates the reflected light from the incident light and deflects it to the detector. Before the light reaches the detector, the out of focus sections are blocked by the pinhole in the conjugate focal plane. The in focus light that passes through the pinhole is measured.

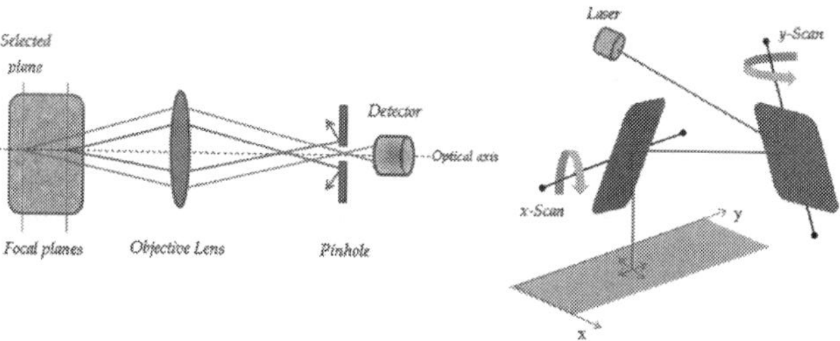

Figure 2. Principle of the confocal (left) and laser scanning (right) microscopy.

The detector, which is usually a photomultiplier tube or avalanche photodiode, amplifies and transforms the intensity of the reflected light signal into an electrical one that is recorded by a computer. In contrast to conventional microscopy, there is never a complete image of the sample at any given instant; rather only one point in the selected plane of the sample is observed. In order to create an image, light from every point in the plane (x-axis, y-axis) must be recorded. This can be done by a raster scanning mechanism which uses two motor driven high-speed oscillating mirrors, which pivot on mutually perpendicular axes. Coordination of the two mirrors, one scanning along the x-axis and the other on the y-axis, produces the rectilinear raster scan (**Figure 2**). During the scanning process, the detected signal is transferred to a computer that collects all the 'point images' of the sample and serially constructs the image pixel by pixel. The brightness of a resulting image pixel corresponds to the relative intensity of the reflected light. The contrast in the images results from variations in the refractive index of microstructures within the tissue. Information can be collected from different focal planes by raising or lowering the objective lens. Then successive planes make up a 'z-stack'. A stack

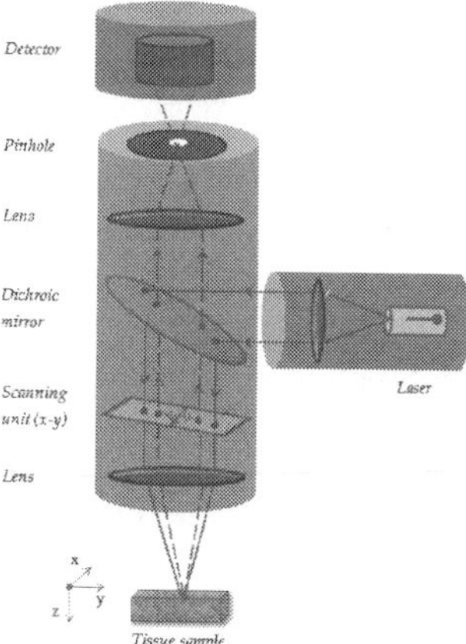

Figure 3. Principle of the confocal laser scanning microscope.

is a sequence of images captured at the same horizontal position (x- and y-axes) at different depths (z-axis). The images are taken enface (horizontally). The confocal laser scanning microscopy is performed with a Vivascope 1000 (Lucid Inc., USA) which uses a diode laser at 830 nm wavelength and a power of <35 mW at tissue level. A ×30 water-immersion objective lens with a numerical aperture of 0.9 is used with water as an immersion medium. The spatial resolution is 0.5–1.0 μm in the lateral and 3–5 μm in the axial dimension.

The images contain a field-of-view of 0.5 × 0.5 mm. Up to 16 layers per lesion can be scanned. All images, stored in BMP file format, are monochrome images with a spatial resolution of 640 × 480 pixels and a grey level resolution of 8 bits.

4. Interpretation of confocal laser scanning microscopic images

The reflectivity of the tissue depends on chemical structures. Melanin and melanosomes have a high refractive index which contributes strongly to the contrast of the resulting image [8–10]. Due to such dominating variations of the refractive index, only a certain part of the in falling light is reflected. This makes the appearance of the tissue in a CLSM image so different from conventional histological views. The power of the 830 nm laser limits the imaging depth to a maximum of 350 μm, corresponding to the papillary dermis (higher power could damage the skin). **Figure 4** shows the views of different skin layers [11]. The stratum corneum shows large polygonal anucleated corneocytes (A). Skin folds and marks appear as dark structures. The next layer is the stratus granulosum (B). The stratum spinosum (C) contains keratinocytes in a honeycomb pattern. In the stratum basalis (D), the basal cells are uniform in size and show higher reflections than spinous keratinocytes and appear very intensively. The dermatological guidelines for the interpretation of melanocytic skin lesions in CLSM views are as follows.

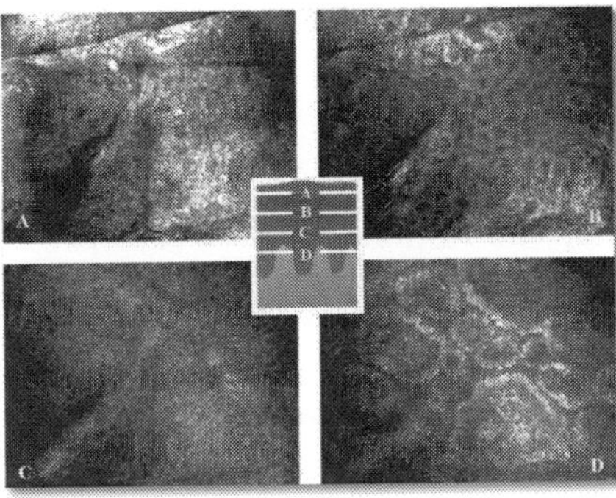

Figure 4. CLSM views of normal skin.

For the diagnosis of CLSM views of benign common nevi and malignant melanoma, architectural structures such as micro-anatomic structures; cell nests, etc., play an important role [12].

Monomorphic melanocytic cells, melanocytic cell nests and readily detected keratinocyte cell boarders are suggestive of benign nevi, whereas polymorphic melanocytic cells, disarray of melanocytic architecture and poorly defined keratinocyte cell borders are suggestive of melanoma (**Figure 5**). The images are taken from the centre of the tumours.

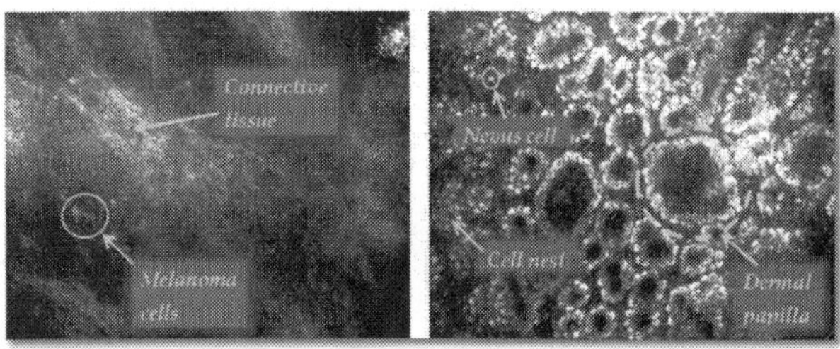

Figure 5. CLSM images of malignant melanoma (left) and common benign nevi (right).

Layers from the plane of the spinous keratinocytes (polygonal cells) to the plane of the basal cells (dermo-epidermal junction) are used for diagnosis.

5. Analysis of tissue structures at different scales

As shown in the previous section, the information at different scales (from coarse structures to details) plays a crucial role in the diagnosis of CLSM images of skin lesions. Wavelet analysis is a method to analyse visual data by taking into account scale information [13].

Figure 6. Scale-space sequence of a successively Laplacian of Gaussian-filtered image.

The multiple resolutions enable a scale invariant interpretation of an image. **Figure 6** illustrates the principle of scale space analysis for four levels of scale (clockwise direction). In the top left image (scale 1), the feature detection responds to fine texture. The images at higher scales are generated by a Laplacian of Gaussian filter (LoG(x, y)), which is also known as Marr-Hildreth operator or Marr wavelet (**Figure 7**), whereby the kernel size (σ) of the Gaussian increases step by step.

$$\mathrm{LoG}(x,y) = \frac{1}{\pi\sigma^4}\left(\frac{x^2 + y^2}{2\sigma^2} - 1\right)\cdot e^{-\frac{x^2+y^2}{2\sigma^2}}$$

The blue and red colours indicate positive and negative values. The images become increasingly blurred and smaller details (or regions) progressively disappear. The detected features are then associated with a larger scale scene structure. The multiresolution analysis is closely analogous to the human vision system which seems to prefer methods of analysis that run from coarse to fine and, repeating the same process, obtain new information at the end of each cycle [14] (**Figure 6** counter clockwise direction). The wavelet decomposition can be realized as a convolution of the image with a filter bank, consisting of high pass and low pass filters [15]. Whereby, for example a first-order derivative can be used as a convolution kernel for the high-pass filter and a moving average as a kernel for the low-pass filter. In our study, the filter coefficients are defined by the Daubechies 4 wavelet transform.

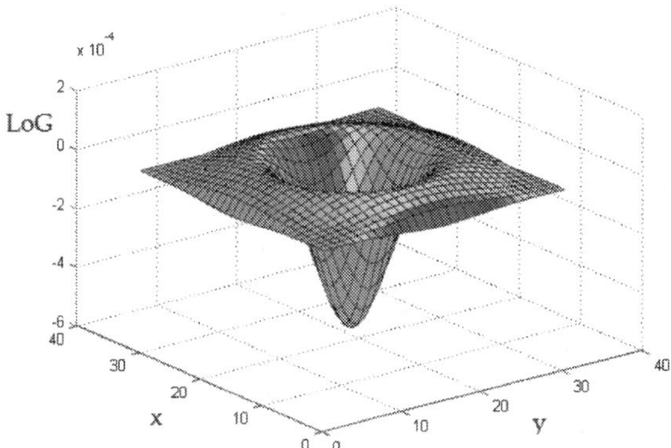

Figure 7. Shape of the Laplacian of Gaussian convolutional filter kernel.

The wavelet decomposition performs a multi resolution analysis, whereby the image is successively decomposed by the filter operations followed by sub-sampling. The (pyramidal) algorithm consists of several steps and operates as follows: at the beginning, the image rows are filtered by the high-pass filter and in parallel by the low-pass filter (**Figure 8**). From both operations result two images (which are called sub-bands), one shows details (high pass) and the other is smoothed out (low pass). The sub-sampling is done by removing every second column in both sub-bands. Subsequently, the columns of both sub-bands are high-pass and independently low-pass filtered. This results in four sub-bands, which differ by the kind of filtering. Again a sub-sampling is done by removing every second row in each sub-band. This is the end of the first step. The mixed filtered (high-low pass, etc.) sub-bands are stored. Only the double low-passed sub-band is processed in the second step (**Figure 8**). The second step repeats the operations of the first step. Again this results in four sub-bands and the fourth smoothed sub-band is used as entry for the following step. At every step, the resulting sub-bands are reduced to half the resolution. The sub-bands with higher spatial resolution contain the detailed information (high pass), whereas the sub-bands with the low-resolution represent the large scale coarse information (low pass). The output of the wavelet decomposition consists of the remaining 'smooth-…-smooth' components and all the accumulated 'detail' components. In other words, via the wavelet decomposition, the image array is decomposed into several sub-bands representing information at different scales. The output of the last low-pass filtering is the mean gray level of the image.

After the dissection of the quadratic sub-bands, they are usually arranged in a quadratic configuration, whereby the three sub-bands of the first step fill 3/4 of the square, the three sub-bands of the second step fill 3/16 of the square, etc. The sub-bands representing successively decreasing scales are labelled with increasing indices (**Figure 9**). Then, the architectural structure information is accumulated along the way of the sub-bands (from coarse to fine). In image processing, it is convenient to display the smoothed image as lowest sub-band in the upper left corner of the quadratic sub-band configuration. The coefficients values in the different sub-bands reflect architectural and cell structures at different scales.

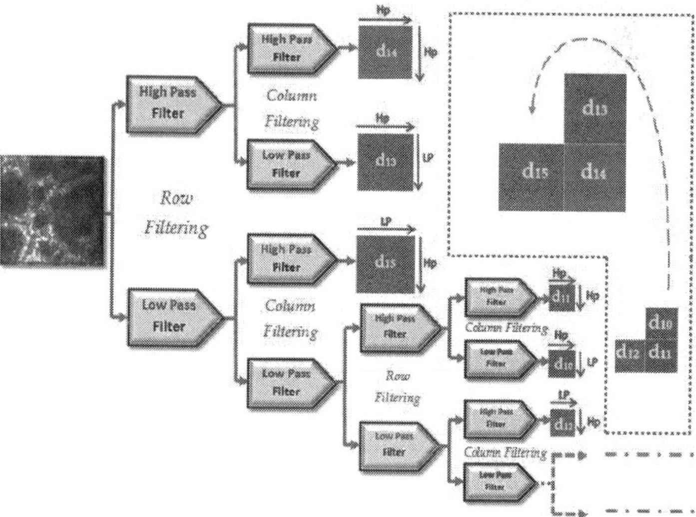

Figure 8. The multiresolution filter bank of the wavelet decomposition.

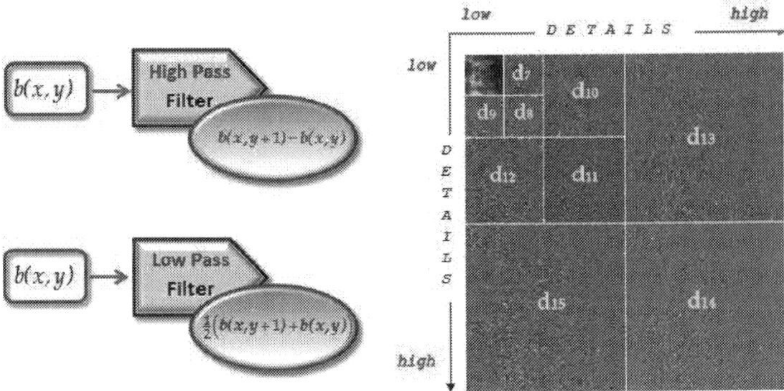

Figure 9. The sub-bands resulting from the successive high and low pass filter operations.

The tissue features are derived from statistical properties of the sub-band coefficients. For the ith sub-band of size N × N, the coefficients are given by:

$$d_i = \left\{ d_i(k,l) \middle| k,l = 1, N \right\}$$

The texture features are based on the variations of the coefficients within each sub-band and the weighted sum of all the coefficients into each sub-band. The standard deviations of the coefficients inside the single sub-bands and the energy and entropy of the different sub-bands are calculated and used as features (for details see: [16]). The standard deviation of the coefficients represents how exposed the tissue structures in the considered sub-band at the given scale are. The total energy of the coefficients in a given sub-band shows to what degree the structures at the corresponding scale contribute to the image. The distribution of the energy of the sub-bands is represented in a power spectrum, enabling an evaluation of their relative contributions.

The next task in automated image analysis is the use of machine-learning algorithms for classification purposes on hand of the feature values [17]. The algorithm learns, by use of a training set, how to assign the tissue images to given classes. Then, in future, the algorithm can apply the gained knowledge to predict the class of unknown tissue. By means of the classification procedure, the primary inhomogeneous set of CLSM samples, consisting of a mix of malignant melanoma and benign common nevi cases, is split into homogeneous subsets, which are assigned to one of the two tumour classes: common benign nevi or malignant melanoma. A homogeneous subset means that it contains only CLSM images with similar feature values, representing one specific kind of tissue. For the discrimination of the CLSM images, the CART (Classification and Regression Trees) algorithm is used [18].

The tree representation consists of different nodes and branches. There is a root node, several leaf (terminal) nodes and inner nodes (**Figure 10**). The first node in the tree is the root node. It contains the feature values of the whole set of CLSM image samples. A leaf node is a homogeneous node which contains only samples belonging to the same class of tissue. The inner nodes contain more or less inhomogeneous sample sets. A branch in the decision tree involves the testing of one particular texture feature (binary tree). Then, the considered node, which is the parent node, is split into two child nodes (**Figure 10**).

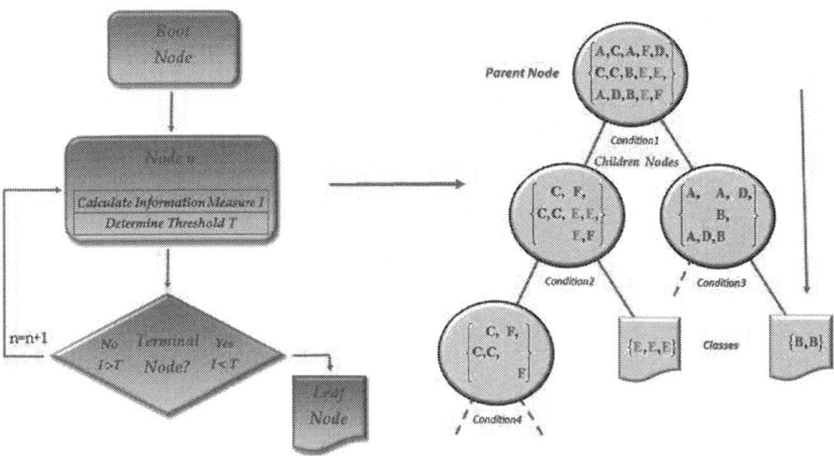

Figure 10. Generation of a decision tree.

The feature is tested by comparing its numerical value with a threshold value that divides the value range. The threshold value is selected automatically by the algorithm in such a way that the subsets of samples in the child nodes are purer than the set in the parent node. To this purpose, an information measure is used which indicates the degree of homogeneity; the value in the leaf nodes is zero and the higher the value of an inner node, the higher is its inhomogeneity. At every branch in the tree, subsets with smaller values of the information measure are generated. The decision tree is generated recursively (details are shown in: [16]). Whereby the algorithm consists in principal of three parts: the determination of the optimal splitting at every node; the decision whether the node is a leaf node or an inner node; the assignment of a leaf node to a specific class (**Figure 10**). To classify an unknown sample, it is routed down the tree according to the values of the different features. When a leaf node is reached, the sample is classified according to the class assigned to the leaf. The tree-based machine-learning algorithm captures the decision structure explicitly. That means the generated decision rules are 'Modus Ponens', with a precondition and conclusion part, and are intelligible in such a manner that they can be understood, discussed and used as diagnostic rules.

$$IF\left(...and.Condition1.and.Condition2\right)THEN\left(Class := A\right)$$

In total, 39 different features are calculated for 16 frequency bands (labelled from 0 to 15). The mean value is calculated from the first four frequency bands; therefore, 13 values result for each feature. The highest frequency bands contain only information about very fine grey level variations, such as noise, and are therefore not considered for the image analysis. The procedure for image analysis (including feature extraction and calculation) was developed with the 'Interactive Data Language' software tool IDL (IDL 7.1, ITT Visual Information Solutions). The tree classification is done by the CART analysis software from Salford Systems, San Diego, USA.

6. Biological motivation for neural networks

A neuron is an electrically excitable cell that receives, processes and transmits information as electrochemical signals. It consists of several dendrites, the soma and an axon (**Figure 11**). The soma is the cell body which contains the nucleus and all the necessary cytoplasmic cell structures. The dendrites are cytoplasmic extensions of the cell body with many branches allowing the cell to receive signals from other neurons. The axon is a special extension which carries signals away from the soma. At its terminal, the axon undergoes extensive branching, enabling communication with many target cells. The neurons maintain voltage gradients across their membranes. Ion channels, embedded in the membrane, enable the generation of intracellular-extracellular ion migrations. The resulting changes in the cross-membrane polarization generate an electrochemical pulse, known as the action potential. These changes in the cross-membrane potential are transferred as a wave of successive depolarization and repolarisation processes along the cell's axon. The axon terminal contains synapses, specialized connections to target neurons, where neurotransmitter chemicals are released. Synaptic signals may be excitatory or inhibitory. Once the pulse from the soma along the axon reaches the synapses, a neurotransmitter is released at the synaptic cleft. The neurotransmitter molecules bound at the receptors in the post-synaptic membrane (of the target neuron) and opens ion channels. Then, the electrochemical pulse is transmitted to the target neuron.

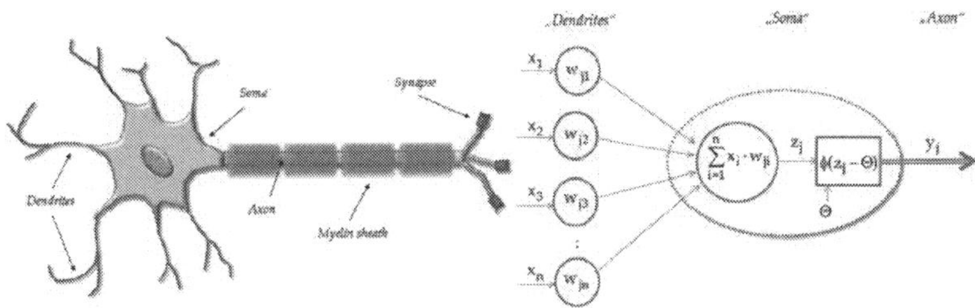

Figure 11. Microanatomy of a natural neuron (left), principle of an artificial neuron (right).

An artificial neuron is a mathematical model of a biological neuron. Artificial neurons mimic the behaviour of the biological neurons. The input of the artificial neuron is represented by a vector:$x = (x_1, x_2, ..., x_n)$, whereby its dimension reflects the number of contributing dendrites (**Figure 11**). In the mathematical model, each 'dendrite' contributes individually through a weighted signal to the input signal. The weight factor $w_j = (w_{j1}, w_{j2}, ..., w_{jn})$ simulates the ratio of synaptic neurotransmitters, whereby positive values represent excitatory and negative values inhibitory behaviour (a weight value zero means that there is no connection between the involved neurons). The summation function represents the soma of the neuron j. The exciting and inhibiting signals are added in the function:

$$z_j = \sum_i x_i w_{ji}$$

The firing behaviour of the neuron is represented by the activation function. Its activation depends on the output of the summation function z_j and a threshold value Θ. If the summation function exceeds the threshold, the neuron is firing and transmits an output signal y_j:

$$y_j = \phi\left(z_j - \theta\right)$$

The biological motivation of the activation function is the threshold potential in natural neurons. Step and sigmoid functions are often used as transfer functions.

7. Artificial neural networks

Artificial neural networks consist of a number of artificial neurons, the computational units, which are interconnected. Each unit performs some small calculation based on inputs it receives from other units, whereby the associated weight factors can be tuned. This tuning occurs by allowing the network to analyse many examples of previously observed data. The most common type of neural network is the feed forward neural network (containing no loops), and in such networks, the computational units are organised into layers from an input layer, where data are fed into the network, to an output layer, where the result of the network's computation is outputted in the form of a classification result or regression result (**Figure 12**). Traditionally, each neuron in a layer is connected to all other neurons in the previous or subsequent layers (fully connected network). Between the output and input layers are hidden layers, and networks that consist of more than one hidden layer are known as *deep learning* algorithms. Such feed forward neural networks have been shown to be universal approximators, that is to say they can learn to approximate any continuous function to arbitrary precision, given enough hidden neurons [19]. Neural networks must be trained. The training data are previous observations that have been collected, and the task of the network is to learn a function which should map new input data to a classification label.

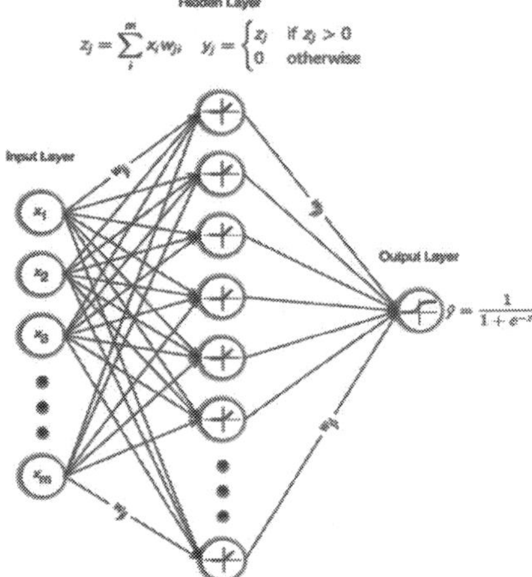

Figure 12. Structure of a feed forward artificial neural network.

In general, feed forward neural networks are supervised machine-learning algorithms. **Figure 12** shows a network with three layers: (1) an input layer, where data are fed in, (2) a hidden layer consisting of neurons that each contain an activation function that reads in data from the input neurons, performs some calculation, and outputs a value, and (3) an output layer that reads data from the hidden layer and makes a prediction based on this input. All connections between neurons have independently adjustable weights (Section 6). All layers are fully connected meaning that each neuron in the input layer is connected to every neuron in the hidden layer. The network learns by adjusting the weights between each of the connected neurons until the network makes good predictions by minimising an error function (backpropagation algorithm).

Fully connected neural networks are useful where individual features of a dataset are not very informative. In image data, where an individual pixel is not likely to be very informative taken on its own, a local combination of pixels may very well be informative and represent an object of interest. However, neural networks are also far more computationally intensive than many other machine-learning algorithms, with the number of tuneable parameters quickly growing into the millions as the network increases in depth or size. Also, neural networks typically work on image data directly, without feature reduction, meaning the dimensionality of the data being analysed by neural networks is much higher than that of other algorithms, which often work on extracted features. One could therefore summarise that neural networks are most useful for high m high n problems—problems where there exist many observations (n) of high dimensional data (m). Of late, neural networks algorithms have re-emerged as a popular technique in machine learning, especially in the field of image analysis. This re-emergence has come due to a number of recent developments in neural network design as well as independent hardware developments. In real-world applications, their usage has grown beyond image analysis and has also been shown to be useful for other tasks, such as natural language processing and artificial intelligence [20, 21]. Nevertheless, a number of advancements in recent years resulted in an upsurge in the usage of neural networks.

First, hardware advancements have made it feasible for larger neural networks to be trained in reasonable amounts of time. As mentioned previously, neural networks that learn on very high-dimensional data require many neurons and layers, meaning networks can consist of many millions of parameters that need to be tuned. This results in large network architectures that have, for a long time, been unfeasibly difficult to train on standard desktop workstations. However, computational enhancements have meant this is no longer the case. These computational advancements are the result of rapid developments in graphics processing unit (GPU) technology due to the ever increasing requirements of the gaming industry, resulting in great improvements in the parallel processing power of GPUs. In 3D gaming, the vast majority of processing power is spent on matrix multiplications, such as transforms and perspective calculations, in order to depict the 3D worlds of games in 2D to the user. Such calculations are, for the most part, performed using matrix and vector multiplications. Such matrix calculations can be performed in parallel, and hence gaming GPUs have evolved to be particularly suited to such parallel processing tasks. To this end, GPUs typically consist of boards with many small, less powerful cores that can perform highly parallel computations. While CPUs tend to possess 2–4 large and fast cores, GPUs possess many hundreds of smaller cores. Crucially, almost 90% of the computational effort required to train a neural network is spent on vector, matrix, and tensor operations, meaning they can benefit from all the recent technological advancements in GPU technology. Indeed, with Moore's Law no longer holding, parallelised algorithms may, in future, be the only way to analyse very large data [22]. Second, empirical data have shown that neural networks with large numbers of hidden layers outperform many algorithms at several machine-learning tasks, especially in computer vision, object recognition and object detection. Deeper and deeper neural networks, with larger and larger numbers of neurons, have achieved human-level performance at very human-like tasks, such as playing video games [23] and playing the game of Go [24]. Deeper networks, however, contain more neurons, each of which needs to perform some calculation, and have its associated weight tuned, resulting in longer training times and larger memory requirements. Again, advances in hardware and optimisation techniques have meant that ever deeper networks are now trainable within reasonable timeframes [25]. Third, more and more data are permanently stored, archived and saved than ever before. This is especially true in fields such as medicine, where large amounts of data are accumulated during routine activities. In the past, these data might have been archived or stored in offline tape drives, or even discarded. However, this is no longer necessarily true as the cost per GB of storage has declined so rapidly, meaning easier access to more data and less likelihood of data being discarded. Deep learning algorithms require large amounts of data to train and access to very large datasets, and the ability for individuals to store large amounts of data has meant they are being applied to such problems more often.

Traditional feed forward neural networks consist of layers, where each neuron is connected to every other neuron in the layers above and below it. These are known as fully connected, or affine, layers. Fully connected neural networks do not consider the spatial relation between pixels in an image. Pixels which are close together are treated exactly like pixels which are far apart when being processed by the network. For the learning of high-level features, this is suboptimal. In terms of image analysis, one particular type of neural network algorithm has stood out as being especially adept at image classification and object recognition. This is the convolutional neural network. The idea behind convolutional neural networks is to restrict the network to take inputs only from spatially nearby neurons. In other words, the layers are not fully connected, as in the example in **Figure 12**.

8. Convolutional neural networks

In the fields of image analysis, object detection and pattern recognition, convolutional neural networks are the state of the art algorithm for practical applications. Following on from our previous work, where we applied multiresolution analysis and CART as tree-based machine-learning method (Section 5), we decided to test the applicability of convolutional neural networks at a similar classification task. Because neural networks learn their own discriminatory, high-level features, the dataset requires no pre-processing or feature extraction, with the exception of image resizing and pixel value normalisation. This is in direct contrast to our previous efforts, where a dedicated feature extraction phase was necessary. Convolutional neural networks (CNN), in effect, emulate the way in which classical pattern recognition works, where local features (edges, corners, etc.) are extracted and combined to generate higher level representations that can be used for object recognition. Convolutional neural networks are locally connected, where each neuron is connected only to those that are spatially close (local receptive fields) in the previous layer, mimicking the visual cortex of some animals. Pixels that are closer to each other are more strongly correlated than those which are further away from each other, and this is something which the convolutional neural network has been designed to be able to account for through its architecture [26].

Network architectures with fully connected layers do not take into account the spatial structure of the images. Instead of using a network architecture which is tabula rasa, convolution neural networks (CNN) try to take advantage of spatial structures in images. They use three basic ideas: local receptive fields, shared weights and pooling. It is helpful to represent the input image as a square of neurons, whose values correspond to the pixel intensities. Then, only small, localized regions of the input image are connected to a neuron in the first hidden layer. Such a region in the input image is called the local receptive field for the corresponding hidden neuron. In other words, the hidden neuron learns to analyse its particular local receptive field. If the receptive field has a size of 5 × 5 pixels, then the hidden neuron is connected by 5 × 5 weights, which are adjusted during learning. The input of the hidden neuron is given by the summation function:

$$y_j = \sum_{l=0}^{4} \sum_{m=0}^{4} w_{l,m}^{j} b_{j+l,k+m}$$

The value $b_{x,y}$ denotes the input activation at position (x, y). The output of the hidden neuron is given by the activation function, for example the sigmoid function. The convolutional operation can be considered as a sliding window, which travels over the image, with the window centre moving one or more pixel a time. This is defined by the stride length. If the window is moved by one pixel, the stride length is 1. For each position of the local receptive field, there is a different hidden neuron in the first hidden layer. The map from the input layer to the hidden layer (convolutional layer) is called a feature map. The weights $w_{l,m}$ defining the feature map are the shared weights. The shared weights define the convolution kernel (convolution is generally the workhorse of image processing). The pixels in the local receptive field are multiplied element-wise with the kernel. Features maps are generated using only neurons which are spatially close to each other, known as spatial connectivity. Each feature map is defined by a specific set of shared weights enabling the network to detect different kinds of features (edges, corners, etc.). The CNN therefore learns objects related to their spatial structure. For image analysis purposes, more than one feature map are required. Therefore, a complete convolutional layer consists of several different feature maps. In addition to the convolutional layers, CNNs also contain pooling layers which usually follow immediately after the convolutional layers. Pooling layers simplify the information in the output from the convolutional layer by generating a condensed feature map (this removes the positional information of the features learned, meaning the learned features are position invariant). For example, each unit in the pooling layer may summarize a region of 2 × 2 neurons in the previous convolutional layer. Pooling is done for each feature map separately. The final layer in the convolutional network is a fully connected layer. This layer connects every neuron from the last pooling layer to every one of the output neurons.

A depiction of a typical 7-layer convolutional neural network can be seen in **Figure 13**. Images are read into the network in the input layer. From this input, a number of feature maps (4) are generated, which are subsampled in a max-pooling phase. Then, both phases are repeated once more, before connecting to a conventional fully connected layer which is finally connected to the output layer. CNNs often contain multiple fully connected layers before the final output layer, and modern CNNs can contain many convolution/max-pooling pairs.

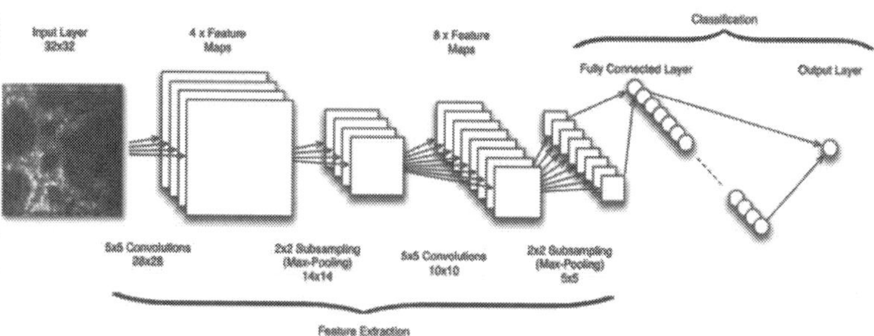

Figure 13. The structure of a typical seven-layer convolutional neural network.

Figure 14 describes the convolutional layer and max-pooling layer in more detail. The input into the convolutional neural network is a vector $x \in R^{1 \times m}$, and the input layer has one neuron per feature. However, the layers can be thought as having their neurons arranged as depicted in **Figures 11** and **12**. In the case above, a 5 × 5 kernel is used, with a stride of 1, which results in a feature map of size 32 − 5 + 1 = 28 × 28. Typically, a convolutional layer is followed by a max-pooling layer, which acts as a type of sub-sampling, in this case halving the size of the previous feature map (**Figure 13**).

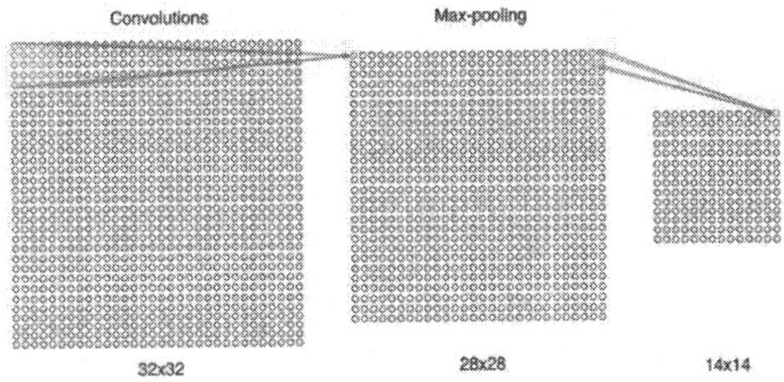

Figure 14. Principle of the convolutional layers and max-pooling layers [27].

Convolutional neural networks possess several characteristics that make them very suitable for the analysis of histological images. First, convolutional neural networks are capable of building models which are translation invariant and robust to transformations in the images, such as rotation, and they can learn features which are robust to scaling. They also generate models which are position invariant. This is especially important for microscopy imagery, where a lesion, for example, has no 'right way up', and cannot even be rotationally normalised.

9. Deep learning analysis of a CLSM image dataset

As stated previously, the goal was to train a model which would classify newly seen images as either malignant or benign. The neural network that was designed was based on the structure of the LeNet-5 convolutional neural network structure and was developed using the Keras deep learning library for Python [26]. The network consisted of a total of eight layers: the input layer, two pairs of convolutional and max-pooling pairs, two fully connected layers, and the output layer. The rectified linear unit (ReLU) was used throughout as the neuron nonlinearity. The ReLU is a computational unit which uses a ramp function [the rectifier $f(x) = \max(0, x)$] and is currently the most popular activation function for deep neural networks. Because of the depth of network, a graphics processing unit (GPU) was used, which greatly increases the speed at which the network can train. In terms of hardware, a mid-range NVidia gaming GPU with 2 GB of dedicated video memory and 640 cores was used for training the network. The card is capable of 1306 GFLOP/s and has a memory bandwidth of 86.4 GB/s. At the time of writing, the card can be purchased for under \$150. The card was installed in a Linux workstation with 32 GB of RAM and a 3.5 GHz 6-core AMD processor running the Xubuntu 14.04 operating system. To illustrate the differences in computational power between a GPU and CPU, and to demonstrate the enormous impact using a GPU can have on training times, we benchmarked our code. Training the network over 20 epochs required 2 min 4 s of time, averaged over three runs, when using the GPU. When using the CPU, this time was 57 min 59 s for 20 epochs (also averaged over three runs), nearly 30 times slower. Experimenting with different parameters, or testing new network structures, can become very tedious when hours of computational power are required per run or experiment. The GPU reduces this time to minutes.

Dropout was used to control overfitting at two points in the network's structure: once after the convolutional and max-pooling pairs, and once again after the first fully connected layer. Dropout helps to control overfitting by randomly setting a certain set percentage of the neurons' weights to zero, effectively forcing the network to relearn those weights, with the intention of mitigating the learning of noise. The output of the network is finally determined by a sigmoid logistic function, squashing the results of the entire network to a value between 0 and 1. Values closer to 1 are therefore classified as being malignant, while values closer to 0 refer to a benign prediction. Such an output can also be used examine the network's confidence at a classification, with a value of 0.99 meaning a highly confident malignant prediction and a value of 0.51 representing an unconfident malignant prediction.

9.1. Input into the neural network

Images are read directly by the neural network. The only pre-processing which was performed was to resize the images from 640×480 to 64×64 pixels. Images are read by the neural network as a series of pixel values stored in a vector. Therefore, a single image is stored as a vector x, so that one instance of an image $x^{(i)} \in \mathbb{R}^{1 \times m} = \left[x_1^{(i)} x_2^{(i)} x_3^{(i)} \dots x_m^{(i)} \right]$. The dataset consisted of $n = 6897$ images, each 64×64 pixels in size, representing a dimensionality $m = 4096$. The entire dataset is therefore stored in an $n \times m$ matrix:

$$X \in \mathbb{R}^{n \times m} = \begin{bmatrix} x_1^{(1)} & \dots & x_m^{(1)} \\ \vdots & \ddots & \vdots \\ x_1^{(n)} & \dots & x_m^{(n)} \end{bmatrix}$$

To reduce the memory footprint, neural networks are typically trained using mini-batches, which are randomly selected subsets of X. Targets, or labels, are stored in an n-dimensional column vector:

$$y = \begin{bmatrix} y^{(i)} \\ \vdots \\ y^{(n)} \end{bmatrix} \left(y \in \{0, 1\} \mid 0 = \text{Benign}, 1 = \text{Malignant} \right).$$

Therefore, to input an image into a neural network, it must first be converted into a vector of pixel values. Each image vector's label is stored numerically in a separate target vector, y. Once these have been prepared, a training matrix X_{train}, a test matrix X_{test}, and their corresponding target vectors y_{train} and y_{test} must also be generated.

9.2. Keras

Recently, a number of frameworks have been developed for deep learning, ranging from low-level, general purpose math expression compilers, such as Theano, to higher level frameworks such as Torch. For this analysis, the Keras framework was used. Keras is written in Python and is based on the Theano framework. It offers a high level control over network construction, abstracting the low-level Theano code, making it possible to design neural network structures in a layer-wise, modular fashion. Layers and functionality are added to the network piece by piece and are finally compiled into a complete network once the desired structure has been built. Users of Python can install Keras using pip, by typing pip install keras at the comment prompt. Keras has a number of requirements, including Theano (which can also be installed using pip install Theano at the command prompt). Briefly, once Keras has been correctly installed and successfully imported into the environment, a convolutional neural network is created by instantiating an object of the Sequential class, and then by adding layers to this object until the desired network is complete. For example, a convolutional layer can be added to the network using the add function: model.add(Convolution2D(…)). Configuring network properties, such as when to use dropout or specifying which activation function should be used, is also performed using the add function of the model object. The network is built in this way until the desired structure has been defined, and is then compiled using the model object's compile function. As Keras is based on Theano, the model is generated into Theano code, which itself is compiled into CUDA C++ code, and subsequently run on the GPU. Upon successful compilation the model, it can be trained on a dataset using the fit function, which takes the training data set as one of its parameters. A trained model can then be tested using the held back test data, using the trained model's evaluate function. Full Python source code for the generation of the model can be found in this book chapter's GitHub repository under https://github.com/mdbloice/CLSM-classification. This source file contains a complete implementation of the network, including the generation of all the plots and figures shown in the Section 10.

10. Results

10.1. Multiresolution analysis

Overall, 857 images of benign common nevi (408 images) and malignant melanoma (449 images) were used as study set [29]. To get more insights into the classification performance, a percentage split was performed by using 66% of the dataset for training and the remaining instances (34%) as the test set (**Table 1**). The classification results of 572 cases (276 benign common nevi, 296 malignant melanomas) in the training set and 285 cases (132 benign common nevi, 153 malignant melanomas) in the test set.

CART	Training set			Test set		
	% Correct	Benign	Malignant	% Correct	Benign	Malignant
Benign	96.6	267	9	78.0	103	29
Malignant	98.0	6	290	84.1	24	129

Table 1. Classification results for features based on multiresolution analysis.

The CART classification shows a correct mean classification of 97.3% samples in the training set and a correct mean classification rate of 81.1% in the test set. In this study, the images were resized to 512 × 512 pixels. To illustrate the differences in the wavelet sub-bands of both tissues, the spectra of the wavelet coefficient standard deviations are shown for typical views of benign common nevi and malignant melanoma (**Figure 15**). The image of benign common nevi show pronounced architectural structures (so called tumour nests), whereas the image of malign melanoma show melanoma cells and connective tissue with few or no architectural structures. These visual findings are reflected by the wavelet coefficients inside the different sub-bands. The standard deviations of the wavelet coefficients in the lower and medium frequency bands (4–10) show higher values for the benign common nevi than for malignant melanoma tissue, indicating more pronounced structures at different orders of magnitude. The tissue of malignant melanoma appears more homogeneous (due to a loss of structure), and the cells are larger as in the case of benign common nevi. The standard deviations in the sub-bands with higher indices (representing finer and more pronounced structures) are lower than in the case of benign common nevi.

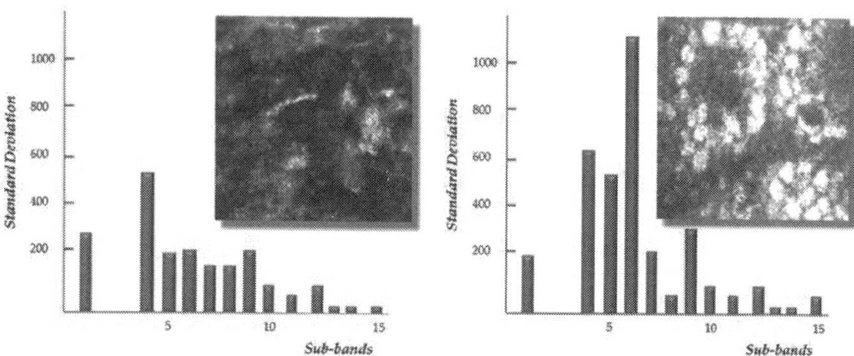

Figure 15. Sub-band spectra for benign common nevi (right) and malignant melanoma (left).

The analysis of the classification tree shows that seven classification nodes indicate benign common nevi and six nodes malignant melanoma. The visual examination of the selected nodes demonstrates characteristic monomorphic melanocytic cells and melanocytic cell nests for benign common nevi [28, 29]. Contrary polymorphic melanocytic cells, a disarray of melanocytic architecture and poorly defined or absent keratinocytic cell borders are characteristic for malignant melanomas.

10.2. Convolutional deep learning neural network

For this study, a dataset consisting of 6897 CLSM images of skin lesions was obtained from our university hospital. The dataset consisted of images of skin lesions in layers of various depths. Before training, the images were randomised and placed into a training set and test set, with the training set consisting of 5000 images and the test set consisting of 1897 images (**Table 2**). It is important to note that, in the case of this project, each image was treated individually, and not treated as belonging to one particular patient or even lesion. The test set, therefore, contained different layers or lesions from potentially the same patient as the training set, as a single patient may have had several scans or may have been examined on multiple occasions.

	Full Dataset	Training Set	Test Set
Total	6,897	5,000	1,897
Benign	3,607	2,655	952
Malignant	3,290	2,345	945

Table 2. The distribution of the classes in the whole dataset and in the training and test set.

Class imbalance occurs when a training set has far more samples of one particular class than another. For example, a small class imbalance existed in the dataset analysed in this chapter, with the samples of benign nevi slightly outnumbering the samples of malignant melanoma (there existed 317 more samples of the former compared to the latter). There are a number of techniques which can be employed to address class imbalance, such as data augmentation (generating synthetic data from your original dataset) or simply by discarding samples to better balance the dataset. In the case of our dataset, class imbalance was not at the degree as to make it problematic. When the training set and test sets were split, however, we ensured that the test set was largely balanced. Class imbalance can also affect how results, such as accuracy and precision/recall, should be perceived when analyzing a trained model on a highly imbalanced test set.

The network, after training for 20 epochs, achieved 93% accuracy on the unseen test set. The model's accuracy on the test set during training, as well as the model's error rate on the training set through each of the 20 epochs is shown in **Figure 16**.

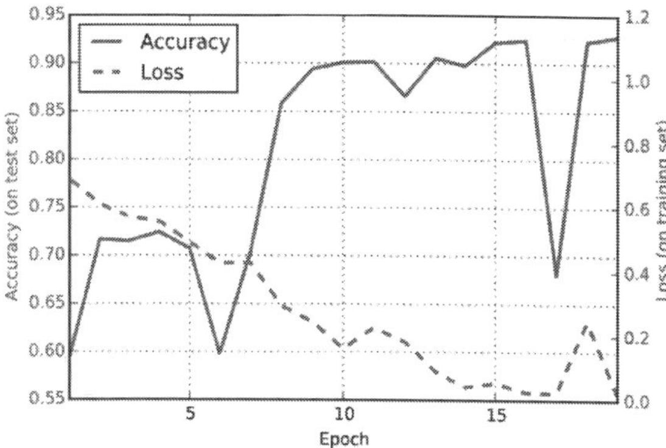

Figure 16. The model's accuracy on the test set and its logistic loss against the training set.

Loss on the training set eventually reduces to almost 0 (meaning it is at this point overfitting heavily), while the accuracy of the model on the unseen test set fluctuates but is tending towards an accuracy of approximately 90%. The accuracy of the final model after epoch 20, when training was terminated, was 93%. A confusion matrix, shown in **Figure 17**, describes the model's accuracy on the test set, in terms of absolute numbers of predicted and actual labels for both the benign and malignant classes.

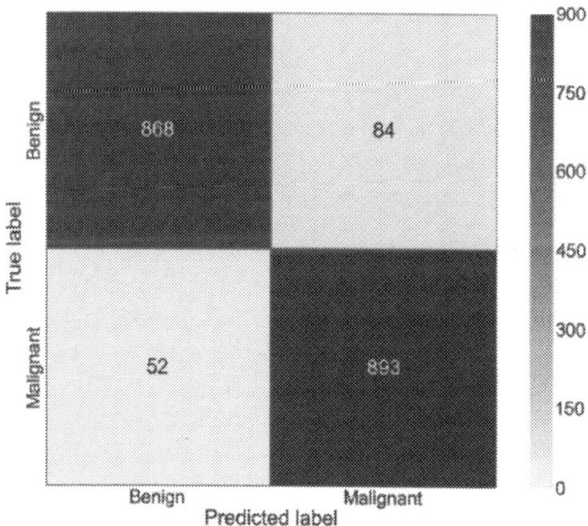

Figure 17. Confusion matrix.

Here, all true/false positives and true/false negatives can be seen. From these values, the precision, recall (sensitivity), and F_1 score (a weighted average of the precision and recall, given by $F_1 = 2 \cdot \frac{\text{precision} \cdot \text{recall}}{\text{precision} + \text{recall}}$) were calculated, as shown in **Table 3**.

	Precision	Recall (Sensitivity)	F_1 score	Support
Benign	0.94	0.91	0.93	952
Malignant	0.91	0.94	0.93	945
Avg/total	**0.93**	**0.93**	**0.93**	1897

Table 3. The generated model's precision, recall and F_1 score measured against the test set.

Table 4 describes the results of the model in absolute terms, with results for the model's predicted labels for both classes versus the actual labels for each class. As well as this, the total number of actual and predicted labels is shown.

	Actual		
Predicted	**Benign**	**Malignant**	**Total**
Benign	**868**	84	952
Malignant	52	**893**	945
All	920	977	1897

Table 4. The generated model's predicted labels versus the actual labels, measured on the test set.

10.3. Transfer learning

Transfer learning is a term that can be applied to several aspects of machine learning. In the case of neural network-based machine-learning approaches, transfer learning often refers to the act of using a pre-trained network as the starting point for a learning procedure, rather than starting with a network which has been initialized with random weights. This is often performed as a time-saving measure, but can also be done when the new data to be classified is scarce. Also, it can be performed only when the data used for pre-training is similar to the new data which should be classified. Furthermore, it constrains the practitioner into using a network which has the same architecture of the pre-trained model. Therefore, it is not useable in all situations, and it does not make sense to use, say, a network pre-trained on the ImageNet dataset (a commonly used benchmarking dataset, containing millions of samples of 1000 classes of images) in the context of CLSM lesion classification.

However, there exist several types of laser scanner-based approaches to skin lesion analysis, where the use of transfer learning may be beneficial. Other methods in the field include two photon excitation fluorescence microscopy, second harmonic imaging microscopy, fluorescence-lifetime imaging microscopy and coherent anti-stokes Raman microscopy. Whether or not transfer learning could indeed be implemented in this context would depend entirely on how well the features learned during pre-training match the features that exist in the new data (in other words, whether the learned features transfer well from one domain to the other). For example, several new methods produce colour images, which would mean the features learned in the analysis described here would likely not transfer well to this new domain (of course, colour images could be converted to greyscale). However, it is conceivable that other technologies, that also produce greyscale images, could make use of a pre-trained network, and thus benefit from pre-trained weight initialisation and therefore transfer learning.

The machine-learning community often makes available pre-trained networks for others to use, such as in the Model Zoo (https://github.com/BVLC/caffe/wiki/Model-Zoo). Some of the networks available on the Model Zoo took many weeks to train on powerful hardware, and is considered a very useful resource by many who do not have the time or the computational resources available to them for such an involved learning task. Of course, a pre-trained network could be made available for the CLSM or skin lesion analysis community, if the network was trained on a sufficiently large dataset and if indeed the learned features would transfer well to other domains.

11. Discussion

Confocal laser scanning microscopy is a technique for obtaining high-resolution optical images with depth selectivity. It enables the noninvasive examination of skin cancer in real-time. This makes CLSM very suitable for screening and early recognition of skin tumours, which augment the success of the therapy. The training of pathologists to acquire and refine their visual diagnostic skills is very time-consuming. To implement diagnostic capabilities on a computer, it is of considerable interest to understand how the diagnostic process unfolds and which texture features are critical for a successful diagnosis. For medical diagnosis, it is important to duplicate the automated diagnostic process.

The multiresolution approach with wavelets features mimics the diagnostic guidelines of the dermatopathologist, as they use multiscale features for the examination of CLSM views. The decision rules generated by machine-learning algorithms, such as CART, represent explicit knowledge that can be used to analyse and refine the diagnostic process. The generated rules can be implemented in viewer software which enables a visual evaluation of the diagnostic performance by the dermatologist. This can be used as a training aid for ongoing dermatologists in education. As shown in the Section 10, the algorithm performance allows a correct classification of 78.0% of the benign common nevi cases and 84.1% of the malignant melanoma in the test set. In contrast, sensitivity and specificity of 85.5 and 80.1% are reached by the human observer (overall performance 82.8%).

Although the CART algorithm discriminates the training set automatically (unsupervised), the feature extraction algorithm is predefined. Algorithms based on artificial neural networks do not perform or require hand-defined analyses of the image features with predefined (filtering) methods. Instead, they use neural computation inspired by the visual system of mammals. Neural networks process an image by use of a hierarchical processing architecture which mimics the way the visual cortex processes visual stimuli from the primary cortex (V1) to different layers (V2–V8) which are selective for different components of the visual stimuli such as orientation, colour, size, depth and motion. Neural networks are well suited for detecting similarities in images. However, the distributed representation of the acquired knowledge complicates the extraction of the diagnostic information. They deliver nothing about the inference mechanism leading to a classification in a form that is easy readable for the human observer. Nevertheless, we can demonstrate a real example as to why artificial neural networks will play an ever more important role in automated medical diagnostic systems. A recent work reported that pigeons (columba livia) proved to have a remarkable ability at discriminating benign from malignant human breast histopathology images and at detecting cancer relevant micro calcifications in mammogram images after differential training with food reinforcement [30]. The discrimination was done by the pigeons via two distinctively coloured response buttons. For a correct discrimination, food was immediately provided by a dispenser. The pigeons proved not only to be capable of image memorization but were able to extend the learned skills to novel tissue images. It results that their diagnostic skills are like that of trained humans. It should be noted that the capabilities were acquired without the benefit of verbal instructions as in the case with human education. The low-level vision capabilities of pigeons appear to be equivalent to those in humans; feedforward and hierarchical processing seem to dominate. It can be assumed that pigeons do not explicitly analyse the images with predefined criteria and explicit instructions as humans do. The reinforcement training of the pigeons resembles the training of artificial neural networks. Given the high diagnostic accuracy of the pigeons they may serve as a model for the development and amelioration of artificial networks (or vice versa). We still do not know in detail how pigeons differentiate such complex visual stimuli but colour, size, shape, texture, and configurational cues seem to participate. Their visual discrimination performance may guide the basic research in artificial neural networks in order to develop computer-assisted image diagnostic systems. Experienced dermatopathologists reported that a beginner (a person in education) examines the CLSM views strictly according to the dermatological guidelines (Section 4), as the computers do by multiresolution analysis. Based on the large amount of previously viewed specimens, an experienced person reports the CLSM views more by its visual appearance (personal communication). This is similar to the image analysis performed by a trained neural network. The receptive field of a sensory neuron is a particular region in the visual system in which a stimulus will trigger the firing of that neuron. In vision research, it is known that a cat's visual cortex only develops its receptive fields if it receives visual stimuli in the first months of life [31]. The receptive fields in the primary visual cortex can be thought as 'feature detectors' or 'flexible categorizers'. This means that they learn the structure of the input patterns and become sensitive to combinations that are frequently repeated [14]. This also demonstrates the importance of convolutional neural networks in image processing and analysis.

In this work, and given the relatively small dataset size, the performance of the trained neural network model is encouraging. However, the results must be considered as a proof of concept, and not a model that could be used in a clinical setting, despite the good accuracy of the trained model. For example, the images were collected from a single department, at one hospital in a single region in Austria. To judge the potential real-world accuracy of a trained model would require a far larger dataset, collected from several regions worldwide, and carefully curated to ensure no unintentional bias is introduced (by only collecting data from patients of a certain age range, for example). By training a model on a far larger dataset such a model could be used in real-world clinical settings as a diagnosis aid.

The work here shows that deep layer neural networks have the capacity to learn the high-level discriminatory features required to classify malignant and benign skin lesions. This can be achieved without any dedicated feature engineering phase, data pre-processing or a priori domain knowledge. In the case of the CLSM image classification task presented here, all that was required was a labelled dataset of previous observations. However, what is also true is that neural networks require far more training data than traditional machine vision methods that work on extracted features. This is due to the very high dimensionality of the data, which in our case was $\mathbb{R}^{4096}$, in contrast to the analysis of the extracted features where the dimensionality was $\mathbb{R}^{39}$. To compensate for a far higher dimensionality, a much larger dataset is, therefore, a necessity. In other words, deep learning neural networks are most suitable for situations where you encounter data with 'high m, high n' properties—high dimensional data, like images, of which many samples exist—such datasets are common in the medical domain, meaning deep learning should be of especial interest to researchers in the area of healthcare informatics.

As parallelized hardware advances, Moore's law begins to plateau, and the amounts of data being stored increases, algorithms that take advantage of this perfect storm will become more and more relevant. We have shown in this chapter that classical approaches to image classification can indeed be emulated by deep neural networks fed with large amounts of observed data. In fields such as medicine, where data are in such abundance, highly parallelized algorithms may be the only approach that can deal with such large data sources in a meaningful way. Fortunately, this is no longer the domain of specialized research institutes with access to cluster computing: such algorithms are trainable without large investments in hardware and can be performed on a standard desktop workstation equipped with a modestly priced GPU.

Author details

Marco Wiltgen* and Marcus Bloice

*Address all correspondence to: marco.wiltgen@medunigraz.at

Institute for Medical Informatics, Statistics and Documentation, Medical University of Graz, Graz, Austria

References

[1] Schaller A, Sattler E, Burghof W, Röken M: Color Atlas of Dermatology. 1te ed. Thieme Verlag. Stuttgart, New York. 2012; 978–3131323415

[2] Sterry W, Paus R: Thieme Clinical Companions Dermatology. Thieme Verlag. Stuttgart, New York. 2006 3–13–1359110

[3] Bolognia J.L, Jorrizo J.L, Schaffer J.V: Dermatology: Expert Consult Premium Edition. Saunders. 3 edition UK. 2012; 978–0723435716

[4] Markovic S.N, Erickson L.A, Flotte T.J, Kottschade L.A. Metastatic malignant melanoma. G Ital Dermatol Venereol. 2009;144(1):1–26.

[5] Oliveria S, Saraiya M, Geller A, Heneghan M, Jorgensen C. Sun exposure and risk of melanoma. Arch Dis Child. 2006;1(2):131–138.

[6] Friedman R, Rigel D, Kopf A. Early detection of malignant melanoma: the role of physician examination and self-examination of the skin. CA Cancer J Clin. 1985;35(3): 130–151.

[7] Pawley J.B. Handbook of Biological Confocal Microscopy. 3rd ed.. Springer, Berlin; 2006. 0-387-25921-X.

[8] Paoli J, Smedh M, Ericson M.B. Multiphoton laser scanning microscopy—a novel diagnostic method for superficial skin cancers. Semin Cutan Med Surg. 2009;28(3):190–195.

[9] Patel D.V, McGhee C.N. Contemporary in vivo confocal microscopy of the living human cornea using white light and laser scanning techniques: a major review. Clin Exp Ophthalmol. 2007;35(1):71–88.

[10] Rajadhyaksha M. Confocal microscopy of skin cancers: translational advances toward clinical utility. Conf Proc IEEE Eng Med Biol Soc. 2009;1:3231–3233.

[11] Hofmann-Wellenhof R, Pellacani G, Malvehy J, Soyer H.P. (eds). Reflectance Confocal Microscopy for Skin Diseases. Springer, Berlin Heidelberg; 2012; 978-3-642-21996-2.

[12] Pellacani G, Cesinaro A.M, Seidenari S. In vivo assessment of melanocytic nests in nevi and melanomas by reflectance confocal microscopy. Mod Pathol. 2005;18:469–474.

[13] Prasad L, Iyengar S.S. Wavelet analysis with applications to image processing. CRC Press, Boca Raton; 1997.

[14] Marr D. Vision. W.H. Freeman, New York, 1982.

[15] Strang G, Nguyen T. Wavelets and Filterbanks. Wellesley-Cambridge Press. MA USA; 1996.

[16] Wiltgen M. Confocal laser scanning microscopy in dermatology: Manual and automated diagnosis of skin tumours. In: Chau-Chang Wang editor. Laser Scanning, Theory and Applications, Intech Publisher. Croatia. 2011;133–170. ISBN 978–953–307–205–0)

[17] Murphy, K.P. Machine learning: a probabilistic perspective. MIT press, USA. 2012.

[18] Breiman L, Friedman J, Olshen R.A, Stone C.F. Classification and Regression Trees. Chapman & Hall, New York, London; 1993.

[19] Hornik K, Stinchcombe M, White H. Multilayer feedforward networks are universal approximators. Neural Netw. 1989;2(5):359–366.

[20] Kalchbrenner N, Grefenstette E, Blunsom P. A convolutional neural network for modelling sentences. *arXiv preprint arXiv:1404.2188*, 2015.

[21] Collobert R, Weston J, Bottou L, Karlen M, Kavukcuoglu K, Kuksa P. Natural language processing (almost) from scratch. J Mach Learn Res. 2011;12:2493–537.

[22] Waldrop M.M. The chips are down for Moore's law. Nature. 2016;530:144–147.

[23] Mnih V, Kavukcuoglu K, Silver D, Rusu AA, Veness J, Bellemare MG, Graves A, Riedmiller M, Fidjeland AK, Ostrovski G, Petersen S. Human-level control through deep reinforcement learning. Nature. 2015;518(7540):529–33.

[24] Silver D, Huang A, Maddison CJ, Guez A, Sifre L, van den Driessche G, Schrittwieser J, Antonoglou I, Panneershelvam V, Lanctot M, Dieleman S. Mastering the game of Go with deep neural networks and tree search. Nature. 2016;529(7587):484–9.

[25] Williams DR, Hinton GE. Learning representations by back-propagating errors. Nature. 1986;323:533–6.

[26] LeCun Y, Bottou L, Bengio Y, Haffner P. Gradient-based learning applied to document recognition. Proc IEEE. 1998;86(11):2278–324.

[27] Michael A. Nielsen, Neural Networks and Deep Learning, Determination Press, USA 2015. URI: http://neuralnetworksanddeeplearning.com/

[28] Lorber A, Wiltgen M, Hofmann-Wellenhof R, Koller S, Weger W, Ahlgrimm-Siess V, Smolle J, Gerger A. Correlation of image analysis features and visual morphology in melanocytic skin tumours using in vivo confocal laser scanning microscopy. Skin Res Technol. 2009;15:237–241.

[29] Wiltgen M, Gerger A, Wagner C, Smolle J. Automatic identification of diagnostic significant regions in confocal laser scanning microscopy of melanocytic skin tumours. Methods Inf Med. 2008;47:14–25.

[30] Leveson R.M, Krupinski E.A, Navarro V.M, Wasserman A. Pigeons (Columba livia) as trainable observers of pathology and radiology breast cancer images. Plos One 10(11). 2015; 1–21.

[31] Hubel DH, Wiesel TN. Period of susceptibility to physiological effects of unilateral eye closure in kittens. J Physiol. 1970;206(2):419–436.

Super-Resolution Confocal Microscopy Through Pixel Reassignment

Longchao Chen, Yuling Wang and Wei Song

Additional information is available at the end of the chapter

http://dx.doi.org/10.5772/63192

Abstract

Confocal microscopy has gained great popularity in the observation of biological microstructures and dynamic processes. Its resolution enhancement comes from shrinking the pinhole size, which, however, degrades imaging signal-to-noise ratio (SNR) severely. Recently developed super-resolution method based on the pixel reassignment technique is capable of achieving a factor of $\sqrt{2}$ resolution improvement and further reaching twofold improvement by deconvolution, compared with the optical diffraction limit. More importantly, the approach allows better imaging SNR when its lateral resolution is similar to the standard confocal microscopy. Pixel reassignment can be realized both computationally and optically, but the optical realization demonstrates much faster acquisition of super-resolution imaging. In this chapter, the development and advancement of super-resolution confocal microscopy through the pixel realignment method are summarized, and its capabilities of imaging biological structures and interactions are represented.

Keywords: super resolution, confocal microscopy, pixel reassignment, computational realization, optical realization

1. Introduction

Better understanding of biological processes at the cellular and subcellular level is closely dependent on the direct visualization of the cellular microstructures. Among the various microscopic techniques, fluorescence microscopy takes advantage of the abilities to observe in real-time the molecular specificities in living biological samples down to the cellular and/or subcellular scale, and thus has found broad applications in the investigations of cell biology and neuroscience. However, the spatial resolution of conventional microscopy is optically diffrac-

tion-limited, restricting its lateral resolution to be ~250 nm and axial resolution to be ~600 nm (primarily determined by the numerical aperture of microscopic objective), respectively. As a result, it is very challenging to resolve the subcellular structures by the conventional microscopic technologies because their microstructures are comparable to (even finer than) the diffraction-limited resolution.

Fortunately, a number of novel fluorescence microscopic techniques with super-resolution capability have been established to break down the optical diffraction limitation in recent years, allowing the observation of many cellular and subcellular structures that are always not resolvable by the conventional fluorescence microscopy. For example, by sharpening the point-spread function of the microscope with the suppression of the fluorescence emission on the rim of a focused laser spot, stimulated emission depletion (STED) microscopy breaks the optical diffraction limitation and achieves resolution as high as ~30 nm [1]. Localization-based techniques, such as stochastic optical reconstruction microscopy (STORM) and photoactivated localization microscopy (PALM), enable imaging at a resolution of ~20 nm [2, 3]. Structured illumination microscopy (SIM) applies spatially structured light illumination for shifting the high spatial frequency to the low-frequency range, which thus can be collected by microscopy [4]. These methods achieve an order of magnitude improvement in spatial resolution over the conventional fluorescence microscopy. Therefore, the super-resolution microscopic technology opens up new windows for observing the previously unresolved cellular structures and provides great potentials for elucidating biological processes at the subcellular and molecular scale [4].

Among these high-resolution fluorescence microscopic techniques, confocal microscopy, the first super-resolution imaging technique, is one of the most widely used imaging approaches with moderately enhanced spatial resolution. Utilizing a focused laser as an excitation source in combination with a pinhole in front of the detector for blocking out out-of-focus signals, confocal microscopy is able to improve the spatial resolution by a factor of $\sqrt{2}$ in principle. However, instead of its super-resolution capability, the sectioning capability is more impressed because the spatial resolution with a factor of $\sqrt{2}$ improvement is hardly accessible in the standard confocal microscopy. The resolution of confocal microscopy relies on the pinhole diameter, that is, higher resolution comes from the smaller sized pinhole filter. Such a small pinhole rejects the unwanted out-of-focus light, while parts of the desired in-focus emission are filtered out simultaneously. As a result, the signal-to-noise ratio (SNR) is drastically decreased as the pinhole size shrinks, which, in turn, practically deteriorates the spatial resolution. Instead, the fluorescence efficiency within the biological samples is often weak, so a relatively large pinhole diameter is typically chosen concerning the imaging SNR. Therefore, the standard confocal microscopy is practically unable to provide super-resolution imaging.

In order to achieve spatial resolution improvement and better imaging SNR simultaneously in confocal microscopy, light/fluorescence signals should be detected with a nearly closed pinhole array instead of a single pinhole [5]. The images acquired by each pinhole within the array have the same resolution but different SNR levels [6]. To overcome this limitation, a method applying the pixel reassignment technique is proposed by reasonably summing the signals from each nearly closed pinhole together, which enables simultaneous improvement of resolution and SNR. In this chapter, we present the state-of-the-art super-resolution techniques based on the pixel reassignment. Section 2 gives the principle of pixel reassignment firstly, and then two different operations realizing the pixel reassignment. Also, some representative super-resolution images in biological specimens are summarized in this section. At last, some advances in super-resolution confocal microscopy through the pixel reassignment will be discussed.

2. Super resolution by pixel reassignment

The concept of pixel reassignment is firstly proposed more than two decades ago to solve the drawbacks in standard confocal microscopy [5]. As we know, the reduction of the pinhole diameter down to zero allows the finest lateral resolution in confocal microscopy in theory, which, however, generates fluorescent images with a very low SNR due to the dramatically

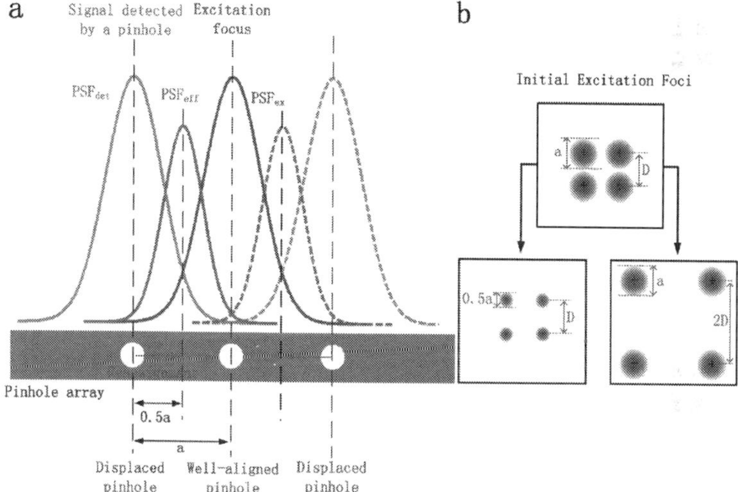

Figure 1. Schematic diagram illustrating the principles of pixel reassignment. (a) One-dimensional representation of pixel reassignment. Two pinholes (left and right) within an array displaces by a distance of 'a' from the excitation focus, which detect light signals mostly originated from the location of the peak of the product of PSF$_{det}$ (x-a) and PSF$_{ex}$ (x). In the case that PSF$_{det}$ and PSF$_{ex}$ are identical (i.e. neglecting the Stokes shift), the maximum in PSF$_{eff}$ occurs at the position with a distance of $a/2$ from the excitation focus. Thus, the detected light signals from the displaced pinholes are reassigned to the well-aligned pinhole that is at the center of the excitation focus and the original detection spot. (b) Pixel realignment operation. Top panel shows the excitation foci (blue circles) created by scanning illuminating laser across the sample, where four excitation foci are with the distance of D and diameter of a. Bottom: Two pixel realignment operations for increasing the image resolution. Lower left panel represents twofold reduction of the foci without altering their distance. Lower right panel displays the increase of the foci distance to 2D, while maintaining all foci sizes. These two implementations produce an equivalent imaging reconstruction, with only different global scaling factor.

degraded light collection efficiency. Although the pinhole size can be adjusted to one Airy unit for better imaging SNR, the lateral resolution is sacrificed. Instead of a single pinhole, a pinhole array is used for the light detection, followed by a reconstruction algorithm for the image formation. As a result, the standard confocal microscopy with the pixel reassignment operation is capable of enhancing its lateral resolution simultaneously with higher imaging SNR.

2.1. Principle of pixel reassignment

Pixel reassignment demonstrates great potentials for improving both lateral resolution and imaging SNR. Instead of summing the signals directly as the conventional imaging technologies, each signal is reassigned to a particular location where the signal most probably comes. **Figure 1(a)** gives the principle of the pixel reassignment in terms of excitation and detection point-spread function (PSF) [7]. The excitation PSF (PSF_{ex}, labeled by blue line) represents the distribution of the corresponding excitation focus. At a displaced pinhole, detection PSF (PSF_{det}, labeled by green line) is centered on the detection axis with a distributed probability of signal detection around that pinhole. The effective PSF (PSF_{eff}, labeled by red line) is contributed from the overlap (multiplication) of PSF_{det} and PSF_{ex}. The well-aligned pinhole is coaxial with the excitation focus, realizing the maximal signal detection probability. As the pinhole detector is far away from the axis of the excitation focus, the signal acquisition probability decreases because of their less overlying; consequently, these nearly closed pinhole detectors induce lower-SNR image.

In the pixel reassignment implementation, a camera (similar with a pinhole array), rather than a point detector, is commonly employed because its individual pixels are considered as infinitely narrow pinhole. Neglecting Stocks shift in single-photon fluorescence and assuming identical PSF_{det} and PSF_{ex}, a maximal probability of signal acquisition (i.e. PSF_{eff}) is at the midway of the peaks of PSF_{det} and PSF_{ex}. **Figure 1(b)** gives two methods for the pixel reassignment operation, either twofold local contraction of the excitation focus without altering the distance between them (panel in lower left of **Figure 1(b)**), or twofold increasing the distance between the foci while maintaining their original size (panel in lower left of **Figure 1(b)**) [8]. By reassigning the signals from all pixels within the detector array (i.e. all displaced pinholes as shown in **Figure 1(a)**) to the particular location, a sharper and higher-SNR image is eventually achieved.

Pixel reassignment technique is able to improve the resolution to a factor of $\sqrt{2}$ without sacrificing SNR, and the resolution can be further improved by deconvolution algorithm up to a factor of 2 [9, 10]. Although the spatial resolution of the pixel reassignment technique is still lower compared with other super-resolution methods, such as STED and STORM [1–3], it overcomes some of their shortcomings. This technique inherits all advantages of the standard confocal microscopy, including high-speed imaging rate, acceptable excitation intensity, optical sectioning capability, and a broad choice of fluorescent dyes and/or proteins, making it a readily accessible technology in a variety of biological investigations.

The pixel reassignment can be considered as an alternative method of SIM, theoretically achieving the same spatial resolution improvement compare with standard SIM through point-like illumination feature. In contrast, the technique demonstrates better feasibility over the standard SIM, that is, the pixel reassignment operation can be easily implemented both computationally and experimentally (optical system adaptation). Unlike computational mode that is always time-consuming in raw data processing, the pixel reassignment realized with optical means is capable of obtaining super-resolution images with fast imaging acquisition. More details on these two different methods for realizing the pixel reassignment are represented as below.

2.2. Computational realization of pixel reassignment

2.2.1. Image scanning microscopy

Image scanning microscopy (ISM), proposed by C. Müller and J. Enderlein in 2009, is a super-resolution microscopic technique based on the pixel reassignment [11]. This system is modified from a standard confocal microscopy that replaces the point detector (normally a photomultiplier tube) with an Electron multiplying CCD (EMCCD) camera (labeled 9) as shown in **Figure 2(a)**. The camera takes an image of each spatial position of the scanning focus, and then an algorithm of the pixel reassignment processing is utilized by summing the raw images to reconstruct an ISM image, which improves the resolution from 244 nm to 198 nm laterally.

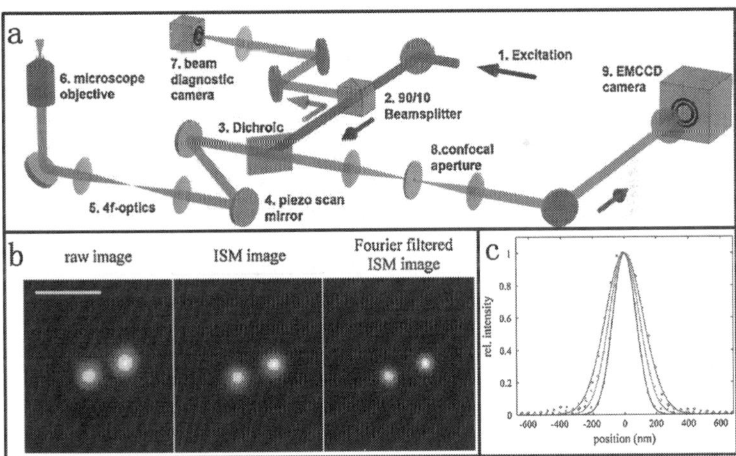

Figure 2. Super-resolution image scanning microscopy (ISM) with computational pixel reassignment. (a) The schematic diagram of ISM system. Fluorescence excitation (1); a super-continuum white light laser equipped with an acousto-optic tunable filter; nonpolarizing beam splitter cube (2); dichroic mirror (3); piezo scanning mirror (4); 4f telescope configuration (5); microscope objective (6); beam diagnostic camera (7); confocal aperture with 200 μm diameter (8); EMCCD camera for fluorescence detection (9). (b) Super-resolution imaging fluorescent beads with 100-nm diameter. Left panel: Confocal microscopy image; middle panel: ISM image; right panel: Fourier-weighted ISM image. Scale bar: 1 μm. (c) Linear cross-sectional distribution along the horizontal axis of an individual bead image in (b). Adapted with permission from reference [11].

Further, deconvolution function is used to improve its lateral resolution up to 150 nm, 1.63-fold better than the image from raw data, as shown in **Figure 2(b)** and **(c)**, respectively. Note that the pinhole in ISM (labeled 8) filters the out-of-focus light signals, maintaining the optical sectioning capability as the standard confocal microscopy. In this work, the realization of the lateral resolution improvement up to 198 nm does not entirely rely on the pinhole because of its relatively large diameter, which, however, gives a high imaging SNR. Therefore, with the computational pixel realignment ISM is able to provide images with optimization of both spatial resolution and imaging SNR.

2.2.2. Multifocal structured illumination microscopy

ISM demonstrates multiple advantages, including the optical sectioning capability as the standard confocal microscopy, the enhanced lateral resolution, and the high fluorescence collection efficiency [11]. However, it is subjected to slow frame rate due to the EMCCD camera (imaging acquisition of 10 ms with each scanning position), and is time-consuming for visualizing the three-dimensional (3D) microstructures.

In order to speed up the imaging acquisition, Shroff et al. developed multifocal structured illumination microscopy (MSIM) by using a sparse lattice of excitation foci (similar to swept-field or spinning disk confocal microscopy) in 2011 [9]. As shown in **Figure 3**, MSIM applies a digital micromirror device (DMD) for generating the sparse lattice illumination patterns.

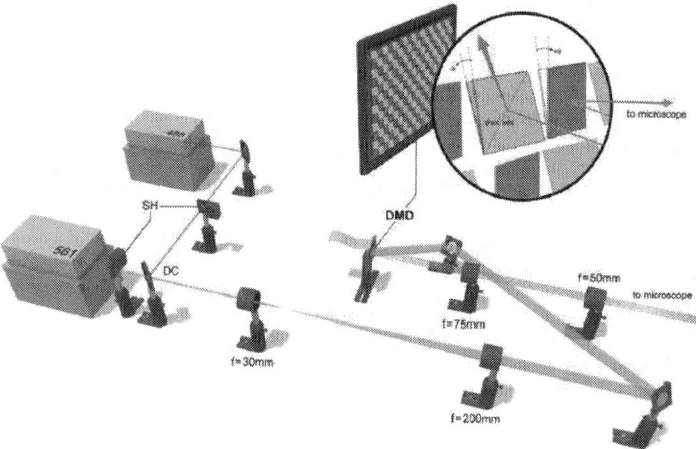

Figure 3. The schematic of multifocal structured illumination microscopy (MSIM). Lasers with 561 and 488 nm serve as illumination sources. Both laser outputs are combined with a dichroic (DC). After beam expanding, both lasers are directed onto a digital micromirror device (DMD). The resulting pattern is de-expanded by a pair of lenses, and is subsequently delivered by the tube lens and microscopic objective inside the microscope (not shown) into the samples. Mechanical shutters (SH) placed in front of the laser output are used for switching illumination on or off. Adapted with permission from reference [9].

After a series of reconstruction steps (open-source software), MSIM enables 3D subdiffractive imaging with resolution doubling, indicating a lateral resolution at 145 nm and an axial resolution at 400 nm. Moreover, it provides the capability of significantly fast imaging acquisition at one 2D image per second.

For super-resolution MSIM, the data acquisition and processing are implemented as below (please refer to **Figure 4** for detailed procedures). First, the sample is excited with a sparse, multifocal excitation pattern. Second, the resulting fluorescence image is recorded with a camera, and then the digital pinholes around each fluorescent focus are applied for rejecting the out-of-focus emission. Afterwards, the pixel reassignment with 2× scaling is used to process the resulting image. Repeat the above procedures for the entire imaging region fully illuminated. Eventually, a super-resolution image with $\sqrt{2}$-fold resolution improvement is obtained through the digital summation of all such pinholed and scaled images. Twofold resolution improvement is further achieved with deconvolution.

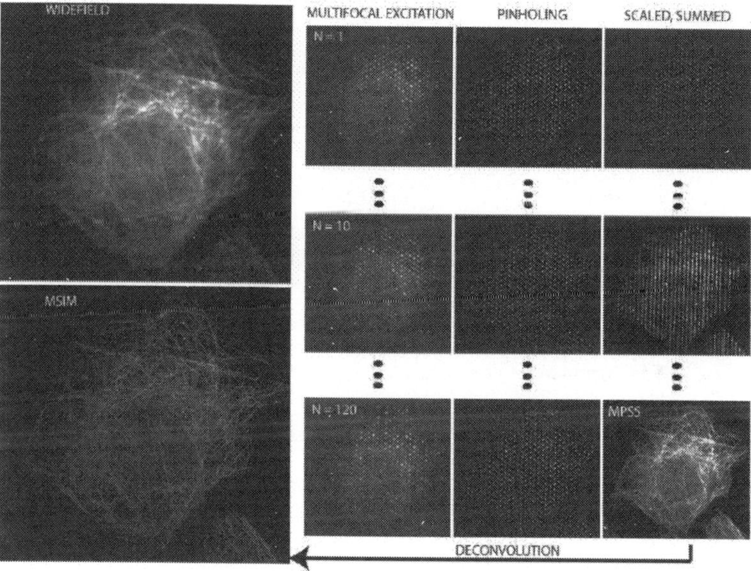

Figure 4. Super-resolution MSIM realization. Top left figure represents a wide-field image produced with a uniformly illuminated pattern onto sample. Right panel provides the reconstructed procedure for the first, tenth, and final raw images of a 120-frame sequence. Lower left figure displays the super-resolution MSIM image by deconvolving the summed image. Adapted with permission from Ref [9].

The resolution improvement of MSIM is demonstrated by imaging antibody-labeled micro-tubules in human osteosarcoma (U2OS) cells embedded in Fluoromount as shown in **Figure 5**. Compared to the wide-field images, the multifocal-excited, pinholed, scaled, and summed (MPSS) images have both higher resolution and better contrast (**Figure 5(b)**). In **Figure 5(d)**, the full-width at half maximum (FWHM) of light intensity of microtubules is estimated at about 145 nm in MSIM images, giving a twofold resolution enhancement compared with the image from wide-field microscopy (~299 nm). Moreover, the frame rate of acquiring an image with field of view at 48 × 49 μm is up to 1 Hz in MSIM, indicating more than 6500-fold faster acquisition over the ISM technology [11].

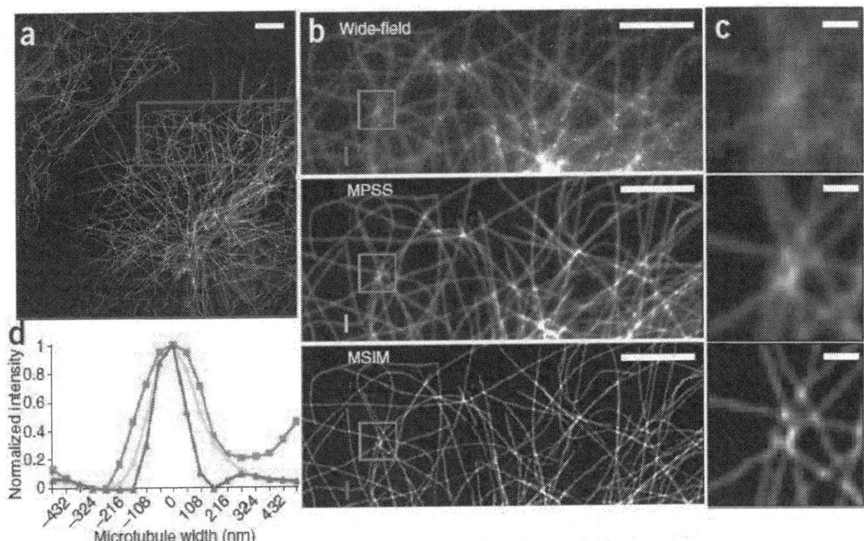

Figure 5. Resolution doubling of MSIM by imaging antibody-labeled microtubules in human osteosarcoma (U2OS) cells. (a) MSIM imaging microtubules labeled with Alexa Fluor 488 in a fixed cell. MSIM image is formed from 224 raw images taking ~1 s total acquisition time with 4.5 ms for each image. Scale bar: 5 μm. (b) Magnified images from the boxed region in (a). Top panel showing a wide-field image, middle panel showing an MPSS image, and bottom panel showing an MPSS and deconvolved (MSIM) image. Scale bars: 5 μm. (c) Close-up images of the boxed regions in (b). Scale bars: 1 μm. (d) Intensity profiles along the colored lines in (b), giving FWHM values at 299 nm in wide-field microscopy, 224 nm in MPSS, and 145 nm in MSIM, respectively. Adapted with permission from reference [9].

2.3. Optical realization of pixel reassignment

The pixel reassignment implemented by the computational means is capable of doubling the resolution than wide-field imaging [9, 11]. The limitation, however, is that the methods are fundamentally time-consuming compared to the standard conventional microscopy because a large number of raw images are essentially acquired and processed. Recently, optically realized pixel reassignment has been developed to overcome the limitations by adapting the optical imaging system instead of digital data-processing operations, which produces images with comparable improvement in the spatial resolution [8, 10, 12].

2.3.1. Instant structured illumination microscopy

Instant structured illumination microscopy (ISIM) is developed by Shroff et al. in 2013 that is analogous to MSIM, while its pixel reassignment process operates optically instead of the digital computation procedures [10]. As shown in **Figure 6**, the DMD used in MSIM is replaced with a converging microlens array. As a result, a multifocal excitation pattern is generated in ISIM. Correspondingly, a matched pinhole array is added to physically reject the out-of-focus emissions. With this modification, the optical pixel reassignment is realized based on the matched microlens array for twofold local contraction of each fluorescent focus. The fluorescence emission pattern is imaged onto a camera by galvanometer scanning. Eventually, the pinholed and scaled images are optically summed, enabling $\sqrt{2}$-fold resolution enhancement.

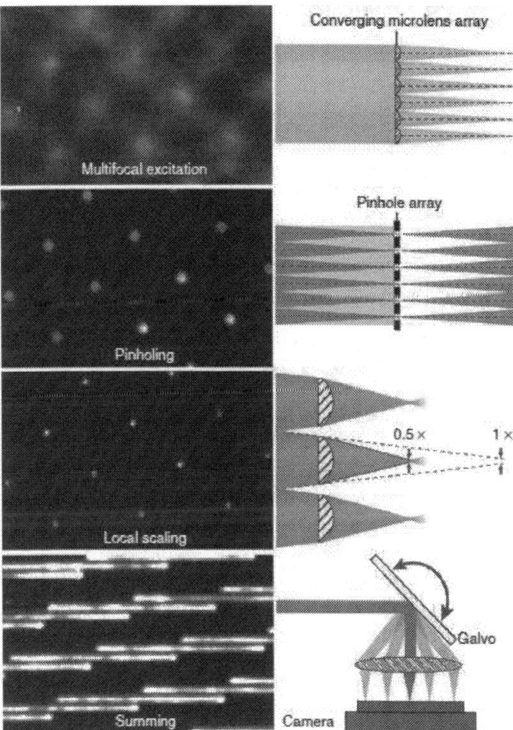

Figure 6. Principles of implementing instant structured illumination for super-resolution realization. A multifocal excitation pattern is produced with a converging microlens array. For fluorescence detection, a pinhole array that matches the microlens array rejects the out-of-focus fluorescence signals. Afterwards, a second, matched microlens array allows a twofold local contraction of each pinholed fluorescence emission. A galvanometer serves as raster scanning of multifocal excitation and summation of multifocal emission, which thus produces a super-resolution image during each camera exposure. Adapted with permission from reference [10].

ISIM demonstrates 3D super-resolution imaging with a lateral resolution of 145 nm and an axial resolution of 350 nm, nearly comparable with MSIM. Moreover, the 100 Hz frame rate comes from the optical operation of pixel realignment in ISIM, allowing super-resolution real-time imaging (almost 100-fold faster than MSIM). Taking into account the data processing duration, the speed-up factor exceeds 10000. In addition, the low illumination power in ISIM ($\sim$5–50 W/cm^2) mitigates photobleaching. As a result, ISIM can perform imaging over tens of time points without obvious photobleaching or photodamage. In **Figure 7**, the rapid growth ($\sim$3.5 µm/s) of endoplasmic reticulum (ER) is monitored by ISIM even though less than 140 ms in the formation and growth of new ER tubules. The biological processes blur in previously developed technologies, such as MSIM and ISM [9, 11]. The capabilities make ISIM a powerful tool for time-lapse super-resolution imaging in living biological samples.

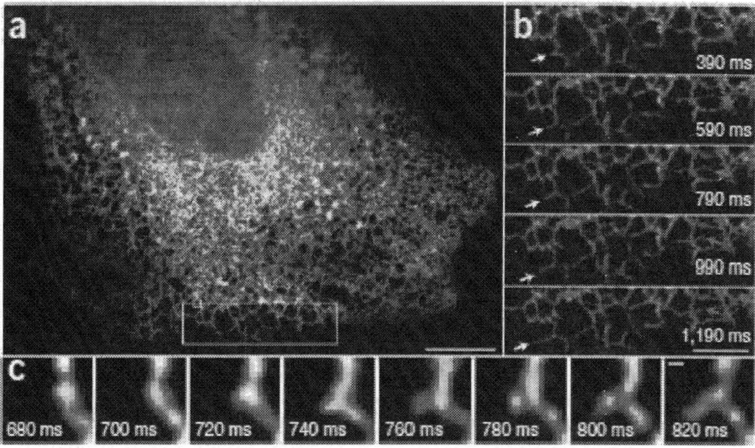

Figure 7. ISIM demonstrates high frame rate of imaging endoplasmic reticulum (ER) at 100 Hz. (a) The first image from 200 time points. ER labeled with GFP-Sec61A within MRL-TR-transformed human lung fibroblasts. Scale bar: 10 µm. (b) Magnification of image with the large white box in (a). White arrows point out the growth process of an ER tubule; blue arrows represent the remodeling of an ER tubule. Scale bar: 5 µm. (c) Magnification of the image with the small white box in (a), displaying the dynamic formation of a new tubule within 140 ms. Scale bar: 200 nm. Adapted with permission from reference [10].

2.3.2. Re-scan confocal microscopy

Rescan confocal microscopy (RCM) is another optical realization of the pixel reassignment technique, proposed by Luca et al. in 2013 [12].Compared with ISIM, it is more easily accessible to build an RCM because this system can be readily modified from a standard confocal microscopy as shown in **Figure 8**. The optical pixel reassignment in RCM is realized as below. The focal length of the lenses L2 and L3 is adapted for twofold local contraction of the fluorescent focus spot. Alternatively, the final fluorescence image is twofold magnified while maintaining the original fluorescence foci size.

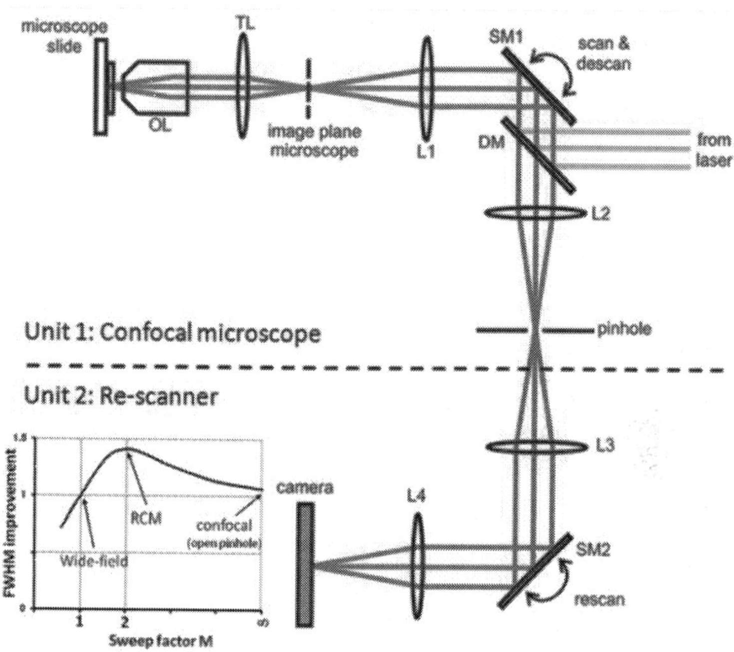

Figure 8. The schematic of rescan confocal microscopy (RCM). Unit 1: A standard confocal microscopy with a set of scanning mirrors for scanning the excitation light and de-scanning the emission light. Unit 2: A re-scanning configuration for 'writing' the light that passes the pinhole onto the CCD-camera. Although the pinhole is in a relatively large diameter, the resolution is $\sqrt{2}$ times improved, which thus gives much more photo-efficient advantage compared to conventional confocal microscopes with the similar resolution. Adapted with permission from reference [12].

This process is accomplished by reasonably changing the angular amplitude of the rescanner. The ratio of angular amplitude of the two scanners, expressed by the sweep factor M, changes the properties of the rescan microscope. For M = 1 the microscope has the same lateral resolution with a wide-field microscope, defined by the well-known optical diffraction limit; it achieves the super resolution for M = 2. The rescanner is used to deliver the fluorescence emission onto the camera pixels. The camera is in the exposure status for optical summation of the fluorescent focus during rescanning.

The lateral resolution improvement of RCM is quantified by imaging 100-nm fluorescent beads. FWHM is found to reduce from 245 nm (15 nm) in wide-field imaging to 170 nm (±10 nm) in RCM imaging, indicating an improvement by a factor of $\sqrt{2}$ without deconvolution. Also, the resolution improvement is concluded by visualizing fluorescently labeled microtubules of HUVEC cell in **Figure 9(a)–(f)**. To demonstrate the capability of RCM for monitoring dynamics, the time-lapse imaging of living HeLa cells expressing EB3-GFP with the growing end of microtubules is observed by RCM. As shown in **Figure 9(g)**, RCM is able to track the fast dynamics (0.5 μm/s) with multiple advantages of improved resolution, high sensitivity, and sufficient imaging rate (1 fps).

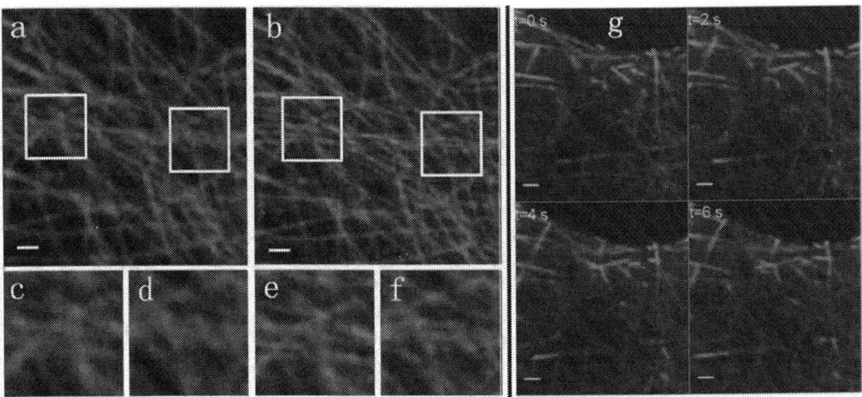

Figure 9. Fluorescently labeled microtubules in HUVEC cells imaged by RCM with sweep-factor M=1 (a), which gives an image with resolution of a wide-field fluorescence microscope determined by the diffraction limit. In double-sweep mode (sweep-factor M=2) (b) RCM gives resolution improvement by a factor of $\sqrt{2}$. Junctions of microtubules (c, e) and parallel microtubules (d, f) are unresolved with wide-field resolution (c, d), but distinguished by RCM in double sweep mode (e, f). (g) Screenshots from an RCM time lapse series of living HeLa cells at M=2 demonstrate the monitoring of fast dynamic structures (0.5 μm/s). Scale bars: 1 μm. Adapted with permission from reference [12].

2.3.3. Two-photon instant structured illumination microscopy

RCM improves resolution by a factor of $\sqrt{2}$ compared with wide-field imaging while possessing optical sectioning capabilities as the traditional confocal microscope [8]. Two-photon excitation offers better optical sectioning capability based on the nonlinear effect. Infrared excitation light minimizes the optical scattering in the tissue, and the fluorescent signals come only from two-photon absorption. These advantages effectively increase the penetration depth and simultaneously suppress the background signal, making the two-photon excitation technique an ideal imaging tool for the thick samples.

Two-photon instant structured illumination microscopy (2P ISIM) is a combination of RCM and two-photon excitation technique, presented by Shroff et al. in 2014, as shown in **Figure 10(a)** [8]. Similarly, an additional scanning component is introduced in 2P ISIM for the optical realization of pixel reassignment. In **Figure 10(b)–(d)**, 2P ISIM provides better resolution than the diffraction-limited two-photon excitation mode by imaging the microtubules. Applying the deconvolution, the lateral resolution is further improved in **Figure 10(c)**. 2P ISIM is quantified by ~150 nm in the lateral resolution and by ~400 nm in the axial resolution, respectively, with 100-nm diameter fluorescent beads as imaging targets. A factor of 2 (with deconvolution) resolution enhancement is obtained compared with the conventional two-photon wide-field imaging (~311 nm).

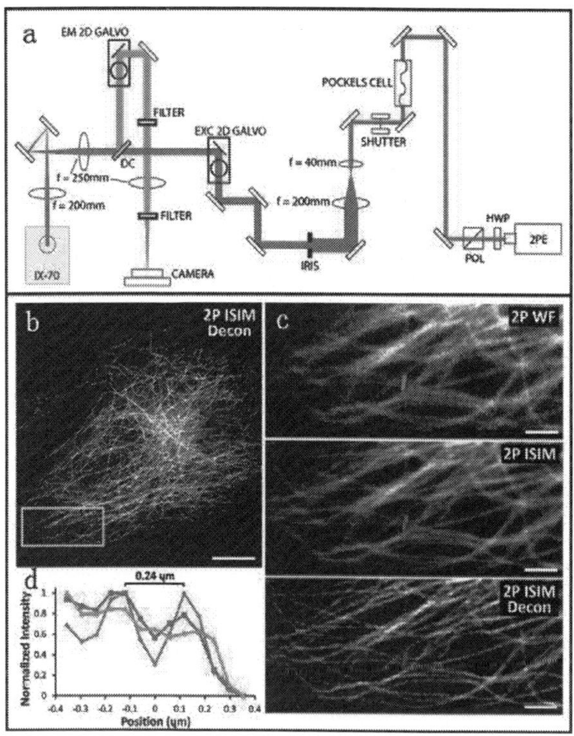

Figure 10. Schematic diagram of two-photon instant structured illumination microscopy (2P ISIM) and its imaging capabilities. (a) Pulsed femtosecond laser (2PE) serves as a two-photon excitation source (labeled by red line). Fluorescence (labeled with green line) is collected and delivered onto a camera. HWP: half-wave plate; POL: polarizer; EXC 2D GALVO: galvanometric mirror for scanning the excitation laser; DC: dichroic mirror; IX-70: microscope part housing objective and sample (not shown); EM 2D GALVO: galvanometric mirror for rescanning the fluorescence emission. (b)–(d) Resolution enhancement of 2P ISIM. (b) 2P ISIM image of immunolabeled microtubules in a fixed U2OS human osteosarcoma cell after deconvolution processing. (c) Magnified view of the yellow rectangular region in (b), indicating the resolution improvement in deconvolved 2P ISIM compared with both 2P wide-field microscopy (2P WF) and 2P ISIM. (d) Fluorescence intensity profiles of microtubules highlighted with green, red, and blue lines in (c). Scale bar: 10 µm in (b) and 3 µm in (c). Adapted with permission from reference [8].

To demonstrate the enhanced penetration ability of 2P ISIM in living thick samples, embryos of transgenic *Caenorhabditis elegans* expressing GFP-H2B are imaged in **Figure 11**. Both imaging resolution and contrast severely degrade at depths of more than ~15 µm from the coverslip surface in 1P illumination due to strong scattering in deep tissue (**Figure 11(a), (b)**). The degradation is not compensated by increasing of the exposure time, which, however, mainly leads to high background noise. Two-photon excitation of 2P ISIM effectively suppresses the out-of-focus emission. Thus, the subnuclear chromatin structures are clearly observed up to the depth of ~30 µm in **Figure 11(c), (d)**, where the fluorescence signals slightly reduce as the depth increases.

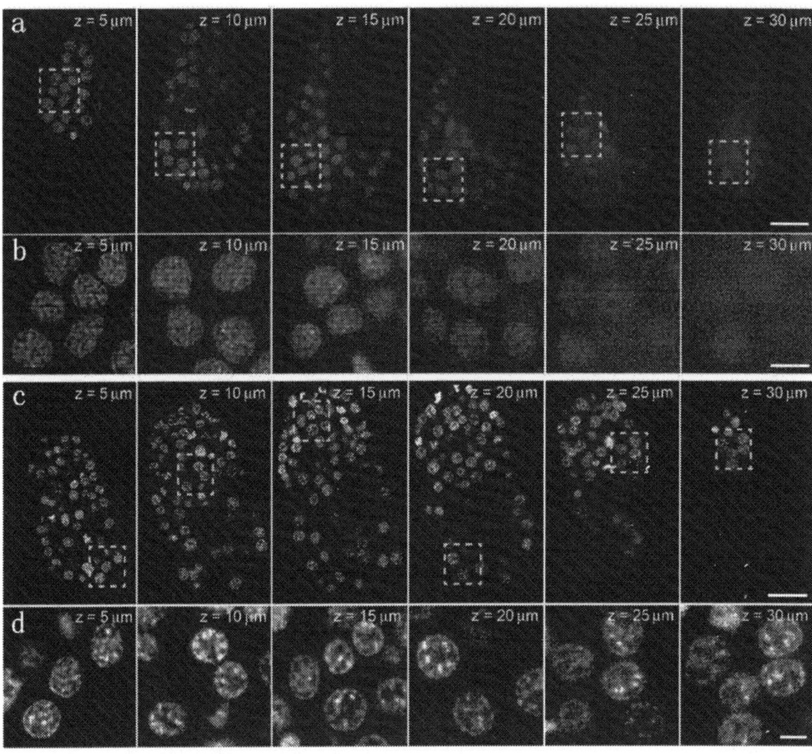

Figure 11. Enhanced penetration ability in 2P ISIM. (a, b) 1P ISIM images of a nematode embryo expressing GFP-H2B in nuclei. (a) Cross sections of the worm embryo at different axial positions. Scale bar: 10 μm. (b) Magnifications of the yellow rectangular regions in (a). Scale bar: 3 μm. The degradation in imaging contrast is observed as the depths increase. (c, d) 2P ISIM visualizes the subnuclear chromatin structure throughout nematode embryos. (c) Cross sections at the representative axial position. Scale bar: 10 μm. (d) Magnifications of yellow rectangular regions in (c), indicating better resolution, higher contrast, and larger imaging depth compared with 1P ISIM. Scale bar: 2 μm. Adapted with permission from reference [8]

3. Conclusion

In this chapter, we represent the super-resolution confocal microscopy (and two-photon microscopy) realized through the pixel reassignment methods computationally and optically. These demonstrate multiple advantages of resolution improvement, high fluorescence collection efficiency, optical sectioning capability, and fast imaging acquisition, which thus is able to investigate biological structures and processes at the cellular and even macromolecular level with 3D spatial scale. Additionally, because the method is directly established based on the standard confocal microscopy and/or two-photon microscopy, it mitigates the requirements in fluorescent probes and/or labeling methods that are always indispensable in some super-resolution fluorescence microscopic technologies, such as STORM and PALM [2, 3].

More importantly, the development of these techniques is not limited in the laboratorial stage. In 2015, the first commercial setup, LSM 800, is established by Carl Zeiss [13], which, in principle, is based on ISM but replaces the EMCCD camera with a 32-channel linear GaAsP-PMT array (i.e. Airyscan detector as shown in **Figure 12**). The highest imaging speed of LSM 800 with 512×512 pixels is up to 8 Hz, tremendous faster than ISM. Therefore, we expect that the super-resolution microscopy based on the pixel reassignment technique has great potentials for boosting imaging acquisition speed, and therefore further provides better understanding in intracellular molecular interactions and dynamic processes within living biological specimens.

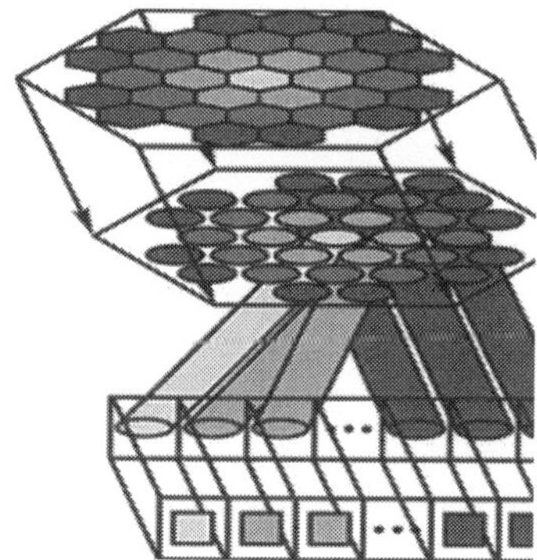

Figure 12. Schematic diagram of Airyscan detector in LSM 800. In brief, a hexagonal microlens array (a) collects incident light, which is in direct connection with the ends (b) of a fiber bundle (c). The other ends (d) of the fibers are in contact with a linear GaAsP-PMT array (e) serving as a detector. Thus, an area detector is created, onto which the Airy disk is imaged via a zoom optic configuration. Note that the single detector element, replacing the classical pinhole, acts as the separate pinholes in Airyscan detection. Adapted with permission from reference [13].

In addition to the issue of imaging acquisition speed, multicolor fluorescence microscopy is desired for investigating the interactions between different structures or biomolecules via labeling them with distinct colors. The possible interactions can be revealed by the co-localization of the different dyes and/or proteins. The standard fluorescence microscopy, however, might give inaccurate co-localization due to the diffraction-limited resolution. In combination with the pixel reassignment, the multicolor imaging technique is anticipated to provide a high-resolution imaging of the biological interaction within live cells.

In MSIM and ISIM based on the pixel reassignment approach [9, 10], both super-resolution imaging capability and color differentiation have been demonstrated, which have the advantages of easily configured optical system and weak cross-talk effect between the different colors. Switching laser lines for the excitation of different fluorophores might induce spatial mismatch in the images. Therefore, it is more preferable for simultaneously exciting all fluorophores and synchronously collecting their fluorescence signals. Multiple detectors with appropriate dichroic mirrors and emission filters can be used to collect the different fluorescence signals with different detection channels. Alternatively, an imaging spectrometer can be applied to record the spectral feature of these fluorophores.

Synchronous imaging decreases the fluorescence photobleaching probability due to low light exposure, benefiting to long-term monitoring of living samples. However, cross-talk of the different fluorophores always occurs because of the broad and overlapping excitation and emission bands of fluorophores. Although the cross-talk effects can be removed by selecting dyes with appropriately wide and non-overlapping emission spectra, the dyes are often inaccessible, which thus restricts its application in multicolor imaging. Linear spectral unmixing analysis is a solution to eliminate the cross-talk effect in spectral imaging [14]. The spectrum of the mixed fluorescent signal is expressed as a linear integration of the component dye spectra [15], and therefore the concentration or intensity of the fluorescence from each dye can be precisely analyzed. Based on the data analysis, both spatial mismatch and cross-talk effect are mitigated in multicolor imaging of live cells.

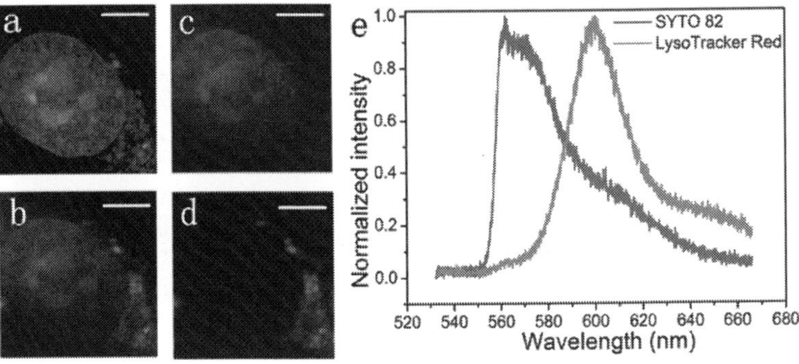

Figure 13. Multicolor RCM reveals the cellular microstructures labeled with different dyes. (a) Simultaneous RCM imaging of nucleus and lysosomes labeled with SYTO 82 and LysoTracker Red in a live bEnd.3 cell, respectively. Based on the linear spectral unmixing analysis, nucleus (c) and lysosomes (d) are differentiated according to their corresponding spectral features (e), respectively. (b) Overlaid image of the RCM images from (c) and (d). Scale bar: 5 μm.

In **Figure 13**, we establish a multicolor RCM with simultaneous excitation of different fluorophores and synchronous collection of their fluorescence. Linear spectral unmixing analysis is implemented for the spectral differentiation of the live cells stained with different dyes. SYTO 82-labeled nucleus and LysoTracker Red-stained lysosomes within live bEnd.3 cells are imaged by RCM with a spectrometer as the spectral detector. The nucleus and lysosomes are captured simultaneously, followed by the linear spectral unmixing analysis based on the known spectral features of these two dyes (severely overlapping as shown in **Figure 13(e)**). **Figure 13(b)–(d)** gives a clear separation of the two kinds of subcellular organelles. This approach is very powerful in investigation of the dynamic interactions of the subcellular structures.

Acknowledgements

This work was supported in part by the National Natural Science Foundation of China grant Nos. 61505238 and 11504042, Daqing Normal University Youth Foundation No. 12ZR12, Daqing Normal University doctor Foundation No. 15ZR03 doctor Foundation, Natural Science Foundation Project of Heilongjiang Province Nos. A200506 and QC2015066, Science and Technology Research Project of Heilongjiang Province Education Department No. 12543002, and Guidance of Science and Technology Plan Projects of Daqing City No. szdfy-2015-59.

Author details

Longchao Chen[1], Yuling Wang[2] and Wei Song[1*]

*Address all correspondence to: weisong1220@gmail.com

1 Shenzhen Institute of Advanced Technology, Chinese Academy of Sciences, Shenzhen University Town, Shenzhen, China

2 School of Mechatronics Engineering, Daqing Normal University, Ranghulu District Xibin Road, Daqing, China

References

[1] V. Westphal and S. W. Hell, "Nanoscale resolution in the focal plane of an optical microscope," Phys. Rev. Lett. 94(14), 143903 (2005).

[2] M. Bates, B. Huang, G. T. Dempsey and X. Zhuang, "Multicolor super-resolution imaging with photo-switchable fluorescent probes," Science 317(5845), 1749–1753 (2007).

[3] M. J. Rust, M. Bates and X. Zhuang, "Sub-diffraction-limit imaging by stochastic optical reconstruction microscopy (STORM)," Nat. Methods 3(10), 793–796 (2006).

[4] B. Huang, M. Bates and X. Zhuang, "Super resolution fluorescence microscopy," Annu. Rev. Biochem. 78, 993 (2009).

[5] C. Sheppard, "Super-resolution in confocal imaging," Optik 80(2), 53–54 (1988).

[6] I. J. Cox, C. J. Sheppard and T. Wilson, "Improvement in resolution by nearly confocal microscopy," Appl. Opt. 21(5), 778–781 (1982).

[7] J. McGregor, C. Mitchell and N. Hartell, "Post-processing strategies in image scanning microscopy," Methods 88, 28–36 (2015).

[8] P. W. Winter, A. G. York, D. Dalle Nogare, M. Ingaramo, R. Christensen, A. Chitnis, G. H. Patterson and H. Shroff, "Two-photon instant structured illumination microscopy improves the depth penetration of super-resolution imaging in thick scattering samples," Optica 1(3), 181–191 (2014).

[9] A. G. York, S. H. Parekh, D. Dalle Nogare, R. S. Fischer, K. Temprine, M. Mione, A. B. Chitnis, C. A. Combs and H. Shroff, "Resolution doubling in live, multicellular organisms via multifocal structured illumination microscopy," Nat. Methods 9(7), 749–754 (2012).

[10] A. G. York, P. Chandris, D. Dalle Nogare, J. Head, P. Wawrzusin, R. S. Fischer, A. Chitnis and H. Shroff, "Instant super-resolution imaging in live cells and embryos via analog image processing," Nat. Methods 10(11), 1122–1126 (2013).

[11] C. B. Müller and J. Enderlein, "Image scanning microscopy," Phys. Rev. Lett. 104(19), 198101 (2010).

[12] G. M. De Luca, R. M. Breedijk, R. A. Brandt, C. H. Zeelenberg, B. E. de Jong, W. Timmermans, L. N. Azar, R. A. Hoebe, S. Stallinga and E. M. Manders, "Re-scan confocal microscopy: scanning twice for better resolution," Biomed. Opt. Express. 4(11), 2644–2656 (2013).

[13] J. Huff, "The Airyscan detector from ZEISS: confocal imaging with improved signal-to-noise ratio and super-resolution," Nat. Methods 12(12), (2015).

[14] T. Haraguchi, T. Shimi, T. Koujin, N. Hashiguchi and Y. Hiraoka, "Spectral imaging fluorescence microscopy," Genes. Cells 7(9), 881–887 (2002).

[15] H. Tsurui, H. Nishimura, S. Hattori, S. Hirose, K. Okumura and T. Shirai, "Seven-color fluorescence imaging of tissue samples based on Fourier spectroscopy and singular value decomposition," J. Histochem. Cytochem. 48(5), 653–662 (2000).

Chapter 5

Second Harmonic Generation Microscopy: A Tool for Quantitative Analysis of Tissues

Juan M. Bueno, Francisco J. Ávila, and Pablo Artal

Additional information is available at the end of the chapter

http://dx.doi.org/10.5772/63493

Abstract

Second harmonic generation (SHG) is a second-order non-linear optical process produced in birefringent crystals or in biological tissues with non-centrosymmetric structure such as collagen or microtubules structures. SHG signal originates from two excitation photons which interact with the material and are "reconverted" to form a new emitted photon with half of wavelength. Although theoretically predicted by Maria Göpert-Mayer in 1930s, the experimental SHG demonstration arrived with the invention of the laser in the 1960s. SHG was first obtained in ruby by using a high excitation oscillator. After that starting point, the harmonic generation reached an increasing interest and importance, based on its applications to characterize biological tissues using multiphoton microscopes. In particular, collagen has been one of the most often analyzed structures since it provides an efficient SHG signal. In late 1970s, it was discovered that SHG signal took place in three-dimensional optical interaction at the focal point of a microscope objective with high numerical aperture. This finding allowed researchers to develop microscopes with 3D submicron resolution and an in depth analysis of biological specimens. Since SHG is a polarization-sensitive non-linear optical process, the implementation of polarization into multiphoton microscopes has allowed the study of both molecular architecture and fibrilar distribution of type-I collagen fibers. The analysis of collagen-based structures is particularly interesting since they represent 80% of the connective tissue of the human body. On the other hand, more recent techniques such as pulse compression of laser pulses or adaptive optics have been applied to SHG microscopy in order to improve the visualization of features. The combination of these techniques permit the reduction of the laser power required to produce efficient SHG signal and therefore photo-toxicity and photo-damage are avoided (critical parameters in biomedical applications). Some pathologies such as cancer or fibrosis are related to collagen disorders. These are thought to appear at molecular scale before the micrometric structure is affected. In this sense, SHG imaging has emerged as a powerful tool in biomedicine and it might serve as a non-invasive early diagnosis technique.

Keywords: second harmonic microscopy, biomedical imaging, collagen, polarization, adaptive optics

1. Principles of second harmonic generation

Non-linear optical microscopy refers to all microscopy techniques based on non-linear optics, in which light-matter interactions violate the linear superposition principle. These techniques can be divided in two main categories: incoherent and coherent. Although in the former, the phase of the emitted optical signal is random, in coherent techniques it depends on a wide variety of factors, including those related to the exciting light or associated with the geometric distribution of the radiating molecules. One of these phenomena is the so called Second Harmonic Generation (SHG).

SHG is a coherent non-linear process where two incident photons at their fundamental frequency interacting with a medium are directly converted into a single photon of exactly the same total energy at double of frequency, without absorption or reemission of photons [1]. This process is carried out via an intermediate virtual state in a single quantum event. **Figure 1** compares this SHG process with the typical linear fluorescence phenomenon.

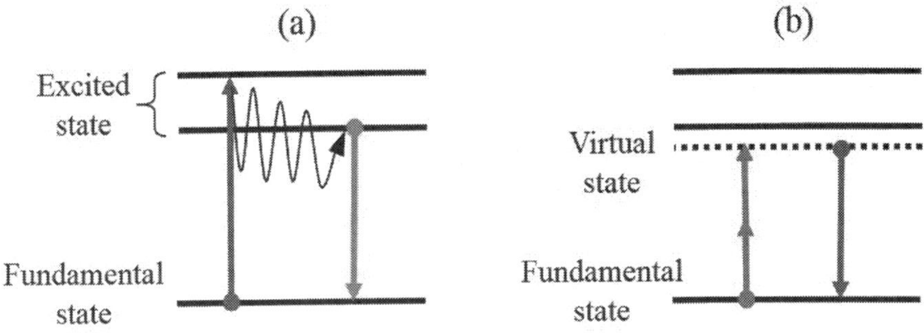

Figure 1. Diagrams of linear fluorescence (a) and SHG (b).

According to the non-linear optics theory, an incident electric field $\vec{E}_{\omega}$ with a frequency ω induces a second-order polarization $\vec{P}_{2\omega,i}$ at 2ω in the ith direction given by [1]:

$$\vec{P}_{2\omega,i} = \varepsilon_0 [\chi^{(1)} \vec{E} + \chi^{(2)} \vec{E}\vec{E} + \chi^{(3)} \vec{E}\vec{E}\vec{E} + ...] \tag{1}$$

where ε_0 is the vacuum permittivity and $\chi^{(n)}$ is the nth-order non-linear susceptibility tensor. The first term of Equation (1) describes the normal absorption and reflection of light, and the second the SHG. Since the term $\chi^{(2)}$ depends on the polarization of the excitation source, the SHG emission is sensitive to polarization [2]. In a medium with hexagonal symmetry, the non-null coefficients $d_{ij} = \frac{1}{2}\varepsilon_0\chi_{ij}^{(2)}$ are d_{31}, d_{33}, d_{15} and d_{14} [3].

These coefficients contain the non-linear optical properties of the material and sum zero for inversion symmetry [4]. Moreover, if Kleinman and cylindrical symmetries are assumed, $d_{14}=0$ and $d_{15} = d_{31}$ [5].

SHG was demonstrated in crystalline quartz in 1962 by Kleinman, and since then this has been commonly used to frequency-double pulsed lasers to obtain shorter wavelengths [5]. Moreover, SHG signal is sensitive to bulk non-centrosymmetric spatial arrangements such as collagen structures or birefringent crystals [6]. The lack of a center of symmetry in an organized material strongly affects the second-order susceptibility and therefore the efficiency of SHG signals [7].

Unlike two-photon excitation fluorescence, SHG is energy conserving (it does not involve an excited state), strongly directional and preserves the coherence of the laser light [8]. Then, a medium is able to provide efficient SHG signal when its structure is organized at the scale of the laser wavelength and lacks a center of symmetry. Further details on the advantages of using SHG imaging (especially for biological applications) will be presented along this chapter.

2. SHG microscopy of biological samples

In 1971, Fine and Hansen proposed that SHG signal could also be produced by biological tissues [9]. In 1974, Hellwarth and Christensen implemented SHG into an optical microscope to visualize the microscopic crystal structure of polycrystalline ZnSe [10]. Later, Gannaway and Sheppard presented SHG images of a lithium niobate crystal by using a laser scanning microscope [11]. These images showed features and contrast levels not seen in regular (linear) microscopy.

However, to our knowledge, the first biological SHG image was reported by Freund in 1986. He imaged a rat tail tendon with high resolution SHG scanning microscopy [12]. More than a decade was necessary to consistently apply SHG microscopy to visualize biological specimens. In particular, a laser scanning microscope was combined with a Ti-Saphire femtosecond laser to acquire live cell images based on SHG [13].

Other biological specimens imaged using SHG microscopy include membranes [7], proteins [13] and collagen-based structures [14] among others. **Figure 2** shows a sample containing starch grains imaged with SHG microscopy.

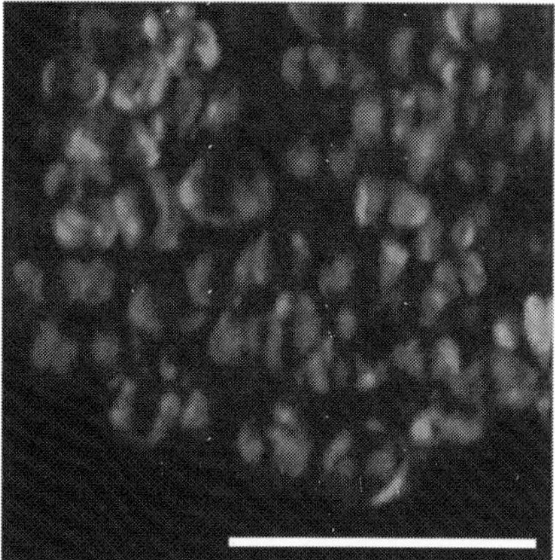

Figure 2. SHG image of starch grains. These are plant polysaccharides with a convenient radial structure. It can be found in both a non-organized (dry) and an organized (hydrated) form. Scale bar: 50 μm.

According to the non-linear nature of harmonic generation, the intensity of the SHG signal depends on the square of the excitation laser intensity, and occurs intrinsically confining the focus of the microscope objective. Since this event takes place in both transversal and axial directions, SHG imaging microscopy provides intrinsic 3D sectioning capabilities with excellent Z-resolution. This property allows optical sectioning of biological samples with reduced out-of-plane photo-toxicity (see **Figure 3**).

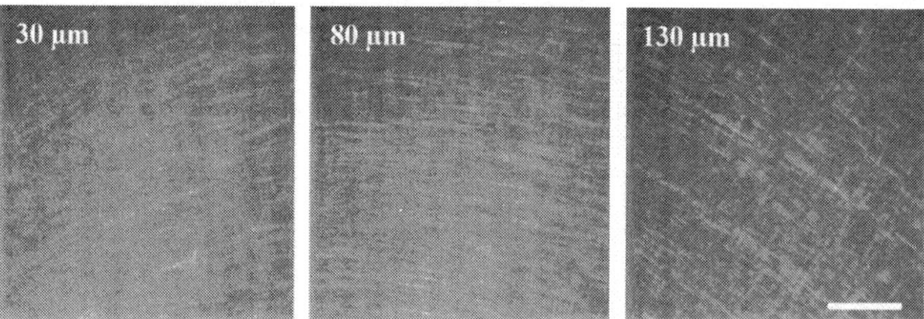

Figure 3. SHG images of an eagle cornea acquired at three different depth locations. Scale bar: 50 μm.

Apart from this inherent property, during the last ten years SHG microscopy has gained increasing popularity in biomedicine mainly due to the provided possibilities for endogenous contrast imaging (staining procedures are not required), reduced tissue damage, sensitivity to molecular architecture organization, preservation of phase information and polarization dependence [15].

SHG is restricted to molecules with non-centrosymmetric organization and is emitted by different biological tissues containing collagen [1], myosin [16] or tubulin (which polymerizes into microtubules) [17]. Type-I collagen is the most abundant structural protein of the human body [18], and due to its presence in connective tissue, SHG signal can be effectively obtained from the cornea, the skin, bones or tendons [19–27]. Collagen plays an important role within the human body and has been studied under many different experimental conditions. Its presence in connective tissues constitutes 6% of the dry weight of the body [28].

The basic structural unit of collagen is the molecule of tropocollagen, presenting a helical structure, formed by three polypeptide chains coiled around each other to form a spiral (see **Figure 4**). These molecules are cross-linked to form collagen fibrils.

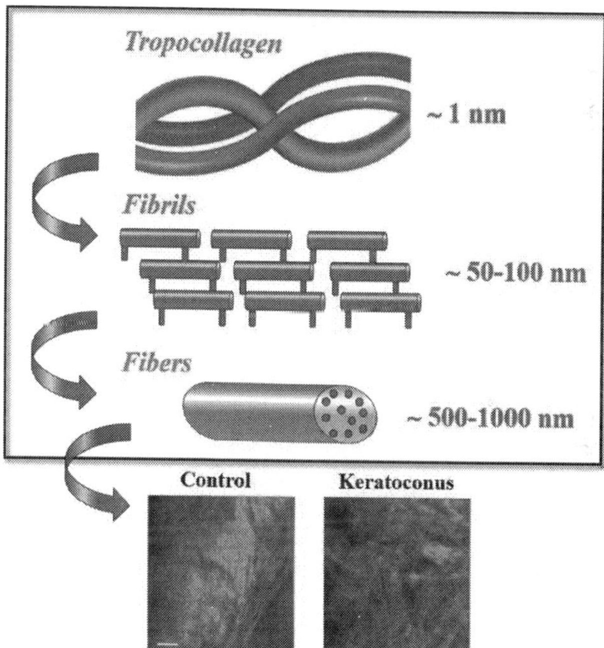

Figure 4. Schematic representation of the collagen structure and SHG images of two human corneas with different collagen distributions. Bar length: 50 μm.

These fibrils are assembled into parallel beams to form collagen fibers [18]. The origin of the SHG signal in collagen arises from its molecular chirality, where the molecules possess permanent dipole moments with high order alignment, ensuring the generation of harmonic signal as a consequence of the non-symmetrical oscillation of the electrons [15]. On the other hand, the intrinsic chirality of the triple-helix of molecular collagen increases the asymmetry of the assembly increasing the non-linear response.

As above explained, the SHG signal can be characterized by the non-linear susceptibility tensor [4]. This depends on the induced dipole moment of the molecules and therefore on the organization of the collagen tissue. If a collagen triple-helix is excited along its main axis, each bond contribution is summed coherently to the SHG signal [29]. When the molecules with the same orientation—or polarity—are assembled to form fibrils, the SHG signal is amplified significantly [30]. Then, the total intensity is the coherent sum of the signal from individual collagen fibrils.

However, in some collagen-based tissues the polarity of the fibrils within the fiber varies randomly [31]. In those cases the organization presents contributions to the axial momentum altering the coherent process of the SHG. Theoretically, the phase matching condition for which the non-linear process is strictly coherent is given by [15]:

$$\vec{\Delta k} = \vec{k_{2\omega}} - \vec{2k_{\omega}} = 0 \tag{2}$$

where $\vec{k_{2\omega}}$ is the wave vector for the SHG emission and $\vec{2k_{\omega}}$ is the wave vector of the incident light. Then, a second harmonic conversion is maximum if $\vec{\Delta k} = 0$. Experimentally, only birefringent crystals have been found to verify this condition [32]. In biological samples, and particularly in collagen-based tissues, the SHG signal is a quasi-coherent process which SHG efficiency conversion depends on how aligned within the fiber the fibrils are.

SHG images at the bottom of **Figure 4** compare the collagen distribution in a normal healthy human cornea and another affected by a pathology called keratoconus. This is a real example on how the coherence of SHG signal is an efficient detector of collagen organization within a sample. Although the control tissue displays a fairly regular distribution of collagen fibers along a preferential direction, these are randomly distributed in the pathologic case. The

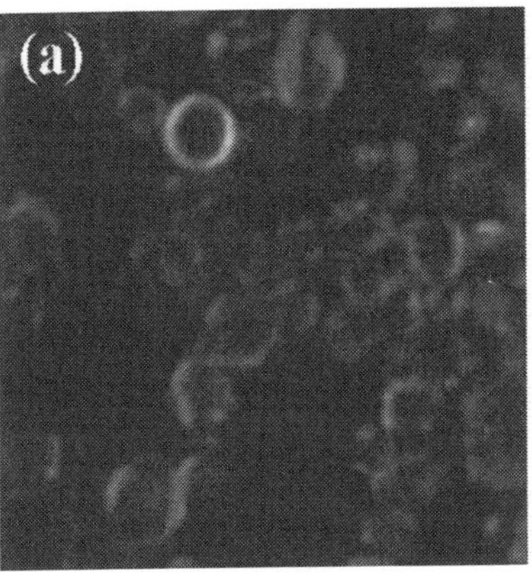
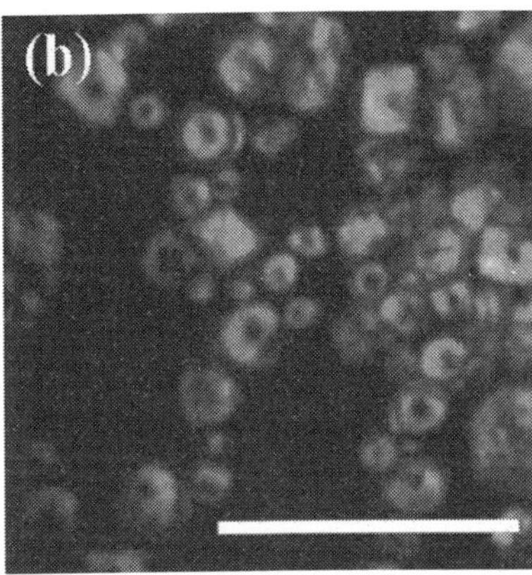

Figure 5. Comparison of SHG images acquired in the backward (a) and forward (b) directions. The differences between the two acquisition directions are readily visible due to the behavior of the SHG emission. Scale bar: 50 µm.

control tissue provides a more efficient SHG conversion because Δk is significantly lower than in the pathologic case. Changes in collagen morphology are firstly produced at molecular level (tropocollagen), which number of aligned dipoles coherently sum, affecting the fibril organization and therefore the distribution of the fibers finally imaged with SHG microscopy.

On the other hand, the emission directionality has an effect in both SHG signal conversion efficiency and observed morphology. This is easy observed in the starch grains presented in **Figure 5**.

In collagen-based tissues, this effect is due to the size and the organization of the collagen fibers, since both affect the phase mismatch and the amplification of the SHG intensity [15]. In general, SHG directionality depends on the distribution of the induced dipoles in the focal volume where the non-linear process takes place [33]. In this sense, the directionality effects of the SHG signal can be appreciated in the visualization of continuous structures (fibers) in the backward channel (i.e. backscattered emission), and a higher segmentation or discrete distribution in the forward directed emission [34]. This implies that the choice of the detection direction in the experimental device will depend on the desired scale of observation. However, since changes in collagen fibers can be observed in the backward configuration [1], this configuration has become suitable for biomedical imaging.

3. Imaging ocular tissues with SHG microscopy

As stated above, the human body is plenty of tissues composed of collagen. These tissues are often the main component of different organs. In particular, the eye is one of them. Although this is not a vital organ (such as the heart or the liver, for instance), it is necessary to have a regular way of living (both humans and animals). Since the middle of nineteenth century, there have been a number of instruments to visualize ocular structures in order to improve the diagnosis and treatment of eye's diseases. Since most ocular elements are transparent, staining procedures are usually required and, under certain experimental conditions, the existing clinical techniques are sometimes not totally appropriate. In that sense, SHG microscopy might be used as a new tool to improve the imaging of some ocular tissues.

The sclera and the cornea are the two structures are the eye's outer tunic, mainly composed of type-I collagen. The former is an opaque connective tissue acting as protective element that gives stability to the ocular globe. Unlike the sclera, the cornea presents high transparency which originates from particular arrangements of the collagen fibers (localized within the stroma, which occupies about 90% of whole corneal thickness). Corneal collagen assembles to form long fibrils with a diameter of approximately 25 nm (in humans) [35]. These are uniformly spaced forming larger bundles or fibers termed "corneal lamellae" (tens of microns in size). In contrast, scleral fibrils have various diameters ranging between 25 and 230 nm [35]. These collagen fibrils also form bundles, however these are not parallel arranged but entangled in individual bundles.

Since type-I collagen is an effective second harmonic generator, both ocular elements can be visualized with SHG imaging microscopy without using labeling techniques. However, the sclera does not have any contribution to the vision function and this is probably the reason because SHG studies on the sclera are scarce. To our knowledge, Han and colleagues were the first to show SHG images of the sclera [36]. The sclera collagen distribution was analyzed through these images in forward and backward directions. They concluded that the sclera presents inhomogeneous, tube-like structures with thin hard shells, maintaining the high stiffness and elasticity of the tissue. SHG imaging was used by Teng and co-authors to resolve the difference in structural orientations between the collagen fibers of the cornea and the sclera: the corneal collagen is organized in a depth-dependent fashion, whereas the sclera collagen is randomly packed [37]. As an example, **Figure 6** shows SHG images of healthy tissues corresponding to a human cornea and a sclera. A simple visual inspection reveals the evident difference in collagen distribution.

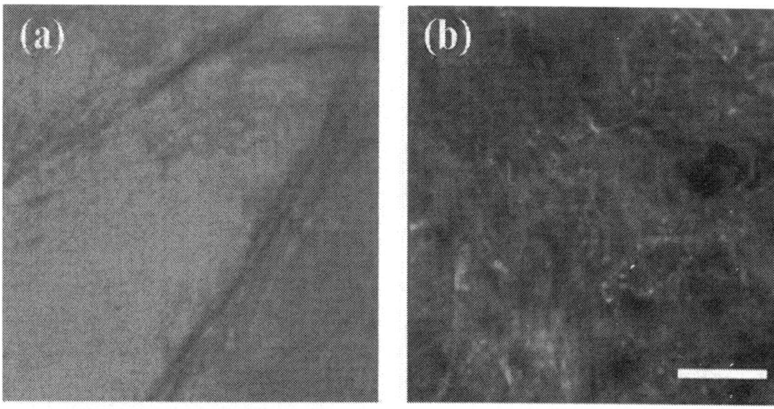

Figure 6. Comparison of collagen distributions in SHG images of a human cornea (a) and a piece of human sclera (b). Although the fibers in the sclera present always a non-organized pattern, in the cornea the arrangement depends on a number of factors as explained below. Scale bar: 50 μm.

SHG images of the sclera as a function depth have also been analyzed. At shallow planes collagen bundles were roughly aligned parallel to the limbus. At deeper locations the fibers did not have a specific orientation of alignment [38]. At the posterior pole this arrangement differed. On the external surface undulating thick bundles without a notable major orientation were found. These also had interwoven structures with various orientations. On the internal surface of the sclera fine collagen bundles were observed. These bundles were frequently branched and intermingled. As already mentioned, the sclera has not a direct implication in visual function, however its structures are related to ocular biomechanics and the changes with intraocular pressure or surgery might be interesting in clinical applications.

Unlike the sclera, the cornea has been analyzed with SHG microscopy by many different authors. Since Yeh et al. obtained SHG images in rabbit corneas without exogenous dyes [19] the corneal structure has been studied in a number of animal models (including humans) under several experimental conditions (see for instance [39] and references therein). SHG images of the cornea have been compared in both forward and backward scattering directions [36]. Although images showed different information, collagen fibers always showed a regular packaging [36, 37]. However, this regular pattern has been shown to change with pathologies [40, 41] or after surgical procedures [42–44]. The corneal stroma also suffers alterations due to scars [45] or changes in the intraocular pressure [46] that have been explored through SHG imaging.

Although SHG microscopy has been used to image the cornea under different experimental conditions, two aspects are really important: (1) the response of the SHG signal to polarized light and (2) the measurement of the stroma organization. Moreover, an objective characterization of structural abnormalities is of great interest to distinguish normal from pathological corneas and the key for possible clinical applications. The next section deals with this topic.

4. Measurement of collagen organization in ocular tissues

As previously mentioned, the collagen arrangement could be compromised due to pathological processes, mechanical trauma or denaturation (aging). Due to this, both classification and quantification of collagen arrangement might be a powerful tool in biomedicine as well as in medical diagnosis, in particular for those pathologies associated with collagen disorders occurring at early stages of the disease.

The analysis of the collagen organization has usually been carried out in a qualitative manner. A quantitative analysis would lead to understand changes in corneal stroma caused by intraocular pressure [46], pathological processes [47, 48] or surgery [49] among others. Although there are several techniques to analyze the spatial distribution of collagen, the bi-dimensional fast Fourier transform (2D-FFT) has often been used for this goal [50].

In particular, the 2D-FFT method has been used with SHG images to compute the degree of organization in corneal collagen in the presence of pathologies [26] or after physical damage

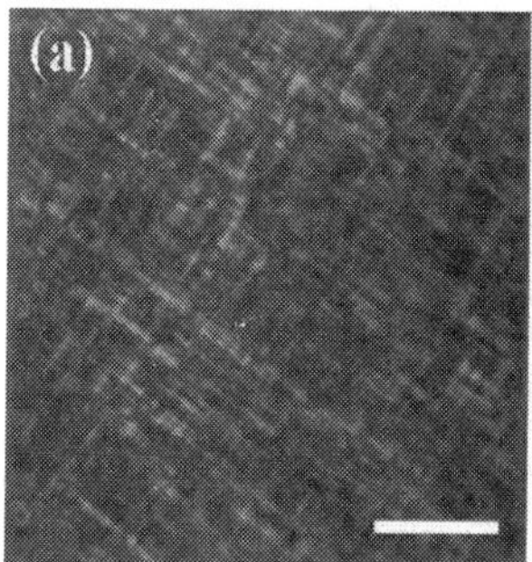
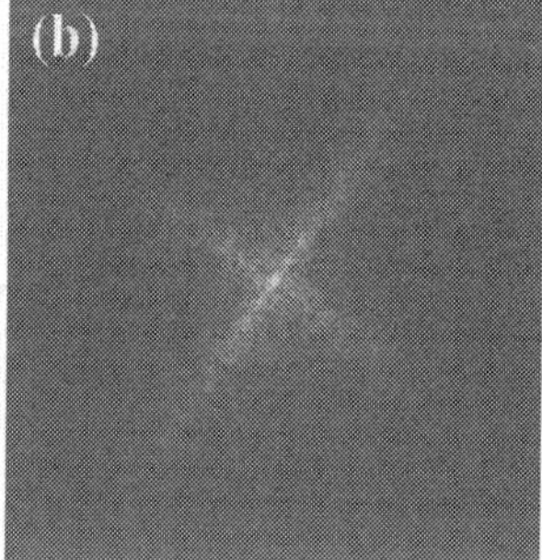

Figure 7. (a) SHG images of an eagle cornea with two preferential directions of the collagen fibers (crosshatched pattern) and (b) the associated 2D-FFT image. As expected, the 2D-FFT presents a cross shape as a result of the distribution of the fibers within the imaged corneal layer. Scale bar: 50 μm.

[51]. If the SHG image shows a structure with a preferential orientation, the spatial frequencies of the 2D-FFT spectrum are aligned along the direction orthogonal to that preferential orientation [52]. Nevertheless, the resolution of the 2D-FFT is limited by the noise of the SHG image and an image filtering is often required. The distribution of the spatial frequencies on the 2D-FFT is also generally fitted by an ellipse, and the ratio between its axes is used as a parameter to quantify the collagen organization. However, when a collagen distribution is arranged in a more complex and heterogeneous pattern (interwoven, crosshatched, …), the best fit is a circle and this operation may lead to erroneous conclusions that require a complicate post-processing [53]. **Figure 7** shows a SHG image with a crosshatched appearance of collagen fibers, together with the corresponding 2D-FFT.

In this sense an alternative procedure based on the structure tensor has recently been presented by these authors as a useful tool to classify the spatial distribution of collagen-based tissues through SHG images [54]. The technique has the advantage of differentiate areas with maximum organization from those locations where the orientation of the collagen fibers is not significant. The structure tensor provides relevant parameters such as the spatially-resolved degree of isotropy (DoI) and the histogram of orientation distribution. The former ranges between 0 and 1 and its value increases with the order of the structure (i.e. the more aligned the fibers the higher the DoI). In the latter, it is verified that the narrower the data, the higher the presence of a dominant orientation. Apart from their quantitative information, a visual inspection of both the DoI map and the orientation histogram permits to discriminate between quasi-aligned and non-organized collagen distributions. For a better understanding of this tool, **Figure 8** shows two SHG images with different spatial distributions and the correspond-

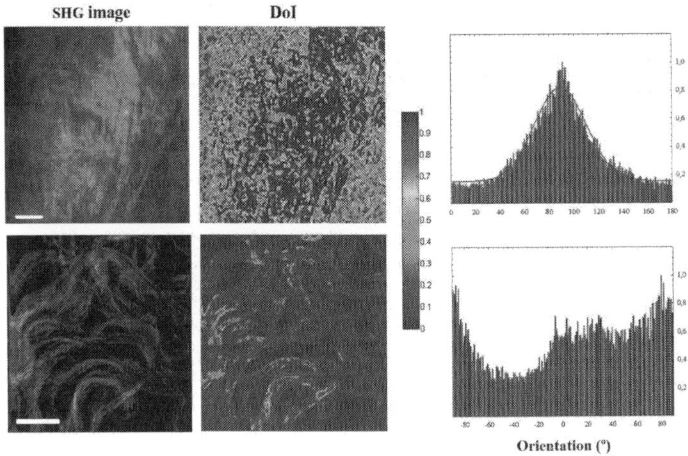

Figure 8. SHG images of a human cornea (upper row) and bovine sclera (bottom row) and their respective spatially-resolved DoI and orientation histogram computed through the structure tensor. Detailed information on how these parameters were computed can be found in [54]. Scale bar: 50 μm.

ing structure tensor parameters. At this point, it is important to notice that the combination of SHG microscopy and the structure tensor could help in the diagnosis of abnormal structures and in the tracking of pathologies related to corneal stroma disorders.

5. Adaptive optics SHG imaging

Despite optical sectioning capabilities and inherent confocality of SHG microscopy, the imaging of thick samples (3D imaging) is limited by the specimen-induced aberrations. When a femtosecond laser beam is focused into a specimen, the deeper the layer to be imaged, the larger the focal spot size. This leads to a reduction in the effectiveness of the SHG process (i.e. lower SHG signal) and a decrease in the quality of the acquired images, both contrast and resolution (**Figure 9**).

To overpass this, adaptive optics (AO) techniques combining a wavefront sensor and an adaptive device (deformable mirror or spatial light modulator) have been used [55–58]. Most authors have been interested in improving two-photon excitation fluorescence images through AO and experiments dealing with AO-SHG imaging are scarce in the literature [59–62].

Although the ideal situation is to compute and correct for the plane-by-plane aberrations, this is experimentally difficult [63]. In that sense, wavefront sensor-less techniques combined with multiphoton microscopy have been reported [58, 59, 61, 62]. With this approach, at a certain plane within the sample the AO element pre-compensates for the "unknown" aberrations without measuring them, but finding the best image according to a pre-defined image quality metric. Although these are time-consuming procedures due to the use of iterative algorithms such as genetic learning, hill-climbing or stochastic, they have provided significantly improved images with more visible details.

Moreover, the dominant aberration term at deeper layers is the spherical aberration [55, 58, 64]. In order to correct for (or minimize) this unwanted spherical aberration, objective correc-

Figure 9. SHG images of a rabbit cornea at different depth locations (10, 100 and 100 µm). The reduction in SHG signal is readily visible. This decrease in SHG effectiveness is mainly due to specimen-induced aberrations and scattering. Scale bar: 50 µm.

tion collars [64, 65] and wavefront sensor-less AO devices [55, 58, 62] have been used. The former is a manual method only valid for a defined set of cover thickness values. The latter is faster than usual since only the spherical aberration term has to be corrected. An alternative technique has recently been reported to improve 3D multiphoton imaging [66]. This is based on the manipulation of the spherical aberration pattern of the incident beam while performing fast tomographic SHG imaging. As expected, when inducing spherical aberration the image quality is reduced at best focus, however at deeper planes a better image quality is obtained. This increases the penetration depth and enables improved 3D SHG images even with non-immersion objectives.

Although these AO techniques can be applied to both non-biological and biological samples, there is a special interest when imaging ocular tissues, the cornea in particular. For this ocular structure the features of interest (especially when analyzing pathologies) might be close to the surface or located deep into it. For shallow planes, SHG images are usually of high quality. However when the plane to be imaged is located at a deeper position AO-SHG can be used as a powerful technique to noticeably enhance SHG images corresponding to those deep corneal layers (see **Figure 10** as an example). Those images will have enough contrast and resolution to observe the collagen bundles [66], and any possible abnormal distribution of them across

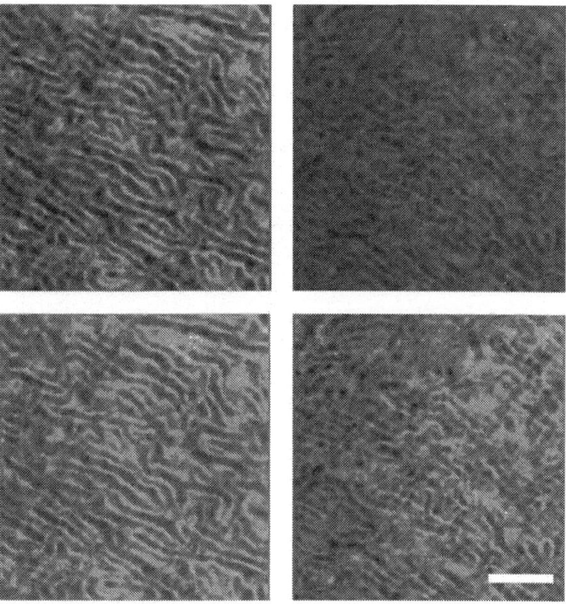

Figure 10. Comparison of SHG images before (upper panels) and after (bottom panels) using AO for two different locations within the sample (porcine cornea). It can be observed how AO improves the quality of the images at both locations. However this is more noticeable for deeper locations within the sample. Scale bar: 50 μm.

the imaged area. Future clinical instruments can also benefit from this implementation which leads to a better visualization of the layered ocular structures.

6. Polarization-sensitive SHG microscopy

The dependence between polarization and SHG signals in collagen is well recognized [67–69]. The combination of non-linear microscopy and polarization allowed detecting changes in collagen arrangement [70] and has been proposed to characterize collagen-based tissues [71, 72]. Type-I collagen fibers exhibit structural anisotropy that can be characterized by the ratio of hyperpolarizabilities or polarization anisotropy $\varrho = \beta_{xxx}/\beta_{xyy}$, which provides information about the internal collagen structure [33, 73].

The polarization anisotropy depends on the orientation of the collagen triple-helix and the orientation of the induced dipoles along the peptide bonds and the values have been reported to be in the range [-3, 3] [33]. Low values of ϱ are associated with immature collagen [67], aging [69] or loss of arrangement in the collagen distribution [1]. Therefore, polarization-sensitive SHG microscopy provides information about the dipolar distribution within the collagen fibers.

SHG intensity has been reported to vary with the angle between the optical axis of the polarizer and the main orientation of the collagen fibers [33, 69]. Moreover, depending on the spatial distribution of collagen fibers the SHG signal will be differently affected by the incoming polarization state [72, 74, 75] (see **Figure 11** as an example). This fact might be of great importance in SHG imaging not only because the total signal varies (for instance) when changing from linear to circular polarization, but also because more details and extra features might be visible for certain polarization states (see **Figure 12**) [72].

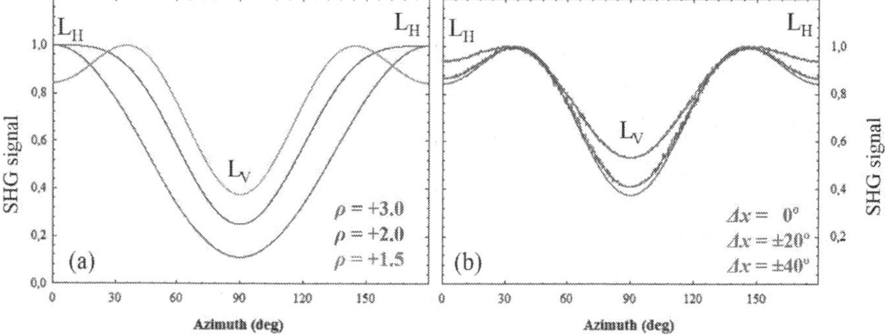

Figure 11. SHG intensity as function of the incident linear polarization for three values of ϱ and parallel-arranged collagen fibers (a), and for three different values of structural dispersion, Δx, and $\varrho=+1.5$ (b). Further details on this can be found in [72].

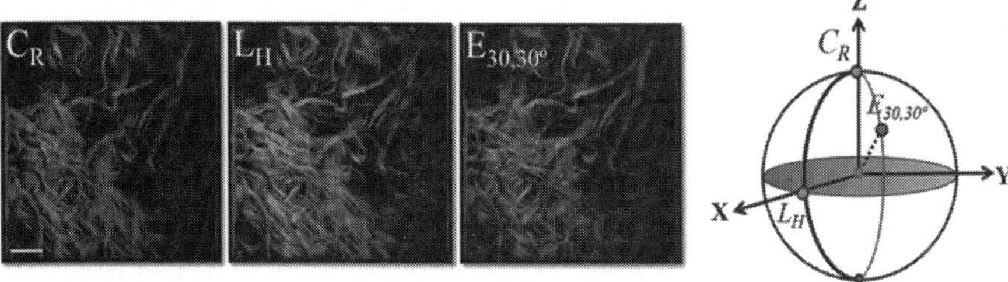

Figure 12. SHG images of a bovine sclera recorded for three different incident polarization states. L_H, linear horizontal; C_R, right circular and $E_{\chi,\psi}$, right elliptical. The location of these polarization states over the Poincaré sphere are shown on the right. Scale bar: 50 μm.

The combination of polarization and SHG signal allows obtaining information about the hierarchical architecture of collagen at molecular scale [76, 77]. This has been used in biopsies to discriminate normal breast from malignant tissue [78], and analyze cancerous ovarian tissues [79].

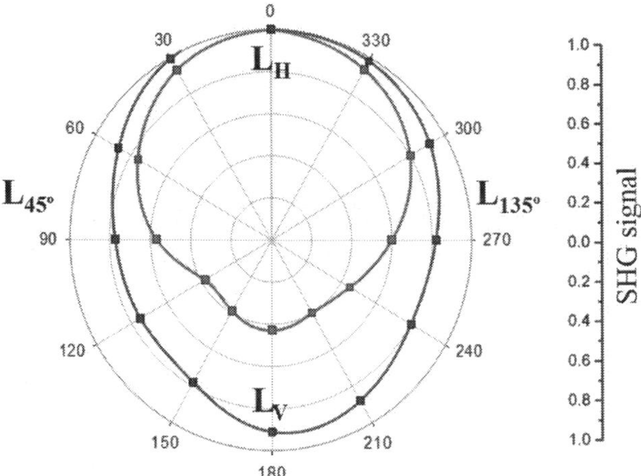

Figure 13. Polar diagram representing the normalized SHG intensity distribution as function of incoming polarization of a control (red) and a keratoconus (blue) cornea. The dependence of SHG intensity with incident polarization is stronger in the former probably due to the presence of a dominant direction of the collagen fibers. This does not exist in the latter and the SHG signal hardly depends on the polarization state.

Pathological alterations of the cornea could seriously compromise vision. In this sense polarization-sensitive SHG microscopy has been proved to be effective to detect structural alterations in keratoconus [40]. It has also been employed to analyze the molecular changes produced by high levels of intraocular pressure and to investigate how these modifications produced within the lamellae affect the stroma thickness [80]. As a possible clinical application,

Figure 13 compares the response to incident linear polarization of a human control and a keratoconus cornea (those presented in **Figure 4**). It can be observed how the SHG signal from the pathological case presents lower dependence with polarization.

7. Conclusions

Along this chapter, the principles of SHG processes and the application to biological imaging have been reviewed. SHG microscopy is a non-linear modality with inherent confocality that allows visualizing non-stained tissues composed of collagen and shows features not seen with regular microscopy. SHG intensity depends on both the size and organization of the collagen fibers. Since SHG directionality depends on the distribution of the induced dipoles within the fibers, the registration of the signal in a backscattered direction can be used to observe the collagen architecture within the specimens (**Figure 13**).

In particular, this is a useful tool to image connective ocular tissues such as the sclera and the cornea with high resolution as function of depth. These tissues can be characterized attending the organization of their collagen fibers. Unlike the sclera that usually presents a non-organized distribution, the arrangement of the corneal fibers depends on numerous factors. These collagen organizations have been discussed and a novel method based on the structure tensor to perform quantitative analyses has been proposed. This permits to classify the spatial distribution of the fibers from the SHG images, and can be used for diagnoses of pathologies related to collagen disorders.

However, the efficiency of SHG imaging of thick samples reduces with depth, as the specimen itself induces aberrations and scattering. To overcome this limitation, AO procedures have been implemented into SHG microscopes. The manipulation of the aberration pattern of the incident laser beam has allowed increasing the image quality of SHG images especially at deeper locations.

On the other hand, SHG signal from collagen structures is polarization dependent. This modulation depends on both collagen internal structure (parameter ϱ) and the arrangement of the fibers (external organization). The combination of polarization and SHG microscopy has been reported to be effective in detecting structural changes in collagen-related pathological processes. The technique could then be a powerful tool in biomedicine and/or in clinical diagnoses.

Author details

Juan M. Bueno*, Francisco J. Ávila, and Pablo Artal

*Address all correspondence to: bueno@um.es

Laboratorio de Optica, Universidad de Murcia, Murcia, Spain

References

[1]   R. M. Williams, W. R. Zipfel and W. W. Webb. Interpreting second-harmonic generation images of collagen I fibrils. Biophys. J. 2005; 88(2): 1377–1386.

[2]   A. Erikson, J. Ortegren, T. Hompland, C. L. Daviesa and M. Lindgren. Quantification of the second-order non-linear susceptibility of collagen I using a laser scanning microscope. J. Biomed. Opt. 2009; 12(4):044002.

[3]   P. N. Butcher and D. Cotter. The elements of nonlinear optics. Cambridge University Press. Cambridge, 1991.

[4]   R. W. Boyd. Non-linear optics. Academic Press. San Diego, USA, 1992.

[5]   D. A. Kleinman. Theory of second harmonic generation of light. Phys. Rev. 1962; 128(4): 1761–1775.

[6]   T. Verbiest, S. Van Elshocht, M. Kauranen, L. Hellemans, J. Snauwaert, C. Nuckolls, T. J. Katz and A. Persoons. Strong enhancement of nonlinear optical properties through supramolecular chirality. Science. 1999; 282(5390):913–915.

[7]   L. Moreaux, O. Sandrea and J. Mertz. Membrane imaging by second harmonic generation microscopy. J. Opt. Soc. Am. B. 2000; 17(10):1685–1689.

[8]   P. J. Campagnola, A. C. Millard, M. Terasaki, P. E. Hoppe, C. J. Malone and W. Mohler. Three-dimensional high-resolution second-harmonic generation imaging of endogenous structural proteins in biological tissues. Biophys. J. 2002; 82(1):493–508.

[9]   S. Fine and W. P. Hansen. Optical second harmonic generation in biological systems. Appl. Opt. 1971; 10(10):2350–2353.

[10]   R. Hellwarth and P. Christensen. Nonlinear optical microscopic examination of structure in polycrystalline ZnSc. Opt. Commun. 1974; 12(3):318–322.

[11]   J. N. Gannaway and C. J. R. Sheppard. Second harmonic imaging in the scanning optical microscope. Opt. Quantum Electron. 1978; 10(5):435–439.

[12]   I. Freund and M. Deutsch. Second harmonic microscopy of biological tissues. Opt. Lett. 1986; 11(2):94–96.

[13]   P. J. Campagnola, M. D. Wei, A. Lewis and L. M. Loew. High resolution nonlinear optical microscopy of living cells by second harmonic generation. Biophys. J. 1999; 77(6):3341–3349.

[14]   P. J. Campagnola and L. M. Loew. Second-harmonic imaging microscopy for visualizing biomolecular arrays in cells, tissues and organisms. Nat. Biotechnol. 2003; 21(11): 1356–1360.

[15]   R. LaComb, O. Nadiarnykh, S. S. Townsend and P. J. Campagnola. Phase matching considerations in second harmonic generation from tissues: effects on emission

directionality, conversion efficiency and observed morphology. Opt. Commun. 2008; 281(7):1823–1832.

[16] S. V. Plotnikov, A. Millard, P. Campagnola and W. Mohler. Characterization of the myosin-based source for second-harmonic generation from muscle sarcomeres. Biophys. J. 2006; 90(2):328–339.

[17] D. A. Dombeck, K. A. Kasischke, H. D. Vishwasrao, M. Ingelsson, B. T. Hyman and W. W. Webb. Uniform polarity microtubule assemblies imaged in native brain tissue by second-harmonic generation microscopy. Proc. Natl. Acad. Sci. U.S.A. 2003; 100(12): 7081–7086.

[18] H. Lodish, A. Berk, C. A. Kaiser, M. Krieger, A. Bretscher, H. Ploegh, A. Amon and M. P. Scott. Molecular cell biology. W.H. Freeman and Company, New York, 2013.

[19] A. T. Yeh, N. Nassif and B. J. Tromberg. Selective corneal imaging using combined second-harmonic generation and two-photon excited fluorescence. Opt. Lett. 2002; 27(23):2082–2084.

[20] G. Cox, E. Kable and M. D. Gorrel. Three-dimensional imaging of collagen using second harmonic generation. J. Struct. Biol. 2003; 141(1):53–62.

[21] S. J. Lin, S. H. Jee and C. Y. Dong. Discrimination of basal cell carcinoma from normal derma stroma by quantitative multiphoton imaging. Opt. Lett. 2006; 31(18):2756–2758.

[22] H. S. Lee, S. W. Teng and C. Y. Dong. Imaging human bone marrow stem cell morphogenesis in polyglycolic acid scaffold by multiphoton microscopy. Tissue Eng. 2006; 12(10):2835–2841.

[23] P. J. Campagnola. Second harmonic generation imaging microscopy: applications to diseases diagnostic. Anal. Chem. 2011; 83(9):3224–3231.

[24] I. Gusachenko, V. Tran, Y. G. Houssen, J. M. Allain and M. C. Schanne-Klein. Polarization-resolved second-harmonic generation in tendon upon mechanical stretching. Opt. Express. 2012; 102(9):2220–2229.

[25] X. Chen, O. Nadiarynkh, S. Plotnikov and P. J. Campagnola. Second harmonic generation microscopy for quantitative analysis of collagen fibrillar structure. Nat. Protoc. 2012; 7(4):654–669.

[26] W. Lo, W. L. Chen, C. M. Hsueh, A. A. Ghazaryan, S. J. Chen, D. H. K. Ma, C. Y. Dong and H. Y. Tan. Fast Fourier transform–based analysis of second-harmonic generation Image in keratoconic cornea. Invest. Opht. Vis. Sci. 2012; 53(7):3501–3507.

[27] H. Yuan-Tan, Y. L. Chang, W. Lo, C. M. Hsueh, W. L. Chen, A. A. Ghazaryan, P. S. Hu, T. H. Young, S. J. Chen and C. Y. Dong. Characterizing the morphologic changes in collagen crosslinked–treated corneas by Fourier transform–second harmonic generation imaging. J. Cat. Refract. Surg. 2013; 39(5):779–788.

[28] R. Seeley, T. Stephens and P. Tate. Anatomy and physiology. McGraw-Hill, Phoenix, 2003.

[29] A. Deniset-Besseau, J. Duboisset, E. Benichou, F. Hache, P. F. Brevet and M. C. Schanne-Klein. Measurement of the second order hyperpolarizability of the collagen triple helix and determination of its physical origin. J. Phys. Chem. B. 2009; 248(4951):73–76.

[30] K. E. Kadler, D. F. Holmes, J. A. Trotter and J. A. Chapman. Collagen fibril formation. Biochem. J. 1996; 316:1–11.

[31] D. A. D. Parry and A. S. Craig. Quantitative electron microscope observations of the collagen fibrils in rat-tail tendon. Biopolymers. 1977; 16(5):1015–1031.

[32] A. Yariv. Quantum electronics. Wiley, New York, 1989.

[33] P. Stoller, P. Celliers, K. Reiser and A. Rubenchik. Quantitative second-harmonic generation microscopy in collagen. Appl. Opt. 2003; 42(25):5209–5219.

[34] I. J. Su, W. L. Chen, Y. F. Chen and C. Y. Dong. Determination of collagen nanostructure from second-order susceptibility tensor analysis. Biophysical J. 2011; 100(8):2053–2062.

[35] Y. Komai and T. Ushiki. The three-dimensional organization of collagen fibrils in the human cornea and sclera. Invest. Ophthalmol. Vis. Sci. 1991; 32(8):2244–2258.

[36] M. Han, G. Giese and J. F. Bille. Second harmonic generation imaging of collagen fibrils in cornea and sclera. Opt. Exp. 2005; 13(15):5791–5797.

[37] S. W. Teng, H. Y. Tan, J. L. Peng, H. H. Lin, K. H. Kim, W. Lo, Y. Sun, W. C. Lin, S. J. Lin, S. H. Jee, P. T. So and C. Y. Dong. Multiphoton autofluorescence and second-harmonic generation imaging of the ex vivo porcine eye. Invest. Ophthalmol. Vis. Sci. 2006; 47(3):1216–1224.

[38] M. Yamanari, S. Nagase, S. Fukuda, K. Ishii, R. Tanaka, T. Yasui, T. Oshika, M. Miura and Y. Yasuno. Scleral birefringence as measured by polarization-sensitive optical coherence tomography and ocular biometric parameters of human eyes in vivo. Biomed. Opt. Express. 2014;5(5):1391–1402.

[39] J. M. Bueno, E. J. Gualda and P. Artal. Analysis of corneal stroma organization with wavefront optimized nonlinear microscopy. Cornea. 2011; 30(6):692–701.

[40] H. Y. Tan, Y. Sun, W. Lo, S. J. Lin, C. H. Hsiao, Y. F. Chen, S. C. Huang, W. C. Lin, S. H. Jee, H. S. Yu and C. Y. Dong. Multiphoton fluorescence and second harmonic generation imaging of the structural alterations in keratoconus ex vivo. Invest. Ophthalmol. Vis. Sci. 2006; 47(12):5251–5259.

[41] C. M. Hsueh, W. Lo, W. L. Chen, V. A. Hovhannisyan, G. Y. Liu, S. S. Wang, H. Y. Tan and C. Y. Dong. Structural characterization of edematous corneas by forward and backward second harmonic generation imaging. Biophys. J. 2009; 97(4):1198–1205.

[42] T.-J. Wang, W. Lo, C. M. Hsueh, M. S. Hsieh, C. Y. Dong and F. R. Hu. Ex vivo multi-photon analysis of rabbit corneal wound healing following conductive keratoplasty. J. Biomed. Opt. 2008; 13(3):034019.

[43] E. J. Gualda, J. R. Vázquez de Aldana, M. C. Martínez-García, P. Moreno, J. Hernández-Toro, L. Roso, P. Artal and J. M. Bueno. Femtosecond infrared intrastromal ablation and backscattering-mode adaptive-optics multiphoton microscopy in chicken corneas. Biomed. Opt. Express. 2011; 2(11):2950–2960.

[44] J. M. Bueno, E. J. Gualda, A. Giakoumaki, P. Pérez-Merino, S. Marcos and P. Artal. Multiphoton microscopy of ex-vivo corneas after collagen cross-linking. Invest. Ophthalmol. Vis. Sci. 2011; 52(8):5325–5331.

[45] S. W. Teng, H. Y. Tan, Y. Sun, S. J. Lin, W. Lo, C. M. Hsueh, C. H. Hsiao, W. C. Lin, S. C. Huang and C. Y. Dong. Multiphoton fluorescence and second-harmonic-generation microscopy for imaging structural alterations in corneal scar tissue in penetrating full-thickness wound. Arch. Ophthalmol. 2007; 125(7):977–978.

[46] Q. Wu and A. T. Yeh. Rabbit cornea microstructure response to changes in intraocular pressure visualized by using nonlinear optical microscopy. Cornea. 2008; 27(2):202–208.

[47] N. Morishige, A. J. Wahlert, M. C. Kenney, D. J. Brown, K. Kawamoto, T. Chikama, T. Nishida and J. V. Jester. Second-harmonic imaging microscopy of normal human and keratoconus cornea. Invest. Ophthalmol. Vis. Sci. 2007; 48(3):1087–1094.

[48] N. Morishige, N. Yamada, X. Zhang, Y. Morita, N. Yamada, K. Kimura, A. Takahara and K. H. Sonoda. Abnormalities of stromal structure in the bullous keratopathy cornea identified by second harmonic generation imaging microscopy. Invest. Ophthalmol. Vis. Sci. 2012; 53(8):4998–5003.

[49] K. Plamann, F. Aptel, C. L. Arnold, A. Courjaud, C. Crotti, F. Deloison, F. Druon, P. Georges, M. Hanna, J. M. Legeais, F. Morin, É. Mottay, V. Nuzzo, D. A. Peyrot and M. Savoldelli. Ultrashort pulse laser surgery of the cornea and the sclera. J. Opt. 2010; 12(8): 084002.

[50] H. G. Adelman. Butterworth equations for homomorphic filtering of images. Comp. Biol. Med. 1998; 28(2):169–181.

[51] P. Matteini, F. Ratto, F. Rossi, R. Cicchi, C. Stringari, D. Kapsokalyvas, F. S. Pavone and R. Pini. Photothermally-induced disordered patterns of corneal collagen revealed by SHG imaging. Opt. Express. 2009; 17(6):4868–4878.

[52] J. C. Russ. The image processing handbook. CRC Press, Boca Raton, 2007.

[53] J. M. Bueno, R. Palacios, M. K. Chessey and H. Ginis. Analysis of spatial lamellar distribution from adaptive-optics second harmonic generation corneal images. Biom. Opt. Express. 2013; 4(7):1006–1013.

[54] F. J. Avila and J. M. Bueno. Analysis and quantification of collagen organization with the structure tensor in second harmonic microscopy images of ocular tissues. Appl. Opt. 2015; 54(33):9848–9854.

[55] L. Sherman, J. Y. Ye, O. Albert and T. B. Norris. Adaptive correction of depth-induced aberrations in multiphoton scanning microscopy using a deformable mirror. J. Microsc. 2002; 206:65–71.

[56] P. Marsh, D. Burns and J. Girkin. Practical implementation of adaptive optics in multiphoton microscopy. Opt. Express. 2003; 11(10):1123–1130.

[57] M. Rueckel, J. A. Mack-Bucher and W. Denk. Adaptive wavefront correction in two-photon microscopy using coherence-gated wavefront sensing. P. Natl. Acad. Sci. U. S. A. 2006; 103(46):17137–17142.

[58] D. Debarre, E. J. Botcherby, T. Watanabe, S. Srinivas, M. J. Booth and T. Wilson. Image-based adaptive optics for two-photon microscopy. Opt. Lett. 2009; 34(16):2495–2497.

[59] A. Jesacher, A. Thayil, K. Grieve, D. Debarre, T. Watanabe and T. Wilson. Adaptive harmonic generation microscopy of mammalian embryos. Opt. Lett. 2009; 34(20):3154–3156.

[60] J. M. Bueno, E. J. Gualda and P. Artal. Adaptive optics multiphoton microscopy to study ex vivo ocular tissues. J. Biomed. Opt. 2010; 15(6):066004.

[61] J. Antonello, T. van Werkhoven, M. Verhaegen, H. H. Truong, C. U. Keller and H. C. Gerritsen. Optimization-based wavefront sensorless adaptive optics for multiphoton microscopy. J. Opt. Soc. Am. A. 2014; 31(6):1337–1347.

[62] M. Skorsetz, P. Artal and J. M. Bueno. Performance evaluation of a sensorless adaptive optics multiphoton microscope. J. Microscopy. 2015; doi: 10.1111/jmi.12325.

[63] M. Skorsetz, P. Artal and J. M. Bueno. Performance evaluation of a sensorless adaptive optics multiphoton microscope. J. Micros. 2015; 261(3):246–258

[64] W. Lo, Y. Sun, S. J. Lin, S. H. Jee and C. Y. Dong. Spherical aberration correction in multiphoton fluorescence imaging using objective. J. Biomed. Opt. 2005; 10(3):034006.

[65] P. A. Muriello and K. W. Dunn. Improving signal levels in intravital multiphoton microscopy using an objective correction collar. Opt. Commun. 2008; 281(7):1806–1812.

[66] J. M. Bueno. Adaptive optics multiphoton microscopy: probing the eye more deeply. Opt. Photo. News. 2014; 25(1):48–55.

[67] S. Roth and I. Freund. Second harmonic generation in collagen. J. Chem. Phys. 1979; 70(4):1637–1643.

[68] S. Roth and I. Freund. Optical second-harmonic scattering in rat-tail tendon. Biopolymers. 1981; 20(6):1271–1290.

[69] I. Freund, M. Deutsch and A. Sprecher. Connective tissue polarity: optical second-harmonic microscopy, crossed-beam summation, and small-angle scattering in rat-tail tendon. Biophys. J. 1986; 50(4):693–712.

[70] P. Stoller, K. M. Reiser, P. M. Celliers and A. M. Rubenchik. Polarization-modulated second harmonic generation in collagen. Biophys. J. 2002; 82(6):3330–3342.

[71] C. H. Lien, K. Tilbury, S. J. Chen and P. J. Campagnola. Precise, motion-free polarization control in second harmonic generation microscopy using a liquid crystal modulator in the infinity space. Biom. Opt. Express. 2013; 4(10):1991–2002.

[72] F. J. Ávila, O. Barco and J. M. Bueno. Polarization dependence of collagen-aligned tissues imaged with second harmonic generation microscopy. J. Biom. Opt. 2015; 20(8): 086001.

[73] I. Gusachenko, G. Latour and M. C. Schanne-Klein. Polarization-resolved Second Harmonic microscopy in anisotropic thick tissues. Opt. Express. 2010; 18(18):19339–19352.

[74] O. Nadiarnykh and P. J. Campagnola. Retention of polarization signatures in SHG microscopy of scattering tissues through optical clearing. Opt. Express. 2009; 17(7): 5794–5806.

[75] O. del Barco and J. M. Bueno. Second harmonic generation signal in collagen fibers: role of polarization, numerical aperture, and wavelength. J. Biom. Opt. 2012; 17(4): 045005.

[76] F. Tiaho, G. Recher and D. Rouède. Estimation of helical angles of myosin and collagen by second harmonic. Opt. Express. 2007; 15(19):12286–12295.

[77] W. L. Chen, T. H. Li, P. J. Su, C. K. Chou, P. T. Fwu, S. J. Lin, D. Kim, P. T. C. So and C. Y. Dong. Second-order susceptibility imaging with polarization-resolved, second harmonic generation microscopy. Proc. SPIE. 2010; 7569(75691):1–7.

[78] R. Ambekar, T. Y. Lau, M. Walsh, R. Bhargava and K. C. Toussaint. Quantifying collagen structure in breast biopsies using second-harmonic generation imaging. Biom. Opt. Express. 2012; 3(9):2021–2035.

[79] K. Tilbury and P. J. Campagnola. Applications of second-harmonic generation imaging microscopy in ovarian and breast cancer. Perspect. Medicin. Chem. 2015; 16(7):21–32.

[80] A. Benoit, G. Latour, S. K. Marie-Claire and J. M. Allain. Simultaneous microstructural and mechanical characterization of human corneas at increasing pressure. J. Mech. Behav. Biomed. Mater. 2015; 60:93–105.

Chapter 6

Nonlinear Microscopy Techniques: Principles and Biomedical Applications

Javier Adur, Hernandes F. Carvalho,
Carlos L. Cesar and Víctor H. Casco

Additional information is available at the end of the chapter

http://dx.doi.org/10.5772/63451

Abstract

Nonlinear optical microscopy techniques have emerged as a set of successful tools within the biomedical research field. These techniques have been successfully used to study autofluorescence signals in living tissues, structural protein arrays, and to reveal the presence of lipid bodies inside the tissular volume. In the first section, the nonlinear contrast technique foundations is described, and also, a practical approach about how to build and combine this setup on a single confocal system platform shall be provided. In the next section, examples of the usefulness of these approaches to detect early changes associated with the progression of different epithelial and connective tissular diseases are presented.

Finally, in the last section, we attempt to review the present-day most relevant analysis methods used to improve the accuracy of multimodal nonlinear images in the detection of epithelial cancer and the supporting stroma. These methods are presented as a set of potential valuable diagnostic tools for early cancer detection and to differentiate clinical subtypes of *osteogenesis imperfecta* disorders, being highly advantageous over present classical clinical diagnostic procedures.

In this chapter, it is proposed that the combination of nonlinear optical microscopy and informatics-based image analysis approaches may represent a powerful tool to investigate collagen organization in skin diseases and tumor cell morphology.

Keywords: nonlinear microscopy, second harmonic generation, third harmonic generation, image analysis, early diagnosis

1. Principles of nonlinear microscopy techniques

The notable advances in cell and molecular biology science have induced the need to imagine cells in an intact-live whole organism. Therefore, the need for real-time observation of cell (and their subcellular components) behavior in whole tissues has become crucial to understanding cellular physiology. During the past decades, imaging techniques have been improved to pursue this goal. One of these techniques is fluorescence imaging. The use of confocal micro-scopy allows the examination of subcellular material with three-dimensional resolution but is restricted by the effective imaging depth (usually less than 200 μm) and phototoxicity, which is caused by using a short wavelength laser (principally continuous wave (CW) laser) [1]. Recent advances in nonlinear optical processes of multiphoton microscopy overcome single-photon linear microscopy technologies, such as confocal microscopy, by their capacity of tissular penetration, clean images production, minimal invasiveness, and chemical selectivity [2]. Therefore, multiphoton fluorescence (MPF) and nonlinear optical (NLO) microscopy in recent year has become one of the key imaging modes and evolved as an alternative to conventional single-photon confocal microscopy. The best-known nonlinear microscopy techniques are two-photon excited fluorescence (TPEF) microscopy, second harmonic gener-ation (SHG) and third harmonic generation (THG) microscopy, and coherent anti-Stokes Raman scattering (CARS) microscopy.

These nonlinear technologies provide several advantages, namely high depth penetration by using a near infrared (pulsed) laser as excitation source, intrinsic tridimensional sectioning and resolution, due to the spatial confinement of the signal to the laser focus, multiple nonlinear processes, and the possibility to use numerous endogenous molecular markers and low phototoxicity that allows the investigation of living processes, without significant perturbation [3]. Together, these advantages allow analyzing the complex relations between tissue and organ function and its structure in complex diseases [4].

To understand this new microscopy instruments, it is advisable to think in classical optical tool. In conventional optical imaging, contrast mechanisms consist of interactions such as absorption, reflection, scattering, and fluorescence, and the response recorded is linearly dependent on the intensity of the incident light. Thereby, there is a linear relationship between the strength of electric field of the light and the induced object polarization. At moderately low incident intensity, the optical response can be approximated to the first-order response as $P = \chi^{(1)} \otimes E$, where $\chi^{(1)}$ is the linear susceptibility, P is the polarization of a material, and E is the strength of an applied optical field. By contrast, nonlinear optical effects occur when a biological tissue interacts with an intense laser beam exhibiting a nonlinear response to the applied field strength. In this situation, the induced polarization vector P of the material subject to the vectorial electric field E can be expressed as $P = \chi^{(1)} \otimes E + \chi^{(2)} \otimes E \otimes E + \chi^{(3)} \otimes E \otimes E \otimes E$ $+...$, where $\chi^{(i)}$ is the $i^{(th)}$ order nonlinear susceptibility tensor and $\otimes$ represents a combined tensor product and integral over frequencies [5]. The bulk nonlinear optical susceptibilities $\chi^{(2)}$ and $\chi^{(3)}$ are obtained from the corresponding high-order molecular nonlinear optical coeffi-cients (hyperpolarizability) β and γ by using a sum of the molecular coefficients over all molecule sites. Typically, materials with conjugated π-electron structures exhibit large optical nonlinearities. The usual linear susceptibility $\chi^{(1)}$ contributes to the single-photon absorption and reflection of the light in tissues. The $\chi^{(3)}$ corresponds to third-order processes such as two-photon absorption, THG, and CARS, while SHG results from $\chi^{(2)}$ [6].

Another way is to think nonlinear optical processes in terms of particles called photons discovered by Einstein, such as processes involving more than one photon. All these processes have some characteristics in common. First, it only occurs if multiple photons coincide in time and space or, in chemical terms, at high concentration of photons. The speed V of a chemical reaction of n elements $f + f + f + f + \ldots \to f_n$ depends on the concentration f elevated to the nth degree, that is, $V \propto [f]^n$ where $[f]$ is the concentration of f. In optics, the concentration is proportional to the laser beam intensity, I = power/area = (energy/time) / area. Thus, the efficiency of a one- photon process is I, and for two-photon would be I^2, and generalizing for n photons I^n. Therefore, it is not surprising that the first nonlinear optical empirical results were materialized only after the advent of pulsed lasers. The NLO microscopes come with the advent of Ti:Sa lasers which produce pulses typically in the range of 100 femtoseconds [fs = 10^{-15} sec], with an average power of 2 watts and repetition time of 12 nanoseconds [ns = 10^{-9} sec]. In this case, the peak power will be 240 kilowatts, although the pulse energy is just 24 nanoJoule. The pulsed laser, therefore, increased the peak power of 2 watts for a CW laser to 240 kilowatts, while keeping the same average power and low energy per pulse. In other words, the very high potencies are obtained by decreasing the pulse duration instead of increasing the pulse energy. Thus, nonlinear effects occur avoiding the potential sample damage. The smaller the temporal pulse duration, greater the efficiency of nonlinear processes, so femtosecond lasers are preferable to picosecond lasers.

The pulsed laser ensures the coincidence of photons in time, but not in the space. The concentration of photons but will be greater the smaller the area of the laser beam, i.e., is maximum in the laser focus. The total generation of events caused by one photon processes is constant, independent of the laser focusing position, because if the process is linear with I, the total number of events is proportional to I multiplied by the area, i.e., N events $\propto$ (power/area) × area = power. As power is constant along the beam, the number of events does not depend on the axial position [along the lens axis—defined as the z-axis]. The processes with more than one photon are proportional to I^n and therefore are inherently confocal. In this case, the amount of events depends very strongly on the beam area, because now N events $\propto I^n$ × area = (power^n/area^{n-1}). Two-photon processes decay inversely with the area, whereas a three-photon does so with the square of the area. This means that events can occur only in the vicinity of the lower area, i.e., the laser focus, that is, the light generated by the nonlinear optical process are generated at the focus of the laser, which becomes intrinsically a confocal microscope. The laser focus is on the operator microscope control and can be used for 3D image reconstruction.

One way to visualize the various nonlinear optical processes is by arrows with length proportional to the photon energy. **Figure 1** schematically depicts a number of nonlinear optical effects, produced for the specimen if the energy density at the focal spot of an objective lens is sufficiently large and also are compared with lineal process produced for a CW laser. Principal contrast mechanisms and characteristics of TPEF, SHG, THG, and CARS modalities are described below.

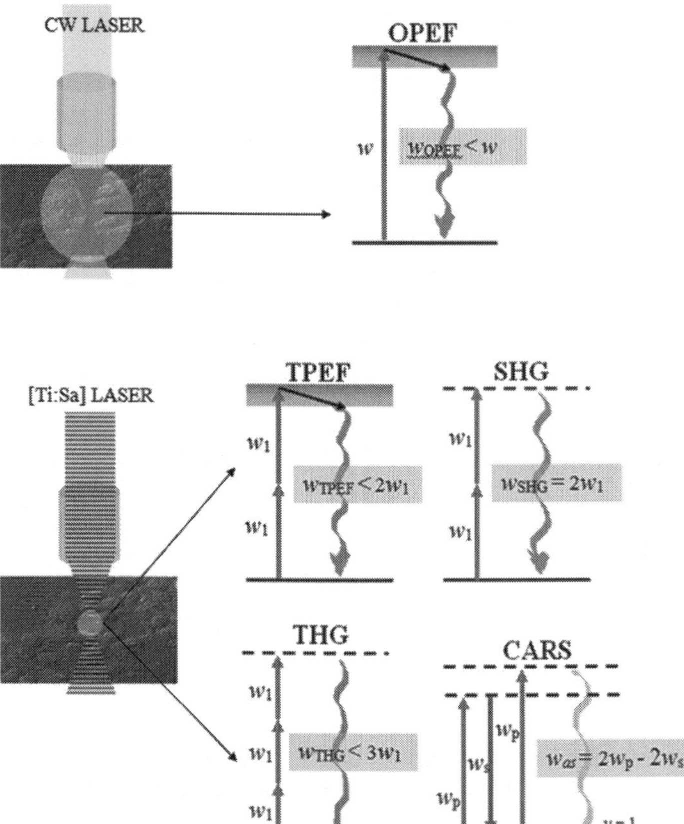

Figure 1. Energy diagrams of linear and nonlinear optical process. Solid lines represent electronic and vibrational states of molecules, while dashed lines denote virtual states. The straight arrows are excitation beams, whereas wavy arrows are the output signal beams. The black arrow represents relaxation in electronically excited states, and red hourglass represents the excitation volume. $\omega 1$ and $\omega 2$ symbolize the two beams available in CARS microscope. $\omega p = \omega 1$ and $\omega s = \omega 2$ in CARS. $v = 0$: vibrational ground state, $v = 1$: vibrational excited state, and Ω is a frequency of vibrational transition between $v = 0$ and $v = 1$. $\omega 1$ = pump beams and $\omega 2$ = Stokes beam from the laser sources. $\omega 3$ is a long wavelength beam for THG obtainable on OPO systems for CARS microscope. Abbreviations: CW: continuous wave, OPEF: one-photon excited fluorescence, OPO: optical parametric oscillator.

TPEF microscopy is a third-order nonlinear optical resonant process where two photons excite an electron from the ground state. It is an inelastic process where photon energy is released at the sample. Two-photon absorption happens only when the energy of the incident photons falls into the two-photon excitation band which is specific for each fluorescent marker. The two-photon excitation band is not exactly half of the one photon excitation band because the selection rules are different. The fact that TPE depends on the square of the incident light provides its confocal features, that is, a process happening only at a focal point volume. Photobleaching is smaller in two-photon excited fluorescence (TPEF) compared to single photon excitation because the excited volume is smaller. Multicolor imaging is allowed to excite different fluorophores simultaneously through different order processes with a single wavelength, in which emissions are spectrally shifted by hundreds of nanometers and uninterrupted for collection. Consequently, multiphoton microscopy is especially appropriate for physiological and pathophysiological studies since it is able to excite endogenous auto-fluorescent components and thus to obtain specific signals such as nicotinamide adenine dinucleotide (NADH) and flavin adenine dinucleotide (FAD) [6].

On the contrary, SHG and THG are coherent second/third-order elastic nonlinear optical processes, respectively. Because two/three photons generate another photon with two/three times the energy of the incident photons, there is no energy released to the medium, meaning out of focus cell photodamages are not expected from these processes. Both SHG/THG can be segregated from fluorescence signals by the wavelength or even by time gating, because the coherent processes are practically instantaneous. The fact that SHG signal is proportional to I^2, while THG signal is proportional to I^3, where I, is the incident light intensity, provides confocality for both techniques. For the same wavelength of the incident light, THG has better optical sectioning resolution than SHG or TPEF but is also more sensitive to changes in the intensity of the light in the focused spot, such as those caused by laser instability or by scattering or defocusing the illumination [7]. The first images of high-resolution SHG were reported in 1998 by Sheppard's group [8], and shortly afterward in 2001, Chu and co-workers showed a multimodal imaging study, including TPEF+SHG+THG [9]. Usually, the third-order nonlinear susceptibility (χ^3), responsible for THG, is much smaller than the second-order one (χ^2), responsible for SHG. In principle, this would mean that THG should be much harder to observe. However, χ^2, as well as any other even susceptibility coefficients, must be null in the presence of inversion symmetry. Therefore, SHG shall be zero in the presence of centro-symmetric molecules, unless an external parameter, such as electric fields or interfaces, breaks the symmetry.

By contrast, all materials have non-zero third-order susceptibility χ^3. Moreover, $\chi 3$ can be several orders of magnitude larger or smaller for different materials. However, THG is null in a homogeneous material, no matter how high χ^3 could be, because the Gouy-phase shift of π across the focus of a Gaussian excitation beam creates a destructive interference between signals generated before and after the focus [10]. Nevertheless, for a nonhomogeneous focal volume, a measurable amount of third harmonic is generated [11]. Since in biological samples, heterogeneity is more common than homogeneity, THG provides an important tool for bioimaging, with the warning that it tends to be brighter at the interface of large granules, lipid droplets, or similar biostructures, compared to the internal signal [6]. SHG imaging modality can probe molecular organization, molecular symmetry, orientation, molecular alignments, and ultrastructure on the micro, as well as the nanoscale. Natural structures are mostly unarranged (optically isotropic) and do not generate any SHG signal. Hardly, some few biological assemblies are ordered and can produce harmonic signal. One of the best known SHG structures in biology is collagen, the major protein of the extracellular matrix. Collagen fibrils often aggregate into larger, cable-like bundles, several micrometers in diameter. This regularly staggered packing order provides the needed structural conditions for efficient SHG [12]. Other examples are acto-myosin assemblies in muscle and microtubule structures in living cells [13, 14]. Discontinuities in refraction index of the optical dispersion properties of biological tissues are able to generate THG signals [15, 16]. The THG can be used to study optical cell interfaces such as those at cell membranes or organelle surfaces. For example, the surface of the erythrocyte can generate significant THG [17]. In contrast to the chemical specificity that characterizes fluorescence images, harmonic generation (SHG and THG) provides an imaging modality specific for structural configuration. In the study of cancer tumors, our experience with both techniques is that SHG is an excellent tool to observe collagen network of extracellular matrix, while THG allows to clearly display the nuclei, which are two key parameters for pathologists [4, 18–20].

Characteristics	TPEF	SHG	THG	CARS
Application in bioscience	1990	1986	1997	1982
Number of photons	2	2	3	3
Susceptibility	χ^3	χ^2	χ^3	χ^3
Advantages	Deeper imaging with less phototoxicity Spatial localization forfluorescence excitation	Coherent process, symmetry selection Probing well-ordered structures, functions of membranes, nonfluorescence tissues No absorption of light	Coherent process, no symmetry requirement No absorption of light Imaging both in bulk and at surfaces for extended conjugation of pi electrons	Coherent process Inherent vibrational contrast for the cellular species, requires no endogenous or exogenous fluorophores Vibrational and chemical sensitivities
Contrast mechanics	Electronics levels of the molecules	Nonlinear properties of the medium	Nonlinear properties of the medium	Vibrational levels of the molecules
Information	Autofluorescence of some biological substances (NADH, FAD, etc.)	Noncentrosymmetric molecules with spatial organization (collagen, elastin, etc.)	Interfaces, optical inhomogeneities, (cell edges, lipids, membranes)	Chemical information (lipids, DNA, proteins)

Table 1. Characteristics of nonlinear optical microscopy.

In addition to harmonic generation microscopy, CARS microscopy is another 3D high-resolution imaging approach that circumvents exogenous probes. CARS is a four-wave mixing process in which a pump beam at frequency w_p, a Stokes beam at frequency w_s, and a probe beam at frequency w'_p are interacted with a sample to result in an anti-Stokes signal at $w_{as} = 2w_p - w_s$. In nearly all experiments, the pump and the probe beams are derived from the resonant oscillation when the beat frequency ($w_p - w_s$) matches the frequency of a particular Raman active molecular vibration mode. Furthermore, due to its coherent nature, CARS signal production only occurs when the field-sample interaction length is less than the coherence length. The generated CARS signal is proportional to $(\chi^{(3)})^2\, I^2_p\, I_s$, having a quadratic dependence on the pump field intensity and a linear dependence on the Stokes field intensity. These properties provide it a 3D-sectioning capability [6]. First CARS microscopy setup was described at the nineties [21] and has now matured into a powerful method for biological imaging. CARS microscopy is more informative than SHG and THG microscopy since it contains rich spectroscopic information about specific molecular species.

To summarize the first section, a description of the physical properties, characteristics, and principal contrast mechanisms of each nonlinear optical imaging method described are summarized in **Table 1**.

2. Nonlinear optical technique implementation

Today, there are multiples ways to assemble a nonlinear microscopy platform. The engineering challenge is to integrate the different modalities on a single platform. In response to this challenge, manufacturers have designed microscopes with multiple input and output ports and increased infinity space for the introduction of customized optics. Coming up next, some laser sources, detectors, and confocal body microscopies routinely used in this technology are enumerated, and a setup configuration by our group is described in some detail.

As was described in the introductory section, for high harmonic generation and multiphoton fluorescence microscopies, short femtoseconds pulses of high peak power are required. While ultrashort few cycle pulses are spectrally very broad, they allow for simultaneous excitation of different chromospheres with spectrally separated absorption bands. Available lasers that conjugate these features are the titanium:sapphire (Ti:Sa) (wavelength range 700–980 nm, pulse width 100 fs, and 76–100 MHz repetition rate), the Cr:forsterite laser (wavelength range 1230–1270 nm, pulse width 65 fs, and 76–120 MHz repetition rate), the Nd:glass laser (wavelength range 1053–1064 nm, pulse width 150 fs, and 70–150 MHz repetition rate), and the femtosecond ytterbium laser (wavelength range 1030 nm, pulse width 200 fs, and 50 MHz repetition rate).

The SHG wavelength excited by a Ti:Sa femtosecond laser operating at 940 nm will be in the blue at 470 nm, and the TPEF will be in the region above 470 nm. While that THG signal generated by a 940 nm principal beam, will be in the UV region at 330 nm. As a result, the THG signal will suffer from the high UV absorption of the principal biological specimens making signal detection difficult. In contrast, using the Cr:forsterite laser operating in the range of 1230 nm allows SHG (615 nm), THG (410 nm), and TPEF (>615 nm), all to fall within the visible spectrum. Additionally, the lowest light attenuation in biological material is generally found in the 1000–1300 nm. In recent years other ultrafast laser systems appeared, such as InSight TM DeepSee TM (wavelength range 680–1300 nm, pulse width <120 fs, and repetition rate 80

MHz), which also can be an excellent light source for multi-modality microscopy. Moving the excitation wavelength to 1200 nm, not only the visible but also the NIR spectrum is open for signal recording.

Nonlinear microscopes share many common features with confocal laser scanning microscopes. In fact, many research groups have implemented multimodal nonlinear platform by coupling source lasers described previously into a confocal scanning microscope. Practically, many scan head models of different manufactures have been used with this proposal, such as Olympus FV300 [22], Olympus FV1000 [23], Zeiss LSM Meta 510 [24], Zeiss LSM 710 [25], Nikon C1 [26], Leica TCS-SP5 [27], and Zeiss LSM 780 [28]. In general, the generated nonlinear signals can be collected with the same microscope objective, separated by a dichroic mirror, which is expressly selected for the given fundamental and fluorescence or harmonic emission wavelengths and focused with a lens through the filter onto detector. If Ti:Sa laser is used, the wavelengths fall within the sensitivity range of high quantum efficiency (QE) silicon-based detectors and photomultiplier tube (PMT) photocathodes that are the currently used detectors. If the source laser used are Nd:glass or Cr:forsterite, special NIR detectors (i.e., indium gallium arsenide (InGaAs) photodiode) are needed. For more data about the optical characteristics of the different detectors, readers can find excellent information in [7].

In our setup (**Figure 2**), we used an inverted Zeiss Axio Observer.Z1 and confocal LSM 780. Briefly, this device is equipped with a UV-lamp, for classical epi-fluorescence operation mode; five lines of CW laser, for confocal studies; and femtosecond (fs) and picosecond (ps) pulsed laser, for nonlinear microscope modalities. The fs laser source is a tunable, Ti:sapphire laser emitting around 690–1040 nm both for efficient TPEF and higher SHG/THG spatial resolution. The picosecond (ps) source is obtained from a synchronously pumped optical parametric oscillator (OPO) system to obtain THG signal in the visible range and high spectral resolution CARS microscopy. The OPO can be easily and continuously tuned over a wide spectral range from 690 to 990 nm for the signal and between 1150 and 2450 nm for the idler output. The fs laser is combined with the scan head through an acousto-optic modulator (AOM) and a collimating telescope (T1) to regulate the beam diameter in the objective back-aperture and the focus position on the microscope focal point. The five wavelengths (signal and idler for each OPO plus the fundamental 1064 nm) are controlled independently with dedicated telescopes (T2, T3, T4, T5, and T6). Delay lines on the five beam paths ensure temporal overlap between the beams. These beams necessary for CARS microscopy are temporally synchronized, recombined (P), and sent onto the backward excitation port of the scan head. The scan head of the LSM780 has a spectral gallium arsenide phosphide (GaAsP) detector with 32 in-line elements and 2 adjacent PMTs. The motorized collimators, the scanners, and the pinhole precisely positionable and the highly sensitive detectors are arranged to provide optimum specimen illumination and efficient collection of the emitted light. The Raman line width is comparable to the spectral width of a picosecond pulse, so that the excitation energy is fully used to take full advantage of the vibrational resonant CARS signal. Working with 1~3 ps spectral pulse widths is possible to obtain the optimal signal-to-background ratio for typical Raman band [29]. A pulse width of a few picoseconds provides a good compromise between the spectral resolution and the peak power and improves the signal-to-background ratio. The

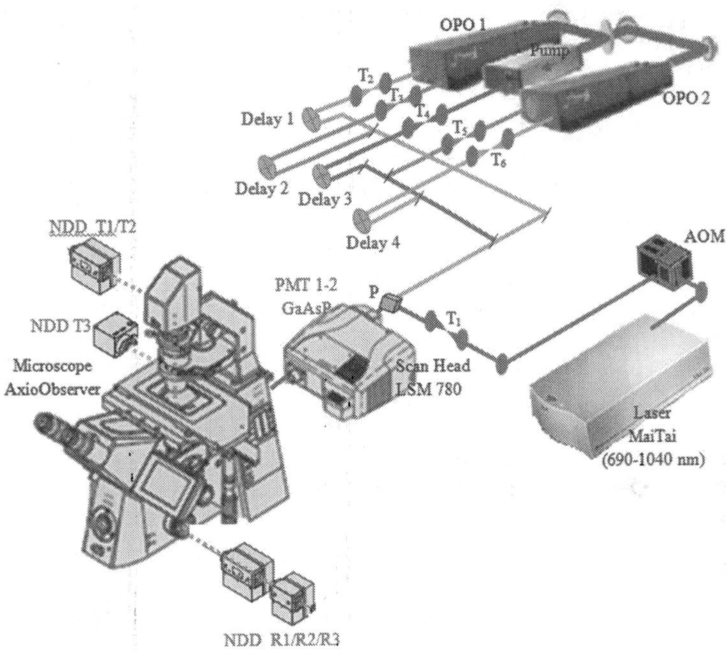

Figure 2. Schematic setup of a multimodal NLO microscope using a femtosecond laser source (Mai-Tai) and picoseconds (OPOs system) pulsed laser. Lasers are coupled to a commercial confocal system platform (inverted Zeiss Axio Observer.Z1 and confocal LSM 780). Principal optics elements are shown with blue letterings (telescope, delay lines, and recombination) and detectors with red letterings. Solid and dashed lines represent excitation and emission paths, respectively. AOM, acousto-optic modulator; T, telescope lens; P, recombine; NDD T, nondescanned detector (transmitted way); NDD R, nondescanned detector (reflective way); PMT, photomultiplier tube.

OPO light source for CARS that we used has already been reported to perform time resolved CARS [30] to improve the sensitivity or to cover the full vibrational Raman spectrum. In our setup, the frequency doubled Nd:YVO generates 8 W of green light (532 nm) and pumps two OPOs with 4 W each. Although the full system can provide up to five wavelengths simultaneously (signal and idler for each OPO plus the fundamental 1064 nm), we normally use three colors. The two signals coming from each OPO are recombined with the fundamental 1064 nm beam from the Nd:YVO oscillator. The power, polarization state, and divergence of each beam are controlled independently with dedicated polarization optics and telescopes. Delay lines on two of the beam paths (signal and 1064 nm) assure temporal overlap between the three beams. The backward detection is achieved with internal GaAsP detector of scan head. A set of dedicated filters is placed before the detectors to select the relevant spectral domains.

The generated signals can be collected with the same microscope objective (reflected or epi-detected), splitted by dichroic mirrors, which were specifically chosen for the given fundamental and fluorescence or harmonic emission wavelengths and focused onto specific detectors. Interference or band pass filters are used in front of the detector for filtering scattered fundamental light and spurious signals outside the desired bandwidth. For epidetection, the system has two internal PMT plus GaAsP avalanche photodiodes inside the scan head (PMT 1-2, **Figure 2**). Also, three non-descanned detectors (NDDs) are available for epi-detection of nonlinear signals (NDD R1/R2/R3). It is also possible to detect the signals in the forward direction. Either one detector with appropriate filters or several detectors recording different signals separated by dichroic mirrors can be used. The system shown here has three NDD detectors (NDD T1/T2/T3). T3 detector was placed right after the sample, holding it as close to the sample as possible. With this system, many configurations can be used and different, linear, and/or nonlinear signals can be simultaneously detected. For example, we can observe CARS signal with internal GaAsP detector of scan head, forward THG, SHG and TPEF with NDD T1/T2/T3, and reflected SHG and TPEF with NDD R1/R2.

3. Biomedical applications

Over the past years, life science interdisciplinary research has routinely used nonlinear microscopy techniques. The combination of SHG, THG, and CARS is used in the production of chemical maps of complex tissues. NLO techniques allow inspecting the assembly of single cells, tissues, and organs as well as monitoring structural and chemical changes related to diverse diseases. Here a few examples of the use of our setup in two applications are shown, epithelial cancer detection and diagnosis of *osteogenesis imperfecta*.

3.1. Cancer detection

Cancer is still a threat to human life [31]. Modern clinically used imaging methods for cancer diagnosis comprise x-ray, CT, MRI, and OCT [32, 33]. The facilities of these technologies are restricted by either low spatial resolution or a lack of chemical specificity, making it difficult to identify the edges of the tumor. Today, new image-based instruments are necessary as diagnostic tool to evaluate structural features with subcellular resolution that are closely linked with tumor malignancy. The combination of different image approaches described in this chapter may represent a powerful combination of tools to study both malignant cells and stromal environment. One of the main examples of such an objective is the collagen organization changes analysis, the remodeling matrix and alterations in epithelial/stromal interface. Highly valuable, structural information revealed by each nonlinear contrast approach can be isolated and analyzed separately, while their superposition allows a better comparison and understanding of the spatial tissue organization. Thus, TPEF and THG can be used to image a variety of well-documented morphologic and architectural alterations, moreover, combined TPEF-SHG can be applied to analyze alterations in epithelial cells and the supporting stroma, and CARS microscopy can be used to understand lipid and proteins composition in tumor tissues.

Figure 3A exemplifies this combination, where NLO techniques were applied to differentiate between normal and malignant (fibroadenoma and invasive lobular carcinoma (ILC)) human breast tissue. The characteristic microscopic appearance of each type of tissue and relationship between cells and stromal compartment can be identified in the SHG/THG combination. The differential orientation and distribution of collagen fibers can be clearly identified in stromal region with SHG image. **Figure 3B** shows a comparative analysis of CARS images of breast tissue. Adipose and fibrous structures of normal tissue possess strong CARS signals. Fibroadenoma exhibits the compressed duct with linear branching pattern, whereas ILC presents single or rows of cells invading into the stroma. Based on examples such as this and our previous work [19], we have established that it is possible to have both qualitative parameters of differences between each kind of breast tumor and to demonstrate the advantage of the integration of as many NLO approaches as possible to analyze breast cancer.

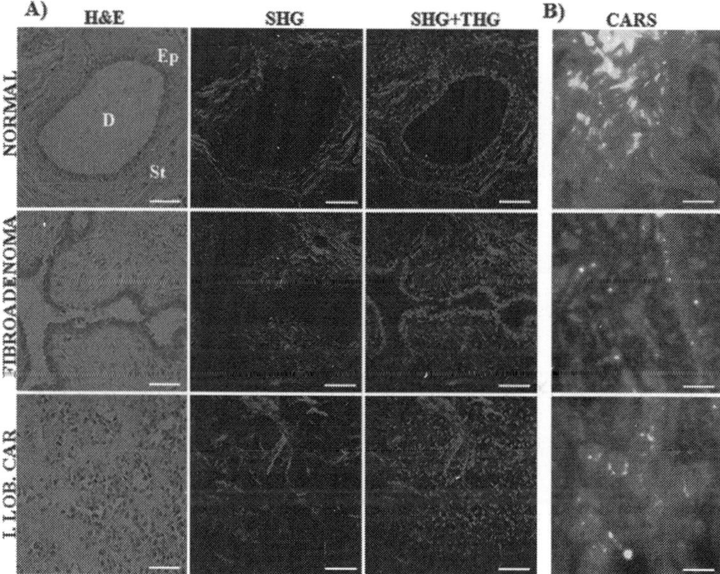

Figure 3. Multimodal NLO approach applied to human breast tumor. (A) Representative H&E-stained and SHG, and SHG+THG cross-sectional images of breast tissues diagnosed as normal (first row), fibroadenoma (second row), and invasive lobular carcinoma (third row). Scale bar = 20 μm. (B) Representative CARS cross-sectional images of breast tissues. Scale bar = 20 μm. D, duct; Ep, epithelium; St, stroma; Fibroad, fibroadenoma; Lob. Carc., lobular carcinoma; I.Lob.Carc, invasive lobular carcinoma. **Figure 3A** from Adur et al., [19].

Human ovarian tumors are shown in **Figure 4**. TPEF signal (green) represents stromal connective tissues. The SHG signal (red) shows collagen fibers, while THG (cyan) enhances the nuclei. The information revealed by each mode can be directly compared, providing a better understanding of the tissue. For example, SHG/THG-merged signals can be used to distinguish the epithelial/stromal interface. It is worth mentioning that these differences and

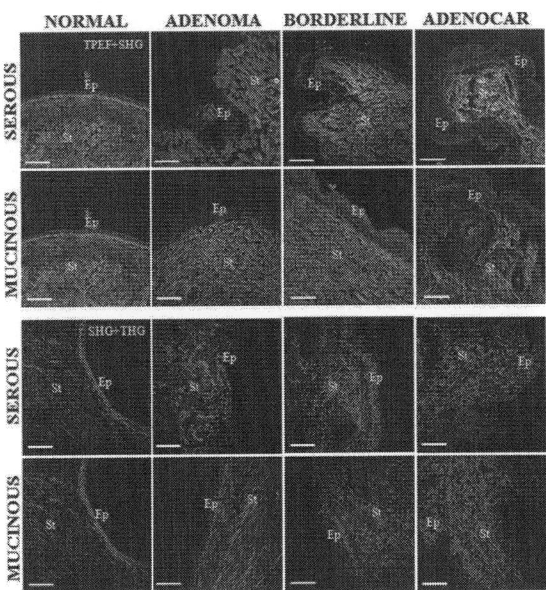

Figure 4. Multimodal NLO approach applied to human ovarian tumor. Representative merges of TPEF (green) and SHG (red) cross-sectional images of ovarian tumor tissues diagnosed as serous-type tumor and mucinous-type tumor; and representative merges of SHG (red) and THG (cyan) cross-sectional images. All scale bar = 20 μm. Ep, epithelium; St, stroma; n, nucleus; c, collagen; Ade, adenoma; Bord, borderline; Adenocar, adenocarcinoma. Reproduced figures from Adur et al. [34] (open access).

contrasts could be automatically and digitally done for quantification (see next section). These data confirm the fact that normal ovaries are more organized tissues than the adenocarcinoma samples. Finally, using THG signal, it was possible to evaluate the differences in the surface epithelium of each tumor type. In normal ovary, cells were arranged in one cell layer. Serous adenoma samples display elevated ciliated and non-ciliated cuboidal epithelial cells with lengthened nuclei, also in one regularly single cell layer. Besides, the serous borderline tumor and serous adenocarcinoma samples are absolutely different from the previous ones, showing epithelial surface with cells of altered sizes lying in multiple layers, including cellular atypia and proliferation. Mucinous tumor samples are similar to borderline/adenocarcinoma, with different size cells forming up multiple layers, but having rich cytoplasmic mucin and basal nuclei.

Figure 5 summarizes the combined uses of these techniques in the analysis of human colon cancer. NLO images clearly demonstrate the circular arrangement pattern of control colonic crypts registered from crypt-cross sections, characterized by epithelial columnar cells and interspersed goblet cells. TPEF (green) revealed the typical foveolar pattern of colon mucosa glands, displaying crypts with rounded luminal openings. SHG (red) specifically traces the collagen scaffold within lamina propria. The evaluation of colonic tissue by SHG microscopies

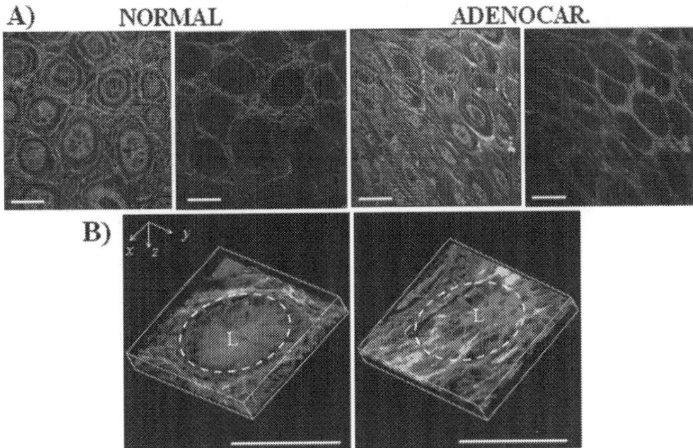

Figure 5. Multimodal NLO approach applied to human colon tumor. (A) Representative TPEF+SHG and SHG cross-sectional images of normal and tumor colon tissues. Scale bars = 20 μm. (B) Maximum projection of 60 images separated 0.5 μm each of normal and adenorcarcinomas colonic tissues. Epithelial-stromal interface is indicated (white outline). Scale bars = 50 μm. L, luminal crypt orifice; Adenocar, adenocarcinoma. Reproduced figures from Adur et al. [37] (open access).

rapidly and clearly allows differentiating between adenocarcinoma and normal tissue states. Whereas individual crypts were easily identified with TPEF, the interspersed connective tissue was detected via SHG of collagen (**Figure 5A**). As was previously stated, in tumoral tissues, SHG images highlight changes in the surrounding fibrous stroma. Another key aspect of NLO techniques in relation to the classical H&E ones is that they have high potential to produce 3D reconstructions and stereological studies. **Figure 5B** shows three-dimensional representations that allow the visualization of indistinguishable features in classical two-dimensional procedures. Using 3D SHG representation, it was possible to detect the tilt and invasiveness of collagen fibers of adenocarcinomas compared to normal colon. Those features are not easily visible in standard two-dimensional H&E-stained sections. In others works, NLO microscopy approaches, especially when combined, can reveal information not distinguishable in H&E stained sections. Different changes in collagen fibers are parameters that can be consistently quantified, which allows to predict an enormous clinical potential in colon cancer. These results show important changes of collagen fiber morphology, alignment, and density in colon tumor tissue, suggesting that collagen fiber inclination angles are a key factor in tumor progression. In agreement with these results, previous reports on human colon and other tissues suggest that the epithelial cells preferentially invade tissues where the collagen fibers became perpendicularly aligned, instead of arbitrarily organized ones [35, 36].

3.2. Osteogenesis imperfecta

Osteogenesis imperfecta (OI) is a heterogeneous disorder of connective tissues (see **Table 2** for types of OI) with an incidence of 1/15000 [38, 39] and disease severity spanning from subclinical osteoporosis to intrauterine lethality. Dominant mutation in collagen type I is the most common cause (>90%). Type I collagen is the most abundant extracellular matrix (ECM) protein in humans and the major structural protein in many organs, for example in skin. It is a heterotrimer consisting of two α1-chains and one α2-chain, encoded by COL1A1 and COL1A2, respectively. Mutations in the genes encoding type I procollagen produce a range of disorders, which include autosomal dominant (AD) OI. Currently, more than 1000 heterozygous COL1A1/2 mutations have been identified (https://oi.gene.le.ac.uk) [40, 41]. Mutation type and position influence the phenotype and as such genotype-phenotype relations exist to some extent. Mainly, two types of mutations in collagen I cause classical dominant OI: quantitative and qualitative collagen defects. These collagen I mutations are reflected in some way on fibril collagen assembly that can be finally observed in an organ, such as the skin. For example, in patients suffering from OI, skin collagen fibers could be smaller and more randomly packed. These disorders in collagen fibrils could be quantified using SHG microscopy.

Type Characteristic	I	II	III	IV	V	VI
Severity	Mild	Perinatal lethal	Severe	Moderate	Moderate	Moderate
Congenital fractures	NO	YES	Usually	Rarely	NO	NO
Bone deformity	Rarely	Very severe	Severe	Mildly moderate	Moderate	Moderate
Stature	Normal	Severely short	Very short	Variable short	Variable	Mildly short
Hearing loss	60% of cases	NA	Common	42% of cases	NO	NO
Respiratory complications	NO	YES	YES	NO	NO	NO

NA: not available

Table 2. Clinical characteristics of osteogenesis imperfecta.

Figure 6 depicts representative images acquired using previous setup, displaying representative TPEF (green) and SHG (red) images. TPEF signals are generated fundamentally by the eosin fluorescence and, in every case, this signal was used to detect just the skin epithelium (dashed white line). The non-contamination confirmation of the SHG signal was established by the wavelength range, half of the excitation, of the signal, by using the avalanche photodiodes (APD) array of the LSM-780 Zeiss scan head CCD. Besides the difference found in the collagen extent, a visual examination of the SHG images of **Figure 6** reveals that the normal skin has thinner collagen fibers that weave in all directions round the hair follicles. The skin from OI patients exhibits changes in collagen fiber thickness when compared to the normal skin. Skin images from the more severe forms of OI result in thicker, broken, and wavy collagen fibers that are firmly packed following the same direction. Moreover, using fresh skin, one can identify a marked reduction in the density of the collagen fibers network in the 3D illustration of SHG images from severe OI patient's samples (**Figure 6B** and **6C**), when compared with 3D SHG images from normal skin fresh biopsies (**Figure 6A**). These skin images are just a basic representative example about how the SHG tool can be used for optical evaluation of OI.

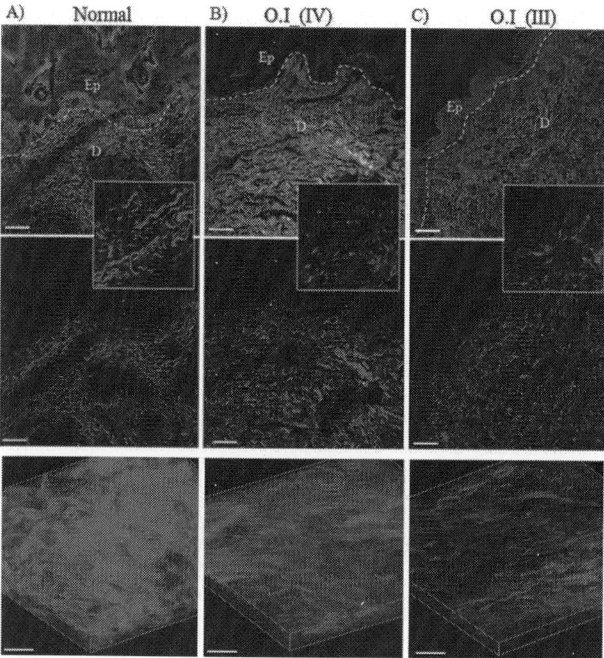

Figure 6. Representative cross-sectional images for TPEF (green) and SHG (red) analyses of normal (A), OI type IV (B), and OI type III (C) skin human tissues. Epidermis/dermis interface is signposted by white outline. Insets highlight visual differences of fiber collagen. Ep, epidermis; D, dermis; and representative 3D maximum projection (40 images at intervals of 1 μm) of SHG images from fresh skin biopsies. Scale bars = 20 μm. Reproduced figures with permission from Journal of Biomedical Optics, 2012 [19].

However, to offer a more accurate diagnostic method, it is necessary to develop reliable quantitative tools that allow discriminating between different OI types. The following section aims to demonstrate that the texture analysis (one of the analyses method presented below), which is the first step to provide SHG image quantitation tool, providing important information about collagen fiber organization.

4. Analysis methods used as diagnostic tools

As was previously mentioned, different processing methods can be used to obtain the relationship between signals of epithelial cells and the collagen matrix obtained with NLO microscopy techniques [4, 42]. Some of the methods that are currently used and others potentially implementable with free software, such as ImageJ (NIH, Bethesda, Maryland, USA), are described below.

4.1. Ratio between collagen and elastic tissue (SAAID)

The second harmonic to autofluorescence aging index of dermis (SAAID) value is a measure of the ratio between collagen and elastic fiber network [4, 43]. As the stroma is composed primarily of collagen and elastic fibers allows the use of nonlinear optical signals to discriminate between altered connective tissue regions near tumor area [44–46]. Specifically, collagen fibers are strong second harmonic signal generators, whereas elastic fibers are only autofluorescent emitters. This parameter can be applied when TPEF and SHG microscopy are simultaneously used [47, 48]. The SAAID index is defined as SAAID = $(I_{SGH} - I_{TPEF})/(I_{SHG} + I_{TPEF})$, where I equals the intensity of each signal, SHG/TPEF are above preselected threshold intensities [43]. For example, to obtain this index, we have used the collagen-elastic tissue ratio map in the whole image of ovarian tissue (**Figure 7A, B**). The whole stroma region was selected as one ROI for each image. It has been demonstrated that collagen content was increased within the tumor stroma. The quantification of these observations is showed by the SAAID bar graph (**Figure 7C**). The corresponding SAAID of adenocarcinoma type exhibits statistically significant ($p<0.05$, t-test) higher values (-0.38 ± 0.03) compared to normal stroma (-0.63 ± 0.06) due to the high SHG (collagen) signal and low TPEF signal in this region. To demonstrate the utility of this index, it was applied to images of ovarian cancer showed in **Figure 4** and represented by bar graph in the **Figure 7D**. The corresponding SAAID of both adenocarcinoma types presented statistically significant ($p<0.05$, t-test) higher values compared to normal stroma due to both the high SHG (collagen) and low TPEF signals in this region [34].

4.2. Tumor-associated collagen signatures (TACS)

This parameter is frequently used to determine the collagen fiber orientation at the tumor stroma boundary [4]. At present, there are three well-characterized TACS. They are reproducible during defined stages of tumor progression: TACS-1 (presence of dense collagen localized around small tumors during early disease), TACS-2 (collagen fibers arranged parallel to the tumor boundary—around 0°), and TACS-3 (collagen fibers disposed perpendicularly to the tumor boundary—around 90°, when the disease becomes invasive) [49]. The collagen-fiber angle calculation (relative to the tumor boundary) is required to know the epithelial zone having abnormal appearance. After this manual selection, fiber angle could be measured using the angle tool option from ImageJ toolbar. This tool measures the angle demarcated by three points. The first is an arbitrary point-guide along the fibril; the second one is the fibril extreme, closer to edge of the tumor; and the third one is any point that connects to the first draws with a path parallel to the epithelium [20]. Using this parameter, for example, to analyze collagen transformation in ovarian cancer, the fiber angle relative to the epithelium has been quantified. SHG images have been used along with collagen orientation, instead of the SHG signal (**Figure 7E, F**). The TACS-2, straightened (taut) collagen fibers, stretched around the epithelium (**Figure 7E**), and TACS-3, identifying radially aligned collagen fibers, that may provide the scaffolding of local invasion (**Figure 7F**), has been found. In normal ovary tissue, collagen fibers were mainly distributed around 0° (see white arrows). Approximately 75% of these fibers are parallel to the epithelium (angle ≤ 20°). In contrast, serous adenocarcinoma exhibits incipient regions of local invasion (TACS-3) with a set of realigned fibers, most of which are disposed around 90° (see white arrows) with respect to epithelium (**Figure 7G**) [4, 20].

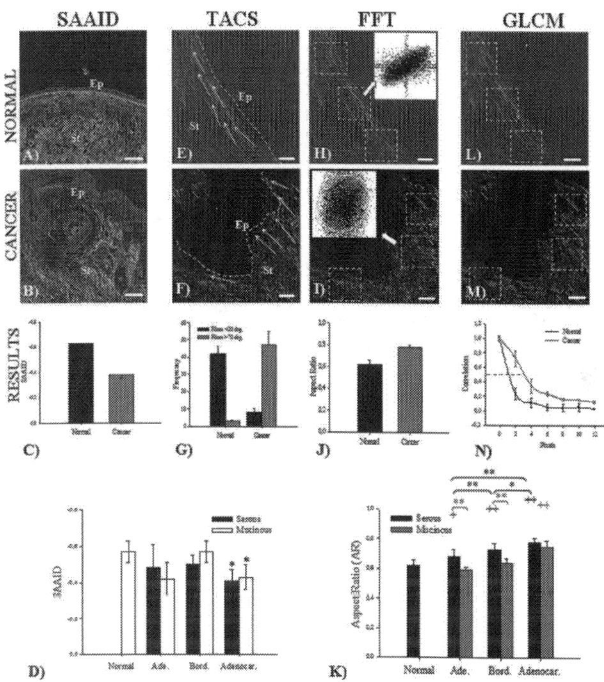

Figure 7. Depicting several applications of different methods to analyze NLO signals. The panel shows representative TPEF (green) and SHG (red) images of (A, E, H, L) normal and (B, F, I, M) cancer ovary. From the TPEF+SHG combination (first column), it is possible to calculate (C) the SAAID ratio. From the SGH image (remaining columns) and using regions near the epithelial/stromal interface (yellow line), it is possible to calculate: (G) TACS (measuring the collagen fiber angle relative to the epithelium); (J) FFT transforms (and fit to ellipse to estimate the anisotropy); and (N) GLCM (correlation value). (D) Bar graphs represent SAAID index quantitative analysis of ovarian tissues. Asterisks indicate a significant increase as compared to the nontumor tissues ($p<0.05$, t-test). (K) Bar graphs represent anisotropy (aspect ratio) quantitative analysis of ovarian tissues. Comparisons with normal tissues are indicated with +. +,* indicates a statistically significant ($p<0.05$) difference and ++, ** indicates a statistically very significant ($p<0.01$) difference following ANOVA. Ep, epithelium; St, stroma; white arrows, collagen fibers; white squares, regions of interest (ROI). Reproduced figures from Adur et al. [4] (open access).

4.3. Fast Fourier transform (FFT) analysis

The FFT has proven to be a good method to assign the degree of image organization [42, 50]. Thereby, the FFT of a set of aligned fibers will have higher values along the orthogonal path to the direction track of the fibers, and its intensity plot seems to have an ellipsoidal shape. If the fibers are perfectly aligned, the ellipse will collapse into a line. For randomly oriented fibers, the intensity plot of the corresponding FFT image looks like a circle. Therefore, the anisotropy of the image can be calculated by performing an elliptic fit on the thresholded FFT images, and then calculating the ratio between its short and long axes, i.e., its aspect ratio (AR) [51]. One sample will be more anisotropic as the AR goes to zero, whereas it will be more isotropic when the AR is closer to one.

To perform anisotropy calculations, squared ROI in the SHG images are usually selected, with the only requirement that they must be placed upon the collagen network around the epithelium, since this is the region responsible for the stroma invasion. The square ROI is required by the FFT procedure of ImageJ, based on an implementation of the 2D Fast Hartley Transform [52]. The FFT can be carried out with the homonymous command of the ImageJ menu [4]. The anisotropy on the ovary SHG image of stromal region has been estimated using this methodology (**Figure 7H, I**). Three ROIs of 150×150 pixels side squared have been used to ensure that the collagen network in the vicinity of the epithelium is registered. **Figure 7J** shows the AR value averaged on all the examined samples. In serous-type tumors, it was found that the AR index turned out to be significantly increased ($p<0.05$, t-test) from normal (0.62 ± 0.04) to adenocarcinoma samples (0.78 ± 0.03).

Using this method, serous and mucinous ovarian cancer samples were analyzed (**Figure 4**). In serous-type tumors, AR increased progressively and significantly ($p<0.05$, ANOVA) from normal to adenocarcinoma, and in mucinous-type tumors (by contrast), AR showed statistically very significant differences only for adenocarcinomas ($p<0.01$, ANOVA) (**Figure 7K**). These results confirm the fact that normal ovaries are more organized tissues as compared to adenocarcinoma. By using this tool, it was possible to discriminate between serous adenoma from mucinous adenoma and serous borderline from mucinous borderline subtypes [48]. Unlike ovarian serous tumors, which are relatively homogeneous in their cellular composition and differentiation degree, mucinous tumors are frequently heterogeneous, with mixtures of benign, borderline, and malignant elements often found within the same neoplasm. The heterogeneity in these mucinous tumors suggests that malignant transformation is sequential and slow, progressing from cystoadenoma to borderline tumor and, finally, to invasive carcinoma [46]. This slow behavior is probably reflected in a more organized stroma [53].

4.4. Gray level co-occurrence matrix (GLCM) analysis

The GLCM analysis method allows the classification of different tissues based on the evaluation of geometrical collagen arrangement [4, 42]. It provides information on the spatial relationships between pixel brightness values in a given image. The GLCM is constructed by counting the number of occurrences of a gray level adjacent to another gray level, at a specified pixel distance "d" and dividing each counting by the total counting number to obtain a probability. The result is a matrix with rows and columns representing gray levels and elements containing the probability $P_d[i,j]$ of the gray-level co-occurrence between pixels. The matrix is usually averaged in opposite and different orientations (0–180°, 45–225°, 90–270°, and 135–315°) unless one-dimensional feature dominates overall possible ones, in which case, the 0–180 average is sufficient. A detailed explanation on how this matrix is created from the original image can be found in [54].

The GLCM analysis can be carried out by different methods; they are commonly classified as contrast methods, orderliness methods, and statistical methods. Contrast methods can be used in order to give quantitative information on the intensity fluctuations in the image [4, 34, 42]. Among the contrast methods, homogeneity is the weighted sum of the GLCM pixel values. The weights are values nonlinearly decreasing according to the distance from the GLCM matrix diagonal increases. The homogeneity parameter gives information about the similitude of two neighboring pixel values, against all the other pair of pixels of the image. Orderliness methods are particularly suitable to images with fibrous structures, such as SHG images of collagen. This approach can be used in order to give a quantitative measurement on the mutual orientation of collagen fiber bundles. The energy parameter, for example, is the root-squared sum of the GLCM pixel values. Considering that it gives higher weight to the hot spots of the GLCM matrix, that parameter can be considered as a measure of the sample orderliness. Statistical methods are based on the statistical analysis of pixel value dependence and can be used for determining repetition of a certain pattern within a tissular image. Among them, the correlation method probably represents the most powerful approach to be applied to SHG images of collagen. Mathematically the parameters are defined in **Table 3**.

Parameters	Interpretation	Mathematical expression
Correlation	Linear dependency of gray levels on those of neighboring pixels	$-\sum_{i,j=0}^{N-1} P_{i,j}\left[\dfrac{(i-\mu)(j-\mu)}{\sigma^2}\right]$
Contrast	Representation of pixels entirely similar to their neighbor	$\sum_{i,j=0}^{N-1} (i-j)^2 P_{i,j}$
Energy	Degree of image's texture directions according to the perception of human eyes	$\sum_{i,j=0}^{N-1} P_{i,j}^2$
Homogeneity	Measure of the amount of local uniformity present in the image	$\sum_{i,j=0}^{N-1} \dfrac{P_{i,j}^2}{1+(i-j)^2}$

Table 3. GLCM parameters and its mathematical expressions.

The texture analyses can be performed with Image-J GLCM Texture plugin, which was described by Walker and collaborators [55]. Also other parameters such Contrast, Entropy, Inertia, and Variance could be estimated from the GLCM approach [42]. Here a characterization of tissues by estimating the typical dimensions in which collagen maintains its organization is showed. For example, the correlation of the image itself with a pixel separation translated from 1 to 12 or 18 pixels (**Figure 7L–N**) was used. The feature was averaged at angles θ = 0, 90, 180, and 270 degrees to take into account the fact that these images do not have a specific spatial orientation. The distance where correlation falls to 1/2 expressed in microns was measured [4]. To perform the calculations, three ROIs (100×100 pixel side squared) in the SHG images near the epithelium were selected. Correlation and entropy were measured using GLCM-Texture plugin from ImageJ, which was previously described by Walker and collaborators [55]. **Figure 7N** shows that the correlation of normal fibrils fall off sharply with distance, indicating distinct, linear fibrils, whereas correlation for the fibrils in adenocarcinomas remained elevated for larger distances, implying less-defined fibrillar structure. Consistent with qualitative appearances, the correlation was found to remain higher in malignant tissues with the Corr50, the pixel distance where the correlation dropped below 50% of the initial value, significantly greater in adenocarcinomas (3.4 pixels) compared with normal ovarian (1.7 pixels) (**Figure 7N**; $p<0.05$, t-test). In the same ROI, the entropy values were 6.26 ± 0.31 and 7.40 ± 0.58 from normal and adenocarcinoma, respectively. This means that normal tissues exhibit a lesser complexity or higher organization than malignant ones [34].

Following the impact of GLCM analysis, the method in the evaluation of patients with OI is showed. Skin samples from healthy and from patients with OI were obtained from the Laboratory of Pediatric Endocrinology, Campinas, SP, Brazil. Biopsies were analyzed and classified as normal (4 cases) or OI (5 cases). OI patients were classified according to clinical observations in mild OI (Type I—1 case), moderately affected and severe OI (Type III or Type IV—4 cases). Normal samples were obtained from eyelid plastic surgery discarded tissue, and patient biopsies were obtained from growing skin. Fresh skin samples in Phosphate Buffer Solution (PBS) were analyzed by 3D SHG representations within 6 hours of the excision. From the mounted SHG pictures, images located in the dermis were taken. Nonsymmetric GLCMs were computed using 256 gray levels. Because collagen fiber orientation changed from sample to sample, four orientations average were used. This scoring method was competent to satisfactorily discriminate the different OI patients according to their clinical severity [56]. Using fresh biopsies, one could detect a marked density decrease of the collagen fibers network in the 3D representations of SHG images from severe OI patient's samples (**Figure 8B** (Type III) and **Figure 8C** (Type IV)), when compared with the 3D representation of SHG images taken from normal skin fresh biopsies (**Figure 8A**). Furthermore, energy value of GLCM texture analysis could not only discriminate type I and type III OI samples from normal skin (**Figure 8D**), but it could also differentiate (with statistical significance) between patients with varying degrees of OI, including Type IV OI (**Figure 8E**). It is well known that dermis collagen fibers have diameter ranges around 0.5 to 3 µm. Therefore, it is expected that GLCM analysis would show a repeating structure with distances of about 1.5–8 pixels corresponding to the 0.5–3 µm range. Considering that OI patients exhibit thicker fibers than normal skin samples, GLCM correlation signals could be to drop on a longer scale. The values of decay length are obtained by fitting the correlation data with a double exponential decay function. The decay length values using the Corr50 (the pixel distance where the correlation dropped below 50% of the initial value) obtained are between 1.3 and 2.5 µm (3.8–7 pixels), confirming that patient D had thicker collagen fiber ≈ 2.5 µm (7 pixels). Using this pixel distance as comparison, the energy parameter shows values of 0.15 ± 0.02 (normal), 0.13 ± 0.04 (Patient (Pat.) A), 0.20 ± 0.03 (Pat. B), 0.24 ± 0.04 (Pat. C), 0.29 ± 0.03 (Pat. D), and 0.33 ± 0.04 (Pat. E), with significant differences ($p<0.05$) when normal skin was compared with OI patient's (B, C, D, and E) skin. This means that this texture parameter clearly allows the identification of each patient pattern. Interestingly, by using this method, it was possible to discriminate one case of type IV patient, exhibiting a more severe phenotype (Patient D) than the others. SHG images of these patient skins display a more compacted collagen pattern (thicker collagen fiber), intermediate between type III, and the two remaining type IV patients. The preliminary results allow auguring that these nonlinear microscopy techniques in association with specific scoring method (energy-GLCM) will be an excellent diagnostic tool to clinically distinguish different types of OI in human skin [56].

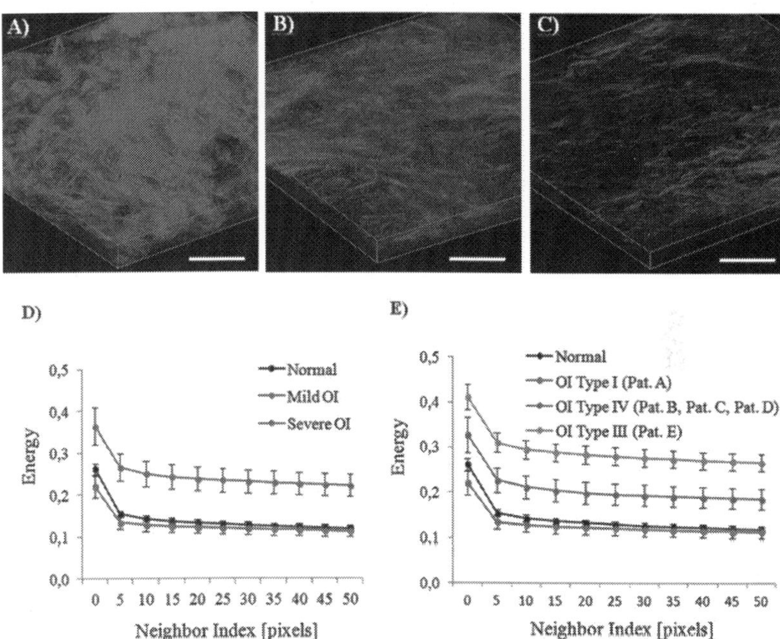

Figure 8. Representative 3D maximum projection (40 images at 1μm interval) of SHG images from fresh skin biopsies, (A) normal skin, (B) OI type III (Patient E), and (C) OI type IV (Patient D). Texture analysis (D, E) using GLCM. Energy values were calculated in dermis tissues versus distance pixels; ranging from 1 to 50 pixels (0.35 μm–17.30 μm) in 0, 45, 90, and 135° directions of image (d): n = 12 normal, n = 3 mild OI, and n = 12 severe OI; and (e) n = 12 normal, n = 3 OI type I, n = 9 OI type IV, n = 3 OI type III. Pat: patients. Reproduced figures from Adur et al. [56] (open access).

5. Conclusions

This chapter summarizes several nonlinear microscopy techniques that can be combined and the images acquired analyzed by a set of quantitative tools. This may allow the implementation of new diagnostic procedures for early detection of various diseases. The integration of a set of microscopy techniques is one of the evolving areas in bioimaging that promises to have a strong impact on the understanding and early detection of diverse pathologies. As has been described and exemplified in the sections of this chapter, the advantages of the techniques are numerous, namely high depth penetration (due to Near Infrared (NIR) laser), intrinsic 3D sectioning and 3D resolution (due to the spatial confinement of the signal to the laser focus), multiple nonlinear processes summed to the possibility to detect several endogenous molecular markers, and low phototoxicity. Additionally, these techniques allow the investigation of living processes in the native environment without major perturbations. All these advantages allow us to postulate that in the near future the NLO techniques together with nonlinear signal processing methods can be very useful in the field of medical diagnosis. In combination with sophisticated animal models and computer-assisted data analysis, NLO microscopy techniques and image processing methods are opening new doors to the study of tumor biology, facilitating the development of new strategies for early tumor diagnosis and other diseases.

As is shown in this chapter, by integrating the strengths of each NLO imaging modality, different structures, and their interactions in a complex biological system can be simultaneously visualized. Additionally, the possibility of obtaining images at high speed and with chemical specificity makes NLO microscopy a powerful tool to evaluate the dynamic behavior of *in vivo* disease progression [57]. CARS microscopy should allow longitudinal studies of lipid metabolism in the same living model organisms over time. Other auspicious application is the label-free imaging of organogenesis and drug delivery. Also, CARS can provide structural information and it has been useful in analyzing molecular orientations in myelin [58], single lipid bilayer [59], cellulose fiber [60], and crystal of clean fourth-order symmetry [61].

In translational research, NLO microscopy has demonstrated the ability of diagnosing diseases of live organisms [62]. Recently, it has been found that changes occurring in collagen deposit and arrangement, in early tumor development and during their progression, can be used as predictable tools of the disease status. The ovary examples in this work demonstrate that AR and correlation analysis have the ability to predict the disease degree in human patients. Therefore, if more experiments are successful, SHG may eventually provide a more rapid, real-time substitute for traditional histopathological processing and analyses. For this disease, mortality rates are elevated because an efficient screening test does not exist presently. Approximately, 15% of ovarian cancers are found before metastasis has occurred. If ovarian cancer is found and treated before this process is triggered, the 5-year survival rate will be around 94%. Thus, an early diagnostic test to detect premalignant changes would save many lives. In this sense, the unique attributes of NLO microscopy described here render these methods as a promising imaging modality for disease diagnostics in the clinic. Also, the medical utility of these optical methods could be improved by the continuous development and refinement of methods to obtain objective, quantitative information. These will be in the form of analysis algorithms such as Helmholtz analysis, wavelet analysis, and with numerical parameters relating to image frequency content and second-order gray-level statistics. Further, a classification scheme could be developed by using a support vector machine.

The effort to develop new diagnosing methods that could better identify early lesions and consequently lead to an early diagnosis is a challenge and a stimulus for research in this area. The outcomes of different works indicate that the combination of diverse image analysis approaches summarized here represent a combination of powerful tools to investigate epithelial cells transformation, collagen organization, and extracellular matrix remodeling in epithelial tumors and *osteogenesis imperfecta* in skin. About OI, it was demonstrated in this chapter that nonlinear microscopy techniques, in combination with image analysis approaches, represent a powerful tool to investigate the collagen organization in skin dermis in patients with OI and having the potentiality to distinguish the different types of OI. The procedure used here requires a skin biopsy, which is almost painless as compared to the bone biopsy commonly used in conventional methods. The data presented here are complementary of existing clinical diagnostic approaches and can be used as a procedure to confirm the disease and evaluate its severity and treatment efficacy.

In cancer diagnosis, there is a growing need for the development of a multimodal imaging-based diagnostic tool to objectively evaluate morphological features with subcellular resolution and molecular compositions that are closely associated with tumor malignancy. With this perspective, NLO microscopy has proven to be useful in cancer research. These techniques have recently emerged as a valuable tool for high-resolution, nondestructive, chronic imaging of living tumors. Moreover, multimodal microscopy can provide a powerful tool for investigating the dynamics of structure-function relationships both at the subcellular and molecular levels.

Today, the application of multimodal nonlinear imaging is recognized in basic research in the biological and biomedical sciences; however, regular applications in clinics are still rare, mainly because of their high cost. Multimodal platforms are still complex and require specialized personnel for its operation. So, mainly technological progresses are required for miniaturization, enhancement of the ease of control, automated data processing, and extraction of significant information. To achieve this goal, the modification of typical clinical endoscopes for in vivo multimodal nonlinear imaging is necessary. The development of nonlinear optical endoscopy, which allows imaging under conditions in which a conventional nonlinear optical microscope cannot be used, will be the primary goal to extend applications of nonlinear optical microscopy toward clinical ones. There are several key challenges involved in the pursuit of in vivo nonlinear optical endoscopy. A few of them are the necessity of obtaining efficient ultrashort pulse laser liberations into a remote place, the need to enhance scan rates for monitoring biological processes, and the miniaturization of the laser-scanning mechanisms to the millimeter scale. Finally, the design of a nonlinear optical endoscope based on micro-optics with great flexibility, and compact enough to be incorporated into endoscopes, will become an evolution of these microscopy platforms. With the continuous advancement in this endoscopic techniques and new laser sources, we have reason to believe that these particularly promising techniques in conjunction with efficient image analysis algorithm will open up many new possibilities for the diagnosis and treatment of different diseases in the near future.

Acknowledgements

HFC acknowledges funding from FAPESP (Grant no. 2009/16150-6). The authors are grateful to FAPESP (Grant no. 2011/51591-3), CEPOF (Optics and Photonics Research Center, FAPESP), INFABIC (National Institute of Photonics Applied to Cell Biology, FAPESP, and CNPq), and PICTO UNER-INTA-CAFEGS (Grant no. 2009-209).

Author details

Javier Adur[1,2,3*], Hernandes F. Carvalho[3,4], Carlos L. Cesar[3,5,6] and Víctor H. Casco[1,2]

*Address all correspondence to: jadur@bioingenieria.edu.ar

1 CITER—Centro de Investigación y Transferencia de Entre Ríos, UNER-CONICET, FI-UNER, Argentina

2 Microscopy Laboratory Applied to Molecular and Cellular Studies, School of Bioengineering, National University of Entre Ríos, Argentina

3 INFABiC—National Institute of Science and Technology on Photonics Applied to Cell Biology, Campinas, Brazil

4 Department of Structural and Functional Biology, Biology Institute, State University of Campinas, Brazil

5 Biophotonic Group, Optics and Photonics Research Center (CEPOF), Institute of Physics "Gleb Wataghin", State University of Campinas, Brazil

6 Department of Physics of Federal University of Ceara (UFC), Brazil

References

[1] Provenzano PP, Eliceiri KW, Keely PJ. Multiphoton microscopy and fluorescence lifetime imaging microscopy (FLIM) to monitor metastasis and the tumor microenvironment. Clin. Exp. Metastasis. 2009; 26:357–370. doi: 10.1007/s10585-008-9204-0

[2] Masters BR, So PTC. The genesis of nonlinear microscopies and their impact on modern developments. In: Master and So, Editors. Handbook of Biomedical Nonlinear Optical Microscopy. New York: Oxford University Press, 2008. p. 5–28.

[3] Meyer T, Schmitt M, Dietzek B, Popp J. Accumulating advantages, reducing limitations: multimodal nonlinear imaging in biomedical sciences – the synergy of multiple contrast mechanisms. J. Biophotonics. 2013; 6:887–904. doi: 10.1002/jbio.201300176

[4] Adur J, Carvalho HF, Cesar CL, Casco VH. Nonlinear optical microscopy signal processing strategies in cancer. Cancer Inf. 2014; 13:67–76. doi: 10.4137/CIN.S12419

[5] Shen YR. The principles of nonlineal optics. New York, John Wiley, 1984.

[6] Gu M, Bird D, Day D, Fu L, Morrish D. eds. Nonlinear optical microscopy. In: Femtosecond Biophotonics Core Technology and Applications. New York: Cambridge University Press, 2010. p. 9–34.

[7] Cheng PC, Sun CK. Nonlinear (harmonic generation) optical microscopy. In: Pawley JB, Editor. Handbook of Biological Confocal Microscopy, 3rd edn. 2006. Springer. New York. p. 703–721.

[8] Gauderon R, Lukins PB, Sheppard CJ. Three-dimensional second-harmonic generation imaging with femtosecond laser pulses. Opt. Lett. 1998; 23:1209–1211.

[9] Chu SW, Chen IH, Liu TM, Chen PC, Sun CK, Lin BL. Multimodal nonlinear spectral microscopy based on a femtosecond Cr:forsterite laser. Opt. Lett. 2001; 26:1909–1911.

[10] Boyd R. Nonlinear Optics, 1st edn. New York, Academic Press, 1992.

[11] Debarre D, Beaurepaire E. Quantitative characterization of biological liquids for third harmonic generation microscopy. Biophys. J. 2007; 92:603–612.

[12] Williams RM, Zipfel WR, Webb WW. Interpreting second-harmonic generation images of collagen I fibrils. Biophys. J. 2005; 88:1377–1386.

[13] Campagnola PJ, Millard AC, Terasaki M, Hoppe PE, Malone CJ, Mohler WA. Three-dimensional high-resolution second-harmonic generation imaging of endogenous structural proteins in biological tissues. Biophys. J. 2002; 82:493–508.

[14] Chu SW, Chen SY, Chern GW, Tsai TH, Chen YC, Lin BL, Sun CK. Studies of chi(2)/chi(3) tensors in submicron-scaled bio-tissues by polarization harmonics optical microscopy. Biophys. J. 2004; 86:3914–3922.

[15] Müller M, Squier J, Wilson KR, Brakenhoff GJ. 3D microscopy of transparent objects using third-harmonic generation. J. Microsc. 1998; 3:266–274.

[16] Moreaux L, Sandre O, Blanchard-Desce M, Mertz J. Membrane imaging by simultaneous second-harmonic generation and two-photon microscopy. Opt. Lett. 2000; 25:320–322.

[17] Sun CK, Chu SW, Chen SY, Tsai TH, Liu TM, Lin CY, Tsai HJ. Higher harmonic generation microscopy for developmental biology. J. Struct. Biol. 2004; 147:19–30.

[18] Adur J, Pelegati VB, Costa LF, Pietro L, de Thomaz AA, Almeida DB, Bottcher-Luiz F, Andrade LA, Cesar CL. Recognition of serous ovarian tumors in human samples by multimodal nonlinear optical microscopy. J. Biomed. Opt. 2011; 16:096017. doi: 10.1117/1.3626575

[19] Adur J, Pelegati VB, de Thomaz AA, D'Souza-Li L, Assunção Mdo C, Bottcher-Luiz F, Andrade LA, Cesar CL. Quantitative changes in human epithelial cancers and osteogenesis imperfecta disease detected using nonlinear multicontrast microscopy. J. Biomed. Opt. 2012; 17:081407-1. doi: 10.1117/1.JBO.17.8.081407

[20] Adur J, Pelegati VB, de Thomaz AA, Baratti MO, Andrade LA, Carvalho HF, Bottcher-Luiz F, Cesar CL. Second harmonic generation microscopy as a powerful diagnostic imaging modality for human ovarian cancer. J. Biophotonics. 2014; 7:37–48. doi: 10.1002/jbio.201200108

[21] Duncan MD, Reintjes J, Manuccia TJ. Scanning coherent anti-Stokes Raman microscope. Opt. Lett. 1982; 7:350–352.

[22] Pelegati VB, Adur J, De Thomaz AA, Almeida DB, Baratti MO, Andrade LA, Bottcher-Luiz F, Cesar CL. Harmonic optical microscopy and fluorescence lifetime imaging platform for multimodal imaging. Microsc. Res. Tech. 2012; 75. doi: 10.1002

[23] Hu W, Zhao G, Wang C, Zhang J, Fu L. Nonlinear optical microscopy for histology of fresh normal and cancerous pancreatic tissues. PLoS One. 2012; 7:e37962. doi: 10.1371

[24] Hompland T, Erikson A, Lindgren M, Lindmo T, de Lange Davies C. Second-harmonic generation in collagen as a potential cancer diagnostic parameter. J. Biomed. Opt. 2008; 13. doi: 10.1117. 054050, doi: 10.1117/1.2983664

[25] Sivaguru M, Durgam S, Ambekar R, Luedtke D, Fried G, Stewart A, Toussaint KC Jr. Quantitative analysis of collagen fiber organization in injured tendons using Fourier transform-second harmonic generation imaging. Opt. Express. 2010; 18:24983–93. doi: 10.1364/OE.18.024983

[26] Mouras R, Rischitor G, Downes A, Salter D, Elfick A. Nonlinear optical microscopy for drug delivery monitoring and cancer tissue imaging. J. Raman Spectrosc. 2010; 41:848–852.

[27] Bianchini P, Diaspro A. Three-dimensional (3D) backward and forward second harmonic generation (SHG) microscopy of biological tissues. J. Biophotonics. 2008; 1:443–450. doi: 10.1002/jbio.200810060

[28] Adur J, Carvalho HF, Cesar CL, Casco VH. Nonlinear imaging microscopy: Methodological setup and applications for epithelial cancers diagnosis. In: Fred Wilkins Ed, Nova Science Publishers. Nonlinear optics: fundamentals, applications and technological advances. New York, Nova Publishers, 2014. p. 97–136.

[29] Cheng JX, Xie XS. Coherent anti-Stokes Raman scattering microscopy: instrumentation, theory, and applications. J. Phys. Chem. B. 2004; 108:827–840.

[30] Volkmer A, Book LD, Xie XS. Time-resolved coherent antiStokes Raman scattering microscopy: imaging based on Raman free induction decay. Appl. Phys. Lett. 2002; 80: 1505–1507.

[31] American Cancer Society. Cancer Facts & Figures 2009. Atlanta, American Cancer Society, 2009.

[32] Tsuboi M, Ueda T, Ushizawa K, Ezaki Y, Overman SA, Thomas GJ. Raman tensors for the tryptophan side chain in proteins determined by polarized Raman microspectroscopy of oriented N-acetyl-L-tryptophan crystals. J. Mol. Struct. 1996; 379:43–50.

[33] Tearney GJ, Brezinski ME, Bouma BE, Boppart SA, Pitris C, Southern JF, Fujimoto JG. In vivo endoscopic optical biopsy with optical coherence tomography. Science. 1997; 276:2037–2039.

[34] Adur J, Pelegati VB, de Thomaz AA, Baratti MO, Almeida DB, Andrade LA, Bottcher-Luiz F, Carvalho HF, Cesar CL. Optical biomarkers of serous and mucinous human ovarian tumor assessed with nonlinear optics microscopies. PLoS One. 2012; 7:e47007. doi:10.1371/journal.pone.0047007

[35] Zhuo S, Zhu X, Wu G, Chen J, Xie S. Quantitative biomarkers of colonic dysplasia based on intrinsic second-harmonic generation signal. J. Biomed. Opt. 2011; 16:120501.

[36] Birk JW, Tadros M, Moezardalan K, Nadyarnykh O, Forouhar F, Anderson J, Campagnola P. Second harmonic generation imaging distinguishes both high-grade dysplasia and cancer from normal colonic mucosa. Dig. Dis. Sci. 2014; 59:1529–1534. doi: 10.1007/s10620-014-3121-7

[37] Adur J, Bianchi M, Pelegati VB, Viale S, Izaguirre MF, Carvalho HF, Cesar CL, Casco VH. Colon adenocarcinoma diagnosis in human samples by multicontrast nonlinear optical microscopy of hematoxylin and eosin stained histological sections. J. Cancer Ther. 2014; 5:1259–1269. http://dx.doi.org/10.4236/jct.2014.513127

[38] Kuurila K, Kaitila I, Johansson R, Grenman R. Hearing loss in Finnish adults with osteogenesis imperfecta: a nationwide survey. Ann. Otol. Rhinol. Laryngol. 2002; 111:939–1946.

[39] Stoll C, Dott B, Roth MP, Alembik Y. Birth prevalence rates of skeletal dysplasias. Clin. Genet. 1989; 35:88–92.

[40] Dalgleish R. The human type I collagen mutation database. Nucleic Acids Res. 1997; 25:181–187.

[41] Dalgleish R. The human collagen mutation database 1998. Nucleic Acids Res. 1998; 26:253–255.

[42] Tilbury K, Campagnola PJ. Applications of second-harmonic generation imaging microscopy in ovarian and breast cancer. Perspect. Med. Chem. 2015; 7:21–32. doi: 10.4137/PMC.S13214

[43] Cicchi R, Sacconi L, Pavone F. Nonlinear imaging of tissues. In: Tuchin VV, Editor. Handbook of photonics for biomedical science. New York, CRC Press, 2010. p. 509–545.

[44] Cicchi R, Massi D, Sestini S, Carli P, De Giorgi V, Lotti T, Pavone FS. Multidimensional non-linear laser imaging of basal cell carcinoma. Opt. Express. 2007; 15:10135–10148.

[45] Cicchi R, Sestini S, De Giorgi V, Massi D, Lotti T, Pavone FS. Nonlinear laser imaging of skin lesions. J. Biophotonics. 2008; 1:62–73. doi: 10.1002/jbio.200710003

[46] Lin SJ, Jee SH, Kuo CJ, Wu RJ, Lin WC, Chen JS, Liao YH, Hsu CJ, Tsai TF, Chen YF, Dong CY. Discrimination of basal cell carcinoma from normal dermal stroma by quantitative multiphoton imaging. Opt. Lett. 2006; 31:2756–2758.

[47] Koehler MJ, König K, Elsner P, Bückle R, Kaatz M. In vivo assessment of human skin aging by multiphoton laser scanning tomography. Opt. Lett. 2006; 31:2879–2881.

[48] Lin SJ, Wu R Jr, Tan HY, Lo W, Lin WC, Young TH, Hsu CJ, Chen JS, Jee SH, Dong CY. Evaluating cutaneous photoaging by use of multiphoton fluorescence and second-harmonic generation microscopy. Opt. Lett. 2005; 30:2275–2277.

[49] Provenzano PP, Inman DR, Eliceiri KW, Knittel JG, Yan L, Rueden CT, White JG, Keely PJ. Collagen density promotes mammary tumor initiation and progression. BMC Med. 2008; 6:11. doi: 10.1186/1741-7015-6-11

[50] Sivaguru M, Durgam S, Ambekar R, Luedtke D, Fried G, Stewart A, Toussaint KC Jr. Quantitative analysis of collagen fiber organization in injured tendons using Fourier transform-second harmonic generation imaging. Opt Express. 2010; 18:24983–24993. doi: 10.1364/OE.18.024983

[51] Matteini P, Ratto F, Rossi F, Cicchi R, Stringari C, Kapsokalyvas D, Pavone FS, Pini R. Photothermally-induced disordered patterns of corneal collagen revealed by SHG imaging. Opt Express. 2009; 17:4868–4878.

[52] Arlo RA. Thesis for the degree of Master of Science by Thayer School of Engineering [master's thesis]. New Hampshire, Dartmouth College Hanover, 1990.

[53] Zhuo S, Zhu X, Wu G, Chen J, Xie S. Label-free monitoring of colonic cancer progression using multiphoton microscopy. Biomed. Opt. Express. 2011; 2:615–619. doi: 10.1117/1.3659715

[54] English RS, Shenefelt PD. Keloids and hypertrophic scars. Dermatol Surg. 1999; 25:631–638.

[55] Walker RF, Jackway PT, Longstaff ID. Genetic algorithm optimisation of adaptive multi-scale in GLCM features. Int. J. Pattern Recogn. 2003; 17:17–39.

[56] Adur J, DSouza-Li L, Pedroni MV, Steiner CE, Pelegati VB, de Thomaz AA, Carvalho HF, Cesar CL. The severity of Osteogenesis imperfecta and type I collagen pattern in human skin as determined by nonlinear microscopy: proof of principle of a diagnostic method. PLoS One. 2013; 8:e69186. doi: 10.1371/journal.pone

[57] Huff TB, Cheng JX. In vivo coherent anti-Stokes Raman scattering imaging of sciatic nerve tissue. J. Microsc. 2007; 225:175–182.

[58] Kennedy AP, Sutcliffe J, Cheng JX. Molecular composition and orientation of myelin figures characterized by coherent anti-Stokes Raman scattering microscopy. Langmuir. 2005; 21:6478–6486.

[59] Potma EO, Xie XS. Detection of single lipid bilayers with coherent anti-Stokes Raman scattering (CARS) microscopy. J. Raman Spectrosc. 2003; 34:642–650.

[60] Zimmerley M, Younger R, Valenton T, Oertel DC, Ward JL, Potma EO. Molecular orientation in dry and hydrated cellulose fibers: a coherent anti-Stokes Raman scattering microscopy study. J. Phys. Chem. B. 2010; 114:10200–10208. doi: 10.1021/jp103216j

[61] Munhoz F, Rigneault H, Brasselet S. High order symmetry structural properties of vibrational resonances using multiple-field polarization coherent anti-Stokes Raman spectroscopy microscopy. Phys. Rev. Lett. 2010; 105:123903.

[62] Wang HW, Langohr IM, Sturek M, Cheng JX. Imaging and quantitative analysis of atherosclerotic lesions by CARS-based multimodal nonlinear optical microscopy. Arterioscler. Thromb. Vasc. Biol. 2009; 29:1342–1348. doi:10.1161/ATVBAHA. 109.189316

Chapter 7

Skin Wound Healing Revealed by Multimodal Optical Microscopies

Gitanjal Deka, Shi-Wei Chu and Fu-Jen Kao

Additional information is available at the end of the chapter

http://dx.doi.org/10.5772/64088

Abstract

Skin is the largest organ of our body serving as the first line defense against pathogens and toxicity. The skin can heal itself if any damage in it occur. Wounds, if not taken care properly, can become chronic and can even cause death. In the field of cosmetics and plastic reconstructive surgery, wounds, are major cause of trauma and costs, which demand proper diagnosis that can help in appropriate treatment. In conventional medicine, wound diagnosis mostly relied on the expertise and experience of physicians on the basis of non-quantitative observation of clinical signs, or invasive histochemical assessment of biopsies.

Methodologies based on light-matter interaction can provide quantitative, noninvasive and real time assessment of a tissue section based on imaging. Depending on the nature of interaction, various contrasts can be achieved by either absorption, scattering, or fluorescence, enabling observation of structural or molecular components of tissue sections. Development of multiphoton nonlinear optical detection techniques provide better resolution and tissue penetration depth with optical sectioning ability by using molecular and structural contrasts simultaneously. This chapter discuses and evaluates various optical approaches with special emphasis on multimodal multiphoton imaging of skin tissue components in correlation to physiological processes that affects the wound healing.

Keywords: skin, wound healing, optical microscopy, NADH, collagen, fluorescence, second harmonic generations

1. Introduction

Skin wounds and their treatment lead to major medical expenses in cosmetic surgery. Chronic wounds, including diabetic ulcers and pressure ulcers, present a significant health and economic concern for individual patients as well as the healthcare system. The diabetic ulcer is a major complication of diabetes mellitus, a disease which afflicts more than 350 million people worldwide. Among them foot ulceration is the leading cause for hospitalization [1]. Acute cutaneous burn wounds are also a serious health-related issue in the global community. Nearly 11 million flame burns occur annually and burn deaths rank in the top 15 causes of death for individuals 5–29 years of age. Around 60% of these burn patients heal with debilitating hypertrophic or keloid scarring [2]. Additionally, the cutaneous burn wound can left deep and large scar in comparison to normal wound after healing. All these scars are formed due to over deposition of collagen fibers to fill up the wound gaps, which are structurally and molecularly different to each other and need different approaches for their diagnosis and management. Improper management of wound may cause serious tissue disfigurement that may cause serious physical and psychological problems in patients.

Wound healing is a widely studied biomedical problem regarding tissue systems. To address the situation of wounds and their assessment of healing potential requires insight of what occurs to the components of skin at cellular and molecular level. Specifically, epithelial cell migration and collagen regeneration by fibroblast cells in the skin were found to have great effects on accelerated wound healing [3]. The entire wound healing process is a complex series of events that starts at the moment of injury and can continue for months even years in a few sequential yet overlapping phases. The characterization of wounds, their healing, and also the timeline of these sequential phases have major clinical significance in assessing severity, healing potential, and determining the correct treatment for all wound types. Traditionally, wound assessment has relied on visual evaluation by trained clinicians, with techniques based on laboratory biopsies providing objective assessment modalities [4, 5]. Currently, histological analysis of the tissue remains the gold standard for precise quantitative and qualitative assessment of wound depth and status. However, the biopsy process is invasive, can be painful, and in some cases can cause additional trauma and worsen scarring [4, 5]. Additionally, the processing required for histochemical observation usually distorts the structural integrity of the tissue.

In contrast to the abovementioned traditional wound assessment procedures, noninvasive imaging by optical means does not require destructive tissue sectioning; it preserves all layers of the skin. By collecting the information through light and tissue interaction, optical imaging assesses wound severity, healing potential, and progress in a rapid, objective, and noninvasive manner. Optical microscopic techniques use various biomolecules as marker to observe skin and its physiology. Various imaging modalities detect scattering and absorption of light by these markers, aiding the qualitative and quantitative evaluation of cell regeneration, metabolic activity, collagen remodeling, blood flow, inflammation, vascular structure, and water content. For example, absorption by hemoglobin provides contrast of veins in technique such as laser Doppler imaging (LDI); reflection and scattering by extracellular matrix provide structural contrast achieved by optical coherence tomography (OCT) and reflectance confocal laser scanning microscopy (RCLM) and fluorescence from molecules such as NADH, FAD [6], and tryptophan [7] provide molecular contrasts for fluorescence imaging of cells that constitute the epidermal layers of the skin. Some molecules, such as collagen and elastin found in the skin dermis, are also known to have autofluorescence [8, 9]. Along with fluorescence, collagen is better known as a strong SH generator that can provide structural contrast while imaging the dermis [9].

In clinical setting, optical imaging with these contrast mechanisms has been or has the potential to study skin wound healing noninvasively. Spectrally resolved tissue imaging with confocal or multiphoton microscopy enables 3D imaging of tissues through depth sectioning and can be used to study skin wound healing [10]. In comparison to other conventional optical microscopies, multiphoton microscopy offers a number of advantages. Nonlinear excitation limits the sample excitation to the focal volume and optical scanning with very small excitation volume results in high-contrast images. Lower scattering of IR light enables deeper penetration in tissue. Large spectral separation between the multiphoton excitation and emission provides easier discrimination of entire emission spectrum [11]. Among the modalities of multiphoton microscopy, two-photon fluorescence (2PF) and SHG present as the most effective ones in tissue imaging for diagnosis and prognosis in skin wound healing.

In this chapter we will discuss recent advancements in optical microscopic techniques for imaging skin tissue and its regeneration during wound healing. We will put forward a comparative idea of various techniques in their specific objectives of skin observations. In doing so we will briefly discuss the wound healing processes at various phases and the corresponding molecular components involved that can be used as biomarkers. Our main emphasis of the chapter is on the analysis of wound healing enabled by multiphoton microscopy (MPM), mainly 2PF and SHG imaging, and their prospects in clinical settings. However, we will also cover other popular methodologies for optical imaging of skin, highlighting their potentials in wound healing study.

2. Skin and wound healing phases

Skin is the largest organ of our body which protects us from excessive water loss and invasion of outside pathogens, senses changes in environment, etc. Before discussing the diagnostic methodologies of skin wound healing, it is very important to understand the anatomy and the molecular basis of skin and how the healing processes are related. In this section we will introduce the skin's anatomic layers and the molecules present in these layers that are potential markers for optical imaging. Various phases of wound healing and the molecular components involve in the process are also discussed in the later part of the section.

From an anatomic point of view, skin is a multilayered tissue as represented in **Figure 1(b)**. It weighs about 10% of our total body weight and thickness is approximately ranged from .5 to 2 mm [12]. The thickness of skin varies in deferent region of the body. Skin is composed of three primary layers:

(1) The epidermis, which preserves body fluid and serves as a barrier to infection, is mainly a stratified squamous epithelium composed of proliferating basal and differentiated keratinocytes. Keratinocytes are the major cells in epidermis constituting 95% of it. It is composed of five stratified layers, namely, stratum corneum, stratum granulosum, stratum spinosum, and stratum germinativum, ranging from 0.05 to 1.5 mm thick [13]. As cells possess autofluorescing chromophores, such as NADH, FAD, and tryptophan, the epidermal physiology can be observed through fluoresce microscopy.

(2) *The dermis* is the layer beneath the epidermis separated by the basement membrane and consists of connective tissues deposited in a space of 0.3–3.0 mm thickness that cushions the body from stress and strain. The connective tissues in dermis are composed mainly of extracellular matrix fibers such as elastin and collagens, ground substances, and specialized cells such as fibroblasts and adipocytes. The main components of extracellular matrix (ECM), collagen, and elastin have autofluorescence, which makes them very useful markers for wound diagnosis. However, collagen being a noncentrosymmetric molecule, SHG is the more popular way of collagen imaging. Additionally, the fibroblasts in dermis play an important role in the wound healing process, which can also be monitored with fluorescence imaging techniques [13].

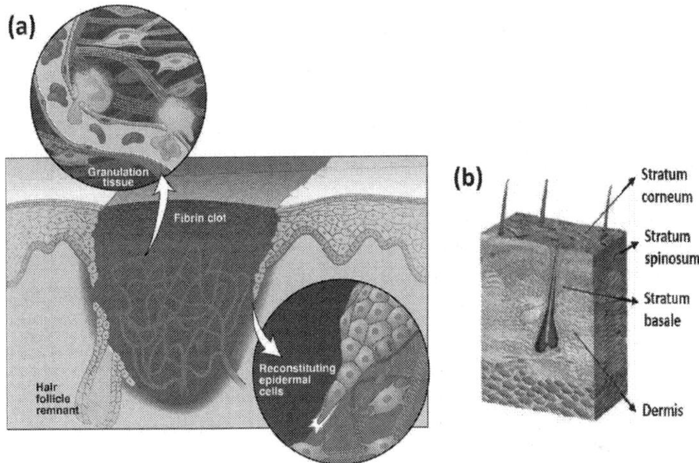

Figure 1. (a) Representative image of the key players in the healing of a skin wound [3]. The wound gap is temporarily plugged with fibrin clot that protects the wound from outside environment along with providing provisions for a dense capillary plexus of new granulation tissue and inflammatory as well as fibroblast cell migration. Reconstructing epidermal cells migrates under the fibrin clot to construct the wound bed that helps in granulation tissue formation to fill the wound gap. (b) Schematic representation of skin anatomy with its different layers [25].

(3) *Hypodermis:* Hypodermis is the layer that lies below the dermis. However, it is not considered to be a part of the skin. It helps joining the skin with the underlying muscles, bones, and blood vessels, as well as nerves. It is mainly composed of loose connective tissues and elastin.

Fibroblasts, macrophages, and adipocytes are the main cells that comprise the hypodermis. Body fat that serves as the insulator of the body also lies in this layer [13].

In normal skin, the epidermis and dermis exist in steady-state equilibrium, forming a protective barrier against the external environment. Once the protective barrier is broken, the normal process of wound healing starts immediately. Wound healing involves sequential phases of cellular initiation and secretion of molecules triggered by specific growth factors and signaling molecules [3]. Initially, a fibrin clot is formed that plugs the defects, which provides a provisional platform for cell migration as depicted in **Figure 1(a)**. In subsequent days, the wound heals completely by forming a dynamic scar tissue rich in collagen [3]. Classical model of wound healing divides the processes into several vital sequential yet overlapping stages, such as (1) hemostasis, (2) inflammation, (3) proliferation, and (4) remodeling.

Hemostasis starts immediately after the wound formation. At this stage, blood changes from liquid state to solid state to stop excessive blood loss, which is termed as blood clotting [14, 15], followed by bacteria and cell debris at the wound site being phagocytosed and removed by macrophages and white blood cells. During this phase the wound site appears red and hotter than the adjacent area marking the onset of inflammation [16–18]. Additionally, at this stage tissue matrix metalloproteinase enzymes start to degrade surrounding ECM proteins such as collagen and necrotic cellular macromolecules to provide a platform for epithelial cells migration [3]. The proliferative phase begins only when the wound is covered by re-epithelium which will migrate to central region of the wound to cover the wound defect. Angiogenesis, collagen deposition, granulation tissue formation, epithelialization, and wound contraction are the signatures of the proliferative phase [19]. The final phase of wound healing is remod eling. It is characterized by the maturation of collagen by rearrangement, intermolecular cross-linking, and alignment along the wound tension line [16]. The remodeling phase may last for a year or even longer with respect to wound size and type [20]. As the wound maturation progresses, the tensile strength of the wound increases, ultimately becoming as strong as 80% of normal tissue [20]. The wound scar gradually flattens and becomes less prominent and more pale and supple. Since activity at the wound site is reduced, consequently blood vessels that are no longer needed are removed by apoptosis and the scar loses its red appearance [21]. The wound healing normally progresses in a predictable, timely manner if not interrupted by any means; otherwise healing may progress inappropriately to transform into a chronic wound or pathological scarring such as a keloid [22, 23]. These scars consist mainly of poorly reconstructed thick parallel bundles of collagens [24]. There are mainly three different kinds of scar tissues depending upon the deposition of ECM [24]: (a) Keloids, (b) hypertrophic scar, and (c) normal scar.

Clinically, keloids are defined as scars growing beyond the confines of original wounds, which rarely regress over time. Hypertrophic scars, on the other hand, are raised scars that remain within the boundaries of the wound and frequently regress spontaneously. Histologically, collagen bundles in the dermis of normal scar tissues appear relatively relaxed and arranged in random arrays, but keloids and hypertrophic scars have collagen bundles that appear much stretched and aligned on the same plane as the epidermis.

3. Current methodologies of wound diagnosis

In clinical practice wound diagnosis is carried out by clinical signs based on the practice and expertise of the physician. For more quantitative and qualitative assessment, histochemical biopsies are employed. Some methodologies commonly used by the clinicians for wound diagnoses are as follows:

(a) *Use of clinical signs:* The clinical signs of infection are erythema, edema, heat, pain, foul odor, and wound breakdown [4]. A physician has to make a firm decision of a wound status on the basis of his or her experience. However, there are several intrinsic limitations to diagnosing a wound through these techniques as all of them are non-quantitative.

(b) *Clinical biopsies:* A biopsy is a medical examination commonly performed by a pathologist involving slice of tissue sections from a diseased or inflamed body part for insights into possible cancerous and inflammatory conditions. It is the medical removal of tissue from a living subject, which is processed into thin slices stained for observation under a microscope or analyzed by biochemical means. This kind of diagnosis has the disadvantage that the method can enlarge the wound. Moreover the time required from biopsy collection to analysis can influence the data and interfere with the wound [5, 26].

(c) *Needle aspiration:* It is a diagnostic procedure used to investigate superficial inflammation, lumps, or masses. In this technique, a thin, hollow needle is inserted into the mass and a portion of the tissue is recovered. The recovered tissue with the cells is stained and examined under a microscope. Needle aspiration biopsies are minor surgical procedures and safe. In spite of being considered as the next best method for microbiological culturing of abscesses and closed wounds, there is a risk, because the biopsy is very small (only a few cells), that the problematic cells may be missed, resulting in a false-negative result [5, 27] that prohibits a definitive diagnosis.

Recently noninvasive approaches have been brought in for assessing skin lesions that include magnetic resonance, ultrasound, and photoacoustic and optical techniques with which intravital imaging of the alterations or aberrations in the skin below the surface has been materialized [28]. Among them optical microscopic techniques provide cost effective and wider range of applications of skin tissue imaging.

4. Optical techniques used in skin observations

Optical imaging techniques are based on the principles of light and tissue interaction for collecting information that is further analyzed to reconstruct an image of the respective tissue section. Depending on the nature of interactions, such as scattering, absorption, or fluorescence, various information can be extracted to reveal anomalies in the tissue sections. The physiological events associated with such structural anomalies also determine the choice of optical modality needed to address the problem noninvasively. In a few excellent reviews, various optical approaches in skin imaging are listed and discussed depending on the skin conditions [28–31].

Optical modalities are comparatively advantageous for their low-cost, easy to use, non-ionizing, mostly noninvasive, and non-contacting attributes. Some of the optical methodologies can provide 3D imaging capability by optical sectioning with high resolution [32–35]. Optical techniques may also be useful in real-time functional imaging regarding skin physiology [36]. Additionally, most of the skin optical imaging techniques use near-infrared (NIR) or infrared (IR) wavelengths, which are less absorbed in tissue, hence penetrating deeper, enabling the imaging of the whole skin layer [37–39]. Most common optical imaging modalities include LDI, tissue spectral imaging (TSI), and OCT, which are useful in imaging macro-masses in skin (macro-imaging modalities). Optical techniques that are useful in imaging at molecular domain or micro contrasts (micro-optical modalities) are RCLM, Raman spectro-microscopy, and laser scanning fluorescence and SHG microscopy. In this section, we are going to discuss the recent advancements in optical techniques that have been applied to evaluate skin wound-related problems noninvasively or hold potential in this regard. The following is separated into two subsections: the macro-optical modalities and micro-optical modalities.

4.1. Macro-optical imaging modalities

4.1.1. Dermoscopy

Dermoscopy or dermatoscopy, also known as epiluminescence microscopy, is the most common basic handheld magnifying tool that aids in first-line optical observation of morphological abnormalities. Recent dermoscopes use polarized light to illuminate the tissue section to visualize horizontal morphological features that are not visible to naked eye [40]. Dermoscopy has been useful in qualitative visualization of skin-related abnormalities such as rosacea [41], diagnosis of hair and scalp diseases [41–43], diagnosis of warts caused by human papillomavirus [44], and determination of the surgical margin of hard to define skin cancers [41].

This method has been widely used in observing skin lesions based on the presence of certain architectural characteristics of the lesion, which provide promising possibilities in skin wound healing study, mainly in collagen regeneration during wound remodeling phase [45]. A dermoscope is easy to use and represents a relatively low-cost first-line diagnostic tool for skin-related issues; however it is not quantitative and requires expertise and experience to have fair diagnostic judgment [46]. Its resolution is only enough to see small lumps and lesions in the skin. Additionally, no functional information can be gathered with this technique. Commercial dermoscopes are available in the market for quite a few time. Companies such as Optilia, WelchAllyn, CALIBER, HYMED, and FotoFinder are manufacturing dermoscopy products of various specifications and models with attached digital cameras to it that are capable of videography also.

4.1.2. Laser Doppler imaging

In LDI, laser light is used to illuminate the tissue section and the backscattered as well as reflected light is collected to image any moving object within the tissue section. With this technique blood flow through superficial skin layer can be calculated based on the Doppler shift introduced by moving blood cells [47, 48]. It is useful in measuring blood perfusion unit [48], which can be applied in extracting useful functional information to assess angiogenesis and endothelial functioning during wound healing [49].

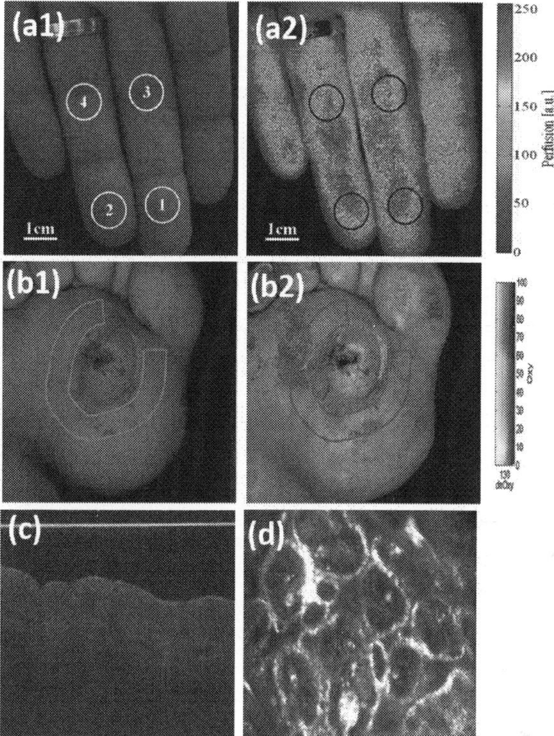

Figure 2. (a1) Visible color image of fingertips and (a2) color-coded blood perfusion map [51]. (b1) Visible and (b2) hyperspectral image of a healing diabetic foot ulcer taken with the HTOM system. HTOM values are 60, 53, and 53% for oxy, deoxy, and StO$_2$, respectively [57]. (c) OCT signals using a super luminescent NIR diode [25] and (d) reflectance confocal microscopy image at wavelength 830 nm of a human nevus detected by VivaScope [25].

LDI is a low-cost, easy to use noninvasive imaging modality compatible with classical medical instrumentation, where discomfort and risk to patients are minimal. A typical LDI system has a resolution of about 2 mm × 2 mm with an average imaging depth of 200–240 μm [50, 51]. **Figure 2(a1)** and **(a2)** depicts a representative comparison of visible **(a1)** and color-coded blood perfusion map **(a2)** [51]. LDI has been reported to be used in imaging microcirculation in burned skin and monitoring blood flow recovery in a skin flap during reconstructive surgery demonstrating its potential for clinical wound assessment applications [47]. There are several other reports of burn wound depth and healing assessment with LDIs in clinical and research settings both on human and animals [50, 52, 53].

Commercial instruments based on the principle of LDI are made available by company such as Moor Instruments for skin perfusion assessments. The moorLDI2-IR laser Doppler blood flow imager can image an area up to 50 cm × 50 cm in one scan in less than 5 minutes. Due to the large area scanning possibility, this method has been very useful in burn wound depth and healing assessment based on angiogenesis.

The main disadvantage of this technique is its limited application only in observation of blood flow, similar to that of laser spackle imaging. It is unable to provide any other functional as well as structural information of skin integrities. The poor resolution in millimeter range is another major limitation of this technique in comparison with other optical techniques. Additionally, the use of visible light illumination in LDI limits its applicability in deep dermal wound assessments.

4.1.3. Tissue spectral imaging

Tissue spectral imaging (TSI) is a technique where a tissue of interest is illuminated by a broadband light and collects the reflected or diffused light through selective narrowband filters in front of the detection unit. This technique thus yields several images of specific wavelengths on the same area, providing quantitative measures of the absorbers or scatterers present [54]. Skin has several chromophores such as hemoglobin, melanin, collagen, and other biomolecules which absorb or scatter light and are responsible for skin physiology.

Diffused multispectral imaging (DMSI) is a type of TSI in which the diffused light through a tissue section was collected with a narrow band-pass filter in front of the detection unit. Based on peak absorption, specific wavelengths were chosen to reconstruct an image that can provide physiological information. It is a widely used technique in imaging hemoglobin as it can discriminate oxygenated and deoxygenated hemoglobin by their spectral signature [28]. Most of the DMSI uses NRI wavelength (700–1,100 nm) for quantitative spectroscopic analysis of structural and chemical integrity of cutaneous tissue, especially oxygen saturation, hemoglobin content, and water content. Careful assessment of the absorption spectra can be applied to monitor the wound severity and healing. Reports at preclinical setting have demonstrated the potential of this technique in differentiating superficial-, intermediate-, deep-, and full-thickness burn wounds based on measurement of water content, oxygen saturation, and total hemoglobin concentration [55, 56]. It has also been found useful in diabetic ulcer assessment and predicting diabetic wound healing as well as differentiating diabetic wounds from nondiabetic wounds [37, 38].

Hyperspectral imaging (HSI) is an effective technique that can capture and map detailed spectroscopic information for every pixel in an image. A single pixel in a hyperspectral image provides information in two-special dimension and one spectral dimension creating a 3D data cube [30]. HSI has been reported to be used in diagnosing diabetic foot ulcers [57, 58] and burn wound edemas [59] where cutaneous tissue oxygenation is observed by assessing oxy- and deoxy-hemoglobin as spectroscopic contrasts. Based on HSI of hemoglobin oxygenation in wound site and areas adjacent to ulcer, wound healing index has been developed [58]. HSI using broadband visible light-emitting diodes has been reported to generate tissue anatomical oxygen map that predicts the risk of diabetic foot ulceration in pre-ulceration tissue [60]. **Figure 2(b1)** and **(b2)** depicts a representative comparison of visible and hyperspectral image of a healing diabetic foot ulcer taken with the hyperspectral tissue oxygenation mapping (HTOM) system [57].

Spectrophotometric intracutaneousanalysis (SIAscope) is a kind of TSI technique based on back-reflected light of wavelength within the range of 400–1000 nm. It is a portable, fast device that predicts burn wound depth by creating a quantitative map of specific chromophores [61]. TSI based on reflectance has also found to have potential in assessing hematomas on the basis of hemoglobin destruction quantification that can determine the age of hematomas in vivo [62]. This information can also be crucial in identifying the layer of skin that sustains the hematoma [63].

Orthogonal polarization spectral imaging (OPSI) uses linearly polarized light to illuminate the skin tissue and collect the emergent depolarized light scattered by the skin components through an analyzer positioned orthogonal to the plane of illumination light polarization [64]. By analyzing the depolarized light, hemoglobin in microcirculation can be visualized to quantify the microvasculature during cutaneous wound healing [65–68].

Thermographic spectroscopy is a spectroscopic imaging technique based on the principle that all the objects including the skin have a heat signature that radiates in IR wavelength. This emission can be detected by using appropriate detector and construct color-coded images that correlate the relative temperature of the specific skin area [69–71]. Usually superficial burn wounds are warmer than uninjured skin due to increased inflammatory processes while deeper burns are cooler than uninjured skin due to structural damage to the vasculature. By using this basic principle, burn wound depth and healing progress over time can be predicted [71, 72].

A number of modalities of TSI have been commercialized by companies such as HySpex and Specim's AisaFENIX. The TSI technique can provide better resolution than LDI, typically up to 0.4–1 μm. Being a wide-field imaging technique, TSI is unable to provide a detailed 2D sectioned image with better resolution; rather it only can provide a molecular map in a certain area. It also suffers from scattering blur and diffraction limitations and has low penetration depth. Additionally, to gather a meaningful spectral information, it requires enough photon information which makes it a relatively slow method. Even with these limitations, this technique holds potential for functional imaging of blood clotting, blood flow during wound inflammation phase, and angiogenesis during superficial skin wounds. Recent advancements in computational methodologies have shown great promises in real-time quantitative functional imaging with improved resolution [54].

4.1.4. Optical coherence tomography (OCT)

OCT is a technique that captures 3D images of a tissue. OCT uses reflected light from tissue to construct cross-sectional images from deeper part of the skin. Most common OCTs use IR illumination, which after scattering from tissue is superimposed with a reference light to generate an interferometric pattern that provides high-resolution 3D depth information by scanning the tissue section in all directions [73–76].

OCT has been an established imaging modality in medical diagnosis and research field. Although it is most popularly used in ophthalmology [77], it has also gripped its root in dermatology [78] study such as keratosis [79], skin cancers [80], skin fibrosis [81], and wound healing. Other than that, it has also been reported to be used in other dermatological problems such as inflammatory diseases and parasitic infection and those of the nails [75, 76]. Recent advancement in OCT allows use of polarized light to image extracellular matrix and other connective tissues in the skin layer that are polarization sensitive [82, 83]. Reports also suggest use of phase-resolved OCT for imaging blood flow in the skin [84]. There are a few excellent reviews that have listed and discussed various applicable possibilities for OCT in dermatology [83, 85, 86].

In diagnosis of wound healing, there are reports of comparing healing assessment of acute wound [87] and superficial wound caused by bacterial infection on mice by OCT to histological findings [88]. A study had reported quantitative evaluation of healing kinetics at real time after fractional laser therapy by OCT demonstrating excellent correlation with findings from histopathological observations [89]. In an in vivo study, OCT has effectively evaluated the various stages of wound healing in 12-day long healing process recognized by re-epithelialization in the early stage, followed by thickening of the epithelial layer around 10th day and formation of scar tissue composed of extracellular matrix along with thickening of epidermal layer in the final stage [90].

OCT's most promising advantage is its ability of axial sectioning and 3D imaging of a tissue mass. OCT techniques using IR light sources are suitable imaging modalities for deep tissue topographical imaging of skin disfigurement. Although the resolution of OCT is lower compared to CLSM or 2PFM or SHG microscopy, the associated resolution degradation with depth is much smaller. OCT cannot produce images at cellular or fibrous molecular resolution; hence it is incapable of imaging a single-cell structure or fibrous collagen structure in the skin dermis [30]. However, in comparison to other macro-optical methodologies, OCT exhibits better resolution. In fact, with sophisticated design, OCT can also achieve a resolution of few tens of micrometer. OCT was also reported to provide even more detailed structural information of a larger mass of tissue than 2PFM at depth of 2–3 mm while imaging thermally injured wounds [91]. OCT is a useful noninvasive technique that has huge potential for wound healing research and assessment. **Figure 2(c)** represents a typical OCT image [25].

OCT has been commercialized by companies such as Optovue, NinePoint Medical, and Thorlabs; two such models from Thorlabs are Ganymede II IR-OCT system and Telesto series spectral domain OCT systems. These systems are mainly operated in IR domain with line scan rate within the range of 5.5–76 KHz.

4.2. Micro-optical modalities

4.2.1. Raman spectro-microscopy

Microscopic imaging based on Raman vibrational spectroscopic contrast provides a useful noninvasive approach for visualizing skin tissues and the corresponding architecture with molecular specificity. A typical Raman microscope detects vibrational scattering changes introduced by the Raman-active molecules in tissue. Molecules rich in CH2 bonds, such as protein and lipid, are good Raman contrast agents and can be interpreted to visualize structural changes occurring in different skin strata [92–94]. An automated Raman micro-spectrometer in confocal settings was reported to be used to determine water concentrations in hydrated and non-hydrated stratum corneum, showing the capacity of this method [95]. However, spontaneous Raman signal is very weak. The Raman detection can be significantly enhanced by CARS. It can visualize structural fibers such as collagen and elastin that constitutes the human dermis along with subcutaneous layer rich in lipids, due to the high density of CH_2 bonds [96–98]. CARS microscopy is the method of choice for studies that require visualization of fat in tissues, which can very effectively characterize obesity in murine skin in vivo [99]. While imaging superficial tissue layers, CARS can provide strong signal from the fat component of the skin that allow video-rate imaging.

Video-rate CARS imaging can be used for imaging lipid lamellae of the stratum corneum, sebaceous glands, and dermal adipocytes, and the fat-containing cells of the subcutaneous layer with imaging depths of up to several hundred micrometers, promising a potential methodology for noninvasive molecular imaging [97]. Recently CARS has also been used in studying transdermal delivery of retinol in mouse ear, a drug with strong CARS signal that stimulates collagen growth in skin and was located in corneocytes of stratum corneum [100].

König and his group have reported a CARS tomography system for skin imaging suitable for clinical environments that is capable of in vivo histology with subcellular resolution and chemical contrast toward patients suffering from psoriasis and squamous cell carcinoma [101]. Their system also has the potential to be used in studying skin wound healing. Although Raman imaging in the form of CARS can provide high-contrast functional imaging with subcellular resolution, it is, however, mostly limited to Raman-active molecules only. In comparison, the Raman scattering cross section is very small which translates to very weak signal intensities, thus requiring very high density of molecules or very long acquisition times in order to acquire a meaningful image.

4.2.2. Laser scanning microscopy techniques:

A commonly used wide-field microscope provides a two-dimensional image, typically in histological observations of biopsies. However it has several drawbacks, including low resolution, low penetration depth, slow imaging rate, and inability to have functional imaging. It delivers poorer image contrast and lacks optical sectioning capability. In contrasts, a laser scanning microscope (LSM) provides a few numbers of platforms for imaging that are improved with respect to all aspects mentioned above. Among them confocal microscopy in linear domain and two-photon fluorescence microscopy (2PFM) and SHG in nonlinear domain are most prominent. Confocal laser scanning microscopy (CLSM) has several advantages over traditional microscopy, including faster data acquisition, optical sectioning of cells and tissues for 3D imaging, and significantly improved spatial resolution [39, 102, 103]. The pioneering work of Minsky, in the year 1957, initiated the development and the first commercialized CLSM was realized in 1987 [104]. However, CLSM has a relatively lower penetration depth compared to MPM, due to the shorter wavelength used. Single-photon confocal microscopy obtains an image section at the expense of photon efficiency, attributing to the spatial filtering pinhole [39, 105, 106]. The overexposure would cause photo bleaching of the sample. As a result, only highly photostable fluorophores work well with this technique. In comparison, MPM uses IR excitation which reduces photo bleaching in a confined way and allows imaging depths of up to ~2 mm. The nonlinear effect forms a virtual pinhole and saves the trouble of precision alignment needed for a physical pinhole [39, 106].

4.2.2.1. Reflectance confocal laser scanning microscopy (RCLSM)

In RCLSM, a pinhole at the confocal image plane eliminates out-of-focus signal to realize optical sectioning for 3D imaging. It uses a focused laser beam for excitation and forms the image by point to point scanning, usually by a pair of computer-controlled galvano mirrors [32, 33]. The reflected light signal is collected by a photo detector after the pinhole. The reflected signal is de-scanned by the same pair of galvano mirrors so the alignment of pinhole is straightforward [107]. The configuration is widely used in commercially available confocal microscopes for skin imaging [33, 107]. It has also been used for assessing and monitoring cutaneous wound healing by evaluating the cellular and morphological parameters of wound bed and wound margins noninvasively over the course of healing [102]. In the reported study, patients with chronic leg ulcers and skin cancers receiving split skin graft were evaluated against healthy individuals, in which various physiological signatures of wound healings at different phases were documented. For example, appearance of inflammatory cells in the epidermis during the early stage of wound healing, proliferative keratinocytes and their migration during granulation and re-epithelialization phases, and the networks of connective tissues during remodeling phases were observed with reflectance CLSM [102].

A commercially available CLSM in reflectance mode is VivaScope®1500 that has planar and axial resolution of 1.25 and 5.0μm, respectively, with an imaging depth up to 200 μm. Its image acquisition speed of 9 frames per second allows real-time videography of wound healing. **Figure 2(d)** represents a reflectance confocal microscopy image by detecting backscattered 830 nm light from a human nevus with the system [25].

Although these instruments are widely used, they are limited to surface imaging only. Therefore, they are not suitable for evaluating deep dermal wounds. Nevertheless, they can image wound margins, which may provide crucial semiquantitative information regarding wound healing with a resolution comparable to that of histological analysis [30]. The reflection contrast-based CLSM is frequently used for structural imaging but is incapable of molecular functional imaging. A typical CLSM has much improved resolution and faster scanning rate than OCT. However, it may be limited by photo bleaching and diffraction blurring when compared to multiphoton techniques.

4.2.2.2. Confocal fluorescence microscopy

Confocal fluorescence microscopy is a technique that allows imaging of living tissue by collecting fluorescence emission from the chromophores present in the tissue. In single-photon fluorescence imaging, a fluorophore absorbs a single photon to be excited into a higher energy state before emitting the fluorescence, and comes down to original lower energy state. The simplest fluorescence imaging instrumentation uses a laser to illuminate the skin at a specific excitation wavelength and collects the filtered fluorescence emission with a detector bearing an optical filter in front of it.

Fluorescence imaging can be done with either staining the tissue by exogenous fluorescent materials or imaging endogenous fluorescence from skin's natural fluorophores. Indocyanine green (ICG) is one commonly used exogenous fluorescence dye that can be located in systemic circulation, which allows the imaging of vascularization and the determination of imaging depth [108]. This technique has been shown to quantitatively measure blood flow in the cutaneous wound that is well correlated with the histological assessment of burn depth [108]. As mentioned in Section 2, endogenous fluorophores, NADH, FAD, and collagen are all important markers in wound healing processes that can be used for wound diagnosis [36, 109].

Along with NADH and FAD, collagen is another abundant molecule present in the skin dermis that is autofluorescing. It can serve as a marker upon exposure to the 325 nm He-Cd laser treatment ($\sim$2 J/cm^2) during skin tissue regeneration, as shown in mouse model by detecting the collagen autofluorescence intensity [110]. In another comparative ex vivo and in vivo study of wound granulation by the same group, normalized NADH/collagen autofluorescence intensity was used to assess collagen deposition during healing [111].

Confocal fluorescence microscopy can provide real-time functional imaging of cells and tissues with improved resolutions. However single-photon imaging may be limited by photo bleaching and low penetration depth. Alternatively, MPF imaging would improve photo stability with deeper penetration.

4.2.2.3. Multiphoton microscopy (MPM)

In multiphoton imaging a simple confocal laser scanning microscope is used with an ultrafast NIR laser source. The pinhole is usually removed and the detection unit is modified with specialized filters. Multiphoton laser technique greatly improves resolution and penetration depth than macro-optical modalities. Its optical sectioning ability does not require a pinhole, which reduces alignment difficulty and the volume of photo bleaching. Additionally, the NIR excitation wavelengths are shown to extend the limit of deep tissue imaging up to 2 mm.

In tissue imaging, commonly used multiphoton techniques are 2PFM and SHG imaging. In 2PFM, the fluorophores absorb two photons simultaneously to be excited to a higher energy real state before emitting the fluorescence, while in SHG, the two photons of the same energy would combine to form a new photon of twice the energy of the incident photon. Biomolecules such as collagen and muscle myosin with noncentrosymmetric molecular structures have the ability to generate SHG signal [8, 112–114]. Skin can be imaged with both fluorescence and SHG contrasts simultaneously with the help of a laser scanning MPM [36, 115]. Zoumie et al.

in their study of a tissue model have described spectrally resolved imaging of different parts of the skin layers by a combined 2PFM and SHG setup [115]. They detected fluorescence from cellular NADH and SHG from collagen. The study of wound healing with fluorescence and SHG is discussed in the following paragraphs.

Cellular NADH autofluorescence in two-photon modality has been used as marker for morphological characterization of epithelia both in vivo [116–118] and ex vivo [119] for animal and human tissues as well as fresh biopsies [120]. It enables optical microscopic imaging being equivalent to histochemical analysis. With the help of 2PFM imaging, various epidermal layers of in vivo skin were discriminated at subcellular spatial resolution based on cellular morphological features [31]. Additionally, the time-correlated single-photon counting technique in conjunction with 2PFM has made functional imaging possible by measuring the lifetime of fluorophores. This technique, termed as fluorescence lifetime imaging (FLIM), is very effective in determining real-time cellular metabolic activity in vivo by measuring the fluorescence lifetime decay of NADH. Cells located in the basal layer exhibit the strongest metabolic activities, while epidermal surface layered cells are found to have lower metabolic activities. FLIM has demonstrated its capacity in characterizing epithelial tissue involved in wound healing and other pathological conditions [31].

NADH, being a metabolic coenzyme, is associated with the cellular metabolic activities through the electron transport chain (ETC) of oxidative phosphorylation. NADH has two functional forms, free and bound. During the process of energy generation, free NADH is bound to mitochondrial membrane proteins [36]. Although the fluorescence emission spectra of both free and bound forms of NADH fall in a very narrow band, their fluorescence lifetimes are well separated. When NADH binds to a protein, its lifetime increases from ~0.4 to ~2.5 ns [121–123]. Therefore by evaluating the contribution of free and bound states to the combined double exponential lifetime, the relative concentrations of individual states can be predicted. In simple words, a cell with higher metabolic activity has a higher concentration of bound NADH than a cell with lower metabolic activity. In addition to that the ratio of bound form NADH to bound form of FAD, termed as cellular redox ratio, can also be a marker for relative metabolic activity determination [124].

The cellular metabolic parameters are viable markers for evaluating wound healing. We have demonstrated on live rat models that the cellular metabolic rate correlates well with wound healing phases [36]. In the study, artificially created incisional wound by punch biopsy was used to evaluate the wound healing from the day of wound formation to scar formation in a 20-day healing course with 2PFM and SHG microscopy. The relative metabolic activities of cells involved in the process of wound healing as time progresses were evaluated by the NADH bound to free ratio, while the changes in collagen concentration are correlated with SHG intensity. These findings suggest the metabolic activities at the wounded sites increase during inflammatory and granulation phases and gradually decrease as wound heals (**Figure 3(b)**). Interestingly, in the beginning of healing, SHG intensity decreases (or collagen concentration), indicating the degradation of collagen in the dermal layer during cell migration. Once new collagens were formed, SHG signal started to increase gradually (**Figure 3(c)**). In general, wounds heal gradually from the edge toward the center; hence the metabolic activities are

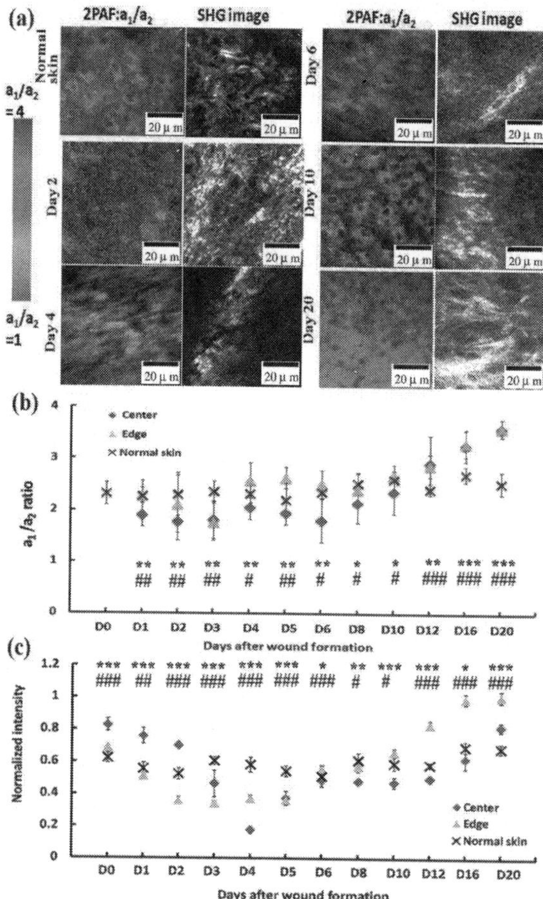

Figure 3. (a) Representative color-coded NADH free/bound (a_1/a_2) lifetime ratio images (left column) and gray-scale SHG intensity images (right column) of collagen regeneration during wound healing [132]. (b) Scatter plot of NADH a_1/a_2 distribution peak value with healing progress both at the center and edge, averaged over 15 wounds at each day of wound observation. The ratio NADH a_1/a_2 is inversely proportional to metabolic activity of cells. Two-side Student's t-test evaluated significant differences of NADH a_1/a_2 values at the center from normal skin that are indicated by *, $P > 0.05$; **, $0.05 > P > 0.001$; and ***, $P < 0.001$. The significant differences of a_1/a_2 values at the edge from the normal skin are designated as #, $P > 0.05$; ##, $0.05 > P > 0.001$; and ###, $P < 0.001$ [132]. (c) Scatter plot of normalized SHG intensity with respect to the maximum intensity observed at the edge on day 20. The changes in SHG intensity elucidate the relative degradation and regeneration of collagen at the center and the edge in the course of wound healing. Two-side Student's t-test evaluated significant differences of SHG intensity at center from normal skin that are indicated by *, $P > 0.05$; **, $0.05 > P > 0.001$; and ***, $P < 0.001$ for center. The significant differences of SHG intensity values at edge from normal skin are designated as #, $P > 0.05$; ##, $0.05 > P > 0.001$; and ###, $P < 0.001$. P value less than 0.05 was considered significant [132].

higher at the edge in the early stages of wound healing, marked by the higher bound to free NADH ratio in lifetime measurement. However, in the proliferative phase the center has higher metabolic activity than the edge since the edge has entered the remodeling phase, in which cell activity decreases and collagen is deposited to fill the wound gap, marked by the increase of SHG intensity. Following the proliferative phase, the whole wound is filled with granulation tissues, mainly collagen, and the cellular metabolism decreases gradually. The wound then heals into a scar, composed of connective tissues marked with higher SHG signal intensity than that from a normal tissue. The lack of cells in scar tissue reduces the need for blood influx, which results in removal of blood vessels by apoptosis and leaves a scar tissue characterized by lower metabolic activity and higher collagen deposition.

The changes of the NADH free to bound ratio (**Figure 3(b)**) and the collagen SHG intensity (**Figure 3(c)**) exhibit as the signature of the various phases in wound healing, which can be used for crucial diagnosis and proper treatment. With the simultaneous measurements of 2PFM and SHG, a correlation between cellular metabolic activities and collagen regeneration can be observed. In **Figure 3(a)**, the morphological features of cells and their gradual appearance in wound region and structural evolution of collagen in a healing wound, acquired by 2PFM and SHG, respectively, are demonstrated. The disordered collagen in normal skin is degraded and more structured collagens are deposited in the process of scar formation as shown in **Figure 3(a)**.

Similar results have also been reproduced by other researchers using combined SHG and 2PFM imaging, where disorganized collagen in fibrin clots and inflammatory cells involved during the early stage of wound healing are distinguished from more organized and aligned collagens in regenerated new skin [125].

SHG is also used in showing the orientation of collagen fibers and their structural changes in the healthy tissues of human dermis [126–129] as well as in in vivo tissue constructs [130]. The efficiency of SHG signal is highly sensitive to the collagen orientation when the incident light is polarized. Along with intensity measurements, polarization-resolved SHG provides information on collagen alignment and orientation during regeneration, which is correlated to wound closure and the way scar tissue forms [131].

Polarization-resolved SHG indicates that collagens are more organized and fibrillary during the proliferative phase, to aid in wound closure when the margins are pulled together by them [132]. In this way, the anisotropic variation of collagen during wound healing can be monitored by collecting the parallel (I_{par}) and the perpendicular (I_{perp}) components of the polarized SHG signals with respect to the incident polarization. **Figure 4** demonstrates representative I_{par} (first row) and I_{par} (second row) polarization-resolved SHG intensity images from biopsy samples taken after discrete days of wound formation along with corresponding anisotropy images (third row) defined by $(I_{par}-I_{perp})/(I_{par}+2I_{perp})$. Anisotropy value equal to 0 corresponds to complete random arrangement of the scatterers, and if it is equal to 1, it corresponds to having a well-aligned, well-structured scattered system [132, 133]. Anisotropic observation of ex vivo rat skin biopsies has revealed maximum anisotropy value of collagen during wound contraction and closure. When the wound gap is filled with matured collagen, the anisotropy decreases gradually [132].

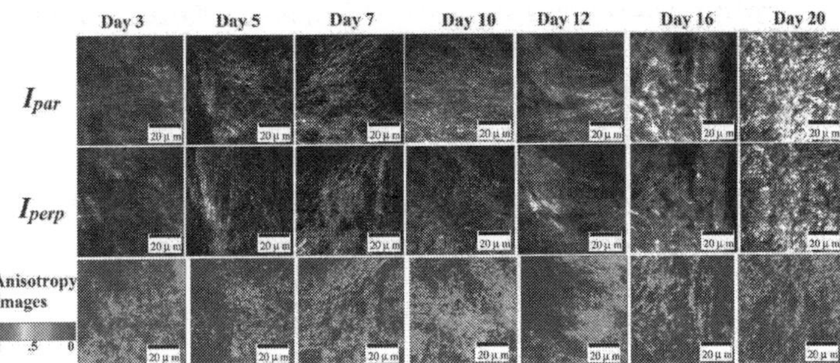

Figure 4. Representative polarization-resolved SHG intensity images of wound biopsy samples taken at different healing stages. The first row depicts the images with the parallel (I_{par}) component. The second row depicts the images with the perpendicular (I_{par}) component. The third row depicts the corresponding anisotropy images defined by (I_{par}-I_{perp})/(I_{par}+2I_{perp}) [132].

In clinical setting, multiphoton imaging of human epidermis and upper dermis has been achived by commercial system such as DermaInspect™ that is able to scan an area of 350 μm×350 μm with special resolution of 1 μm in lateral and 2 μm in axial directions [25]. The system provide non-invasive in vivo optical biopsies of skin at subcellular resolution by detecting autofluorescence from biomolecules such as NADH, flavins, porphyrins, elastin and melanin and SHG signals from collagens.

5. Conclusion

Wound healing is an important physiological process that follows a certain sequential order. Migration of various cells and the involvement of certain molecules at the wound site characterize the various phases during healing progression. Detailed quantitative and qualitative information of these components at a specific time provides critical insights on wound healing. Optical methodologies are versatile and include techniques that can gather a wide variety of information on multiple components noninvasively, which presents tremendous future prospects in terms of clinical implications. The available modalities present enormous potentials to supplement clinical assessment and to aid research in the field of cutaneous healing and skin tissue regeneration.

The versatile optical modalities discussed in this chapter have their own significance in assessing specific wound-related problems. Some modalities are simple and easy to operate, which provide relatively low-cost first-line diagnosis. More complex techniques can provide better resolution and sophisticated structural information. By judicially combining various contrasts from the skin components, these optical techniques can address a wide variety of skin wound-related issues. These can include observations of subsurface morphological features using dermascope, blood flow using LDI, molecular and functional signatures using TSI, structural revelation using OCT, RCLM and SHG microscopy, or molecular identification with Raman and fluorescence imaging. Each technique would provide unique yet complimentary information.

Multimodal MPM presents the most sophisticated approach for quick, qualitative, and quantitative skin wound healing study as it integrates multiple contrast mechanisms for imaging the skin. Specifically, 2PFM and SHG are favorable in wound assessment for their high-resolution, better penetration depth, optically sectioned 3D imaging with the provision of structural and real-time molecular functional signature.

Emerging super resolution imaging based on saturation excitation (SAX) of scattering from metallic nanoparticles may extend the possibilities of super resolving the skin abnormalities [134]. Ointments and sunscreen lotion could effectively carry the nanoparticles into skin epidermis to facilitate the new optical techniques. With the ongoing rapid advancements in photonics and imaging, one can expect new and novel techniques will find unprecedented and enlightening applications in dermatology in the coming future.

Author details

Gitanjal Deka[1*], Shi-Wei Chu[1] and Fu-Jen Kao[2]

*Address all correspondence to: gitanjal@yahoo.com

1 Department of Physics, National Taiwan University, Taipei, Taiwan

2 Institute of Biophotonics, National Yang Ming University, Taipei, Taiwan

References

[1] H. Brem, and M. Tomic-Canic, "Cellular and molecular basis of wound healing in diabetes," J. Clin. Invest. 117, 1219-1222 (2007).

[2] M. Peck, J. Molnar, and D. Swart, "A global plan for burn prevention and care," Bull. World. Health. Organ. 87, 802-803 (2009).

[3] P. Martin, "Wound healing--aiming for perfect skin regeneration," Science 276, 75-81 (1997).

[4] S. E. Gardner et al., "A tool to assess clinical signs and symptoms of localized infection in chronic wounds: development and reliability," Ostomy. Wound Manag. 47, 40-47 (2001).

[5] G. Dow et al., "Infection in chronic wounds: controversies in diagnosis and treatment," Ostomy. Wound Manag. 45, 23-40 (1999).

[6] T. Y. Buryakina, P.-T. Su, W.-J Syu, C. A. Chang, H.-F. Fan, and F.-J. Kao, "Metabolism of HeLa cells revealed through autofluorescence lifetime upon infection with entero-hemorrhagic *Escherichia coli*" J. Biomed. Opt. 17, 101503 (2012).

[7]   B. Banerjee, T. Renkoski, L. R. Graves, N. S. Rial, V. L. Tsikitis, V. Nfonsam, J. Pugh, P. Tiwari, H. Gavini, and U. Utzinger, "Tryptophan autofluorescence imaging of neoplasms of the human colon," J. Biomed. Opt. 17, 016003 (2012).

[8]   W. R. Zipfel, R. M. Williams, R. Christie, A. Y. Nikitin, B. T. Hyman, and W. W. Webb, "Live tissue intrinsic emission microscopy using multiphoton-excited native fluorescence and second harmonic generation," Proc. Natl. Acad. Sci. U. S. A. 100, 7075-7080 (2003).

[9]   X. Jiang, J. Zhong, Y. Liu, H. Yu, S. Zhuo, and J. Chen, "Two-photon fluorescence and second-harmonic generation imaging of collagen in human tissue based on multiphoton microscopy," Scanning 33, 53-56 (2011).

[10]   W. Denk, J. H. Strickler, and W. W. Webb, "Two-photon laser scanning fluorescence microscopy," Science 248, 73-76 (1990).

[11]   P. T. So, C. Y. Dong, B. R. Masters, and K. M. Berland, "Two-photon excitation fluorescence microscopy," Annu. Rev. Biomed. Eng. 2, 399-429 (2000).

[12]   Y. Lee, and K. Hwang, "Skin thickness of Korean adults," Surg. Radiol. Anat. 24, 183-189 (2002).

[13]   L. A. Baden, "Surgical Anatomy of the Skin," Arch. Dermatol. 125, 719-719 (1989).

[14]   G. C. Troy, "An overview of hemostasis," Vet. Clin. North Am. Small Anim. Pract. 18, 5-20 (1988).

[15]   T. A. Springer, "Biology and physics of von Willebrand factor concatamers," J. Thromb.Haemost. 9, 130-143 (2011).

[16]   H. P. Lorenz, and M. T. Longaker, "Wounds: biology, pathology, and management," in Essential Practice of Surgery (Springer, 2003), pp. 77-88.

[17]   F. H. Epstein, A. J. Singer, and R. A. Clark, "Cutaneous wound healing," N. Engl. J. Med. 341, 738-746 (1999).

[18]   S. A. Eming, T. Krieg, and J. M. Davidson, "Inflammation in wound repair: molecular and cellular mechanisms," J. Invest. Dermatol. 127, 514-525 (2007).

[19]   K. S. Midwood, L. V. Williams, and J. E. Schwarzbauer, "Tissue repair and the dynamics of the extracellular matrix," Int. J. Biochem. Cell Biol. 36, 1031-1037 (2004).

[20]   R. F. Diegelmann, and M. C. Evans, "Wound healing: an overview of acute, fibrotic and delayed healing," Front. Biosci. 9, 283-289 (2004).

[21]   D. G. Greenhalgh, "The role of apoptosis in wound healing," Int. J. Biochem. Cell Biol. 30, 1019-1030 (1998).

[22]   R. O'leary, E. Wood, and P. Guillou, "Pathological scarring: strategic interventions," Eur. J. Surg. 168, 523-534 (2002).

[23] A. Desmoulière, C. Chaponnier, and G. Gabbiani, "Tissue repair, contraction, and the myofibroblast," Wound. Repair. Regen. 13, 7-12 (2005).

[24] T.-L. Tuan, and L. S. Nichter, "The molecular basis of keloid and hypertrophic scar formation," Mol. Med. Today. 4, 19-24 (1998).

[25] K. König, "Clinical multiphoton tomography," J. Biophotonics 1, 13-23 (2008).

[26] L. Brentano, and D. L. Gravens, "A method for the quantitation of bacteria in burn wounds," J. Appl. Microbiol. 15, 670 (1967).

[27] N. S. Levine, R. B. Lindberg, A. D. Mason JR, and B. A. Pruitt JR, "The quantitative swab culture and smear: A quick, simple method for determining the number of viable aerobic bacteria on open wounds," J. Trauma. Acute. Care. Surg. 16, 89-94 (1976).

[28] B. Dasgeb, J. Kainerstorfer, D. Mehregan, A. Van Vreede, and A. Gandjbakhche, "An introduction to primary skin imaging," Int. J. Dermatol. 52, 1319-1330 (2013).

[29] D. W. Paul, P. Ghassemi, J. C. Ramella-Roman, N. J. Prindeze, L. T. Moffatt, A. Alkhalil, and J. W. Shupp, "Noninvasive imaging technologies for cutaneous wound assessment: a review," Wound. Repair. Regen. 23, 149-162 (2015).

[30] H. Albert, "A survey of optical imaging techniques for assessing wound healing," Int. J. Intell. Control Syst. 17, 79-85 (2012).

[31] R. Cicchi, D. Kapsokalyvas, and F. S. Pavone, "Clinical nonlinear laser imaging of human skin: a review," BioMed Res. Int. 2014, 903589 (2014).

[32] S. Paddock, T. Fellers, and M. Davidson, *Confocal Microscopy*, Humana Press, Spinger, New York (2001).

[33] M. Rajadhyaksha, S. González, J. M. Zavislan, R. R. Anderson, and R. H. Webb, "In vivo confocal scanning laser microscopy of human skin II: advances in instrumentation and comparison with histology," J. Invest. Dermatol. 113, 293-303 (1999).

[34] J. Fredrich, "3D imaging of porous media using laser scanning confocal microscopy with application to microscale transport processes," Phys. Chem. Earth. Pt. A 24, 551-561 (1999).

[35] J. Lawrence, D. Korber, B. Hoyle, J. Costerton, and D. Caldwell, "Optical sectioning of microbial biofilms," J. Bacteriol. 173, 6558-6567 (1991).

[36] G. Deka, W.-W. Wu, and F.-J. Kao, "In vivo wound healing diagnosis with second harmonic and fluorescence lifetime imaging," J. Biomed. Opt. 18, 061222 (2013).

[37] M. S. Weingarten, E. Papazoglou, L. Zubkov, L. Zhu, G. Vorona, and A. Walchack, "Measurement of optical properties to quantify healing of chronic diabetic wounds," Wound. Repair. Regen. 14, 364-370 (2006).

[38] E. S. Papazoglou, L. Zubkov, L. Zhu, M. S. Weingarten, S. Tyagi, and K. Pourrezaei, "Monitoring diabetic wound healing by NIR spectroscopy," in Conf. Proc. IEEE. Eng. Med. Biol. Soc. (IEEE, 2005) Shanghai, China, pp. 6662-6664.

[39] A. Diaspro, *Confocal and Two-photon Microscopy: Foundations, Applications and Advances*, Wiley-VCH; New York, (2001).

[40] E. Ruocco, G. Argenziano, G. Pellacani, and S. Seidenari, "Noninvasive imaging of skin tumors," Dermatol. Surg. 30, 301-310 (2004).

[41] G. Micali, F. Lacarrubba, D. Massimino, and R. A. Schwartz, "Dermatoscopy: alternative uses in daily clinical practice," J. Am. Acad. Dermatol. 64, 1135-1146 (2011).

[42] S. Inui, T. Nakajima, and S. Itami, "Significance of dermoscopy in acute diffuse and total alopecia of the female scalp: review of twenty cases," Dermatology 217, 333-336 (2008).

[43] M. Wu, S. Hu, and C. Hsu, "Use of non-contact dermatoscopy in the diagnosis of scabies," Dermatol. Sinica. 26, 112-114 (2008).

[44] E.-M. De Villiers, C. Fauquet, T. R. Broker, H.-U. Bernard, and H. zur Hausen, "Classification of papillomaviruses," Virology 324, 17-27 (2004).

[45] G. Argenziano, H. P. Soyer, S. Chimenti, R. Talamini, R. Corona, F. Sera, M. Binder, L. Cerroni, G. De Rosa, and G. Ferrara, "Dermoscopy of pigmented skin lesions: results of a consensus meeting via the Internet," J. Am. Acad. Dermatol. 48, 679-693 (2003).

[46] G. Pagnanelli, H. Soyer, G. Argenziano, R. Talamini, R. Barbati, L. Bianchi, E. Campione, I. Carboni, A. Carrozzo, and M. Chimenti, "Diagnosis of pigmented skin lesions by dermoscopy: web-based training improves diagnostic performance of non-experts," Br. J. Dermatol. 148, 698-702 (2003).

[47] M. Leutenegger, E. Martin-Williams, P. Harbi, T. Thacher, W. Raffoul, M. André, A. Lopez, P. Lasser, and T. Lasser, "Real-time full field laser Doppler imaging," Biomed. Opt. Express 2, 1470-1477 (2011).

[48] K. Wårdell, A. Jakobsson, and G. E. Nilsson, "Laser Doppler perfusion imaging by dynamic light scattering," IEEE Trans Bio-Med Eng 40, 309-316 (1993).

[49] S. Kubli, B. Waeber, A. Dalle-Ave, and F. Feihl, "Reproducibility of laser Doppler imaging of skin blood flow as a tool to assess endothelial function," J. Cardiovasc. Pharmaco. 36, 640-648 (2000).

[50] Z. Niazi, T. Essex, R. Papini, D. Scott, N. McLean, and M. Black, "New laser Doppler scanner, a valuable adjunct in burn depth assessment," Burns 19, 485-489 (1993).

[51] M. Leutenegger, E. Martin-Williams, P. Harbi, T. Thacher, W. Raffoul, M. André, A. Lopez, P. Lasser, and T. Lasser, "Real-time full field laser Doppler imaging," Biomed. Opt. Express 2, 1470-1477 (2011).

[52] S. A. Pape, C. A. Skouras, and P. O. Byrne, "An audit of the use of laser Doppler imaging (LDI) in the assessment of burns of intermediate depth," Burns 27, 233-239 (2001).

[53] E. Droog, W. Steenbergen, and F. Sjöberg, "Measurement of depth of burns by laser Doppler perfusion imaging," Burns 27, 561-568 (2001).

[54] J. M. Kainerstorfer, M. Ehler, F. Amyot, M. Hassan, S. G. Demos, V. Chernomordik, C. K. Hitzenberger, A. H. Gandjbakhche, and J. D. Riley, "Principal component model of multispectral data for near real-time skin chromophore mapping," J. Biomed. Opt. 15, 046007 (2010).

[55] M. G. Sowa, L. Leonardi, J. R. Payette, J. S. Fish, and H. H. Mantsch, "Near infrared spectroscopic assessment of hemodynamic changes in the early post-burn period," Burns 27, 241-249 (2001).

[56] K. M. Cross, L. Leonardi, M. Gomez, J. R. Freisen, M. A. Levasseur, B. J. Schattka, M. G. Sowa, and J. S. Fish, "Noninvasive measurement of edema in partial thickness burn wounds," J. Burn. Care. Res. 30, 807-817 (2009).

[57] A. Nouvong, B. Hoogwerf, E. Mohler, B. Davis, A. Tajaddini, and E. Medenilla, "Evaluation of diabetic foot ulcer healing with hyperspectral imaging of oxyhemoglobin and deoxyhemoglobin," Diabetes care 32, 2056-2061 (2009).

[58] L. Khaodhiar, T. Dinh, K. T. Schomacker, S. V. Panasyuk, J. E. Freeman, R. Lew, T. Vo, A. A. Panasyuk, C. Lima, and J. M. Giurini, "The use of medical hyperspectral technology to evaluate microcirculatory changes in diabetic foot ulcers and to predict clinical outcomes," Diabetes Care 30, 903-910 (2007).

[59] M. A. Calin, S. V. Parasca, R. Savastru, and D. Manea, "Characterization of burns using hyperspectral imaging technique – a preliminary study," Burns 41, 118-124 (2015).

[60] D. Yudovsky, A. Nouvong, K. Schomacker, and L. Pilon, "Assessing diabetic foot ulcer development risk with hyperspectral tissue oximetry," J. Biomed. Opt. 16, 026009 (2011).

[61] H. Tehrani, M. Moncrieff, B. Philp, and P. Dziewulski, "Spectrophotometric intracutaneous analysis: a novel imaging technique in the assessment of acute burn depth," Ann. N Y. Acad. Sci. 61, 437-440 (2008).

[62] V. Hughes, P. Ellis, T. Burt, and N. Langlois, "The practical application of reflectance spectrophotometry for the demonstration of haemoglobin and its degradation in bruises," J. Clin. Pathol. 57, 355-359 (2004).

[63] M. Bohnert, R. Baumgartner, and S. Pollak, "Spectrophotometric evaluation of the colour of intra-and subcutaneous bruises," Int. J. Dermatol. 113, 343-348 (2000).

[64] W. Groner, J. W. Winkelman, A. G. Harris, C. Ince, G. J. Bouma, K. Messmer, and R. G. Nadeau, "Orthogonal polarization spectral imaging: a new method for study of the microcirculation," Nat. Med. 5, 1209-1212 (1999).

[65] S. Langer, F. Born, R. Hatz, P. Biberthaler, and K. Messmer, "Orthogonal polarization spectral imaging versus intravital fluorescent microscopy for microvascular studies in wounds," Ann. N Y. Acad. Sci. 48, 646-653 (2002).

[66] O. Goertz, A. Ring, A. Köhlinger, A. Daigeler, C. Andree, H.-U. Steinau, and S. Langer, "Orthogonal polarization spectral imaging: a tool for assessing burn depths?" Ann. N Y. Acad. Sci. 64, 217-221 (2010).

[67] S. M. Milner, "Predicting early burn wound outcome using orthogonal polarization spectral imaging," Int. J. Dermatol. 41, 715-715 (2002).

[68] S. M. Milner, S. Bhat, S. Gulati, G. Gherardini, C. E. Smith, and R. J. Bick, "Observations on the microcirculation of the human burn wound using orthogonal polarization spectral imaging," Burns 31, 316-319 (2005).

[69] A. Renkielska, A. Nowakowski, M. Kaczmarek, and J. Ruminski, "Burn depths evaluation based on active dynamic IR thermal imaging—a preliminary study," Burns 32, 867-875 (2006).

[70] N. J. Prindeze, P. Fathi, M. J. Mino, N. A. Mauskar, T. E. Travis, D. W. Paul, L. T. Moffatt, and J. W. Shupp, "Examination of the early diagnostic applicability of active dynamic thermography for burn wound depth assessment and concept analysis," J. Burn. Care. Res. 36, 626-635 (2015).

[71] R. N. Lawson, G. Wlodek, and D. Webster, "Thermographic assessment of burns and frostbite," Can. Med. Assoc. J. 84, 1129-1131 (1961).

[72] A. Renkielska, A. Nowakowski, M. Kaczmarek, M. K. Dobke, J. Grudziński, A. Karmolinski, and W. Stojek, "Static thermography revisited—an adjunct method for determining the depth of the burn injury," Burns 31, 768-775 (2005).

[73] A. F. Fercher, "Optical coherence tomography–development, principles, applications," Zeitschrift Med. Phys. 20, 251-276 (2010).

[74] D. Huang, E. A. Swanson, C. P. Lin, J. S. Schuman, W. G. Stinson, W. Chang, M. R. Hee, T. Flotte, K. Gregory, and C. A. Puliafito, "Optical coherence tomography," Science 254, 1178-1181 (1991).

[75] E. Sattler, R. Kästle, and J. Welzel, "Optical coherence tomography in dermatology," J. Biomed. Opt. 18, 061224 (2013).

[76] A. Pagnoni, A. Knuettel, P. Welker, M. Rist, T. Stoudemayer, L. Kolbe, I. Sadiq, and A. Kligman, "Optical coherence tomography in dermatology," Skin. Res. Technol. 5, 83-87 (1999).

[77] L. M. Sakata, J. DeLeon-Ortega, V. Sakata, and C. A. Girkin, "Optical coherence tomography of the retina and optic nerve–a review," Clin. Experiment. Ophthalmol. 37, 90-99 (2009).

[78] S. Chua, "High-Definition Optical Coherence Tomography for the Study of Evolution of a Disease," Dermatol. Bullet. 26, 2-3 (2015).

[79] V. R. Korde, G. T. Bonnema, W. Xu, C. Krishnamurthy, J. Ranger-Moore, K. Saboda, L. D. Slayton, S. J. Salasche, J. A. Warneke, and D. S. Alberts, "Using optical coherence tomography to evaluate skin sun damage and precancer," Lasers Surg. Med. 39, 687-695 (2007).

[80] M. Mogensen, L. Thrane, T. M. Jørgensen, P. E. Andersen, and G. B. Jemec, "OCT imaging of skin cancer and other dermatological diseases," J. Biophotonics 2, 442-451 (2009).

[81] O. Babalola, A. Mamalis, H. Lev-Tov, and J. Jagdeo, "Optical coherence tomography (OCT) of collagen in normal skin and skin fibrosis," Ann. N Y. Acad. Sci. 306, 1-9 (2014).

[82] S. Sakai, M. Yamanari, Y. Lim, N. Nakagawa, and Y. Yasuno, "In vivo evaluation of human skin anisotropy by polarization-sensitive optical coherence tomography," Biomed. Opt. Express 2, 2623-2631 (2011).

[83] M. C. Pierce, J. Strasswimmer, B. H. Park, B. Cense, and J. F. de Boer, "Advances in optical coherence tomography imaging for dermatology," J. Invest. Dermatol. 123, 458-463 (2004).

[84] Y. Zhao, Z. Chen, C. Saxer, S. Xiang, J. F. de Boer, and J. S. Nelson, "Phase-resolved optical coherence tomography and optical Doppler tomography for imaging blood flow in human skin with fast scanning speed and high velocity sensitivity," Opt. Lett. 25, 114-116 (2000).

[85] J. Welzel, "Optical coherence tomography in dermatology: a review," Skin. Res. Technol. 7, 1-9 (2001).

[86] T. Gambichler, A. Pljakic, and L. Schmitz, "Recent advances in clinical application of optical coherence tomography of human skin," Clin. Cosmet. Investig. Dermatol. 8, 345-354 (2015).

[87] N. Greaves, B. Benatar, S. Whiteside, T. Alonso-Rasgado, M. Baguneid, and A. Bayat, "Optical coherence tomography: a reliable alternative to invasive histological assessment of acute wound healing in human skin?" Br. J. Dermatol. 170, 840-850 (2014).

[88] K. Sahu, Y. Verma, M. Sharma, K. Rao, and P. Gupta, "Non-invasive assessment of healing of bacteria infected and uninfected wounds using optical coherence tomography," Skin. Res. Technol. 16, 428-437 (2010).

[89] E. C. E. Sattler, K. Poloczek, R. Kästle, and J. Welzel, "Confocal laser scanning microscopy and optical coherence tomography for the evaluation of the kinetics and quantification of wound healing after fractional laser therapy," J. Am. Acad. Dermatol. 69, e165-e173 (2013).

[90] M. J. Cobb, Y. Chen, R. A. Underwood, M. L. Usui, J. Olerud, and X. Li, "Noninvasive assessment of cutaneous wound healing using ultrahigh-resolution optical coherence tomography," J. Biomed. Opt. 11, 064002 (2006).

[91] J. G. Fujimoto, C. Pitris, S. A. Boppart, and M. E. Brezinski, "Optical coherence tomography: an emerging technology for biomedical imaging and optical biopsy," Neoplasia 2, 9-25 (2000).

[92] G. Zhang, D. J. Moore, C. R. Flach, and R. Mendelsohn, "Vibrational microscopy and imaging of skin: from single cells to intact tissue," Anal. Bioanal. Chem. 387, 1591-1599 (2007).

[93] H. Lui, J. Zhao, D. McLean, and H. Zeng, "Real-time Raman spectroscopy for in vivo skin cancer diagnosis," Cancer. Res. 72, 2491-2500 (2012).

[94] Y. Li, R. Chen, H. Zeng, Z. Huang, S. Feng, and S. Xie, "Raman spectroscopy of Chinese human skin in vivo," Chin. Opt. Lett. 5, 105-107 (2007).

[95] P. Caspers, G. Lucassen, H. Bruining, and G. Puppels, "Automated depth-scanning confocal Raman microspectrometer for rapid in vivo determination of water concentration profiles in human skin," J. Raman Spectro. 31, 813-818 (2000).

[96] H.-W. Wang, T. T. Le, and J.-X. Cheng, "Label-free imaging of arterial cells and extracellular matrix using a multimodal CARS microscope," Opt. Commun. 281, 1813-1822 (2008).

[97] C. L. Evans, E. O. Potma, M. Puoris' haag, D. Côté, C. P. Lin, and X. S. Xie, "Chemical imaging of tissue in vivo with video-rate coherent anti-Stokes Raman scattering microscopy," Proc. Natl. Acad. Sci. U. S. A. 102, 16807-16812 (2005).

[98] M. N. Slipchenko, T. T. Le, H. Chen, and J.-X. Cheng, "High-speed vibrational imaging and spectral analysis of lipid bodies by compound Raman microscopy," J. Phys. Chem. B 113, 7681-7686 (2009).

[99] D. Haluszka, R. Szipocs, N. Wikonkal, and A. Kolonics, "Characterization of obesity in murine skin in vivo by CARS and SHG microscopy using a cost efficient, fiber laser based wavelength extension unit," in 6th EPS-QEOD Europhoton Conference. Neuchatel, Switzerland (2014).

[100] P. D. Pudney, M. Mélot, P. J. Caspers, A. Van Der Pol, and G. J. Puppels, "An in vivo confocal Raman study of the delivery of trans retinol to the skin," Appl. Spectrosc. 61, 804-811 (2007).

[101] M. Weinigel, H. Breunig, M. Kellner-Höfer, R. Bückle, M. Darvin, M. Klemp, J. Lademann, and K. König, "In vivo histology: optical biopsies with chemical contrast using clinical multiphoton/coherent anti-Stokes Raman scattering tomography," Laser Phys. Lett. 11, 055601 (2014).

[102] S. Lange-Asschenfeldt, A. Bob, D. Terhorst, M. Ulrich, J. Fluhr, G. Mendez, H.-J. Roewert-Huber, E. Stockfleth, and B. Lange-Asschenfeldt, "Applicability of confocal

laser scanning microscopy for evaluation and monitoring of cutaneous wound healing," J. Biomed. Opt. 17, 076016 (2012).

[103] M. Rajadhyaksha, M. Grossman, D. Esterowitz, R. H. Webb, and R. R. Anderson, "In vivo confocal scanning laser microscopy of human skin: melanin provides strong contrast," J. Invest. Dermatol. 104, 946-952 (1995).

[104] M. Minsky, "Memoir on inventing the confocal scanning microscope," Scanning 10, 128-138 (1988).

[105] T. Wilson, "The role of the pinhole in confocal imaging system," in *Handbook of Biological Confocal Microscopy*, Springer, New York, pp. 167-182 (1995).

[106] F. Helmchen, and W. Denk, "Deep tissue two-photon microscopy," Nat. Methods 2, 932-940 (2005).

[107] P. Calzavara-Pinton, C. Longo, M. Venturini, R. Sala, and G. Pellacani, "Reflectance confocal microscopy for in vivo skin imaging," Photochem. Photobiol. 84, 1421-1430 (2008).

[108] J. Still, E. Law, K. Klavuhn, T. Island, and J. Holtz, "Diagnosis of burn depth using laser induced indocyanine green fluorescence: a preliminary clinical trial," Burns 27, 364-371 (2001).

[109] M. E. Gschwandtner, E. Ambrózy, S. Marić, A. Willfort, B. Schneider, K. Böhler, U. Gaggl, and H. Ehringer, "Microcirculation is similar in ischemic and venous ulcers," Microvasc. Res. 62, 226-235 (2001).

[110] V. Prabhu, S. B. Rao, E. M. Fernandes, A. C. Rao, K. Prasad, and K. K. Mahato, "Objective assessment of endogenous collagen in vivo during tissue repair by laser induced fluorescence," PloS One 9, e98609 (2014).

[111] V. Prabhu, S. B. Rao, S. Chandra, P. Kumar, L. Rao, V. Guddattu, K. Satyamoorthy, and K. K. Mahato, "Spectroscopic and histological evaluation of wound healing progression following Low Level Laser Therapy (LLLT)," J. Biophotonics 5, 168 (2012).

[112] S.-J. Lin, R.-J. Wu, H.-Y. Tan, W. Lo, W.-C. Lin, T.-H. Young, C.-J. Hsu, J.-S. Chen, S.-H. Jee, and C.-Y. Dong, "Evaluating cutaneous photoaging by use of multiphoton fluorescence and second-harmonic generation microscopy," Opt. Lett. 30, 2275-2277 (2005).

[113] T. Luo, J. Chen, S. Zhuo, K. Lu, X. Jiang, and Q. Liu, "Visualization of collagen regeneration in mouse dorsal skin using second harmonic generation microscopy," Laser Phys. 19, 478-482 (2009).

[114] E. Brown, T. McKee, A. Pluen, B. Seed, Y. Boucher, and R. K. Jain, "Dynamic imaging of collagen and its modulation in tumors in vivo using second-harmonic generation," Nat. Med. 9, 796-800 (2003).

[115] A. Zoumi, A. Yeh, and B. J. Tromberg, "Imaging cells and extracellular matrix in vivo by using second-harmonic generation and two-photon excited fluorescence," Proc. Natl. Acad. Sci. U. S. A. 99, 11014-11019 (2002).

[116] B. R. Masters, P. T. So, and E. Gratton, "Multiphoton excitation microscopy of in vivo human skin: functional and morphological optical biopsy based on three-dimensional imaging, lifetime measurements and fluorescence spectroscopy," Ann. N Y. Acad. Sci. 838, 58-67 (1998).

[117] B. Masters, P. So, and E. Gratton, "Optical biopsy of in vivo human skin: multi-photon excitation microscopy," Lasers Med. Sci. 13, 196-203 (1998).

[118] J. A. Palero, H. S. De Bruijn, H. J. Sterenborg, and H. C. Gerritsen, "Spectrally resolved multiphoton imaging of in vivo and excised mouse skin tissues," Biophys. J. 93, 992-1007 (2007).

[119] L. H. Laiho, S. Pelet, T. M. Hancewicz, P. D. Kaplan, and P. T. So, "Two-photon 3-D mapping of ex vivo human skin endogenous fluorescence species based on fluorescence emission spectra," J. Biomed. Opt. 10, 024016 (2005).

[120] J. Paoli, M. Smedh, A.-M. Wennberg, and M. B. Ericson, "Multiphoton laser scanning microscopy on non-melanoma skin cancer: morphologic features for future noninvasive diagnostics," J. Invest. Dermatol. 128, 1248-1255 (2008).

[121] W. R. Zipfel, R. M. Williams, and W. W. Webb, "Nonlinear magic: multiphoton microscopy in the biosciences," Nat. Biotechnol. 21, 1369-1377 (2003).

[122] A. Gafni, and L. Brand, "Fluorescence decay studies of reduced nicotinamide adenine dinucleotide in solution and bound to liver alcohol dehydrogenase," Biochemistry 15, 3165-3171 (1976).

[123] M. Wakita, G. Nishimura, and M. Tamura, "Some characteristics of the fluorescence lifetime of reduced pyridine nucleotides in isolated mitochondria, isolated hepatocytes, and perfused rat liver in situ," J. Biochem. 118, 1151-1160 (1995).

[124] M. C. Skala, K. M. Riching, A. Gendron-Fitzpatrick, J. Eickhoff, K. W. Eliceiri, J. G. White, and N. Ramanujam, "In vivo multiphoton microscopy of NADH and FAD redox states, fluorescence lifetimes, and cellular morphology in precancerous epithelia," Proc. Natl. Acad. Sci. U. S. A. 104, 19494-19499 (2007).

[125] J. Chen, C. Guo, F. Zhang, Y. Xu, X. Zhu, S. Xiong, and J. Chen, "In-vivo monitoring rat skin wound healing using nonlinear optical microscopy," Proc. of SPIE BiOS (International Society for Optics and Photonics, 2014) San Francisco, California, United States, 894820-894827.

[126] R. Cicchi, S. Sestini, V. De Giorgi, D. Massi, T. Lotti, and F. Pavone, "Nonlinear laser imaging of skin lesions," J. Biophotonics 1, 62-73 (2008).

[127] P. Stoller, K. M. Reiser, P. M. Celliers, and A. M. Rubenchik, "Polarization-modulated second harmonic generation in collagen," Biophys. J. 82, 3330-3342 (2002).

[128] R. Cicchi, D. Kapsokalyvas, V. De Giorgi, V. Maio, A. Van Wiechen, D. Massi, T. Lotti, and F. S. Pavone, "Scoring of collagen organization in healthy and diseased human dermis by multiphoton microscopy," J. Biophotonics 3, 34-43 (2010).

[129] T. Yasui, Y. Tohno, and T. Araki, "Characterization of collagen orientation in human dermis by two-dimensional second-harmonic-generation polarimetry," J. Biomed. Opt. 9, 259-264 (2004).

[130] B. A. Torkian, A. T. Yeh, R. Engel, C.-H. Sun, B. J. Tromberg, and B. J. Wong, "Modeling aberrant wound healing using tissue-engineered skin constructs and multiphoton microscopy," Arch. Dermatol.L 6, 180-187 (2004).

[131] L. Cuttle, M. Nataatmadja, J. F. Fraser, M. Kempf, R. M. Kimble, and M. T. Hayes, "Collagen in the scarless fetal skin wound: detection with Picrosirius-polarization," Wound. Repair. Regen. 13, 198-204 (2005).

[132] G. Deka, K. Okano, W.-W. Wu, and F.-J. Kao, "Multiphoton microscopy for skin wound healing study in terms of cellular metabolism and collagen regeneration," in SPIE BiOS (International Society for Optics and Photonics, 2014), pp. 894820-894820-894827.

[133] P. J. Campagnola, A. C. Millard, M. Terasaki, P. E. Hoppe, C. J. Malone, and W. A. Mohler, "Three-dimensional high-resolution second-harmonic generation imaging of endogenous structural proteins in biological tissues," Biophys. J. 82, 493-508 (2002).

[134] S.-W. Chu, T.-Y. Su, R. Oketani, Y.-T. Huang, H.-Y. Wu, Y. Yonemaru, M. Yamanaka, H. Lee, G.-Y. Zhuo, and M.-Y. Lee, "Measurement of a saturated emission of optical radiation from gold nanoparticles: application to an ultrahigh resolution microscope," Phys. Rev. Lett. 112, 017402 (2014).

Chapter 8

Automated Identification and Measurement of Haematopoietic Stem Cells in 3D Intravital Microscopy Data

Reema Adel Khorshed and Cristina Lo Celso

Additional information is available at the end of the chapter

http://dx.doi.org/10.5772/64089

Abstract

Image analysis and quantification of Haematopoietic stem cells (HSCs) position within their surrounding microenvironment in the bone marrow is a fast growing area of research, as it holds the key to understanding the dynamics of HSC-niche interactions and their multiple implications in normal tissue development and in response to various stress events. However, this area of research is very challenging due to the complex cellular structure of such images. Therefore, automated image analysis tools are required to simplify the biological interpretation of 3D HSC microenvironment images. In this chapter, we describe how 3D intravital microscopy data can be visualised and analysed using a computational method that allows the automated quantification of HSC position relative to surrounding niche components.

Keywords: intravital microscopy, 3D image analysis, bone marrow visualisation, haematopoietic stem cell niche, object segmentation and classification

1. Introduction

Somatic stem cells have the extraordinary ability to maintain their own pool, while replenishing dead cells and regenerating tissues after injuries throughout our lifetime. Dividing stem cells have the potential to differentiate into other cell types such as blood cells, skin cells or brain cells, or maintain themselves through a process called "self-renewal". The stem cells that form blood and immune cells are called haematopoietic stem cells (HSCs). These stem cells are responsible for turnover and maintenance of red blood cells, platelets and immune cells. During differentiation, HSCs generate multi-potent and lineage-committed progenitor cells prior to attaining maturity, which results in the generation of billions of new blood cells every day [1].

The bone marrow (BM) is the main site, where HSCs and their immediate progeny reside, precisely contained in a regulated and a very complex environment called the niche. The niche microenvironment has a direct impact on the function of HSCs, determined by the interaction of the HSCs themselves with particular cellular and molecular components in their surroundings [1, 2].

To observe and understand such interactions, single-cell resolution intravital microscopy of fluorescent HSCs and niche components is a vital tool, allowing the study of stem cells, their behaviour and interactions during steady state, aging and disease [3, 4]. This method has been successfully utilised to directly observe the HSCs with their niche in mouse bone marrow. Fluorescent dyes were used to label HSCs prior to transplantation into transgenic reporter mice that expressed GFP under the control of an osteoblast-specific promoter. Osteoblast cells reside on the bone marrow endosteal surface and contribute to the formation of bone [5]. This kind of fluorescent niche component served as a fundamental tool for the visualisation of the HSC endosteal niche and resulted in multiple observations indicating that functional HSCs localise near osteoblastic cells [5–7]. Intravital microscopy of HSCs injected into wild type and other bone marrow stroma/haematopoietic reporters indicated the importance of GalphaS receptor subunit on homing HSCs [8], of interactions with nestin-positive mesenchymal stem and progenitor cells [9] and with regulatory T cells [10]. Time-lapse imaging has revealed that infection-exposed HSCs interact with larger endosteal niches [7, 11] and that aged HSCs interact with the bone marrow differently than young ones [12]. However, a clear understanding of how the coordinated action of multiple niche components regulates HSC fate is not clear, and considering the localization of HSCs relative to multiple surrounding cellular and structural constituents of the bone marrow microenvironment is the first step towards unravelling HSCs dynamics and signal exchanges.

Even though manual distance measurement of HSCs to other niche components is possible [1, 6, 13], it suffers from various limitations including the extensive time spent on conducting such manual analysis as well as intra- and inter-researcher inconsistencies and human error. Therefore, a specialised image analysis tool capable of completing such tasks will add the great benefits of producing consistent unbiased results, and will simplify the interpretation of the biological microenvironment events of the bone marrow niche.

During the last decade, several attempts have been made to develop image analysis approaches to automate the quantification of medical and biological images, from cell segmentation and detection to tumour diagnosis and classification [14–16]. However, the direct application of such methods to three-dimensional (3D) intravital microscopy images of the bone marrow is not possible. This is due to the complex structure of the tissue, decreased signal with increased depths, unpredictable scattering of both the excitation and emission photons and the irregular encasing bone structure, which restricts the resolution of intravital microscopy images when compared to *ex vivo* and soft tissue imaging.

Consequently, we developed an image analysis pipeline, to specifically tackle all the main challenges associated with intravital microscopy 3D images of bone marrow. The approach starts by detecting and segmenting HSCs and surrounding niche components, for instance, the bone and osteoblast. Then it classifies each segmented HSCs and assigns them into one of three classes, and finally it quantifies them based on their location to surrounding niche components. This analysis will allow the extraction of variable quantitative features, which could potentially reveal novel aspects of HSC biology, and especially their relation to the bone marrow microenvironment [11].

This chapter is organised as follows: first an introduction of the bone marrow and haematopoietic stem cells niche highlighting the importance of intravital microscopy and automated image analysis is provided (Section 2), followed by a detailed description of intravital microscopy of the bone marrow and the main challenges faced during the image analysis stage (Section 3). The automated image analysis tool is then described (Section 4), detailing how the three main segmentation parameters can be optimised for a particular cell or niche component (Section 5). Finally, a description of how a supervised machine learning classifier can be used to classify both HSCs and vasculature is presented (Section 6), followed by a brief description of how to obtain 3D positional measurements of HSCs to other niche components (Section 7).

2. Intravital microscopy of mouse bone marrow

In vivo imaging of stem cells is a growing field, providing unique insights of their behaviour and especially of their interaction with their surrounding microenvironment. This approach has been instrumental in producing new hypotheses and revealed a number of novel findings concerning the regulation of fate decision of multiple somatic stem cells [17, 18], including HSCs [1, 3, 4].

Nevertheless, how multiple niche components affect and regulate HSCs fate is still a significant question that scientists are attempting to answer. Recording the localization of HSCs relative to multiple surrounding cellular and structural constituents of the bone marrow microenvironment is the first step towards understanding such phenomena.

Confocal fluorescence and second harmonic generation microscopy of HSCs and their niche components in the mouse calvarium (top of the skull) bone marrow provides a powerful tool to observe cellular interactions and has been successfully used to detect and study fluorescently labelled HSCs and GFP expressing osteoblasts and other niche components in transgenic reporter mice [7, 11].

Confocal microscopy provides a better 3D optical resolution than epifluorescence microscopy because it restricts the light that reaches the photomultiplier through a pinhole. While in epifluorescence microscopy in-focus image objects are mixed with out-of-focus image information arising from regions outside the focal plane, in confocal microscopy the pinhole blocks out-of-focus signal. This technology has allowed major advances in the field of biological imaging, due to its cost-effective solution and ease of use. In addition, confocal microscopes have improved sensitivity, resolution and intensity compared to epifluorescent microscopes. Confocal microscopy is also less phototoxic and can therefore be used to generate not only static but also dynamic data of living cells within their tissues, through time-lapse acquisition. All these qualities make confocal microscopes an ideal instrument for *in vivo* studies.

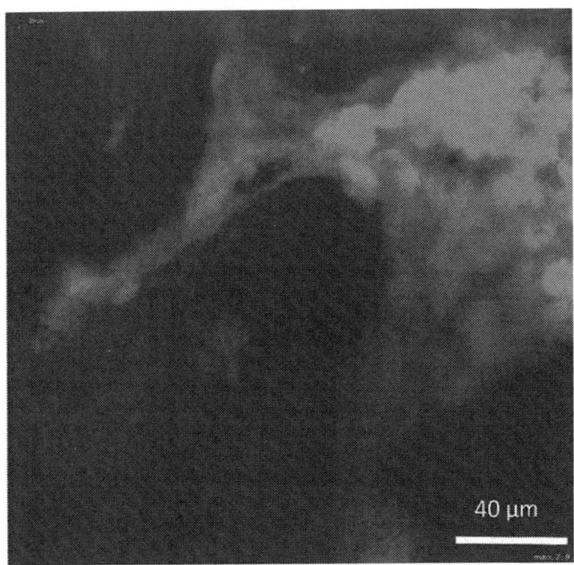

Figure 1. Maximum intensity projection of a 3D stack example of raw bone marrow *in vivo* images including DiD signal (red), GFP signal from osteoblastic cells (green) and bone collagen SHG signal (blue). Confocal microscopy was used to generate the signal for both DiD-labelled HSCs and osteoblasts using 633 and 488 nm lasers, respectively. Second harmonic generation (SHG) signal was obtained using 840-nm excitation of collagen to resolve bone structures.

Two-photon excitation of a fluorophore provides 3D optical sectioning similar to confocal imaging. However, it uses a wavelength roughly twice the length of the absorption peak of the specimen being imaged. In this chapter, we describe a particular application of two-photon excitation, named second harmonic generation (SHG). This takes advantage of a characteristic specific to certain molecules, such as collagen, which is the emission of photons at half of the wavelength of the exciting photons. Because collagen is one of the main components of bone, SHG signal allows detection of calvarium bone without the need to use specific transgenic fluorescent reporter mouse strains. An example of raw bone marrow *in vivo* images containing HSCs, osteoblast and SHG bone signal is presented in **Figure 1**. Two-photon excitation is also widely used to detect signal from multiple fluorophores, including GFP and tomato fluorescent protein [17], and signal obtained from two-photon excitation fluorescence microscopy can be analysed with the same computational algorithm we describe below.

3. Application and challenges of bone marrow intravital microscopy image analysis

The combination of confocal and SHG microscopy has greatly simplified high-resolution fluorescence imaging of animal tissues and organs. In Section 2, we have briefly described the advantages of the combination of confocal/SHG microscopy. Here, we describe the experimental set-up used and the challenges posed by the intravital images when they have to be further analysed. In the following sections, we describe in details how HSCs and bone marrow components can be automatically segmented, classified and measured.

To visualise HSCs, they were first labelled *ex vivo* using lipophilic membrane dyes such as 1,1′-Dioctadeciyl-3,3,30,30-Tetramethylindodicarbocyanine (DiD) to generate a bright fluorescent signal [1, 5]. DiD-labelled HSCs were then injected into Col2.3GFP recipient mice, which allow the visualisation of osteoblasts as GFP-positive cells. Confocal microscopy was used to generate the signal for both DiD-labelled HSCs and osteoblasts using 633 and 488 nm lasers, respectively. Second harmonic generation (SHG) signal was obtained from 840-nm excitation of collagen to resolve bone structures. Acquisition setting (e.g. gain, laser powers, step and stack size) can vary between multiple users, leading to overall brighter/dimmer images and a range of field of view sizes and depths. Such variance imposes a challenge for automated image analysis. Therefore, developing an image analysis tool that can deal with the variability of setting preferences is a crucial point.

Another challenge imposed on image analysis from *in vivo* microscopy data is the decreased signal at increased depths. This results in non-uniform intensity and brightness of objects. In addition, light scattering caused by the surrounding tissue and bone restricts the resolution of *in vivo* microscopy of bone marrow compared to that of other soft tissues or *ex vivo* techniques.

To address these issues, we developed a local heterogeneity-based image segmentation (LH-SEG) approach [11] that employs multi-resolution segmentation [19] and mean intensity difference to neighbour thresholding. The approach works by comparing local morphological and intensity characteristics of objects, which most often are smaller than the cells and structures recorded. These detected objects are then grouped based on their homogeneity with other neighbouring objects within a defined distance. Because the method is applied to each 2D slide, it ensures reliable edge detection and segmentation across cells and structures with high intensity heterogeneity, despite the decreased signal (intensity) at increasing depths.

DiD labelling produces a bright fluorescent signal of HSCs [1, 5]. However, this and other related dyes have a number of limitations when associated with intravital microscopy:

1. The dye often diffuses from the labelled cells into the surroundings, and is diluted upon cell division, resulting in a loss of brightness and intensity of signal.

2. Labelling HSCs does not always provide a homogenous staining, which can result in some HSCs being brighter than others.

3. The dye often leads to background noise as result of cell debris and aggregate signal.

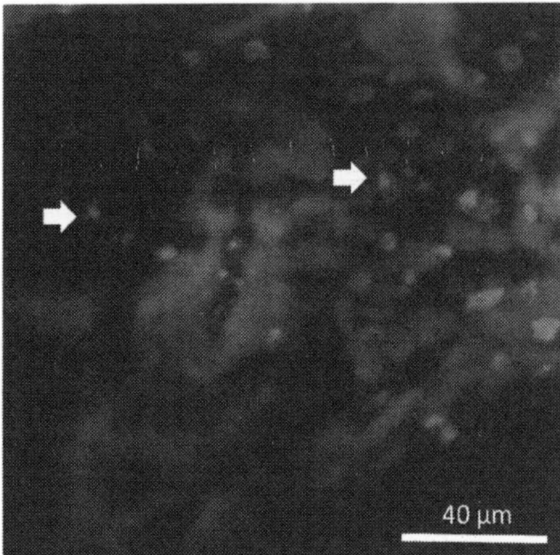

Figure 2. Maximum intensity projection of a 3D stack example representing the challenges for automated analysis of raw bone marrow *in vivo* images. DiD signal is in red. Arrows point at single DiD-labelled HSCs with varying intensity levels and sizes; the remaining red signal represents debris and background noise. Osteoblast cells (green) show non-uniform intensity levels too.

Figure 2 shows examples of the points listed above.

To overcome these issues, LH-SEG detects cells with variable intensities and signal brightness. Moreover, a machine learning classification protocol based on morphological and textural features recognises and classifies all segmented DiD signal, to differentiate real HSCs from cell debris and aggregate.

4. Image analysis

Image analysis is the process of extracting meaningful information from images, using manual or automated methods (the latter known as computer vision techniques) [20, 21]. The selection of the appropriate image analysis method for a certain type of images determines the success rate of the analysis. Therefore, understanding the challenges associated with the images acquired is the first step towards developing an effective image analysis solution.

Automated image analysis has a number of advantages over the human manual analysis. Human vision can easily be biased by pre-conceived concepts, affecting the output results and the rigorous testing of hypothesis. Manual analysis can also be time-consuming compared to automated analysis tasks, in which large datasets can be batch processed without the need of human monitoring, allowing users to perform other tasks while the analysis is being carried out.

Microscopy images are often complex, with a range of artefacts and background noise, which require variable image processing steps before any meaningful quantification can be extracted from objects and region of interests. An overall image analysis protocol, including image processing, analysis and data output, needs to be designed and tailored according to the targeted image datasets. A general outline of the image analysis pipeline described in this chapter is provided in **Figure 3**.

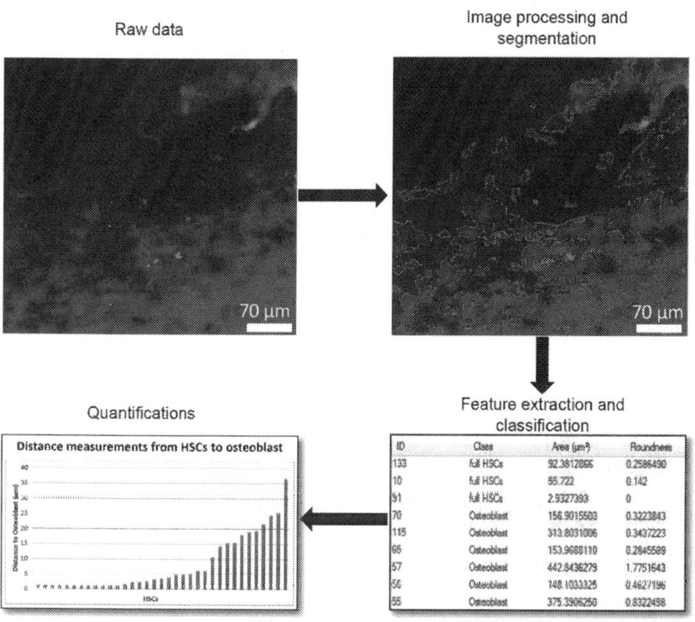

Figure 3. Outline of our image analysis protocol, starting from raw data and applying image processing and segmentation, feature extraction and classification, and finally distance measurements.

4.1. Image acquisition and processing

Image acquisition is the process of obtaining the raw image data using a microscope. Such data usually contain a number of imperfections, for example, due to oversaturation, out of focus signal and uneven excitation due to the irregular structure of the tissue itself.

To overcome these issues, filtering techniques are usually introduced to the image analysis as a pre-processing step. Smoothing and de-nosing filters such as the median filter and Gaussian filter are applied to enhance image quality and reduces the noise introduced by the image acquisition process.

In the case of our *in vivo* bone marrow images, we applied Gaussian blur [22] to handle fluorescence intensity heterogeneity. The blur is typically used to smooth images detail and reduce noise from uneven signal intensity. A small kernel size of (3 × 3) was used to reduce noise while avoiding over-smoothing of image objects.

4.2. Image segmentation

Object detection is a crucial step, as it determines the objects of interest in an image. This step ultimately governs the quality of quantifications extracted from each object. A very effective method of object detection is image segmentation. Segmentation is the process of dividing an image into smaller meaningful segments, by selecting a group of pixels to represent a region, or an object, contained within a border. Objects are segmented such that the pixels enclosed within one border share certain characteristics that define them as an object.

Successful segmentation is dependent on the method used, which critically needs to be tailored for each particular set of images. Images with a high level of contrast between objects of interests and background and with uniform intensities can be segmented using simple thresholding and edge detection approaches. More complex images with high intensity heterogeneity and intricate structures such as *in vivo* single cell resolution images require a more sophisticated segmentation approach.

The main challenges imposed upon segmentation of bone marrow *in vivo* images are the heterogeneity of fluorescent intensity as represented in GFP and DiD signal in **Figure 2**, the loss of signal with increasing depths and the unpredictable shape of stroma components. To overcome these issues, we have developed a two-step method of segmentation (LH-SEG). This method combines two powerful approaches: the multi-resolution segmentation and mean intensity difference to neighbourhood thresholding (MDN).

This segmentation approach was mainly developed to minimise artefacts due to loss of signal with increased depths. The first step of LH-SEG, multi-resolution segmentation, starts by dividing each 2D slice of the 3D stack into smaller segments grouped by their homogeneity in shape and texture. Homogeneity for each pixel is then calculated by selecting a scale parameter α. The parameter α is optimised for each object type (HSC, osteoblast and bone), taking into consideration their morphological and textural characteristics. Optimisation of this parameter for each object type is described in detail in Sections 4.3 and 5.

Adjacent homogeneous image segments are then merged based on their mean intensity difference to neighbourhood (MDN) threshold. MDN threshold calculates the difference between an image segment and its neighbouring image segments using the mean intensity values described in [11] as follows:

$$T_{\bar{\Delta}_k}(v) = \frac{1}{W} \sum_{u \in N_v(d)} w_u (\bar{c}_k(v) - \bar{c}_k(u)) \tag{1}$$

where w is the image channel weight. Images are weighted by the distance between the segmented image objects, defined as follows:

$$w = \sum_{u \in N_v} (d)^{w_u} \tag{2}$$

where v and u are two segmented image objects, N_v is the direct neighbour to the segmented image object v. u is defined as a direct neighbour to v if the minimum distance between them is less than or equal to d. d is the distance between neighbouring segments and defined as the radius of the segmented image object perimeter in pixel. w_u is the weight of the segmented image object defined by the difference of the mean intensity value between v and u in a given distance d. $\bar{c}_k$ is the mean intensity value of channel k. The appropriate MDN threshold for $T_{\bar{\Delta}_k}$ and distance feature d need to be selected for effective segmentation of each image object category (HSC, osteoblast and bone) (from [11]).

Figure 4 shows how, at the end of this process, segmented images on each z plane are then linked together to generate a 3D rendering of the whole stack.

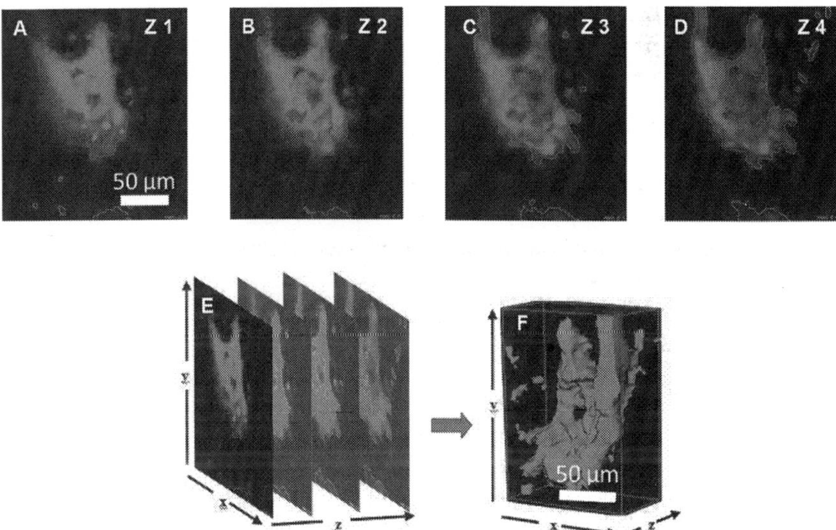

Figure 4. (A–D) segmented regions of osteoblast 2D slices using the LH_SEG method before they are merged (E) for the 3D rendering (F).

4.3. Parameter optimisation

Image segmentation is a fundamental step of image analysis, which ultimately determines the level of success of the overall quantification. Automated image analysis approaches can only perform sufficiently well when their parameters are optimally set for a certain type of images or dataset. In this section, we describe how the optimal settings can be selected for datasets containing intravital microscopy bone marrow 3D stacks. Three fundamental parameters are targeted: (1) the multi-resolution segmentation scale parameter α; (2) MDN threshold $T_{\bar{\Delta}_k}$; and (3) the MDN distance between neighbouring image segments d. The optimisation of parameters will be discussed in detail for each image object category (HSCs, osteoblast and bone) in Section 5.

The selection of the scale parameter α depends on the physical and textural structure of the objects. Different values of parameter α for the multi-resolution segmentation influence the output of the segmentation process. Selecting high scale parameters results in fewer, larger segments that can be bigger than the object observed, while lower scale parameter values result in smaller objects. To illustrate the consequences of using a range of parameter values for α on the resulting segmentation, we will present parameter values that would represent excessive cases of error when increasing or decreasing the scale parameter, as shown in **Figure 5**.

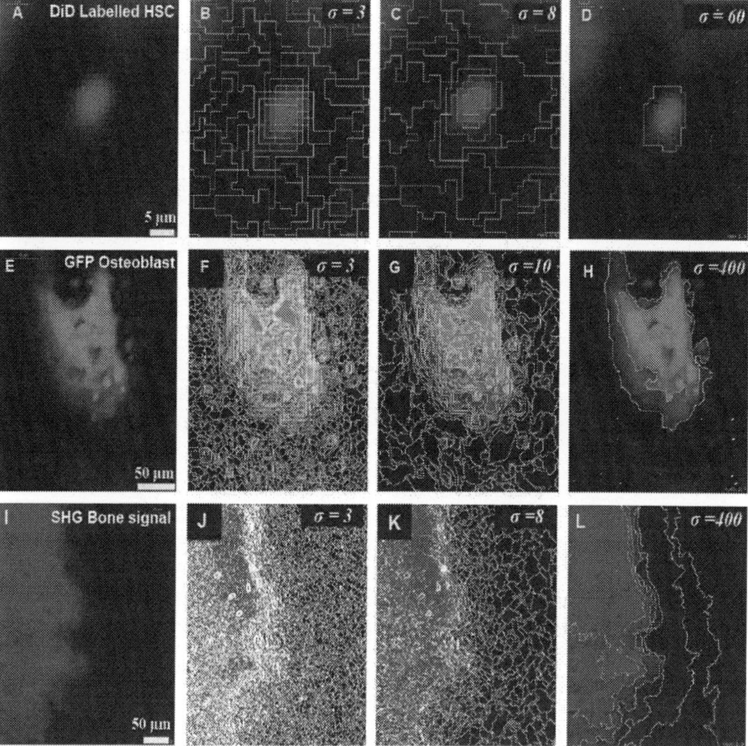

Figure 5. Optimisation of the multi-resolution segmentation parameter α for each image object category. Column (A–I) represent the data of each object category after smoothing with the convolution filter. Column (B–J) represent region segmentation error when decreasing the α parameter. Column (C–K) represent correct region segmentation with the optimal parameter α. Column (D–L) represent region segmentation error when increasing the α parameter.

MDN threshold has two parameters: MDN threshold value $T_{\bar{\Delta}_k}$ and the distance to neighbouring image segments d. Results of all three parameters optimisation for each component type (HSCs, osteoblast and bone), together with examples of the consequences of using excessively high or low parameter values, are described in Section 5 and presented in **Figures 6 and 7**.

5. Parameter optimisation

In this section, we present examples of the results obtained when different parameter values for α, $T_{\bar{\Delta}_k}$ and d are selected, to help the reader apply the same decision process to his/her own image dataset.

5.1. Multi-resolution parameter α

5.1.1. α for HSCs

Selecting a small value such as $\alpha = 3$ results in the region of the image corresponding to an HSC being divided in small segments (irregular boxes delimited by white lines, **Figure 5**) and increases the running time of the overall process. On the other hand, a scale parameter of $\alpha = 8$ leads to larger and fewer segments, covering the area occupied by the HSC and, once combined, describing its edges, therefore, serves as the optimal parameter for HSCs. Increasing the scale parameter α will result in even fewer, larger segments comprising areas of heterogeneous intensity and causing some background signal to be included in the HSC segment region (**Figure 5A–D**).

5.1.2. α for osteoblasts

Selecting a small value such as $\alpha = 3$ results in the osteoblastic cell region being divided in very small segments (irregular boxes delimited by white lines) and increases the running time of the overall process. A scale parameter of $\alpha = 10$ leads to fewer, larger segments, which cover the area occupied by the osteoblastic cells and combined describe their edges. Having fewer segments reduces the processing time. Therefore, $\alpha = 10$ serves as the optimal parameter for osteoblasts. Increasing the scale parameter α marginally did not have a major change in the segments produces, however, increasing the parameter extensively to e.g. $\alpha = 400$ results in larger segments comprising areas of heterogeneous intensity and causing some background signal to be included in the osteoblastic segment region (**Figure 5E–H**).

5.1.3. α for bone

Selecting a small value such as $\alpha = 3$ results in the bone region being divided into small segments (irregular boxes delimited by white lines) and increases the running time of the overall process. Similarly to osteoblasts, a scale parameter $\alpha = 8$ leads to fewer, larger segments covering the area occupied by the bone and bone cavities and, once combined, describing its edges. Again, dealing with fewer segments reduces the processing time. Therefore, $\alpha = 8$ serves as the optimal parameter for bone. Increasing the scale parameter α marginally did not have a major impact on the segments produced, however, increasing the parameter extensively to e.g. $\alpha = 400$ results in larger segments comprising areas of heterogeneous intensity and causing some background signal to be included in the bone segment region **Figure 5I–L**.

5.2. MDN threshold value $T_{\bar{\Delta}_k}$

5.2.1. $T_{\bar{\Delta}_k}$ for HSCs

The optimisation of the MDN threshold value depends on the correct detection of HSC edges. To observe the effects of changing the MDN threshold on HSC segmentation, we set $\alpha = 8$ and $d = 30$, then change the parameter $T_{\bar{\Delta}_k}$. Selecting $T_{\bar{\Delta}_k} \geq 10$ resulted in some background signal being included in the HSC segment region, while $T_{\bar{\Delta}_k} \geq 68$ resulted in correct segmentation of the HSC region. On the other hand, selecting a threshold value of $T_{\bar{\Delta}_k} \geq 200$ resulted in restricting the selection of segments to only those which have $T_{\bar{\Delta}_k} \geq 200$ mean intensity difference value to their neighbouring segments and excluded HSC segments of lower intensity from the final HSC object. Hence selection of low threshold values for the MDN will result in increased segment region sizes, while selecting higher MDN threshold values will restrict the selection of segments to regions of smaller size (**Figure 6A–C**).

5.2.2. $T_{\bar{\Delta}_k}$ for osteoblasts

To observe the effects of changing the MDN threshold on osteoblastic cell segmentation, we set $\alpha = 8$ and $d = 30$, then change the MDN threshold parameter $T_{\bar{\Delta}_k}$. $T_{\bar{\Delta}_k} \geq 0$ resulted in some background signal being included in the osteoblastic cell segment regions, while $T_{\bar{\Delta}_k} \geq 8$ resulted in correct segmentation of the osteoblastic cell regions. $T_{\bar{\Delta}_k} \geq 100$ resulted in restricting the selection of segments to only those which have $T_{\bar{\Delta}_k} \geq 100$ mean intensity difference to their neighbouring segments and excluded osteoblast segments of lower intensity from the final osteoblastic cell objects. Hence selection of low threshold values for the MDN will result in increased segment regions, while selecting higher MDN threshold values will restrict the selection of segments (**Figure 6D–F**).

5.2.3. $T_{\bar{\Delta}_k}$ for bone

To observe the effects of changing the MDN threshold on bone segmentation, we set $\alpha = 8$ and $d = 70$, then change the MDN threshold parameter $T_{\bar{\Delta}_k}$, while $T_{\bar{\Delta}_k} \geq 0$ resulted in some background signal being included in the bone segment region. $T_{\bar{\Delta}_k} \geq 4$ resulted in correct segmentation of the bone region, $T_{\bar{\Delta}_k} \geq 80$ resulted in restricting the selection of segments to only those which have ≥ 80 mean intensity difference to their neighbouring segments and excluded bone segments of lower intensity from the final bone object. Hence selection of low threshold values for the MDN will result in increased bone segment regions, while selecting higher MDN threshold values will restrict the selection of segments (**Figure 6G–I**).

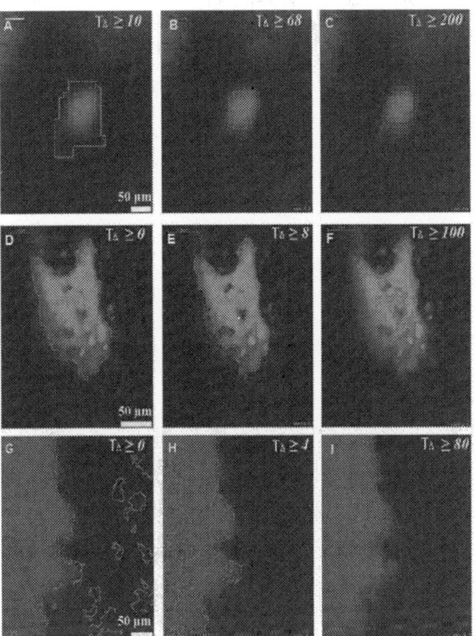

Figure 6. Optimisation of the MDN threshold parameter $T_{\bar{\Delta}}$ for each image object category. Column (A–G) represent object segmentation error when decreasing the $T_{\bar{\Delta}}$ parameter. Column (B–H) represent correct object segmentation with the optimal parameter $T_{\bar{\Delta}}$. Column (C–I) represent object segmentation error when increasing the $T_{\bar{\Delta}}$ parameter.

5.3. Distance to neighbour d parameter

5.3.1. d for HSCs

To illustrate the process of selecting the optimal distance to neighbour d parameter, we use varying distance d between neighbouring segments as follows:

We set the optimal multi-resolution scale parameter $\alpha = 8$ and MDN $T_{\bar{\Delta}_k} \geq 68$, then select the neighbourhood size. Small neighbourhood size such as $d = 2$ restricts the selection to fewer segments, specifically to those within a two pixels radius from the centre of the HSC region. As a result, the edges of the HSC are excluded from the final object. On the other hand, increasing the neighbouring distance to $d = 30$ resulted in correct selection of segments belonging to the HSC and elimination of segments containing background signal. Increasing the neighbouring distance further to about $d = 120$ resulted in some background signal being included in the final HSC region (**Figure 7A–C**).

5.3.2. d for osteoblasts

To show the effects of changing the neighbourhood size d on osteoblast segmentation, we use the optimal multi-resolution scale parameter $\alpha = 8$ and MDN threshold to $T_{\bar{\Delta}_k} \geq 8$, then select the neighbourhood size d. Selection of small neighbourhood size such as $d = 2$ restricted the selection of segments, resulting in fragmented segments scattered across the osteoblastic cell regions and failed to detect actual osteoblast regions. On the other hand, $d = 2$ resulted in correct selection of segments belonging to the osteoblasts and elimination of segments containing background signal. Increasing the neighbouring distance to $d = 120$ resulted in some background signal being included in the final osteoblast region (**Figure 7D–F**).

To demonstrate the effect of the selection of the value of parameter d on low intensity osteoblast regions, we selected images from lower slices of the same 3D stack of images (**Figure 7G–I**).

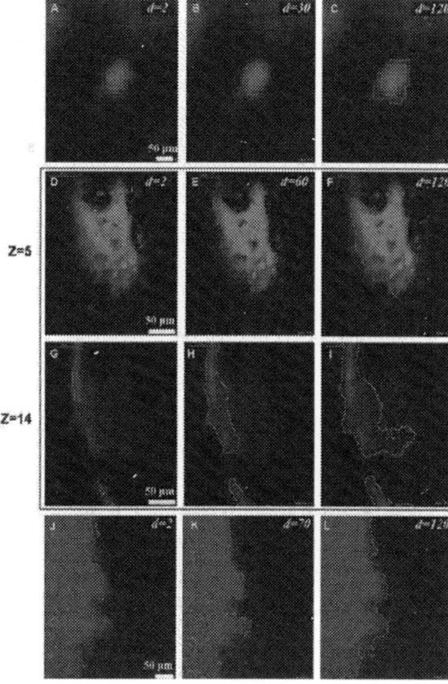

Figure 7. Optimisation of Distance to neighbour d parameter for each image object category. Column (A–J) represent object segmentation error when decreasing the d parameter. Column (B–K) represent correct object segmentation with the optimal parameter d. Column (C–L) represent object segmentation error when increasing the d parameter. Row (G–I) represent images from lower slices of the same 3D stack of images (D–F).

5.3.3. d for bone

To show the effect of changing the neighbourhood size d on bone segmentation, we use the optimal multi-resolution scale parameter $\alpha = 8$ and MDN threshold to $T_{\bar{\Delta}_k} \geq 4$, then select the neighbourhood size d. Selection of small neighbourhood size such as $d = 2$ restricted the selection of segments, resulting in fragmented segments scattered across the bone regions and failed to detect actual bone regions. As a result, the edges of the bone are excluded from the final object. On the other hand, $d = 70$ resulted in correct selection of segments belonging to the bone and elimination of segments containing background signal. Increasing the distance to $d = 120$ resulted in some background signal being included in the final bone region and missing small detail of bone cavities (**Figure 7J–L**).

Note: Parameter selection can be further observed on the YZ and XZ dimensions to ensure accurate selection of the optimal parameters (**Figure 8**). To optimise the parameters for further niche components and cell types (e.g. vasculature), the same steps can be followed as described for the HSCs, osteoblasts and bone.

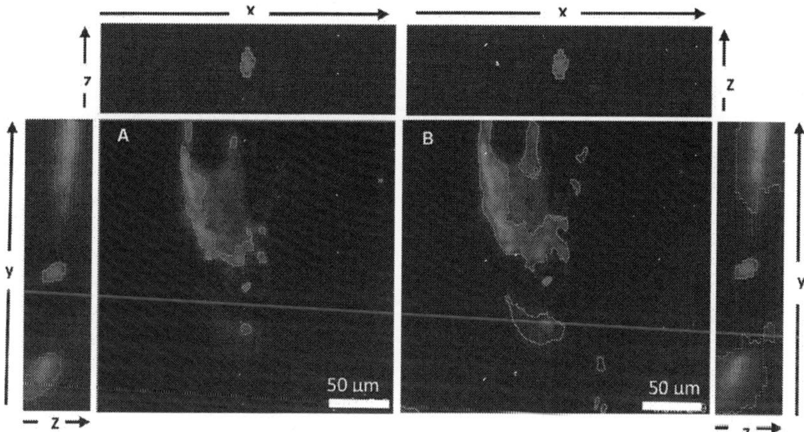

Figure 8. Examples of different dimensions view to observe the segmentation results following selecting different parameter values for α, $T_{\bar{\Delta}_k}$ and d.

6. Machine learning

Machine learning (ML) has become a valuable artificial intelligence tool, increasingly used for analysis of complex image data [23, 24]. ML serves two main objectives: classification and regression. Classification approaches are performed when a computer is given a set of options and is expected to divide them into a subset of categorise. Regression is the process of selecting the appropriate response to a particular situation from a set of possible responses. Furthermore, ML classifiers can be categorised into two main models: supervised and unsupervised classifiers. Unsupervised approaches utilise a clustering technique, where ML looks for resemblances across data, and then splits the same data into clusters. Clusters are then used to define the classes. Conversely, supervised approaches require a training set, where the ML learns information by extracting a set of features from that particular set, and the ML classifier is then expected to classify a new set of data and categorise it into a k number of classed based on the discriminative features extracted from the training set. Supervised classifiers are capable of identifying a set of complex features, suitable for classifying heterogeneous imaging data such as HSCs and vasculature from *in vivo* bone marrow 3D images, and then use them to perform such classification.

6.1. Decision tree classifier

The decision tree (DT) classifier is a structured approach that builds classification models from an input dataset to predict the output of an unknown dataset. The DT classifier starts the classification procedure structurally from the top node and uses the feature vectors to split it into further nodes [25, 26]. The selection of the feature vectors is measured by the purity of a particular subset, such that if a pure subset is produced, the splitting stops, otherwise splitting continues until a pure subset is allocated. Impurity of a specific subset is calculated by the entropy and is defined in [27] as follows:

$$E(S) = -\sum_{c \in C} p(c) \log_2 p(c) \tag{3}$$

where S is the dataset for which the entropy is calculated, c the set of classes in set S, $p(c)$ the proportion of the number of elements belonging to class c to the number of elements in set S. Pure subsets will generate a value of 0, while impure subsets will generate a value of 1.

$$I(S,F) = \sum_i \frac{|S_i|}{|S|} . E(S_i) \tag{4}$$

where F is a feature. When a feature F splits the set S into subsets S_i, the average entropy is computed and the sum compared to the entropy of the original set S; from [28].

$$Gain(S,F) = E(S) - I(S,F) = E(S) - \sum_i \frac{|S_i|}{|S|} . E(S_i) \tag{5}$$

In this chapter, we describe the application of the DT classifier for classifying DiD signal (potential HSCs) and vasculature structures. The DT classifier was selected in this study due to its computational simplicity and illustrative attributes represented in the DT output. The classifier also has the advantage of being able to select discriminative features without a prior step of feature optimisation, allowing the incorporation of a wide range of features for testing at the training stage.

6.2. Training and testing the classifier

6.2.1. HSCs

LH-SEG provides a sufficient level of DiD objects detection. However, not all the DiD objects are HSCs. This is due to cell debris and the diffusion of the dye, which could lead to aggregates producing signal similar in shape and intensity to that of the HSCs.

In order to identify HSCs, we propose the use of DT classifier. We first trained our classifier to distinguish three classes of DiD objects: Class-1 represents HSCs distinguished by their smoother, round surface and high intensity as observed for quiescent HSCs; Class-2 comprises HSCs that have less rounded shapes and present small uropod-like protrusions [27], as previously observed in time-lapse images of migratory HSCs [7]; and Class-3 contains DiD objects that are not HSCs and characterised by their exceedingly uneven morphology. A selection of objects was manually selected for each category and fed into the classifier as training set. The exact number of objects needed for training each class depends on the variability of the structures contained, such that objects with regular, predictable characteristics such as Class-1 HSCs would require fewer objects for training, while irregular objects containing high variability such as class-3 would require a higher number of objects for training to ensure the classifier collects sufficient numbers of characteristics to handle the complexity and inconsistency of objects. After training, the classifier is tested on different datasets from those used for training with a 3-fold cross-validation approach. Classification results provided a high accuracy for all classes and successfully classified the DiD object to the three proposed classes as described in [11] and **Figure 9A** and **B**.

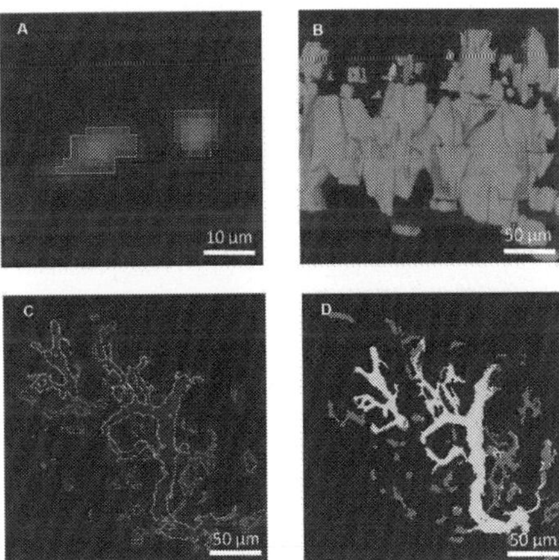

Figure 9. (A) Classification results of DiD objects using our proposed approach: two DiD-labelled HSCs were classified as HSC class 1 (red) and 2 (moccasin). (B) 3D rendering of DiD HSCs represented in (A) inside their microenvironment. (C) Classification results of vasculature using our proposed approach, showing small sinusoids (magenta) and large sinusoids (yellow). (D) 3D rendering of the 3D stack containing the slice shown in (C).

6.2.2. Vasculature

Blood vessels inside the bone marrow are another component that plays a vital role in the maintenance of HSCs. Interactions between vasculature and HSCs have become one of the main headlines for many recent studies [29, 30]. As demonstrated in our pervious study, segmentation of vasculature of *in vivo* images was possible using the LH-SEG [11]. However, the automated identification of the variable components of the BM vascular network is a challenging task due to the complex and interlinked structure of blood vessels. Therefore, the selection of effective thresholds for classifying the vasculature into different regions based on their appearance and morphology is a challenging task. To overcome these issues, we propose a DT Classifier to define distinct features of the different types of vasculature and classify them into four categories based on their morphological and topological characteristics inside the BM space [28]. Blood vessels share mutual features in relation of their complexity and interlinking physical structure. Consequently, we manually select different 2D segmented from the vasculature region to create the training sets for four classes of blood vessels: (1) large sinusoids: larger blood vessels in comparison to other sinusoids and positioned mostly towards the periphery of the BM space; (2) small sinusoids: blood vessels smaller in section when compared to the large sinusoids, spread across bone marrow space; (3) the central sinus: a large venous vessel situated in the centre of the bone marrow space; and (4) the bifurcation of the central sinus, which results from the central sinus branching out towards either sides of the BM space in its frontal area. The number of samples was optimised for each class to avoid under-fitting and over-fitting the classifier. Three-fold cross validation was used in this experiment. Morphological and topological features were fed to the classifier for extraction of discriminative values. Classification results provided a high accuracy for all classes and successfully classified all vessels into one of the four classes as described in [23] and presented in **Figure 9C** and **D**. This demonstrates the applicability of ML classification approaches to 3D *in vivo* images of the bone marrow, which escalates the throughput of intravital imaging and our understanding of the complexity of the HSC and their interaction with multiple niche components.

7. 3D positional measurements

Following segmentation and classification of DiD-labelled HSCs, HSCs position relative to their nearest osteoblast and bone is measured as described in [11], and could be measured to their nearest vessel or other segmented bone marrow components. The overall automated distance measurements from HSCs to osteoblast and bone were equivalent to previously published data obtained using manual distance measurements and described in [5, 6]. Further quantitative measures such as the morphological and textural characteristics of each segmented and classified object can also be obtained using this tool.

8. Conclusion

In this chapter, we describe an approach for the visualisation, analysis and quantification of HSCs within their surrounding microenvironment in the bone marrow using intravital imaging and image analysis approaches.

Confocal fluorescence and SHG microscopy combined facilitate the collection of large amounts of image data of HSCs and their niche components. They specifically allowed the first observations of the cellular interactions of *ex vivo* labelled HSCs with GFP expressing osteo-blasts and bone, and were further used to uncover several other details of HSC-niche interactions.

An image analysis approach combining image segmentation (LH-SEG) and machine learning classification detects and defines the edges of HSCs, osteoblast, bone and blood vessels. Subsequently, detected HSC and vasculature objects are fed into a decision tree classifier for categorising the detected signal into further classes. HSC signal was categorised into three classes representative of their morphological and intensity characteristics, proving high levels of accuracy for detection of HSCs and elimination of DiD debris and background signal. Furthermore, vasculature in the bone marrow was categorised into four classes based on their intrinsic morphological and position features. This will simplify the analysis of this component and allow simultaneous analysis of other components, increasing our understanding of the complexity of the HSC niche. After all the biologically important objects have been segmented and classified, it is then possible to derive quantitative information from these features and further translate it into a deeper understanding of the biological events observed.

This automated image analysis pipeline was particularly designed to address the major challenges imposed on *in vivo* 3D images of the bone marrow using confocal and SHG microscopy and provided high level of accuracy in both segmentation and classification of HSCs and their niche components. Moreover, optimisation of the segmentation parameters is fundamental for extracting resourceful and precise quantitative features from all segmented and classified objects. However, this method can be virtually applied to any complex 3D imaging data, especially those generated by confocal and two photon microscopy, such as *ex vivo* whole mount immunofluorescence of whole bones, and intravital microscopy data from other tissues and organs.

In conclusion, once the correct parameters are selected as we describe, this approach is applicable to the analysis of a broad range of 3D and intravital microscopy images from other tissues and organs, leading to a faster pace of discovery not only in the haematopoiesis and bone marrow field, but also in other somatic stem cell studies, and developmental and cancer biology studies. Further development of the described tool can be expanded to more programming languages and software platforms. This will accelerate the productivity of the method and allow an even wider range of applications.

Author details

Reema Adel Khorshed* and Cristina Lo Celso

*Address all correspondence to: r.khorshed@imperial.ac.uk

Department of Life Sciences, Imperial College, London, UK

References

[1] C. Lo Celso, C. P. Lin, and D. T. Scadden, "In vivo imaging of transplanted haemato-poietic stem and progenitor cells in mouse calvarium bone marrow," *Nat Protoc*, vol. 6, pp. 1–14, Jan 2011.

[2] S. J. Morrison and D. T. Scadden, "The bone marrow niche for haematopoietic stem cells," *Nature*, vol. 505, pp. 327–34, Jan 16 2014.

[3] L. Ritsma, S. I. Ellenbroek, A. Zomer, H. J. Snippert, F. J. de Sauvage, B. D. Simons, et al., "Intestinal crypt homeostasis revealed at single-stem-cell level by in vivo live imaging," *Nature*, vol. 507, pp. 362–5, Mar 20 2014.

[4] A. Kohler, V. Schmithorst, M. D. Filippi, M. A. Ryan, D. Daria, M. Gunzer, et al., "Altered cellular dynamics and endosteal location of aged early haematopoietic progenitor cells revealed by time-lapse intravital imaging in long bones," *Blood*, vol. 114, pp. 290–8, Jul 9 2009.

[5] C. Lo Celso, H. E. Fleming, J. W. Wu, C. X. Zhao, S. Miake-Lye, J. Fujisaki, et al., "Live-animal tracking of individual haematopoietic stem/progenitor cells in their niche," *Nature*, vol. 457, pp. 92–6, Jan 1 2009.

[6] S. W. Lane, Y. J. Wang, C. Lo Celso, C. Ragu, L. Bullinger, S. M. Sykes, et al., "Differential niche and Wnt requirements during acute myeloid leukemia progression," *Blood*, vol. 118, pp. 2849–56, Sep 8 2011.

[7] N. M. Rashidi, M. K. Scott, N. Scherf, A. Krinner, J. S. Kalchschmidt, K. Gounaris, et al., "In vivo time-lapse imaging of mouse bone marrow reveals differential niche engage-ment by quiescent and naturally activated haematopoietic stem cells," *Blood*, May 21 2014.

[8] G. B. Adams, I. R. Alley, U. I. Chung, K. T. Chabner, N. T. Jeanson, C. Lo Celso, et al., "Haematopoietic stem cells depend on Galpha(s)-mediated signalling to engraft bone marrow," *Nature*, vol. 459, pp. 103–7, May 7 2009.

[9] S. Méndez-Ferrer, T. V. Michurina, F. Ferraro, A. R. Mazloom, B. D. MacArthur, S. A. Lira, et al., "Mesenchymal and haematopoietic stem cells form a unique bone marrow niche," *Nature*, vol. 466, pp. 829–834, 2010.

[10] J. Fujisaki, J. Wu, A. L. Carlson, L. Silberstein, P. Putheti, R. Larocca, et al., "In vivo imaging of Treg cells providing immune privilege to the haematopoietic stem-cell niche," *Nature*, vol. 474, pp. 216–219, 06/09/print 2011.

[11] R. A. Khorshed, E. D. Hawkins, D. Duarte, M. K. Scott, O. A. Akinduro, N. M. Rashidi, et al., "Automated Identification and Localization of haematopoietic Stem Cells in 3D Intravital Microscopy Data," *Stem Cell Rep*, vol. 5, pp. 139–53, Jul 14 2015.

[12] H. Geiger and K. L. Rudolph, "Aging in the lympho-haematopoietic stem cell compartment," *Trends Immunol*, vol. 30, pp. 360–5, Jul 2009.

[13] A. Sanchez-Aguilera, Y. J. Lee, C. Lo Celso, F. Ferraro, K. Brumme, S. Mondal, et al., "Guanine nucleotide exchange factor Vav1 regulates perivascular homing and bone marrow retention of haematopoietic stem and progenitor cells," *Proc Natl Acad Sci USA*, vol. 108, pp. 9607–12, Jun 7 2011.

[14] G. Chung and L. Vese, "Image segmentation using a multilayer level-set approach," *Computing and visualisation in Science*, vol. 12, pp. 267–285, 2009/08/01 2009.

[15] S. Y. Yeo, X. Xie, I. Sazonov, and P. Nithiarasu, "Level set segmentation with robust image gradient energy and statistical shape prior," in *IEEE International Conference on Image Processing*, pp. 3397–3400, 2011.

[16] K. Nandy, J. Kim, D. P. McCullough, M. McAuliffe, K. J. Meaburn, T. P. Yamaguchi, et al., "Segmentation and quantitative analysis of individual cells in developmental tissues," *Methods Mol Biol*, vol. 1092, pp. 235–53, 2014.

[17] P. Rompolas, K. R. Mesa, and V. Greco, "Spatial organization within a niche as a determinant of stem-cell fate," *Nature*, vol. 502, pp. 513–518, 10/24/print 2013.

[18] L. Ritsma, S. I. J. Ellenbroek, A. Zomer, H. J. Snippert, F. J. de Sauvage, B. D. Simons, et al., "Intestinal crypt homeostasis revealed at single-stem-cell level by in vivo live imaging," *Nature*, vol. 507, pp. 362–365, 03/20/print 2014.

[19] G. Mallinis, N. Koutsias, M. Tsakiri-Strati, and M. Karteris, "Object-based classification using Quickbird imagery for delineating forest vegetation polygons in a Mediterranean test site," *ISPRS J Photogramm Remote Sens*, vol. 63, pp. 237–250, 3//2008.

[20] F. Long, J. Zhou, and H. Peng, "Visualisation and analysis of 3D microscopic Images," *PLoS Comput Biol*, vol. 8, p. e1002519, 2012.

[21] S. Uchida, "Image processing and recognition for biological images," *Dev Growth Differ*, vol. 55, pp. 523–549, 2013.

[22] B. Moon, "A Gaussian smoothing algorithm to generate trend curves," *Korean J Comput Appl Math*, vol. 8, pp. 507–518, 01 2001.

[23] E. Mjolsness and D. DeCoste, "Machine learning for science: state of the art and future prospects," *Science*, vol. 293, pp. 2051–5, Sep 14 2001.

[24] C. Sommer and D. W. Gerlich, "Machine learning in cell biology—teaching computers to recognize phenotypes," *J Cell Sci*, vol. 126, pp. 5529–39, Dec 15 2013.

[25] C. Agarwal and A. Sharma, "Image understanding using decision tree based machine learning," in *International Conference on Information Technology and Multimedia*, pp. 1–8, 2011.

[26] O. Aydemir and T. Kayikcioglu, "Decision tree structure based classification of EEG signals recorded during two dimensional cursor movement imagery," *J Neurosci Methods*, vol. 229, pp. 68–75, May 30 2014.

[27] M. F. Krummel and I. Macara, "Maintenance and modulation of T cell polarity," *Nat Immunol*, vol. 7, pp. 1143–9, Nov 2006.

[28] R. A. Khorshed and C. Lo Celso, "Machine learning classification of complex vasculature structures from *in-vivo* bone marrow 3d data," *(ISBI), IEEE International Symposium on Biomedical Imaging*, Prague, 2016.

[29] A. Mendelson and P. S. Frenette, "haematopoietic stem cell niche maintenance during homeostasis and regeneration," *Nat Med*, vol. 20, pp. 833–846, 08//print 2014.

[30] Daniel J. Nolan, M. Ginsberg, E. Israely, B. Palikuqi, M. G. Poulos, D. James, et al., "Molecular signatures of tissue-specific microvascular endothelial cell heterogeneity in organ maintenance and regeneration," *Dev Cell*, vol. 26, pp. 204–219, 29 July 2013.

Chapter 9

Microscopic Investigations on Woody Biomass as Treated with Ionic Liquids

Toru Kanbayashi and Hisashi Miyafuji

Additional information is available at the end of the chapter

http://dx.doi.org/10.5772/62721

Abstract

Woody biomass is one of the most promising renewable alternatives to fossil resources. However, some physical and chemical treatment is required to convert their chemical components into biofuels and valuable chemicals because of their low degradative properties. Recently, there has been considerable interest in ionic liquid treatment for biorefinery, and many fundamental studies on the reactivity of wood with ionic liquids have been performed from a chemical and morphological point of view. This chapter highlights the findings regarding morphological and topochemical features of wood cell walls in the degradation process as a result of ionic liquid treatment. Bright-field microscopy and scanning electron microscopy have revealed the swelling behavior of cell walls and the detailed ultrastructural features of wood tissues treated with ionic liquid. Polarized light microscopy and confocal Raman microscopy have clarified the changes in cellulose crystallinity and distribution of chemical compositions such as polysaccharides and lignin during ionic liquid treatment at the cellular level.

Keywords: Biorefinery, Cell wall, Liquefaction, Ionic liquid, Wood

1. Introduction

The efficient use of lignocellulosics has been an important approach to prevent exhaustion of fossil resources and global warming caused by increasing emissions of greenhouse gases. Among the various types of lignocellulosics, woody biomass is regarded as a promising resource because it is carbon neutral, abundantly available in many regions, and does not compete with agricultural production. Wood cell walls decompose to form persistent organic complexes mainly composed of cellulose (40–50%), hemicellulose (25–35%), and lignin (18–35%) [1, 2]. To convert

their chemical components into transportable biofuels and chemical feedstocks, it is necessary to develop the processing technology to break their rigid structure. To date, various conversion methods such as acid hydrolysis [3–5], enzymatic saccharification [6, 7], pyrolysis [8–10], and supercritical or sub-critical fluid treatment [11–13] have been investigated. However, practical methods have not yet been established.

Ionic liquids are defined as organic salts with low melting points, and have many advantages including negligible vapor pressures, chemical and thermal stability, non-flammability, low viscosity, and reusability [14–16]. In addition, they can dissolve a wide range of organic and inorganic substances [14]. Ionic liquid treatment is attractive as a new conversion technology for woody biomass. **Figure 1** shows the typical cations and anions found in ionic liquids. There are an infinite number of combinations of cations and anions, and their physical properties, such as melting point, viscosity and dissolving power, can be easily changed by altering the combination. This is why ionic liquids are called "designer solvents" [17].

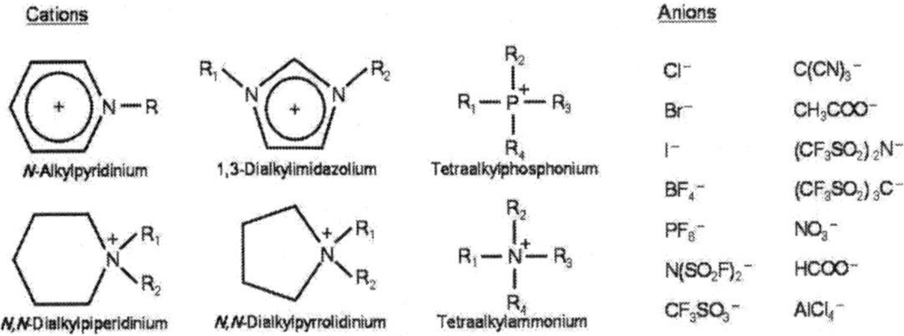

Figure 1. The structures of typical cations and anions in ionic liquids.

Recent studies revealed that certain types of ionic liquids can liquefy cellulose [18–20], lignin [21], and even wood cell walls [22–26]. Using ionic liquids as the solvent to process woody biomass, many fundamental studies on the reaction behavior of wood in ionic liquids have been carried out focusing on the chemical processes [27–31]. However, woody biomass is a very inhomogeneous composite at the cell level. Wood comprises various types of tissues such as the tracheid, wood fibers, vessels, and parenchyma. In addition, wood cell walls consist of several layers: a compound middle lamella (middle lamella + primary wall; CML) and a secondary wall (S), which is generally composed of S_1, S_2, and S_3 sublayers. The chemical components and distribution vary depending on the wood species, types of tissues and their layers [32]. Therefore, to improve the chemical conversion process using ionic liquids, a better understanding of the effects of ionic liquid treatment of wood, such as the interaction of wood with ionic liquids at the cell level and the deconstruction behavior of various types of tissues in ionic liquids, are required.

In this chapter, we focus on morphological and topochemical studies on the liquefaction of wood in ionic liquids, especially 1-ethyl-3-methylimidazolium chloride ([C2mim][Cl]) and 1-ethylpyridinium bromide ([EtPy][Br]), using various microscopy techniques. [C2mim][Cl] and [EtPy][Br] are known as the ionic liquids which can preferentially liquefy cellulose [25] and lignin [31], respectively. Bright-field microscopy and polarized light microscopy were employed to determine the swelling and decomposition behaviors of wood cell walls and the state of cellulose crystallinity during ionic liquids treatment. Scanning electron microscopy (SEM) was used to observe the detailed ultrastructural changes in various wood tissues treated with ionic liquids. Confocal Raman microscopy was employed to examine the changes in chemical components including polysaccharides and lignin at the cellular level and to visualize their distribution on the cell walls during ionic liquids treatment.

2. Application of microscopy techniques to examine wood liquefaction

2.1. Light microscopy analysis

Morphological features of wood cell walls during the liquefaction process in ionic liquid were determined by light microscopy [33–38]. **Figure 2** shows bright-field microscopy and polarized light microscopy images of *Cryptomeria japonica*, which is the most common softwood species in Japan, treated with [C2mim][Cl] at 120°C for 0 h and 24 h. All the cell walls swelled after 24 h of [C2mim][Cl] treatment (**Figure 2c**). Although the cell walls of tracheids in latewood (summerwood; formed late in the growing season) were disordered and distorted after the treatment, those in earlywood (springwood; formed early in the growing season) were barely changed. The polarized light microscopy studies show that the brightness from the birefringence of cellulose in both earlywood and latewood was decreased after treatment (**Figure 2d**). These results imply that the crystalline structures of cellulose in wood are amorphized before the wood cell walls liquefy completely during [C2mim][Cl] treatment.

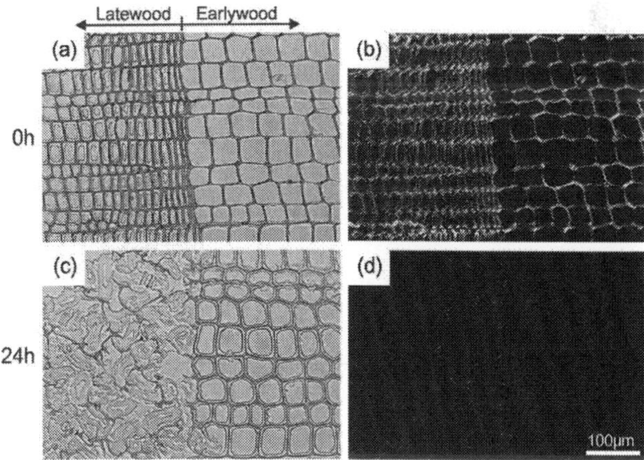

Figure 2. Bright-field microscopy (**a**, **c**) and polarized light microscopy images (**b**, **d**) of transverse sections of *Cryptomeria japonica* before (**a**, **b**) and after treatment with [C2mim][Cl] at 120°C for 24 h (**c**, **d**).

To study the detailed swelling behavior of tracheids arising from [C2mim][Cl] treatment, we performed time sequential measurements of the cell wall area, cell lumen area, and the total of cell lumen and cell wall areas, in earlywood and latewood, in transverse sections (**Figure 3**). The cell wall area in earlywood increased only slightly at an early stage of [C2mim][Cl] treatment, whereas that in latewood increased significantly. After the initial swelling, the cell wall area in earlywood showed no further changes, whereas that in latewood increased gradually with prolonged treatment time. After 72 h of treatment, the cell wall area in earlywood and latewood had increased by 1.5 and 4 times, respectively. These results indicate that the swelling behavior of the tracheids of *Cryptomeria japonica* is different for the two morphological regions. In particular, tracheids in latewood are more prone to swelling and being broken after [C2mim][Cl] treatment.

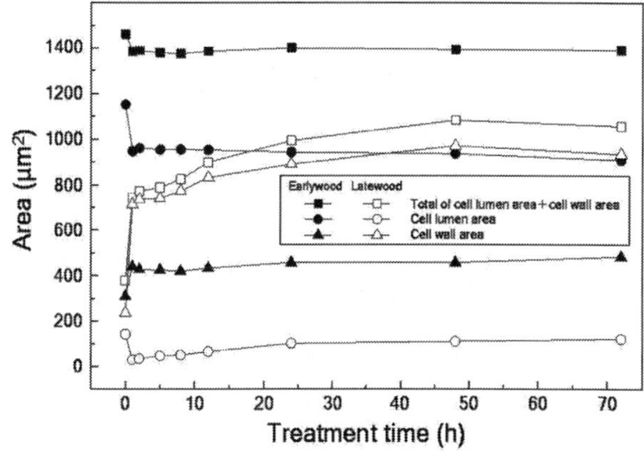

Figure 3. Changes in cell wall area, cell lumen area, and total of cell lumen + cell wall area in earlywood and latewood of *Cryptomeria japonica* during [C2mim][Cl] treatment at 120°C [34].

Changes in the cell wall area of fibrous cells of various Japanese hardwood species during [C2mim][Cl] treatment were also measured (**Figure 4**). At the initial stages of [C2mim][Cl] treatment, the cell wall areas of all species increased rapidly. Thereafter, the cell wall areas of *Fagus crenata* and *Quercus mongolica* in both earlywood and latewood increased continuously with progressing treatment time. After 72 h of treatment, the cell wall areas of *Fagus crenata* in both earlywood and latewood had increased by 4 times, and those of *Quercus mongolica* in earlywood and latewood had increased by 3.5 and 5 times, respectively. However, the swelling behavior of two other species that were tested was different from *Fagus crenata* and *Quercus mongolica*. The cell wall area of *Quercus glauca* in latewood increased gradually with progressing treatment time, but not in the earlywood. The change in cell wall area in *Trochodendron aralioides* levelled off in both earlywood and latewood. After 72 h of treatment, the cell wall areas of *Quercus glauca* in earlywood and latewood had increased by 2 and 3 times, respectively, and those of *Trochodendron aralioides* in earlywood and latewood had increased by 1.5 and 2 times, respectively. Therefore, swelling behavior, including the swelling ratio and time-dependent change of fibrous cells during [C2mim][Cl] treatment, differed according to the wood species and morphological regions, including earlywood and latewood.

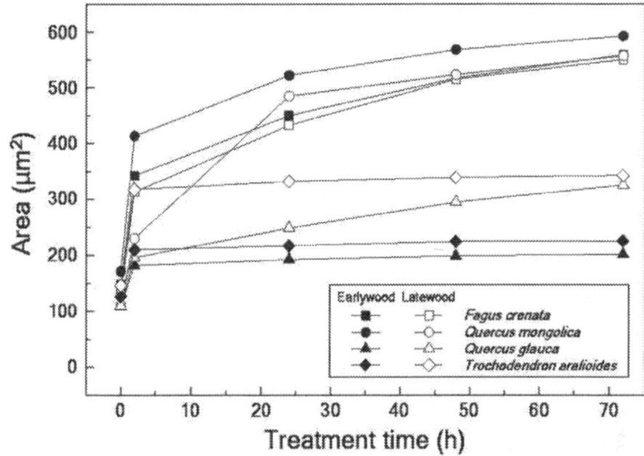

Figure 4. Changes in the cell wall area of fibrous cells in the earlywood and latewood of various wood species during [C2mim][Cl] treatment at 120°C [35, 36].

Figure 5 shows bright-field microscopy and polarized light microscopy images of *Cryptomeria japonica* before and after treatment for 72 h with [EtPy][Br] at 120°C. The cell walls of both earlywood and latewood were swollen by [EtPy][Br] treatment (**Figure 5c**). Although the cell walls in earlywood were well ordered during [EtPy][Br] treatment, those in latewood were partially dissociated. After 72 h of treatment, tracheids in earlywood and latewood were

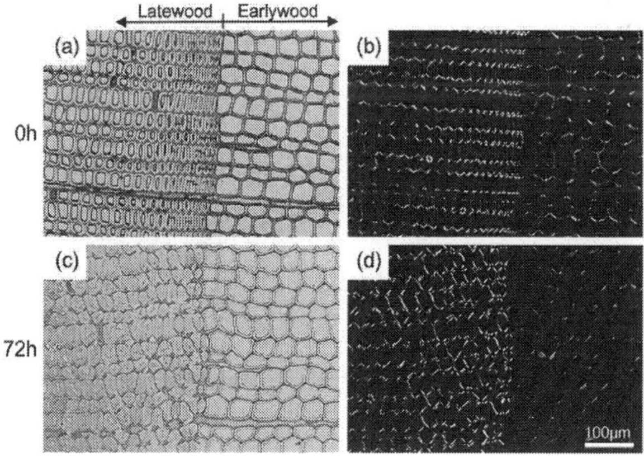

Figure 5. Bright-field microscopy (**a, c**) and polarized light microscopy images (**b, d**) of transverse sections of *Cryptomeria japonica* before (**a, b**) and after treatment with [EtPy][Br] at 120°C for 72 h (**c, d**) [37].

swollen by 1.3 and 2 times, respectively. Therefore, the swelling efficiency of [EtPy][Br] is lower than that of [C2mim][Cl]; and morphological changes in wood cell walls differ for the different types of ionic liquids. In the polarized light micrographs, the brightness from the birefringence of cellulose showed no changes during [EtPy][Br] treatment (**Figure 5d**). This result implies that [EtPy][Br] treatment has no marked effect on the crystalline structure of cellulose in the wood cell walls.

2.2. Scanning electron microscopy observations

Ultrastructural changes in wood cell walls due to ionic liquid treatment were observed by SEM [34–38]. **Figure 6** shows SEM images of various tissues of *Cryptomeria japonica* treated with [C2mim][Cl] and [EtPy][Br]. The dissociations and distortions of cell walls were observed in latewood after 72 h of treatment with both [C2mim][Cl] and [EtPy][Br] (**Figure 6e, i**). These changes were caused by significant swelling of tracheids in latewood. The magnified view of earlywood (**Figure 6b, f, j**) and latewood (**Figure 6c, g, k**), shows the CML (indicated by small arrows) disappeared in both earlywood and latewood after [EtPy][Br] treatment (**Figure 6j, k**) but was preserved after [C2mim][Cl] treatment (**Figure 6f, g**). These changes may be related to lignin distribution in wood cell walls. CML contains more than 50% lignin, whereas S_2 (indicated by arrowheads) contains about 20% lignin [39]. Our recent research revealed that lignin can be preferentially liquefied by [EtPy][Br] [31], which explains why the CML was liquefied more rapidly than S_2 by [EtPy][Br]. **Figure 6d, h,** and **l** shows the changes in bordered pits in earlywood. The torus (indicated by large arrows) in the pits was broken after [C2mim][Cl] treatment, while the warts on the surface of tracheids disappeared after [EtPy][Br]

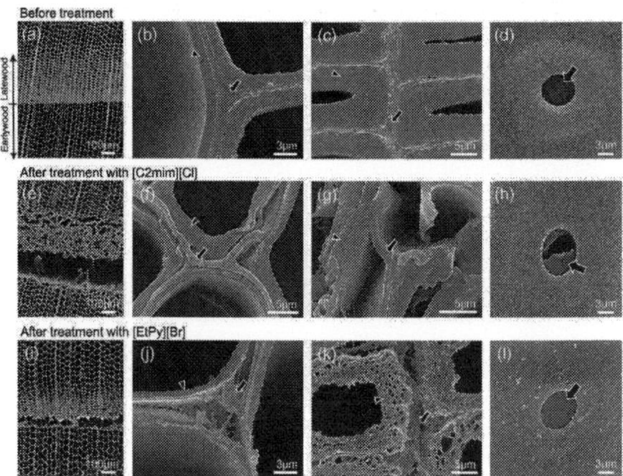

Figure 6. SEM images of various tissues of *Cryptomeria japonica* before (**a–d**) and after treatment with [C2mim][Cl] at 120°C for 48 h (**e–h**) and with [EtPy][Br] at 120°C for 72 h (**i–l**) [37]. (**a, e, i**) Around the annual ring boundary. (**b, f, j**), a magnified view of earlywood, (**c, g, k**) a magnified view of latewood, (**d, h, l**) bordered pits. The S_2, CML, and torus are indicated by the arrowhead, small arrow and large arrow, respectively.

treatment. Harada and Côté have reported that the torus is mainly composed of cellulose microfibrils [40]. In addition, Jansen et al. stated that warts are mainly composed of lignin and hemicellulose [41]. The differences in the reactivity of [C2mim][Cl] and [EtPy][Br] with these wood tissues derive from differences in their chemical components.

Figure 7 shows SEM images of various tissues of *Fagus crenata* treated with [C2mim][Cl] and [EtPy][Br]. Cracks occurred between the ray parenchyma (indicated by large arrowheads) and peripheral tissues after treatment with both ionic liquids. These changes are presumably due to the bonds between the ray parenchyma and tissues adjacent to it weakening after treatment as well as the difference in the swelling behavior. **Figure 7b, f, and j** show magnified views of the morphological features of wood fibers treated with [C2mim][Cl] and [EtPy][Br]which are similar to tracheids in latewood of *Cryptomeria japonica* (**Figure 6g, k**). Various changes were observed at vessel pits. Intervascular pits (vessel–vessel pits) were occluded after [C2mim][Cl] treatment (**Figure 7g**), while ray-vessel pits were not occluded. Instead, their pit membranes (indicated by large arrows) were broken after the treatment (**Figure 7h**). However, the intervascular pits and ray-vessel pits showed no significant changes after [EtPy][Br] treatment (**Figure 7k, l**). These results indicate that morphological changes in the pits differed depending on the types of pits and ionic liquids used.

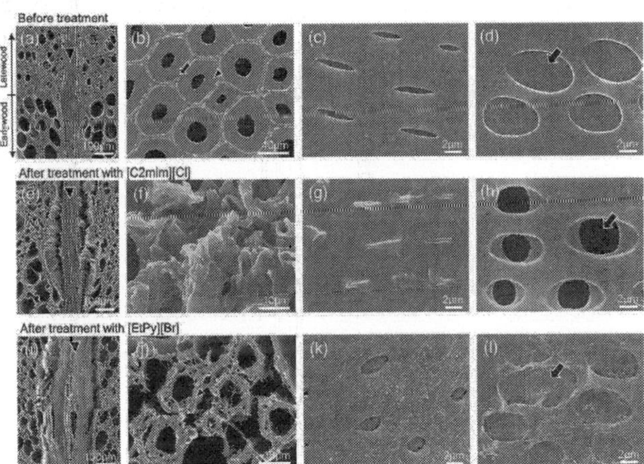

Figure 7. SEM images of various tissues of *Fagus crenata* before (**a–d**) and after treatment with [C2mim][Cl] (**e–h**) and [EtPy][Br] (**i–l**) at 120°C for 72 h [35, 38]. (**a, e, i**) Around annual ring boundary. (**b, f, j**) Wood fibers. (**c, g, k**) Intervascular pits. (**d, h, l**) Ray-vessel pits. The S_2, CML, ray parenchyma and pit membranes are indicated by the small arrowhead, small arrow, large arrow head and large arrow, respectively.

Figure 8 shows SEM images of transverse sections of two hardwood species treated with [C2mim][Cl]. Although wood fibers (indicated by arrowheads) of both *Fagus crenata* and *Quercus mongolica* showed significant deformation, the axial parenchyma cells (indicated by arrows) maintained their shapes (**Figure 8c, d**). These differences in morphological changes between the wood fibers and axial parenchyma cells are mainly attributed to their chemical composition. Fujii et al. reported that the lignin content of the axial parenchyma cells is higher than that of the wood fibers [42]. In our previous papers, we showed that lignin is much more difficult to react with [C2mim][Cl] than cellulose and hemicellulose [25, 28]. The reactivity of [C2mim][Cl] with the axial parenchyma cells is lower than that with the wood fibers, thus the axial parenchyma cells did not collapse after [C2mim][Cl] treatment.

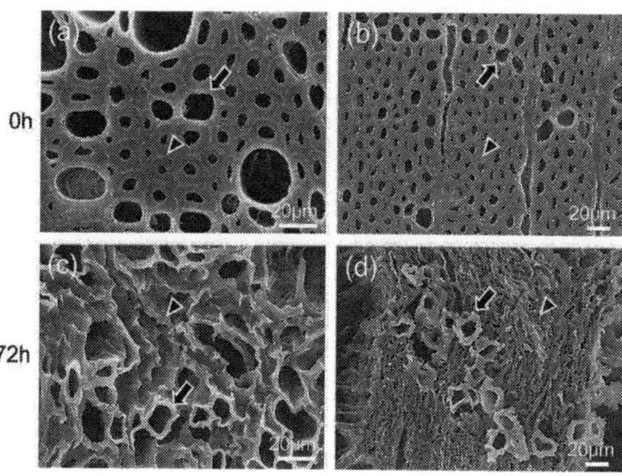

Figure 8. SEM images of transverse sections of *Fagus crenata* (**a, c**) and *Quercus mongolica* (**b, d**) before and after treatment with [C2mim][Cl] at 120°C for 72 h [35, 36]. Wood fibers and axial parenchyma cells are indicated by arrow heads and arrows, respectively.

2.3. Confocal Raman microscopy analysis

Raman spectra can reveal much information about functional groups, hydrogen and chemical bonds, and the surrounding environment. Raman spectroscopy is used to identify the chemical structure of a substance. Confocal Raman microscopy couples Raman spectroscopy with a confocal microscope to perform detailed analysis quickly. In recent years, confocal Raman microscopy has received attention as a new method of spectroscopic analysis for plant cell walls because of its characteristic advantages. It is non-destructive, has a high spatial resolution (approximately 0.3–2 μm), is not hindered by the presence of water [43, 44], and little-to-no sample pre-treatment is required. Several research groups have studied the chemical composition of native wood cell walls using this method [45–51]. In addition, it has been reported that confocal Raman microscopy is an effective tool to investigate topochemical changes in wood after pre-treatment for biorefinery [52–54].

We applied confocal Raman microscopy to determine the changes in chemical components and their distribution in wood at the cellular level during [C2mim][Cl] and [EtPy][Br] treatment [37, 38, 55, 56]. **Figure 9** shows Raman spectra obtained from S$_2$ of the wood fibers of *Fagus crenata* treated with [C2mim][Cl] and [EtPy][Br]. The characteristic Raman bands of lignin were observed at 370, 1331, 1459, 1600, 1657, 2848, and 2943 cm^{-1}, whereas those of the polysaccharides (cellulose and hemicellulose) were observed at 380, 436, 520, 1093, 1118, 1152,

1378, and 2895 cm^{-1}. These band assignments were based on previous studies [57–61]. Although all the band intensities of the polysaccharides sharply decreased during [C2mim] [Cl] treatment, most band intensities of lignin barely changed, except for some bands (**Figure 9a**). For instance, the band intensity at 1657 cm^{-1} was assigned to the ethylenic C=C bond in coniferyl/sinapyl alcohol units and γ–C=O in coniferyl/sinapyl aldehyde units decreased gradually. These results indicate that [C2mim][Cl] prefers to react with polysaccharides in S$_2$ of wood fibers rather than with lignin, despite the molecular structures of lignin changing partially during [C2mim][Cl] treatment. Regarding time-dependent spectral changes during [EtPy][Br] treatment, most band intensities of polysaccharides showed no significant changes, while those of lignin decreased with prolonged treatment time (**Figure 9b**). These spectral changes indicate that lignin is liquefied more readily than polysaccharides in [EtPy][Br].

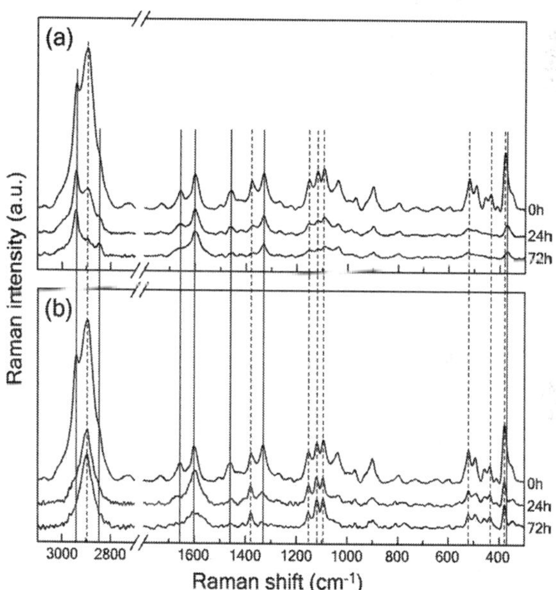

Figure 9. Raman spectra for the S$_2$ of wood fibers of *Fagus crenata* after treatment with [C2mim][Cl] (a) and [EtPy][Br] (b) at 120°C for 0 h, 24 h, and 72 h. Lignin and polysaccharides are indicated by solid lines and dashed lines respectively. [38, 56].

To study the changes in the distribution of chemical components in the wood cell walls over a wide range during ionic liquids treatment, Raman mapping analysis was applied on transverse sections. Raman mapping was performed on tracheids and wood fibers because these tissues are the main elements of *Cryptomeria japonica* and *Fagus crenata*, respectively. In the Raman images, bright areas indicate high concentrations of a specific chemical composition, whereas dark areas indicate low concentrations.

Figure 10 shows the results of time-sequential Raman mapping analysis of the distribution of lignin and polysaccharides of tracheids of *Cryptomeria japonica* and wood fibers of *Fagus crenata* during [C2mim][Cl] treatment. Within both wood species, the lignin concentration was higher in CML than in S_2 while the polysaccharide concentrations were lower in CML than in S_2. After treatment, changes in the distribution of chemical compositions were similar for the tracheids of *Cryptomeria japonica* and wood fibers of *Fagus crenata*. Although the lignin concentration in S_2 decreased with increasing [C2mim][Cl]treatment time, the concentration in CML showed no significant change. These results indicate that lignin in CML has a high chemical resistance to [C2mim][Cl]. The differences in the reactivity of [C2mim][Cl] with lignin in each morphological region are attributed to many factors such as molecular structure, concentration, and penetrability of [C2mim][Cl]. Polysaccharide concentrations decreased significantly after 24 h of treatment. After that, most of the Raman signal derived from polysaccharides disappeared. Therefore, in fibrous cells of wood, polysaccharides were more rapidly liquefied by [C2mim][Cl] than lignin.

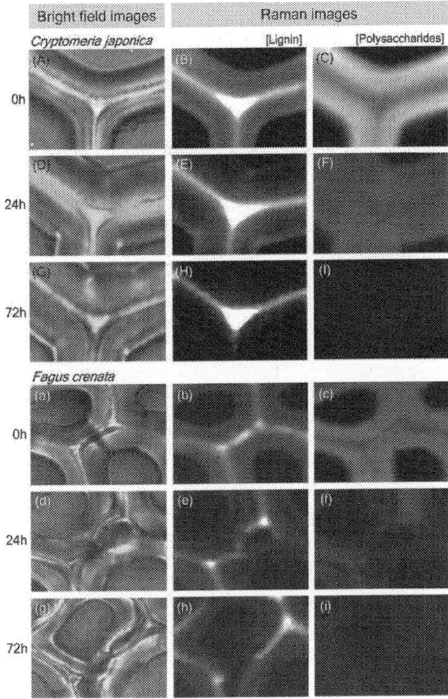

Figure 10. Raman mapping of transverse sections of tracheids of *Cryptomeria japonica* (**A–I**) and wood fibers of *Fagus crenata* (**a–i**) before and after treatment with [C2mim][Cl] at 120°C for 0 h, 24 h, and 72 h [55, 56]. The left panel shows bright field images of the measured position, the middle panel shows the distribution of lignin, the right panel shows the distribution of polysaccharides. Bright regions indicate high concentrations of specific chemical compositions, dark regions indicate low concentrations.

Raman mapping analysis was also performed on the wood samples treated with [EtPy][Br] (**Figure 11**). Although the lignin concentration in S_2 and in CML decreased with prolonged treatment time, the lignin in CML was preserved at relatively high concentration after 72 h of treatment in both tracheids of *Cryptomeria japonica* and wood fibers of *Fagus crenata*. Polysaccharide concentrations in both species decreased after 24 h of treatment. However, after 72 h of treatment, the polysaccharide concentrations in the tracheids of *Cryptomeria japonica* showed no significant changes, while those in the wood fibers of *Fagus crenata* continued to decrease gradually. These results imply that the reactivity of [EtPy][Br] with lignin is higher than with the polysaccharides in S_2 of fibrous cells. In addition, the liquefaction behavior of polysaccharides in fibrous cells during [EtPy][Br] treatment is different for different wood species.

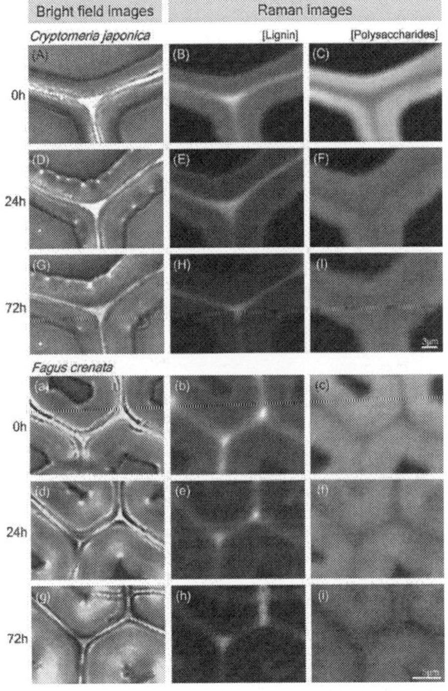

Figure 11. Raman mapping on transverse sections of tracheids of *Cryptomeria japonica* (**A–I**) and wood fibers of *Fagus crenata* (**a–i**) before and after treatment with [EtPy][Br] at 120°C for 0 h, 24 h, and 72 h [37, 38]. The left panel shows bright field images of the measured position, the middle panel shows the distribution of lignin, the right panel shows the distribution of polysaccharides. Bright regions indicate high concentrations of specific chemical compositions, dark regions indicate low concentrations.

To gain insight into the reactivity of [C2mim][Cl] with various morphological regions, Raman spectra were acquired for the S_2 of wood fibers, the cell corner of wood fibers, vessels, and axial parenchyma cells of *Fagus crenata* (**Figure 12**). During [C2mim][Cl] treatment, all the polysaccharide band intensities of S_2 of wood fibers and vessels decreased markedly, whereas those of the axial parenchyma cells decreased slightly. In addition, the intensity of the lignin band at 1657 cm^{-1} decreased significantly for all the measured regions except for the sample of axial parenchyma cells. Overall, the Raman spectra for axial parenchyma cells were not changed by [C2mim][Cl] treatment compared with those for the other morphological regions. This result indicates that both the polysaccharides and lignin of axial parenchyma cells are difficult to react with [C2mim][Cl]. The tendency of the spectral changes agrees with the morphological changes observed by SEM (**Figure 8**).

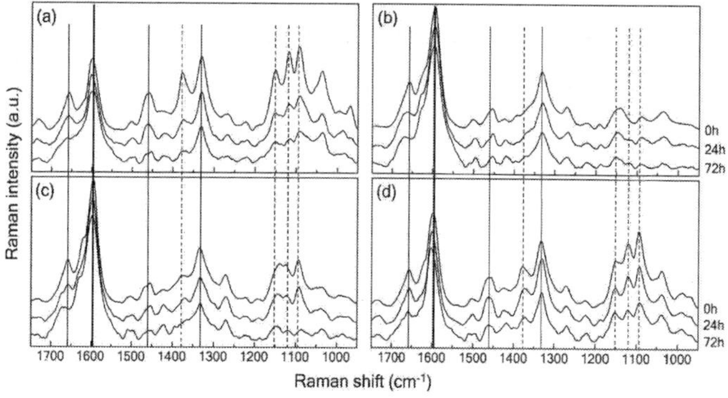

Figure 12. Raman spectra focusing in the spectral region of 950–1750 cm^{-1} for S_2 of wood fibers (**a**), cell corner of wood fibers (**b**), vessels (**c**), and axial parenchyma cells (**d**) of *Fagus crenata* before and after treatment with [C2mim][Cl] at 120°C for 0 h, 24 h, and 72 h; [56]. Lignin and polysaccharides are indicated by solid lines and dashed lines respectively. The spectra were normalized based on the band of an aromatic ring vibration around 1600 cm^{-1} (thick lines), which has superior chemical stability under [C2mim][Cl] treatment conditions, as an internal reference to visualize the spectral differences after treatment with [C2mim][Cl] clearly.

3. Conclusions

Using various microscopy techniques, the morphological and topochemical features of wood cell walls treated with ionic liquids were studied. During the processing of wood liquefaction in ionic liquids, the ultrastructure and chemical compositions of wood showed inhomogeneous changes at the cellular level. The interaction of ionic liquid with wood cell walls was quite different depending on the types of ionic liquids, wood species, tissues, and cell wall layers. These findings will serve to cultivate a better understanding of the liquefaction mechanism of woody biomass in ionic liquids and accelerate development of ionic liquid treatment for wood-based biorefinery.

For research in wood chemistry and anatomy, many microscopy techniques have been applied to investigate the characteristics of the cell walls. However, it is hardly possible to examine the chemical compositions and their distribution with nanoscale spatial resolution while at the same time observing the ultrastructure such as ultrathin layers. The development of sensitive analytical methods in the wood cell walls for chemical information with much higher spatial resolution will open a new field of wood science and technology.

Acknowledgements

This work was partly supported by the "Science and Technology Research Promotion Program for Agriculture, Forestry, Fisheries and Food Industry" (No. 26052A) from the Ministry of Agriculture, Forestry and Fisheries of Japan, a Grant-in-Aid for Scientific Research (C) (No. 25450246) from the Japan Society for the Promotion of Science (JSPS), and a Grant-in-Aid for JSPS Fellows (No. 15J05592).

Author details

Toru Kanbayashi[1,2] and Hisashi Miyafuji[2*]

*Address all correspondence to: miyafuji@kpu.ac.jp

1 Department of Wood Improvement, Forestry and Forest Products Research Institute, Ibaraki, Japan

2 Graduate School of Life and Environmental Sciences, Kyoto Prefectural University, Kyoto, Japan

References

[1] Zhu JY, Pan X, Zalesny Jr. RS. Pretreatment of woody biomass for biofuel production: energy efficiency, technologies, and recalcitrance. Applied Microbiology and Biotechnology. 2010;87(3):847–857. DOI: 10.1007/s00253-010-2654-8

[2] Pettersen RC. The Chemical Composition of Wood. In: Rowell RM, editor. The Chemistry of Solid Wood, Advances in Chemistry Series, Vol. 207. Washington, DC: American Chemical Society; 1984. p. 57–126.

[3] Taherzadeh MJ, Karimi K. Acid-based hydrolysis processes for ethanol from lignocellulosic materials: a review. BioResources. 2007;2(3):472–499.

[4] Taherzadeh MJ, Eklund R, Gustafsson L, Niklasson C, Liden G. Characterization and fermentation of dilute-acid hydrolyzates from wood. Industrial & Engineering Chemistry Research. 1997;36(11):4659–4665. DOI: 10.1021/ie9700831

[5] Parisi F. Advances in Lignocellulosics Hydrolysis and in the Utilization of the Hydrolyzates. In: Fiechter A, editor. Advances in Biochemical Engineering/Biotechnology, Vol. 38. Berlin: Springer-Verlag; 1989. p. 53–87. DOI: 10.1007/BFb0007859

[6] Chang VS, Holtzapple MT. Fundamental factors affecting biomass enzymatic reactivity. Applied Biochemistry and Biotechnology. 2000;84–86:5–37. DOI: 10.1007/978-1-4612-1392-5_1

[7] Zhao Y, Wang Y, Zhu JY, Ragauskas A, Deng Y. Enhanced enzymatic hydrolysis of spruce by alkaline pretreatment at low temperature. Biotechnology and Bioengineering. 2008;99(6):1320–1328. DOI: 10.1002/bit.21712

[8] Piskorz J, Radlein D, Scott DS, Czernik S. Liquid Products from the Fast Pyrolysis of Wood and Cellulose. In: Bridgwater AV, Kuester JL, editors. Research in Thermochemical Biomass Conversion. London: Elsevier Applied Science; 1988. p. 557–571. DOI: 10.1007/978-94-009-2737-7_43

[9] Kwon GJ, Kuga S, Hori K, Yatagai M, Ando K, Hattori N. Saccharification of cellulose by dry pyrolysis. Journal of Wood Science. 2006;52(5):461–465. DOI: 10.1007/s10086-005-0784-x

[10] Hosoya T, Kawamoto H, Saka S. Influence of inorganic matter on wood pyrolysis at gasification temperature. Journal of Wood Science. 2007;53(4):351–357. DOI: 10.1007/s10086-006-0854-8

[11] Oasmaa A, Solantausta Y, Arpiainen V, Kuoppala E, Sipilä K. Fast pyrolysis bio-oils from wood and agricultural residues. Energy & Fuels. 2010;24(2):1380–1388. DOI: 10.1021/ef901107f

[12] Yamazaki J, Minami E, Saka S. Liquefaction of beech wood in various supercritical alcohols. Journal of Wood Science. 2006;52(6):527–532. DOI: 10.1007/s10086-005-0798-4

[13] Xu C, Etcheverry T. Hydro-liquefaction of woody biomass in sub- and super-critical ethanol with iron-based catalysts. Fuel. 2008;87(3):335–345. DOI: 10.1016/j.fuel.2007.05.013

[14] Sheldon R. Catalytic reactions in ionic liquids. Chemical Communications. 2001;23:2399–2407. DOI: 10.1039/B107270F

[15] Seddon KR. Ionic liquids for clean technology. Journal of Chemical Technology and Biotechnology. 1997;68(4):351–356. DOI: 10.1002/(SICI)1097-4660(199704)68:4<351::AID-JCTB613>3.0.CO;2-4

[16] Rogers RD, Seddon KR. Ionic liquids – Solvents of the future? Science. 2003;302(5646):792–793. DOI: 10.1126/science.1090313

[17] Freemantle M. Designer solvents – Ionic liquids may boost clean technology development. Chemical & Engineering News. 1998;76:32–37.

[18] Swatloski RP, Spear SK, Holbrey JD, Rogers RD. Dissolution of cellulose with ionic liquids. Journal of the American Chemical Society. 2002;124(18):4974–4975. DOI: 10.1021/ja025790m

[19] Zhang H, Wu J, Zhang J, He J. 1-Allyl-3-methylimidazolium chloride room temperature ionic liquid: a new and powerful nonderivatizing solvent for cellulose. Macromolecules. 2005;38(20):8272–8277. DOI: 10.1021/ma0505676

[20] Fukaya Y, Hayashi K, Wada M, Ohno H. Cellulose dissolution with polar ionic liquids under mild conditions: required factors for anions. Green Chemistry. 2008;10:44–46. DOI: 10.1039/B713289A

[21] Pu Y, Jiang N, Ragauskas AJ. Ionic liquid as a green solvent for lignin. Journal of Wood Chemistry and Technology. 2007;27(1):23–33. DOI: 10.1080/02773810701282330

[22] Honglu X, Tiejun S. Wood liquefaction by ionic liquids. Holzforschung. 2006;60(5):509–512. DOI: 10.1515/HF.2006.084

[23] Fort DA, Remsing RC, Swatloski RP, Moyna P, Moyna G, Rogers RD. Can ionic liquids dissolve wood? Processing and analysis of lignocellulosic materials with 1-*n*-butyl-3-methylimidazolium chloride. Green Chemistry. 2007;9(1):63–69. DOI: 10.1039/B607614A

[24] Kilpeläinen I, Xie H, King A, Granstrom M, Heikkinen S, Argyropoulos DS. Dissolution of wood in ionic liquids. Journal of Agricultural and Food Chemistry. 2007;55(22):9142–9148. DOI: 10.1021/jf071692e

[25] Miyafuji H, Miyata K, Saka S, Ueda F, Mori M. Reaction behavior of wood in an ionic liquid, 1-ethyl-3-methylimidazolium chloride. Journal of Wood Science. 2009;55(3):215–219. DOI: 10.1007/s10086-009-1020-x

[26] Abe M, Yamada T, Ohno H. Dissolution of wet wood biomass without heating. RSC Advances. 2014;4(33):17136–17140. DOI: 10.1039/C4RA01038H

[27] Sun N, Rahman M, Qin Y, Maxim ML, Rodriguez H, Rogers RD. Complete dissolution and partial delignification of wood in the ionic liquid 1-ethyl-3-methylimidazolium acetate. Green Chemistry. 2009;11(5):646–655. DOI: 10.1039/B822702K

[28] Nakamura A, Miyafuji H, Saka S. Liquefaction behavior of Western red cedar and Japanese beech in the ionic liquid 1-ethyl-3-methylimidazolium chloride. Holzforschung. 2010;64(3):289–294. DOI: 10.1515/hf.2010.042

[29] Nakamura A, Miyafuji H, Saka S. Influence of reaction atmosphere on the liquefaction and depolymerization of wood in an ionic liquid, 1-ethyl-3-methylimidazolium chloride. Journal of Wood Science. 2010;56(3):256–261. DOI: 10.1007/s10086-009-1081-x

[30] Brandt A, Hallett JP, Leak DJ, Murphy RJ, Welton T. The effect of the ionic liquid anion in the pretreatment of pine wood chips. Green Chemistry. 2010;12:672–679. DOI: 10.1039/B918787A

[31] Yokoo T, Miyafuji H. Reaction behavior of wood in an ionic liquid, 1-ethylpyridinium bromide. Journal of Wood Science. 2014;60(5):339-345. DOI: 10.1007/s10086-014-1409-z

[32] Saka S. Chemical Composition and Distribution. In: Hon DNS, Shiraishi N, editors. Wood and Cellulosic Chemistry. New York: Marcel Dekker; 1991. p. 59–88.

[33] Miyafuji H, Suzuki N. Observation by light microscope of sugi (*Cryptomeria japonica*) treated with the ionic liquid 1-ethyl-3-methylimidazolium chloride. Journal of Wood Science. 2011;57:459–461. DOI: 10.1007/s10086-011-1190-1

[34] Miyafuji H, Suzuki N. Morphological changes in sugi (*Cryptomeria japonica*) wood after treatment with the ionic liquid, 1-ethyl-3-methylimidazolium chloride. Journal of Wood Science. 2012;58(3):222–230. DOI: 10.1007/s10086-011-1245-3

[35] Kanbayashi T, Miyafuji H. Morphological changes of Japanese beech treated with the ionic liquid, 1-ethyl-3-methylimidazolium chloride. Journal of Wood Science. 2013;59(5):410–418. DOI: 10.1007/s10086-013-1343-5

[36] Kanbayashi T, Miyafuji H. Comparative study of morphological changes in hardwoods treated with the ionic liquid, 1-ethyl-3-methylimidazolium chloride. Journal of Wood Science. 2014;60(2):152–159. DOI: 10.1007/s10086-014-1389-z

[37] Kanbayashi T, Miyafuji H. Topochemical and morphological characterization of wood cell wall treated with the ionic liquid, 1-ethylpyridinium bromide. Planta. 2015;242(3): 509–518. DOI: 10.1007/s00425-014-2235-7

[38] Kanbayashi T, Miyafuji H. Anatomical and topochemical aspects of Japanese beech (*Fagus crenata*) cell walls after treatment with the ionic liquid, 1-ethylpyridinium bromide. Microscopy and Microanalysis. 2015;21(6):1562–1572. DOI: 10.1017/S1431927615015275

[39] Donaldson LA, Lignification and lignin topochemistry – an ultrastructural view. Phytochemistry. 2001;57(6):859–873. DOI: 10.1016/S0031-9422(01)00049-8

[40] Harada H, Côté WA. Structure of Wood. In: Higuchi T, editor. Biosynthesis and Biodegradation of Wood Components. Orlando: Academic Press; 1985. p. 1–42.

[41] Jansen S, Smets E, Baas P. Vestures in woody plants: a review. IAWA Journal. 1998;19(4):347–382. DOI: 10.1163/22941932-90000658

[42] Fujii T, Shimizu K, Yamaguchi A. Enzymatic saccharification on ultrathin sections and ultraviolet spectra of Japanese hardwoods and softwoods. Mokuzai Gakkaishi. 1987;33(5):400–407.

[43] Fackler K, Thygesen LG. Microspectroscopy as applied to the study of wood molecular structure. Wood Science and Technology. 2013;47(1):203–222. DOI: 10.1007/s00226-012-0516-5

[44] Lupoi JS, Singh S, Simmons BA, Henry RJ. Assessment of lignocellulosic biomass using analytical spectroscopy: an evolution to high-throughput techniques. BioEnergy Research. 2014;7(1):1–23. DOI: 10.1007/s12155-013-9352-1

[45] Agarwal UP. Raman imaging to investigate ultrastructure and composition of plant cell walls: distribution of lignin and cellulose in black spruce wood (*Picea mariana*). Planta. 2006;224:1141–1153. DOI: 10.1007/s00425-006-0295-z

[46] Gierlinger N, Schwanninger M. Chemical imaging of poplar wood cell walls by confocal Raman microscopy. Plant Physiology. 2006;140(4):1246–1254. DOI: 10.1104/pp.105.066993

[47] Röder T, Koch G, Sixta H. Application of confocal Raman spectroscopy for the topochemical distribution of lignin and cellulose in plant cell walls of beech wood (*Fagus sylvatica* L.) compared to UV microspectrophotometry. Holzforschung. 2004;58(5):480–482. DOI: 10.1515/HF.2004.072

[48] Schmidt M, Schwartzberg AM, Perera PN, Weber-Bargioni A, Carroll A, Sarkar P, Bosneaga E, Urban JJ, Song J, Balakshin MY, Capanema EA, Auer M, Adams PD, Chiang VL, James Schuck P. Label-free in situ imaging of lignification in the cell wall of low lignin transgenic *Populus trichocarpa*. Planta. 2009;230(3):589–597. DOI: 10.1007/s00425-009-0963-x

[49] Hänninen T, Kontturi E, Vuorinen T. Distribution of lignin and its coniferyl alcohol and coniferyl aldehyde groups in *Picea abies* and *Pinus sylvestris* as observed by Raman imaging. Phytochemistry. 2011;72(14–15):1889–1895. DOI: 10.1016/j.phytochem.2011.05.005

[50] Sun L, Simmons BA, Singh S. Understanding tissue specific compositions of bioenergy feedstocks through hyperspectral Raman imaging. Biotechnology and Bioengineering. 2011;108(2):286–295. DOI: 10.1002/bit.22931

[51] Zhang Z, Ma J, Ji Z, Xu F. Comparison of anatomy and composition distribution between normal and compression wood of *Pinus bungeana* Zucc. revealed by microscopic imaging techniques. Microscopy and Microanalysis. 2012;18(6):1459–1466. DOI: 10.1017/S1431927612013451

[52] Ma J, Zhang X, Zhou X, Xu F. Revealing the changes in topochemical characteristics of poplar cell wall during hydrothermal pretreatment. BioEnergy Research. 2014;7(4):1358–1368. DOI: 10.1007/s12155-014-9472-2

[53] Ji Z, Ma J, Xu F. Multi-scale visualization of dynamic changes in poplar cell walls during alkali pretreatment. Microscopy and Microanalysis. 2014;20(2):566–576. DOI: 10.1017/S1431927614000063

[54] Zhou X, Ma J, Ji Z, Zhang X, Ramaswamy S, Xu F. Dilute acid pretreatment differentially affects the compositional and architectural features of *Pinus bungeana* Zucc. compression and opposite wood tracheid walls. Industrial Crops and Products. 2014;62:196–203. DOI: 10.1016/j.indcrop.2014.08.035

[55] Kanbayashi T, Miyafuji H. Raman microscopic analysis of wood after treatment with the ionic liquid, 1-ethyl-3-methylimidazolium chloride. Holzforschung. 2015;69(3):273–279. DOI: 10.1515/hf-2014-0060

[56] Kanbayashi T, Miyafuji H. Raman microscopic study of Japanese beech (*Fagus crenata*) as treated with the ionic liquid, 1-ethyl-3-methylimidazolium chloride. Journal of Wood Chemistry and Technology. 2016;36(3):224–234. DOI: 10.1080/02773813.2015.1112404

[57] Wiley JH, Atalla RH. Band assignments in the Raman spectra of celluloses. Carbohydrate Research. 1987;160:113–129. DOI: 10.1016/0008-6215(87)80306-3

[58] Agarwal UP, Ralph SA. FT-Raman spectroscopy of wood: Identifying contributions of lignin and carbohydrate polymers in the spectrum of black spruce (*Picea mariana*). Applied Spectroscopy. 1997;51(11):1648–1655. DOI: 10.1366/0003702971939316

[59] Edwards HGM, Farwell DW, Webster D. FT Raman microscopy of untreated natural plant fibres. Spectrochimica Acta Part A: Molecular and Biomolecular Spectroscopy. 1997;53(13):2383–2392. DOI: 10.1016/S1386-1425(97)00178-9

[60] Agarwal UP. An Overview of Raman Spectroscopy as Applied to Lignocellulosic Materials. In: Argyropoulos DS, editor. Advances in Lignocellulosics Characterization. Atlanta: TAPPI Press; 1999. p. 201–225.

[61] Agarwal UP, McSweeny JD, Ralph SA. FT-Raman investigation of milled wood lignins: Softwood, hardwood, and chemically modified black spruce lignins. Journal of Wood Chemistry and Technology. 2011;31(4):324–344. DOI: 10.1080/02773813.2011.562338

Non-Optical Microscopy

Chapter 10

The New Youth of the *In Situ* Transmission Electron Microscopy

Alberto Casu, Elisa Sogne, Alessandro Genovese,
Cristiano Di Benedetto, Sergio Lentijo Mozo,
Efisio Zuddas, Francesca Pagliari and Andrea Falqui

Additional information is available at the end of the chapter

http://dx.doi.org/10.5772/63269

Abstract

The idea of *in situ* transmission electron microscopy (TEM) and its possible ramifications were proposed at the very dawn of electron microscopy, but the translation from theory to practice encountered many technological setbacks, which hindered the feasibility of the most elaborated approaches until recent times. However, the several technological improvements achieved in the last 10–15 years filled this gap, allowing the direct observation of the dynamic response of materials to external stimuli under a vast range of conditions going from vacuum to gaseous or liquid environment. This resulted in a blossoming of the *in situ* TEM and scanning TEM (STEM) techniques to a new youth for a vast, growing range of applications, which cannot be rightfully detailed in a short span; therefore, this chapter should be intended as a guide highlighting a selection of the most inspiring, recently achieved results.

Keywords: analytical electron microscopy, environmental transmission electron microscopy, *in situ* transmission electron microscopy, scanning transmission electron microscopy, transmission electron microscopy

1. Introduction

The idiom "seeing is believing" is very old and has been often used since its religious inception to the present day as the title of novels, songs, movies, and documentaries. However, it is also the actual utterance of any *in situ* microscopist. So, what is the *in situ* transmission electron microscopy (TEM)? It consists in the ability to look by TEM and in real time at the dynam-

ic behavior of a material as a consequence of an external *stimulus*, so providing an insight into its properties and modifications while they are happening. This kind of information is inaccessible by other techniques at the atomic scale, which is used to perform imaging, structural and/ or compositional analyses by TEM. The main difference with respect to any other TEM observation technique lies in the fact that instead of carrying out experiments on specimens outside the microscope and using the TEM for observations before and after the experiments' completion, the *in situ* TEM permits to image the evolution of materials while it is occurring, thus not losing any information about intermediate states. As briefly mentioned above, the evolution over time of samples under TEM observation takes place under an external *stimulus*. Then, it is possible changing the environment around the TEM sample in several ways: varying the temperature up to very high or down to very low values; keeping the specimen in thermally controlled static or dynamic gaseous environment, in the latter case also controlling the gas pressure; putting the sample in different liquids under static or flowing regime, also to perform local chemical/electrochemical reactions; applying to it local strain, compression, nanoindentation; and putting it in an electrical or magnetic field, as well as irradiating it by electrons or light.

Why we titled this chapter "The new youth of the *in situ* electron microscopy"? Because if on the one hand the possibility to perform *in situ* experiments was always chased resolutely since the beginning of TEM studies, on the other hand it was just the dramatic advancements achieved by TEM in the last 10–15 years that gave the necessary impulse to further develop the *in situ* studies. First, TEM—even in the scanning mode (STEM)—has recently known a new and strong evolution with respect to the past. The two fundamental events that greatly improved the way to perform (S)TEM imaging consisted in the availability of both ultrabright electron sources [1] and spherical aberration correctors [2, 3] for either the condenser or the objective lens. The first improvement made electron beams very bright allowing to work in both TEM and STEM geometries with very low exposition times for both imaging and spectroscopic acquisition. The second one permitted to achieve unprecedented spatial resolution, down to less than 0.1 nm, using conventional electron acceleration voltages. Furthermore, the last decade also saw other important developments: electron energy resolution was improved to less than 0.1 eV by using beam monochromators [4] or cold field emission electron sources [5], while the most recent silicon drift detectors (SDD) with very large solid angle for energy-dispersive spectroscopy (EDS) [6], electron energy loss spectroscopy (EELS) image filters [7], and the ultrafast CMOS cameras [8] allowed to investigate the specimens' evolution in terms of both imaging and elemental analyses down to the atomic scale with very high time resolution and under very low electron dose conditions, respectively. As a consequence, all the (S)TEM-based analytical and imaging techniques largely benefited from these achievements and also pushed up the capability to realize further *in situ* studies mainly by means of novel and dedicated microscopes or specimen holders, which actually work as small "laboratories" providing the means to submit the samples to all the possible external *stimuli* mentioned above. In most of the cases, micro- and nanoelectromechanical systems (MEMS and NEMS, respectively) as well as nanodevices led to the miniaturization of the specimen holders' or stages' components. This allowed performing *in situ* experiments with exceptional minimization of the sample drift, currently reduced to about 1 nm min^{-1} and characterized by an increase in complexity and number of possible *stimuli*, while granting a fine devices' control and the simultaneous recording of massive amounts of data.

Due to the current and continuously growing vastness of the *in situ* (S)TEM field, this chapter shows only some of the most recent and exciting innovations and results achieved in this field. Not all applications are then represented here, but we decided to focus the attention mainly on the *in situ* (S)TEM techniques and experiments that in our opinion well represent the most important, recently achieved novelties with the main aim of showing the improvement in quality of today's scientific results with respect to the past.

The chapter starts with a section dedicated to the *in situ* heating experiments because heating any sample upon the TEM high vacuum conditions is a decades-old kind of experiment that was affected by two concomitant limitations until few years ago. Since the heating was usually realized by a small furnace surrounding the sample, the first unavoidable limitation was due to the thermal dilatation of the metal constituting the TEM grid during all thermal ramps as a consequence of a huge drift suffered by the whole sample that made recording any image or analytical data impossible until the thermal steady state was reached. The second effect was less crucial and due to the high amount of heat produced by the furnace: the increase in temperature was unavoidably accompanied by the emission of highly intense infrared (IR) radiation coming from the heated furnace and TEM grid. This IR emission saturated any EDS detector even at low temperature (i.e., less than 100°C), making impossible to use this spectroscopic technique to qualitatively or quantitatively analyze the elements constituting the samples. However, thanks to the new MEMS-based heating holders, where a heating chip has substituted the furnace that heated the TEM grid, both these limitations have been dramatically reduced. First, it is now possible reaching sample drift as low as 1 nm min^{-1} during the thermal ramps. Second, being the heating element much smaller than the classical furnace, the power needed to reach high temperatures is also much lower than the one needed to increase the temperature of a standard metal TEM grid, resulting in a strong limitation of the produced heat. This in turn allows the use of EDS for analyzing samples with temperatures up to some hundreds of degrees without saturating the EDS detector. In the following section, the attention is then focused on a sort of "variation on a theme" of the *in situ* sample heating. In this case, a different sample environment is taken into account, being it a gaseous one developed to study *in situ* solid-gas reactions. This section explains the two main current routes to perform this kind of experiments, namely, (a) by using a very small and physically isolated volume where the gas could either be sealed or let flow inside the TEM sample holder; (b) by inserting a heating sample holder in a particular kind of microscope, usually known as environmental TEM (E-TEM), featuring a portion of the TEM column filled with gas at a controlled pressure and limited by two zones where a differential pumping system prevents any gas leak in the rest of the TEM column and gun. Section 3 of this chapter is dedicated to a sort of further development of the gas-solid *in situ* (S)TEM techniques and devices shown in the previous section, with the aim to realize *in situ* imaging of samples immersed in a liquid environment and to follow real-time chemical reactions that are occurring in a liquid solution. From a historical point of view, these kinds of experiments had already been attempted just after the TEM invention, between the end of the 1930s and the beginning of the 1940s of the twentieth century, but they became effectively attainable just in very recent years. In this regard, the most interesting aspects of performing *in situ* liquid (S)TEM experiments consist in the fact that, first, this is the only TEM technique that allows the observation of living cells, although with some limitations that are discussed below and, second, that the high electron scattering power of any liquid requires slightly different sample configurations and preparations, depending on the geometry (TEM or STEM) that is used to perform the *in situ* imaging.

2. *In situ* heating TEM experiments in vacuum

Studies involving temperature and its effects on the features and evolution of various systems are most common in a very broad number of different branches, ranging from fundamental research and going to strictly applicative performance issues. Despite the very different focus, all these kind of studies rely on the basic assumption that the goodness and reliability of the results will necessarily go through trying to obtain the most thorough description of the evolution of the system and of the phenomena leading to said evolution. Thus, while TEM stands as the obvious choice for gathering a vast array of accurate direct structural information, the choice of performing *in situ* rather than *ex situ* heating treatment lies in the possibility of obtaining as precise data as possible, thus limiting the temperature ranges where unnoticed transient variations could take place. The *ex situ* approach consists in an out-of-TEM-column annealing and in the subsequent observation by TEM of its results, so it is a simpler setup from the point of view of the actual TEM (only an external furnace/heating source is required), which does not affect the microscope's performance in terms of spatial resolution. Conversely, this approach is inherently limited by the fact that the study of a temperature-dependent system would require the *ex post* reconstruction of the intermediate states by repeating the annealing at different user-defined temperatures. Leaving aside any consideration regarding the samples, this approach in principle allows for a very precise control over temperature, but also retains a degree of uncertainty over the evolution of the system given by the user-defined rate of temperature sampling, which must obviously be a discrete, not continuous, one.

On the other hand, the *in situ* approach implies that the annealing takes place inside the TEM column by using a dedicated heating holder, thus allowing a direct, continuous observation of the temperature-dependent evolution of the sample. The *in situ* approach to annealing seems the most promising one, but in practice it relies heavily on the technological/technical advancements in both design and performance of the heating holders, namely, minimizing the thermal mechanical drift associated with the heating of the holder and extending its temperature range while improving the precision of temperature control and thermal stability of the holder, thus allowing the possibility to run better controlled and more elaborated experiments.

In this framework, the evolution observed in the design of heating holders, going from homemade custom ones [9, 10] to commercial furnace-type ones [11, 12] to the most recent developments involving strongly localized micron-wide heating areas, affected the feasibility of heating experiments.

In furnace-type heating holders, the annealing is achieved by heating a section located at the tip of the TEM holder, where the TEM grid is lodged and clamped. A heating filament in proximity to the TEM grid is responsible for the heating of this "hot zone," while the temperature can be measured by a thermocouple or a temperature-to-current calibration curve (**Figure 1A**).

Obviously, the hot region of the furnace must be mechanically stable within the working temperature range in order to minimize the mechanical drift associated with the thermal expansion of the furnace, while being as thermally isolated as possible from the rest of the holder in order to limit the heat conduction to the rest of the tip/holder and the consequent instability in reaching and measuring the correct temperature. Thus, the improvement of this category of holders goes through finding appropriate materials for the different components of the tip and designing a heating system that maximizes both the heating performance and the thermal and mechanical stabilities while minimizing the undesired side effects, such as heat dissipation to the rest of the holder and high power consumption.

From the materials point of view, this implies lowering the coefficient of thermal expansion for the ones used in the furnace, while those used to connect the furnace to the tip must also guarantee low thermal conductivity, since the connections should insulate the furnace against the heat conduction to the holder. Moreover, the rest of the tip should have high heat capacity in order to minimize the heating by irradiation coming from the furnace, thus keeping the holder as close as possible to room temperature. The results achieved with regard to the materials must be supported by improving the design of the holder in order to increase its ease of use and its efficiency in terms of performance vs. power consumption. This means, for example, trying to minimize the size of the furnace (and consequently increasing the available heating rates while decreasing the power consumption necessary for any heat treatment) or designing tips that decrease the heat dispersion from the furnace. However, one inherent problem of indirect-heating holders, such as the furnace-based ones, lies in the gradient of temperature occurring between the heating furnace and the TEM grid during the annealing: this difference can be minimized in stable temperature conditions, but it still induces a transient state of thermal imbalance and mechanical drift due to thermal expansion.

In this framework, the run for improving the performance of heating holders leads to the introduction of different solutions that abandoned the furnace-based architecture for annealing TEM samples (**Figure 1B** and **C**). Kamino holders [13] adopted heating wires as a direct source of Joule heating and as a support for the TEM sample: this architecture allows for a direct heating of the sample and higher annealing temperatures but lacks versatility, because samples need to be supportable by the heating wire. A further step in this direction is represented by heating holders based on MEMS (microelectromechanical systems) chips technology [14, 15]. These chips are constituted by a semiconductor (usually Si or Si-based material) as main body, overlaid by a thin film and featuring a small, electron-transparent area devoted to the heating and subsequent TEM analysis, thus effectively combining the heating element and the TEM grid in the same object. The presence of a support film allows for greater versatility in terms of possible samples, while dedicated conductive wirings provide the heating and temperature measurements of the heating area in a fashion similar to the furnace-type holders: a heating wire acts as an electrical furnace, while a second one is used as a temperature probe against a resistance-to-temperature calibration curve. Given the small dimensions of the heating area (usually less than 0.1 mm² wide) with respect to those of the whole chip, choosing a material with high thermal conductivity enables the chip itself to act both as a thermal sink with respect to the heating area and as an insulating buffer with respect to the holder. On the other hand, the engineering of a miniaturized heating area with low heat capacity implies using smaller heating currents. The combination of these characteristics allows faster temperature ramps and faster cooling rates by the sole variations in heating current values, resulting in a more precise control over the desired temperatures and an overall expansion of the possibilities of annealing experiments.

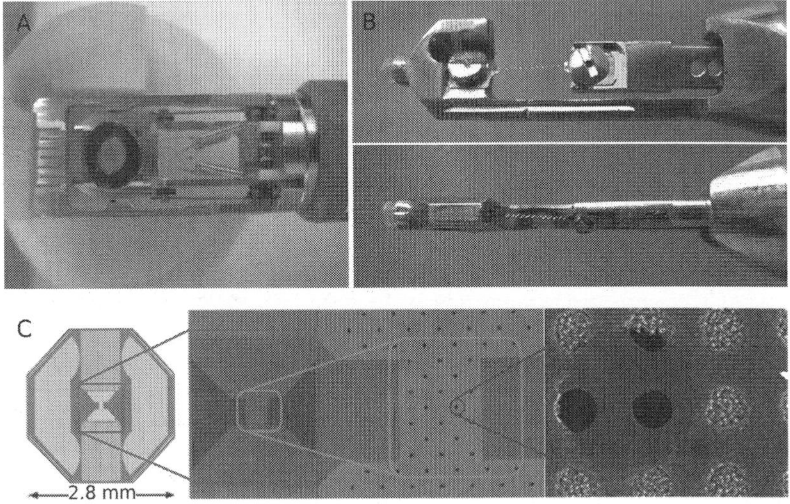

Figure 1. (A) Top view of a furnace-based heating holder. The circular hole at the center of the furnace serves as lodging for the TEM grid. Adapted with permission from [116]; (B) top and side views of a Kamino-type, direct-heating holder displaying a fine tungsten wire heater. Adapted with permission from [117]; (C) top view schematic of MEMS-based heater chip. Backscattered and secondary electron images of the central region of the chip at increasing magnification reveal the structure of its core and the carbon-film-covered holes. Adapted with permission from [15].

Considering these innovations in the more general framework of the recent achievements of TEM, the possibility of performing *in situ* annealing experiments with atomic resolution has become a realistic goal, shedding new light on the fine mechanisms of already well-established areas of interest and on previously impossible experimental routes.

2.1. *In situ* TEM heating in Au nanoparticles

The study of colloidal metallic nanoparticles (NPs) has been a major research subject during the past decades, at first to show, then to study and modify their properties for a vast array of possible applications, such as biomedical, sensor technologies, energy storage and nanocatalysis. The development and improvement of imaging techniques such as TEM allowed a clearer comprehension on the connections between structural characteristics and physical properties at the nanoscale, which are required to understand and optimize their use in diverse applications.

In this general context, Au NPs were a constant object of interest in studies describing different synthetic methods and morphologies, their properties, and their possible applications [16–20]. However, regardless of the particular field of application, any engineering attempt would benefit from a detailed analysis of the morphological and structural transformations happening at the nanoscale. This is the reason for developing and improving a nanoscale phase diagram for Au NPs, i.e., a predictive map of their structural-, morphology-, and temperature-dependent stable configurations. Previous experimental and theoretical studies on nanogold [21, 22] were recently improved thanks to *in situ* annealing studies, taking advantage of the improved temperature control granted by MEMS-based technology and leading from qualitative to quantitative phase diagrams. Barnard [23] provided a comparative study between their calculated phase map and *in situ* annealing high resolution TEM (HRTEM) data. The quantitative phase map was obtained by combining the calculated free energy of different motifs according to a free energy vs. temperature vs. size model and illustrated the stable motifs for gold NPs, while the annealing experiments (up to 1000 K) provided the phase evolution data for NPs of different sizes, showing the temperature-driven evolution from ichosaedral, to decahedral, and finally to a disordered motif with "roughened" surface (**Figure 2A**). These results were further validated and refined by devoting attention to the high temperature regime and surface roughening [24]. In fact, *in situ* HRTEM annealing data showed that despite a modest size variation at lower temperatures, a consistent decrease in size could be observed once the thermal threshold of surface roughening had been exceeded. This experimental trend was consistent with previous theoretical models, but also highlighted which physical phenomena took place in the premelting regime. The occurrence of NPs with an amorphous, irregular surface and a crystalline core excluded an amorphization by melting the whole NP in favor of a coexisting liquid/solid state, where the evaporation of the liquid molten phase is responsible for the size variation.

The relationship between theoretical and experimental data with regard not only to the structural evolution of the NPs but also to their so-called roughened and melting states proved the suitability of this course of action for obtaining a full structural characterization. In fact, a similar approach was recently tried by Baumgardner [25], who studied the structural evolution of a heterogeneous $Au:Fe_2O_3$ NP system at elevated temperatures by *in situ* annealing. In particular, by taking advantage of Z-contrast in high angle annular dark-field scanning TEM (HAADF-STEM), it was possible to follow the evolution from a dual population of Au and Fe_2O_3 NPs to the formation of new mixed Au/Fe_2O_3 NPs via fluid shape transformation. Two possible morphologies emerged, namely, phase-segregated Fe_2O_3/Au structures or surface-alloyed NPs with an Au/Fe layer in between the iron oxide core and the gold shell. The evolution of phase-segregated and surface-alloyed NPs was further monitored by increasing the *in situ* annealing temperature while collecting HAADF-STEM and chemical data (**Figure 2B**). Thus, variations in chemical composition and size of each NP vs. temperature were combined as a nanosized phase diagram, which works as a compositional stability phase map for this binary system.

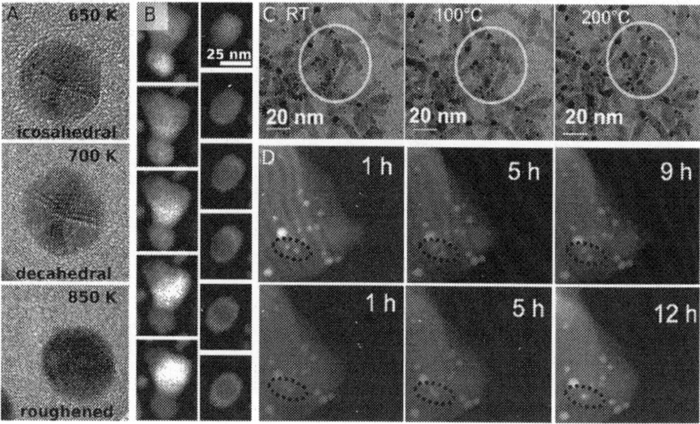

Figure 2. (A) Representative HRTEM images showing the phase variations of a 7.5 nm wide Au NP as a function of temperature. Adapted with permission from [23]; (B) phase transition from alloy to phase segregated (left) or core/shell nanoparticles (right). Adapted with permission from [25]; (C) HRTEM images showing the evolution of Au@CdSe networks during the *in situ* annealing experiment. Adapted with permission from [27]; (D) HAADF-STEM images showing the motion of Au NPs in SBA-15 during *in situ* TEM heating experiments conducted at 550°C (top) and 700°C (down). Adapted with permission from [28].

On the other hand, *in situ* annealing TEM is also a valuable tool for less systematic and more application-based studies, focused on trying to take into account (and possibly take advantage of) the properties of Au NPs for developing nanosized heterostructures for different applications. In this regard, a good example is given by the mobility and sintering of Au NPs. The mobility was a key factor in studying CdSe nanorods (NRs) decorated with Au domains and their structural and morphological evolution under thermal annealing for possible applications in the fields of electronics and photovoltaics [26, 27]. While at room temperature the Au decoration of CdSe NRs consisted in small Au NPs attached to the lateral facets of the rods and larger Au NPs located at their tips, heating the samples above 200°C caused the migration of the small NPs toward the tips and the subsequent ripening of the larger Au NPs. In particular, the ripening of Au NPs at the tips of nearby CdSe rods led to bigger sized NP/NR interfaces, effectively forming a robust network (**Figure 2C**). Further investigations on the charge transport mechanisms showed that the annealing improved the conductance of these networks, thus making them viable candidates in the development of bottom-up electronic devices.

Conversely, the mobility and sintering of Au NPs is a drawback in the case of catalysis, where stability problems caused by deactivation due to NP sintering at high temperatures are well-known problems for metallic NP-based catalysts. Thus, a deeper comprehension of the sintering mechanism of NPs inside a porous matrix is a basic research study with immediate applicative repercussions. Liu et al. [28] studied the migration of Au NPs inside the ordered channels of mesoporous silica via *in situ* TEM annealing (**Figure 2D**). By comparing the migration of Au NPs inside silica mesoporous matrices with different pore diameters and micropore volumes at high temperatures, they observed that in fact the sintering was substantially suppressed by the largest silica matrix with largest micropore volumes, since the Au NPs tend to be embedded in the micropores of the mesochannel walls. This result highlighted the main importance of abundance and size of micropores to hinder the migration and sintering of the NPs. Moreover, by performing *in situ* STEM-EELS during the annealing and observing the local variations in the plasmonic features of a region with two coalescing NPs, they could suggest that an expansion of the conduction electron clouds induced by the annealing could trigger electron-tunneling-induced coalescence of nearby Au NPs.

2.2. *In situ* chemical reactions of semiconductor-based materials

Semiconductors represent another broad research subject in the fields of chemistry and material science, but recent studies conducted on these systems offer a different prospective on the role and possibilities of *in situ* annealing experiments.

In fact, while colloidal synthesis is a well-known route to synthesize and fine-tune NPs with well-defined crystal structure, shape and size [29, 30], *in situ* annealing TEM represents a useful tool not only to investigate (at atomic level and in real time) the properties and stability of promising nanomaterials [31–37] but also to trigger extensive modifications in nano-materials of technological interest.

In this context, the studies by Hellebusch et al. [34] and Hudak et al. [35] represent perfect examples of basic research of immediate applicative interest, since both present stability tests for materials that are already being used as building blocks in a vast range of applications. In particular, Hellebusch et al. [34] investigated the *in situ* sublimation of CdSe NRs in order to understand if the sublimation could influence their facet stability and the possible role of surface ligands. While the sublimation was always anisotropic, a temperature-dependent effect was observed: the mass loss occurred either noncontinuously from both ends of the NRs at lower temperatures or continuously from one end at higher temperatures (**Figure 3A**). In both cases, the NR sublimation and the mass loss occurred along the long c-axis, featuring the least stable facets. After reaching a transition state with a particle length of 2–3 nm (without variations in the NR diameter), the sublimation proceeded along ab-axes until the full sublimation was achieved. Also, an explanation to the dual sublimation trend involving surface ligands and generic contaminants was proposed. Taking into account the presence of a superficial layer coating the NRs, it was proposed that the NR sublimation fronts were hindered by this layer and that the sublimation could start whenever one of the tips was "freed" by nucleation or desorption of the contaminant. Thus, similar sublimation and the contaminant deposition rates at lower temperature would imply a noncontinuous sublimation, while at higher temperatures the sublimation could be favored, leading to a continuous sublimation process.

On the other hand, Hudak et al. [35] investigated the dissolution of Au-decorated SnO_2 nanowires. *In situ* TEM annealing experiments were conducted between 400 and 800°C and etching was observed at $T > 450$°C with Au moving along the wires and consuming them by a solid-liquid-vapor (SLV) dissolution mechanism. In particular, the finer temperature control granted by MEMS-based holders showed that Au motion was indeed a temperature-guided process with the motion/dissolution rate being activated, accelerated, or quenched by reaching, going beyond, or below the threshold temperature. By studying the evolution of the Au/ SnO_2 interface against temperature, the SLV mechanism was observed: reaching the threshold temperature activated the diffusion of Sn in the Au tip and the subsequent melting of the Au/ Sn alloy to a droplet; the interfacial SnO_2 dissolved in the droplet, leading to the supersaturation of Sn and its ejection via evaporation or surface diffusion (**Figure 3B**). Depending on the temperature and the size, the Au/Sn droplet could overcome adhesion to the substrate and move through the NW by the incorporation/ejection of Sn, resulting in a temperature-controlled etching mechanism.

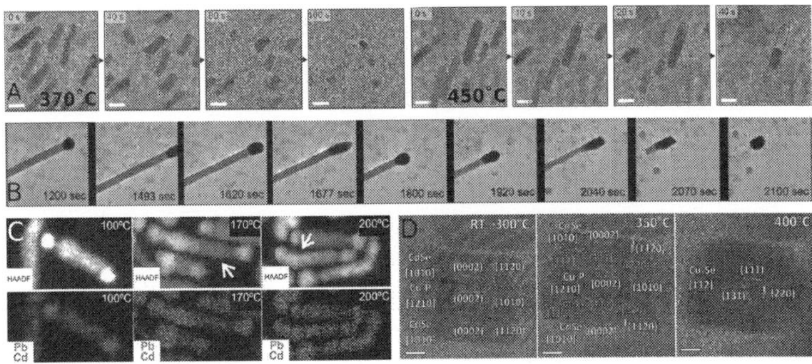

Figure 3. (A) CdSe NRs sublimation during *in situ* annealing experiments at different temperatures. Scale bars are 10 nm. Adapted with permission from [34]; (B) dissolution of the SnO_2 nanowire into the Au catalyst tip during *in situ* heating experiments. Field of view is 870 nm. Adapted with permission from [35]; (C) HAADF-STEM images and the corresponding STEM-EDX elemental maps of dumbbell heteronanostructures collected during *in situ* annealing experiments at different temperatures. Adapted with permission from [36]; (D) HRTEM images of a single CdSe/Cu$_3$P/CdSe heterostructure collected during an *in situ* annealing experiment from RT to 400°C, where the sole Cu$_2$Se phase is observed. Scale bars are 5 nm. Adapted with permission from [37].

An alternative approach to *in situ* experiments takes advantage of the annealing to introduce chemical transformations in candidate materials. Yalcin et al. [36] showed the evolution of PbSe-CdSe nanodumbbell heterostructures (HSs) subject to *in situ* evaporation-induced cation exchange (CE) by a solid-solid-vapor (SSV) epitaxial growth mechanism. The annealing-driven Cd evaporation from the superficial sites and the increased mobility of Cd along the HSs triggered the migration of Cd vacancies from the surface to the PbSe/CdSe interface, where they recombined with Pb atoms, forming Pb vacancies. Subsequent recombination with excess Pb adsorbed on the PbSe domains led to the growth of the PbSe phase at the expense of the CdSe phase (**Figure 3C**), leading to the reduction or the complete substitution of the CdSe phase.

Also, De Trizio et al. [37] obtained a new phase through substitution of both cationic and anionic atomic species by performing *in situ* annealing on a "sandwich"- type CdSe/Cu$_3$P/CdSe HS. Structural and chemical analyses highlighted that the exchange reaction was a two-step process with Cu first diffusing and substituting Cd to form Cu$_2$Se in the peripheral regions of the HS (as proven by the blurring of the Cu$_3$P/CdSe interface and the rearranging of CdSe) and

Se subsequently diffusing to the Cu_3P region to finally form a single-phase Cu_2Se NP (**Figure 3D**). Both studies showed ordinate fronts of atomic diffusion and substitution, resulting in the formation of new phases where the orientations of the starting structures are mimicked, while the shape and size of the starting nanostructures are maintained.

Thus, *in situ* TEM annealing represents a solid tool not only to study the stability of a vast array of nanostructures but also to finely tune the parameters leading to their modification.

2.3. *In situ* TEM heating for graphene studies

Graphene, with its unique physicochemical properties [38–40], has been the subject of various studies voted to better understand its properties in response to external *stimuli* at the nano- or atomic scale. In this context, the use of *in situ* TEM to investigate the graphene response during annealing processes permits to study and comprehend how its behavior at high temperature can be exploited to tune the growth and engineer its properties for possible applications.

In particular, the control over orientation during the growth of the graphene layer has been an ongoing issue since graphene layers were first produced by mechanical exfoliation of bulk graphite by Gemin and Novoselov in 2004 [41] regardless of the growth technique (CVD [42], thermal decomposition of SiC [43] and molecular beam epitaxy (MBE) [44]), and investigating its growth process is a mandatory prerequisite for a successful application. In this perspective, *in situ* TEM annealing provides the chance to study how the temperature can influence the growth process of graphene. Liu reported, through aberration-corrected STEM analysis, a direct observation of graphene growth and domain boundary formation obtained by annealing bilayer graphene (BLG) and using the hydrocarbon residues from the microscope column as a carbon source for in-plane graphene growth at the step edge of the BLG substrate [45]. By keeping the electron beam energy below a threshold of ~50 kV to prevent knock-on damage in graphene structure [46] and annealing to 700°C, an extra growth was observed from the edge of the second layer of a single BLG-crystal following mechanisms determined by step-edge structures with residual Si atoms from the CVD fabrication process acting as catalysts during the graphene growth and sliding over the graphene edges during e-beam scanning (**Figure 4A**). Moreover, it was possible to investigate the relation between growth rate and residual hydrocarbon gas pressure by changing the vacuum conditions inside the TEM column (in the range $1.0–2.9 \times 10^{-5}$ Pa), thus finding faster growth rates when more residual hydrocarbon gases were left in the TEM column. In fact, the step-edge in-plane growth is a combined mechanism of physical absorption of C atoms and catalytic effect of Si atoms present on bilayer edges, where the thermal treatment provides the needed activation energy and the electron beam accumulating hydrocarbon molecules over the scanned area. Graphene growth is strongly dependent on the equilibrium between the number of available C atoms and the time needed for the C-to-graphene bonding to occur: if the C atoms do not have enough time to bond in an energetically favorite site, then graphene will be just contaminated with amorphous C. In fact, graphene observed in STEM mode at room temperature and low vacuum condition (non-zero pressure of gaseous hydrocarbons) presented contamination by amorphous C, while if the graphene substrate was heated up to 500–700°C, the diffusing free C atoms were more likely to form sp^2 bonds with the step-edge C-atoms of graphene, leading to in-plane graphene growth. On the other hand, raising the temperature above 700°C caused the etching effect to be dominant over the carbon adsorption, so that no growth was observed.

Given the significance of the etching and curing effects in the graphene layer at high temperatures [47, 48], Kano et al. [49] investigated the structure and dynamics of Cu atoms embedded in single-layer graphene by aberration-corrected TEM operating at 80 kV while heating the sample through a MEMS-based chip. Cu atoms could replace C atoms of graphene under irradiation by a focused electron beam, more easily when residual oxygen and hydrocarbon contaminations were present. However, the Cu-C substitution was in competition with Cu evaporation: analyzing these processes during *in situ* annealing led to finding a threshold value with substitution being favored up to 300°C and Cu evaporation leading at higher temperatures. Then, choosing adequate electron beam currents and annealing temperatures allowed a control over the density of Cu-C substitutions. In particular, observations at 150 and 300°C showed that the Cu substitution led to local straining of graphene. These observations suggested that Cu atoms tended to combine with SW defects (Stone-Wales defects, i.e., pairs of five- and seven-membered rings at a 30° grain boundary [50]) forming grain boundaries between the reconstructed and original grains and, preferentially, replacing C atoms in defective graphene areas. Thus, the substitution modified the lattice structure by promoting C-C bond rotation, formation and mending of nanopores and rotation of grains mediated by defects (**Figure 4B**). In turn, this caused the disruption of grains in the graphene and its reconstruction as a less-strained lattice. So, unlike metals such as Cr, Ti, Pd, Ni, and Al, whose presence determines an etching of graphene, the presence of individual Cu atoms can catalyze a curing effect of graphene [51].

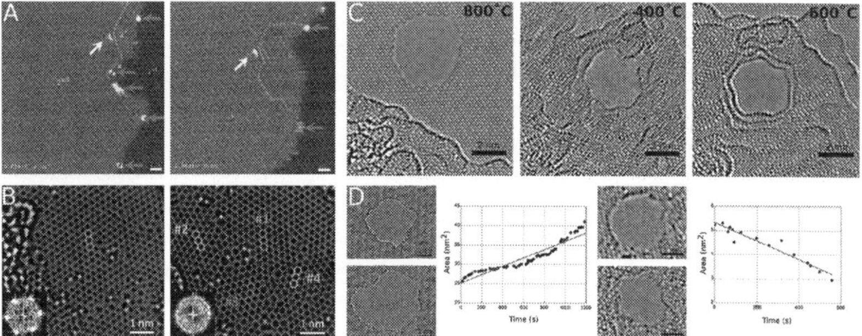

Figure 4. (A) Sequential ADF (annular dark-field) images showing graphene from the step edge of the BLG at 500°C. The dashed magenta line indicates the initial 2nd layer step edge, while the cyan dot line highlights the rotated 2+ layer. Green arrows show the single Si atoms at the step edge and a white arrow indicates the pinned Si atom. Adapted with permission from [45]; (B) representative TEM images taken before (left) and after (right) the reconstruction of graphene at 150°C. Insets show the FFT images. Adapted with permission from [49]; (C) representative TEM images showing the evolution of a graphene nanopore with varying temperatures. Adapted with permission from [55]; (D) nanopore sizes control, showing selected frames of electron beam hole sputtering at 800°C (left) and nanopores shrinkage at room temperature (right). Variations in the nanopores area trend are presented along with a linear fit in the graphs beside the frames. Adapted with permission from [55].

The effects of temperature on graphene are not limited to the cleaning and restoration of the lattice: in fact, those same temperature-dependent effects can be exploited to perform nano-fabrication. Nanopores embedded in thin membranes attracted special interest for their potential application in label-free, single-molecule detection of chemicals or biomolecules [52, 53]. Xu showed that a strongly focused 300 kV electron beam could be used to sculpt free-standing monolayer graphene with close-to-atomic precision in the STEM mode while heating the sample at 600°C [54]. The same electron beam with different scanning dwell times was used for sculpting and imaging: this allowed an immediate switching between sculpting and imaging and consequently fine-tuning the shape of the sculpted lattice. The effect of temperature was clarified by performing sculpting at 20, 400, 600, 700, and 800°C under identical STEM conditions while varying dwell times. The sculpting was successfully performed between 400 and 700°C, but it led to the contamination of the specimen at 20°C and it was not possible, even with large dwell time, at 800°C likely due to self-repair process being faster than C removal. More in detail, the experiment consisted in three steps: preparation, sculpturing, and inspection. The preparation of the sculpting area was performed in imaging mode at 600°C so that any isolated defects created by the electron beam were removed by self-repair of the lattice. Then, in sculpting mode, using a longer dwell time, several adjacent carbon atoms were knocked-out to prevent self-repair of the graphene lattice; extending the initial hole in a predefined direction it was possible to shape the graphene in a pattern with a precise position, size, and orientation. Finally, switching again to the imaging mode the sculpted pattern was inspected without introducing damages. Furthermore, the patterned graphene nanostructures were stable after being cooled to room temperature and stored in air. He also studied the stability of nanopores, developing a method to eliminate the dangling bonds at the pores edge in monolayer graphene and creating the so-called closed edge nanopores by using electron beam irradiation during *in situ* annealing and leading to pores that were stable for up to 2 days in air [55]. The process, taking advantage of the fast temperature ramps of MEMS-based holders, could be divided into four steps: the nanopores were first created by focusing of the electron beam at 800°C, with a subsequent cooling at room temperature to aggregate and fix the hydrocarbon adsorbates to their edges; a subsequent second heating to 800°C led to the crystallization of hydrocarbons to form closed edges around the pores, while a final cooling to room temperature stabilized the newly formed pores. As for the study of Xu, this process was strongly dependent on precise values of temperature (**Figure 4C**) and time (**Figure 4D**), but all these studies demonstrate how recent advances of *in situ* TEM annealing introduce the possibility to investigate in real time and at atomic scale the relationships between structure, properties, and functionality of graphene, which are essential in tailoring or designing devices based on this promising carbon-based material.

2.4. New directions and perspectives of *in situ* TEM annealing

The resurgence of *in situ* techniques due to the technological advancements experienced in recent years allowed an improved/deeper understanding of a vast range of processes at atomic scale, but other than pushing forward along already established routes also offered a chance for branching toward new directions.

A couple of recent studies dealing with the role of carbon with regard to *in situ* reactions from different aspects proved a good point in this matter. Carbon, as a support film of TEM grids or as an undesired presence inside the TEM column (e.g., hydrocarbon, which can be encountered as residues in the TEM column [45] or as a capping agent of NPs) is widely known to be a possible "active player" in annealing reactions. In particular, the role of carbon as a capping agent was the main focus of a study by Asoro, who successfully accounted for its role during sintering [56]. By performing *in situ* annealing experiments on carbon-covered Ag NPs of different sizes and origins and comparing the resulting diffusivity values with analogous experiments involving carbon-covered Ag NPs on a micrometric Ag wire, they observed that the formation of a neck bridging adjacent NPs is opposed by the presence of carbon coating, which acts as a steric barrier to its formation and, consequently, to sintering. In fact, the interparticle diffusion of Ag atoms must occur through the carbon: the formation of the Ag neck determines the out-diffusion of C atoms from the contact region, but also a decrease in diffusivity during ripening processes at the nanoscale (**Figure 5A**). The role of carbon capping to phase transitions during annealing experiments was also the main focus of a study conducted by Chen et al. [57]. By comparing the results of *in situ* annealing experiments performed on carbon- or Cu_2O-supported Cu NPs, different melting conditions were observed. The melting point of Cu NPs on Cu_2O substrates was 100 K lower than its counterpart for Cu NPs on a carbon substrate and the reason of this discrepancy in thermal stability was attributed to the presence of a coating layer of carbon adsorbed on the surface of the NPs and acting as a

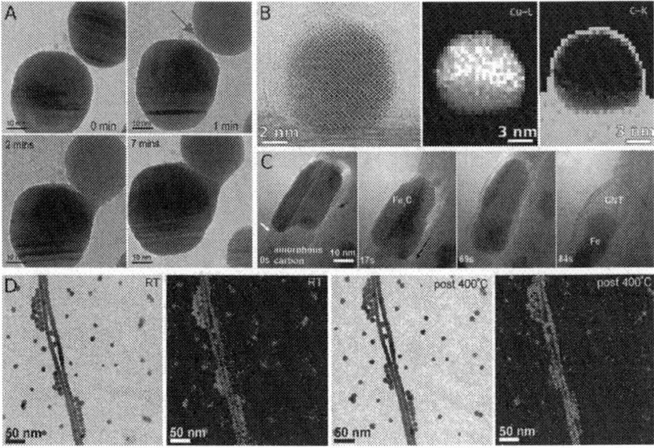

Figure 5. (A) TEM images collected during *in situ* heating at 300°C, showing that the carbon residue on the surface of two sintering Ag NPs can prevent neck growth. Adapted with permission from [56]; (B) a typical Cu NP supported on graphite with a thin amorphous layer on its surface. Core-loss images of C-K and Cu-L edges confirm the presence of a 1 nm thick amorphous layer identified as C. Adapted with permission from [57]; (C) *in situ* TEM observations of the nucleation and growth of a CNT from an Fe_2O_3 catalyst NP, showing the evolution of the NP along with the dissolution of amorphous carbon and the growth of the CNT. Adapted with permission from [59]; (D) sequence of elastic-filtered TEM images and normalized EFTEM maps of $Cu_{2-x}Se$ NPs and CdSe NWs recorded at RT and postthermal treatment at 400°C. Green and red color coding indicates the presence of Cu and Cd in the EFTEM maps, respectively. Adapted with permission from [61].

thermal shield (**Figure 5B**). Core loss EELS mapping of C and Cu proved the presence of the C coating layer, while molecular dynamics simulations confirmed its shielding effect on Cu NPs, thus showing how "external factors" should be taken into account to explain discrepancies usually observed at the nanoscale.

Romankov and Park [58] approached the problem of C contribution to *in situ* annealing experiments from a very different starting point, i.e., structural studies and evolution of nanolaminated composite materials against temperature, and suggested an unconventional use of the carbon layer. By annealing a crystalline/amorphous nanocomposite (CoFeNi/Cu/ZrAlO)/C/W lamellar sample, they took advantage of the intermediate amorphous C layer to observe the out-diffusion of atomic species from all the layers and uses it as an *in situ* "reaction chamber" where those species could form new HSs or grow new phases. Moreover, the complex, liquid-like diffusion of CoFeNi at the C/CoFeNi interface was observed. During the annealing, the interfacial atoms from the highly disordered layers of CoFeNi were attracted by the carbon surfaces of the pores, which led to the lengthening of branches formed by a disordered shell and an ordered core and their diffusion through the carbon layer. Carbon is then a suitable candidate in the fabrication of nanocomposite structures or as a porous matrix for *in situ* observation of liquid-solid reactions.

Recently, *in situ* TEM annealing was also used to shed new light on some fundamental phenomena that have been generally accepted but never investigated in detail at the nanoscale, such as the nucleation, growth, and transformation processes of various well-known nanostructures.

A good example is given by carbon nanotubes (CNTs), which lacked a detailed nucleation study, despite their popularity for many possible applications. While their nucleation was generally ascribed to a generic vapor-liquid-solid (VLS) mechanism, the study conducted by Tang et al. [59] was specifically focused on shedding light on these aspects. The nucleation and growth of CNTs were studied by using an unconventional *in situ* annealing setup composed by poorly crystalized CNTs containing the catalytic NPs (Fe_2O_3 or Au) and acting as crucibles. The CNTs, also acting as C reservoirs for the growth of new nanotubes, were connected to electrodes and the system was annealed by heating the CNTs by Joule effect. During the annealing, the Fe_2O_3 NPs interacted with carbon from the CNT to form the Fe_3C catalyst, which triggered the nucleation of new CNTs and in turn transformed in metallic Fe once the CNT growth had stopped (**Figure 5C**). Moreover, *in situ* TEM showed the actual evolution of the Fe_3C and Au nanocatalysts during CNT nucleation and growth with the NPs going through multiple intermediate physical states that could not be reconducted to the simple, previously derived VLS model. Boston adopted a similar approach with a MEMS-based setup to investigate the nucleation and growth of quaternary Y_2BaCuO_5 nanowires (NWs) by a microcrucible mechanism [60]. Freezing the NW growth by rapidly cooling the system, the molten $BaCO_3$ NPs were observed diffusing on the surface of the porous precursor matrix to form an amorphous region at the base of the NW. This region, containing a mixture of the constituting elements of the NWs, was identified as the actual microcrucible, with the $BaCO_3$ NPs collecting the other cations by solubilization while diffusing on the matrix. Given the dynamic nature of the liquid/solid interface between microcrucible and NW, the shape and size of the microcrucible was expected to change with the NW growth and its evolution was studied by *in situ* annealing. Therefore, the direct TEM observation of the microcrucible creep with the NW growth during the *in situ* annealing showed how variations in the microcrucible modify the shape, size, and crystallinity of the NW, thus offering new insight into the growth of complex NWs.

The study of Casu et al. [61] provides a further example, focusing on modifications occurring during CE reactions in solid state at the nanoscale during *in situ* TEM annealing. In fact, while the out-diffusion and motion of chemical species have previously been accounted for as a generic possibility during *in situ* exchange reactions happening at the nanoscale, it was never actively investigated or used as a tool for *in situ* ionic exchange reactions. Taking advantage of the temperature-dependent copper depletion of Cu_2Se NPs, CE reactions were studied between two distinct populations of metallic Cu/Cu_2Se NPs and CdSe NRs/NWs with the TEM grid substrate acting as a medium for the Cu diffusion. Upon *in situ* annealing of Cu-containing and CdSe NPs, a CE reaction was observed in the latter, with Cu completely substituting Cd (**Figure 5D**). The structural and chemical data collected during the thermal-driven CE reactions proved that the kinetics of solid-state CE was slower than those of CE occurring in solution, therefore allowing the direct observation of the structural modifications occurring to CdSe upon CE. These observations showed a novel route for CE reactions at the solid state, while the slower kinetics offer a chance to study the intermediate steps that cannot be observed for CE reactions performed in solution.

3. *In situ* gas-solid reactions in environmental (S)TEM

Chemical reactions between gaseous phases and solid materials represent a challenging topic not only for the basic scientific research but also for important industrial applications. Since the dawn of the TEM technique, scientists tried to locally modify the environment around the specimen by introducing liquid or gaseous phases in order to study the materials under reactive or real environments [62–64]. These attempts clashed with the "classical" configuration of the TEM, which normally needs high vacuum conditions (1×10^{-7} Torr) [65] to protect the electron gun, to prevent contamination of the sample, and to avoid any blurring effect due to additional scattering of the electron beam with atoms that are not part of the specimen. The challenge was then to increase the gas pressure in a localized and controlled environment around the specimen so that temperature, gas composition, and pressure could be tuned without compromising the performance of the microscope itself. Scientists and manufactures developed two confinement methods for gaseous phases to achieve controlled environment TEM (E-TEM): (i) differentially pumped systems, featuring modified TEM column and vacuum system [66–72], and (ii) window-closed cells designed *ad hoc* to fit in TEM holders [73–77]. Neither of these methods requires special protocols for specimen preparation and both allow investigation by the usual TEM and STEM techniques [78].

Nowadays, coupling these two methods with the exceptional spatial and time resolution capabilities of state-of-the-art (S)TEMs and detectors, such as those used in EDS and EELS, creates an extremely powerful tool for structural and chemical analyses of materials up to the atomic scale and over the reaction time [79]. However, the introduction of gases to create an environmental cell inside the TEM implies some additional limitations, i.e., the degradation of signal and spatial resolutions. The presence of an external gas species implies a direct rise in the number of molecules along the path L of the electron beam in the vicinity of the specimen and, in turn, is responsible for additional scattering events (*extrinsic scattering*) [80]. Although *extrinsic scattering* can present additional contributions coming from the interaction of electrons with the confining thin windows in the case of window-closed cells, the main contribution usually comes from interaction with gases, so the number (m) of *extrinsic scattering* varies proportionally to the gas pressure and the m value can be calculated considering the number of *extrinsic scattering* events within a cylindrical volume, defined by scattering cross section (σ) and length of electrons path (L) as expressed in Eq. (1):

$$m = \frac{LPM\sigma}{k_{\mathrm{B}}T} \tag{1}$$

where σ is the mass density of the gas, P is the gas pressure, M is the number of atoms per gas molecule, k_{B} is the Boltzmann's constant, and T is the temperature. The *extrinsic scattering* causes an overall reduction of the signal-to-noise ratio, followed by a progressive blurring effect and worsening of spatial resolution, since any imaging (textural and structural data) and spectro-scopic (chemical data) information correlated to the specimen will be forwarded only by those electrons which did not undergo *extrinsic scattering* [80]. Hence, the forwarded signals can go through the TEM column toward the different types of detectors for recording and analysis processes.

3.1. The differentially pumped system

The differentially pumped system approach relies on a modified TEM setup to perform environmental studies. In this case, the environment cell (E-Cell), generally intended as the experimental chamber where the specimen comes in contact with user-introduced gaseous species, is constituted by the portion of TEM column around the sample that is limited by a double series of additional apertures with small holes, which are positioned in correspondence with the objective lens pole pieces to confine the gaseous phases. Two additional sets of apertures along the column combined to a differential pumping system create zones of decreasing pressure between the high pressure zone of the E-Cell and the high vacuum of the other parts of the TEM column, thus protecting the electron gun from any gas coming from the E-Cell (**Figure 6A**).

This architecture allows the separate vacuum control of the E-Cell from the rest of the TEM column so that all the volume around the sample delimited by the objective lens can be filled with gases at low pressure and evacuated independently. The introduction of external gases implies that the E-Cell needs to be connected to an external gas manifold and flow controller system equipped with gas reservoir tanks, unreactive pipelines, and pressure gauges for any tuning and monitoring procedures [71] but no modification is necessary with regard to the holder and any type of tilting and heating TEM holders can be used to perform *in situ* experiments.

The main advantage of the differential pumping is that it is an *open* system, meaning that the electron scattering comes only from the sample and the gas molecules, while no containment window is needed, although it is also worth noting that this architecture can reach only limited pressure values (around 20 Torr) without jeopardizing the TEM and electron gun [81]. However, this solution comes with a cost, i.e., the blocking of electrons scattered at high angle by the lower series of differential apertures below the sample [67]. This constraint imposes a strong limitation to the capabilities of different analytical techniques, e.g., STEM imaging by annular dark field (ADF) and high angle annular dark field (HAADF) detectors, electron diffraction of patterns collected at high angle [82, 83], and EELS, since the range of collection semiangle of the spectrometer cannot exceed the angular value of the lower E-Cell and differentially pumped apertures. Moreover, the introduction of a differentially pumped system results in a more complex and more expensive system.

In addition, the imaging resolution is not only affected by the presence of gases, but it is also influenced by additional electro-optical and microscope-related parameters with opposite effects, i.e., the electron dose and the primary electron energy. In fact, under the same conditions of E-Cell gas pressure and primary high tension, reducing the electron-dose increases

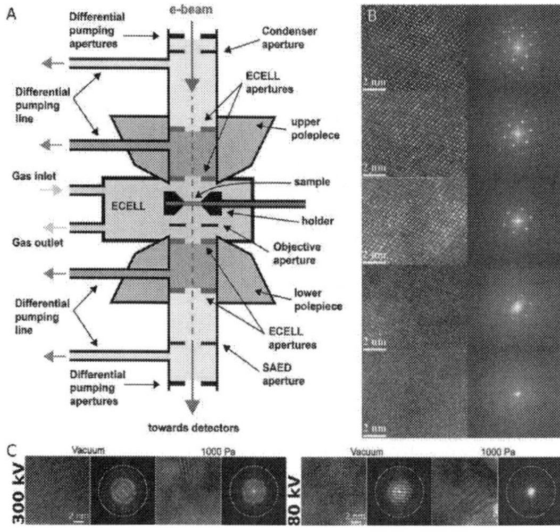

Figure 6. (A) Schema of a differentially pumped TEM system after [71]; (B) sequence of HRTEM images and the corresponding diffractograms of an Si 011 crystal lattice measured under 4 mbar of N_2 gas pressure and different electron doses. Adapted with permission from [84]; (C) HRTEM and the corresponding power spectrum of an Au film on carbon film, showing the pressure dependence of the transmitted information limit under different conditions of acceleration tension and N_2 gas pressure. The white circle corresponds to 10 nm^{-1}. Adapted with permission from [85].

the spatial resolution (**Figure 6B**) [84], while reducing the high tension worsens the spatial resolution (**Figure 6C**) [85]. These effects can be explained thinking that a low electron dose will lower the electron density of the beam and the overall number of *extrinsic scattering* events, while, on the other hand, a lower tension will induce a stronger scattering effect between the electron beam and the atoms of the sample and the gas molecules, the latter resulting in a worsening of the signal-to-noise ratio [86]. Therefore, a proper evaluation of the best setting of the microscope according to the type of material, gas phase, pressure, and chemical reaction to be studied is mandatory before starting any experiment of environmental (S)TEM.

In general, the combination of *open* architecture and limited *extrinsic scattering* contributions due to the use of low working pressures make the open E-Cell suitable for chemical investigations of materials through EELS analysis. For these reasons, many recent basic and applicative studies of catalytic materials took advantage of this combination to study the structural and textural modifications of NPs exposed to reactive gas, following their change of valence state via EELS near-edge fine structure analysis.

In particular, studying the modifications occurring to catalysts in different temperature and pressure regimes is a focal point for assessing the best working conditions and comprehending variations in catalytic performance. This is the case of cuprous oxide (Cu_2O) NPs, which represents an active photocatalyst for hydrogen production by water splitting through direct irradiation with visible light ($\lambda > 400$ nm) [87]. Studying the variations in the oxidation state of copper on Cu_2O nanocubes under water vapor (5 mbar) and light exposure ($\lambda = 405$ nm) (**Figure 7A**) by EELS near-edge structure analysis of the Cu $L_{2,3}$ edge highlighted a variation in the near-edge features throughout the photocatalytic reaction. In fact, the two strong L_2 and L_3 ionization edges — "white lines" — showed by oxidized copper (Cu^+, Cu^{2+}) at 951 and 931 eV, respectively, were substituted by one step-like abrupt onset of L_3 edge followed by weak features, which is typical of metallic copper (Cu^0) and revealed that the photocatalytic reaction also led to a self-degradation process of the NPs to metallic Cu. Another example is offered by the study on the evolution of Ni NPs during oxidation [88]. In fact, Ni is a viable catalyst for a wide range of applications, including energy production in solid fuel cells, but its oxidation leads to a lessening of its catalytic performance. Therefore, analyzing in detail by *in situ* E-TEM the structural and chemical transformations induced by oxidation to crystallites represents a first step in understanding and, possibly, control this process. In particular, the comparison of structural and chemical data showed that the oxidation process progressed as the pervasive formation of a thin NiO layer covering the Ni NPs, and subsequently growing at the expense of the Ni component by increasing the temperature, until the final evolution into porous, polycrystalline NiO structures, which are responsible for the worsened catalytic performance. Xin used a similar approach to monitor the transformation of nanoporous CoO_x/silica nanocomposites under H_2 flux at different temperatures [89]. In fact, while the chemical analysis of the Co $L_{2,3}$ edge of EELS signal exhibits a clear fingerprint of reduction of Co valence from Co^{2+} to Co^0 during the exposure to H_2 at high temperatures, the *in situ* TEM showed a coarsening of the composites, reflected by an increase in size of the Co component at the expense of the composites porosity. This evidence translated in the quantification of a structural trend from a porous nanosized network to solid NPs, which can be used to tailor the catalyst features in its actual working conditions (**Figure 7B**).

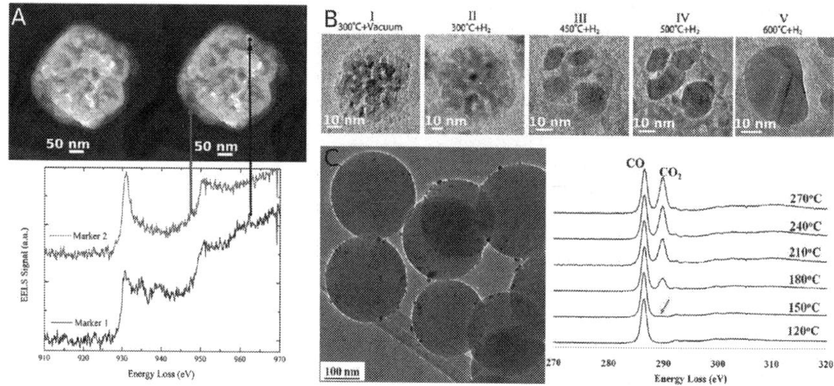

Figure 7. (A) HAADF STEM images of a Cu_2O nanocube before (left) and 30 min after (right) the photocatalytic reaction. The corresponding EEL spectra of the peripheral region (below) show the coexistence of metallic Cu portions (marker 1) in the Cu_2O nanocube (Marker 2). Adapted with permission from [87]; (B) *in situ* observation of structural changes of Co/CoO_x NPs under different reduction conditions and temperatures. Adapted with permission from [89]; (C) representative TEM image of Ru/SiO_2 NPs (left) and background subtracted EEL spectra acquired at different temperatures during the oxidation reaction of CO (right). Adapted with permission from [91].

On the other hand, E-TEM is also a powerful tool for direct observation in more applicative-inclined studies, devoted to test and observe the performance and limits of materials with more immediate practical repercussions. Such is the case of the study on yttria-stabilized zirconia (YSZ) NPs as a treatment device for soot exhaust products by their oxygen channeling capability [90]. *In situ* E-TEM experiments, conducted under high temperature conditions with a 3 mbar pressure of O_2 on a tight mix of YSZ and soot, offered direct evidence of the role of YSZ. In fact, the direct observation showed how the oxidation front of soot was indeed at the surface of the NPs, where the O^{2-} is made available, while soot particles not in contact with YSZ remained unaffected despite being surrounded by gaseous oxygen, thus confirming how the oxidation of carbon soot to CO_2 was due to the capability of YSZ of conducting O^{2-} ions toward the NPs surface and making them available for oxidation. Similarly, Chenna et Crozier [91] were able to detect and quantify the catalytic process in the E-TEM by EELS, showing the feasibility of an *Operando TEM* approach for simultaneous structural and reactivity studies. To achieve this goal, they investigated the oxidation effects of Ru/SiO_2 nanocomposites on CO gas. Taking advantage of the high catalytic performance of Ru/SiO_2 and of the variation of C π^* peak, going from 286.4 eV for CO to 289.7 eV for CO_2, they heated the sample and collected the EELS spectra during the oxidation of CO to CO_2. By monitoring the appearance and growth of the C π^* peak of CO_2 and the variation of its intensity with respect to the C π^* peak of CO, it was possible to confirm that the oxidation reaction was taking place to observe temperature-dependent variations in the oxidation (**Figure 7C**), thus showing the feasibility and power of the *Operando* approach in E-TEM.

3.2. The window-closed E-cell

The window-closed cell is the main alternative to the differentially pumped system for environmental TEM analysis. In this case, the E-cell is incorporated into the tip of a dedicated TEM holder and uses two parallel electron-transparent windows to confine the sample and the gases inside a tiny-volume nanoreactor (**Figure 8A**), while no modification to the TEM column is required. By reducing the volume to a small, closed layer around the specimen, the gaseous phases can reach pressures above one atmosphere [92–94], while the nanoreactor can be safely inserted in the TEM column. The main advantage, other than the higher working pressures, is that this setup can be installed in different TEMs without any modification to the column and vacuum system with a direct saving of costs of purchasing and maintenance.

The confinement diaphragms of the window-closed cell must have some fundamental prerequisites. Namely, they must (i) be electron-transparent, (ii) be strong enough to confine the pressurized gas, and (iii) provide a low diffraction contribution in order to reduce any effects of diffraction contrast and limit the superposition of additional periodic patterns to the "proper" signals coming from the sample [95]. These are the reasons why the diaphragms are made of thin and amorphous materials with low average atomic number (e.g., carbon, silicon nitride), thus also minimizing their *extrinsic scattering* contribution [95]. However, the additional contributions of diaphragms to the *extrinsic scattering* can increase the energy spread of primary electrons, resulting in a worsening of spatial resolution of both the microscope [75] and EELS capabilities because of a stronger damping effect on contrast transfer function and a worse energy spread of the beam entering the specimen, respectively. In a similar fashion, the escape of any X-ray signals generated during the scattering events between primary electrons and sample atoms will be partially or totally absorbed by the diaphragms [92]. The "closed" nature of this setup does not require any additional apertures in the lower part of objective lens and column, thus making available to the detectors the forwarded signals coming from electrons scattered at high angle and removing some of the limitations encountered in the differentially pumped systems. However, the gas path length between the windows is still a crucial parameter: the spatial resolution displays an overall worsening under high-pressure conditions (1 atm) in nanoreactors with thin windows but longer gas paths (more than 1 mm), due to the dramatic increase in the number of *extrinsic scattering* events [96], akin to the differentially pumped systems.

Also, in this configuration the gas pressure and the temperature inside the cell can be monitored and varied continuously, up to the instrumental limit. The holder is generally connected to an external gas line equipped with gas reservoir tanks, unreactive pipelines, and manometer pressure gauges capable to control the pressure from few mbars to few bars, while a dedicated vacuum pump is used for purging operations [93].

In this regard, the introduction of micromechanical system (MEMS) technology led to important improvements in the mechanical characteristics, design, and performance of window materials and in the engineering of holders with respect to previously available setups (**Figure 8B**) [92, 94, 97]. The last generations of closed cell nanoreactors, obtained via MEMS technology, integrate electron-transparent diaphragms made of 10 nm thick amorphous silicon nitride, flow gas pipes of a few tens of micrometer wide, micropressure controlling gauges, and microheaters [98, 99]. Usually, the heater system is a metallic spiral strip (e.g., underdoped Pt or Pt-based alloys) embedded on the lower thin window and the gas path between the windows is in the order of some tens of micrometers, which allows a reduction in the volume of the cell. Moreover, the spiral shape of the heater helps to minimize any spurious magnetic fields interfering with those of the electromagnetic lenses [98] and any thermal gradient upon the sample. The combination of these parameters makes MEMS-based nanoreactors suitable for HRTEM and HRSTEM imaging in Cs-corrected microscopes even at atomic scale, as demonstrated, e.g., by the study of Allard et al. [93] on catalytic nanomaterials, which exhibited atomic columns resolution under vacuum in Au NPs grown on Fe_2O_3 and in Rh nanocrystals on $CaTiO_3$ under pressures up to 1 atm. Ultimately, the improvement in performance over a wide range of working pressures and temperatures and the adaptability to different (S)TEMs without any modification of columns and vacuum systems provide the window-closed cells a high experimental flexibility, which makes them valuable in a variety of scientific studies devoted to the *in situ* observation of chemical reactions. Obviously, given the novelty of the most recent developments of the E-TEM technique, different technical approaches are being explored. While third-party solutions are already available in the market, many groups are also developing their own custom-built setups to perform environmental TEM experiments.

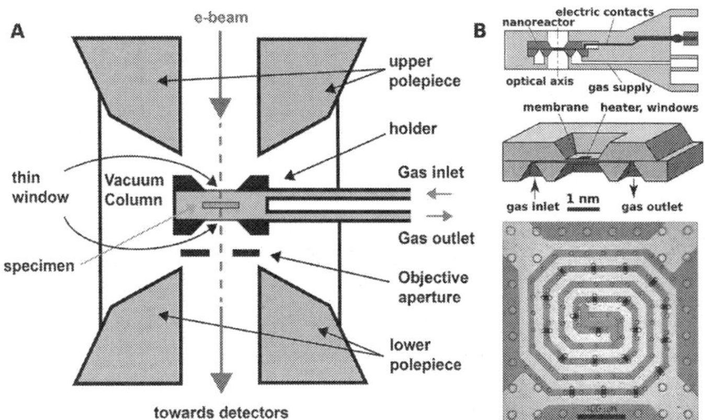

Figure 8. (A) Schema of TEM holder with window-closed cell; (B) MEMS-based nanoreactor. Top: Schematic lateral cross section with the electrical contact for the heater; middle: schematic 3D view of the MEMS nanoreactor; bottom: active area, featuring the microheater spiral (in white) with ovaloid electron-transparent windows and circular spacers. Adapted with permission from [92].

In this context, Sun et al. adapted a wet cell they previously developed into a closed-cell setup for *in situ* E-TEM experiments devoted to observe the oxidation of Ag NPs in air [100]. Due to its ionizing effects on molecular oxygen, the electron beam was used to obtain atomic and ionic oxygen inside the E-Cell, which acted as ionizing agents for Ag. Therefore, variations in the electron beam current intensity were reflected in the oxidation reactions products: at first increases in current density were revealed by the progressive decrease in Ag NPs size accompanied by the nucleation of new Ag and silver oxide grains; further increases in current density resulted in the progressive predominance of oxide over metallic silver with the heating effects of the electron beam having an increasingly prominent role. The melting of crystalline silver oxide due to beam heating led to the formation of a noncrystalline, highly dynamic vapor phase, which swept through the E-Cell either aligning to nucleate clusters of Ag NPs or dissolving NPs (**Figure 9A**).

More recently, Vendelbo et al. [97] used a MEMS-based nanoreactor to study *in situ* the oscillatory variations in the structure of catalytic Pt NPs during CO oxidation. In fact, while the oscillatory behavior of catalytic reactions under stable conditions was already known, the role of nanosized catalysts was still unclear. Experimental data were acquired while 1.0 bar of CO/O_2/He was flowing through the E-Cell containing Pt NPs at a steady temperature of 659 K. The catalytic performance of the NPs was verified by measuring variations in mass spectrometry of the CO and CO_2 phases exiting the nanoreactor, while time-resolved HRTEM imaging recorded variations in the structure of the Pt NPs. Finally, calorimetric information about the chemical reaction was extracted by recording fluctuations in the electrical power consumption of the MEMS nanoreactor due to the heat released by the CO oxidation. By correlating the reaction data with the variations in morphology observed on the catalytic NPs, it was observed that a periodic and reversible faceting of Pt between a spherical and a facetted shape is indeed the cause of the periodic oscillations in these catalytic reactions (**Figure 9B**). In a similar fashion, the *in situ* study performed by Zhang et al. [101] on the evolution of Si-supported Pd@CeO_2 nanocatalysts revealed the presence of dynamical processes that altered drastically the structure of the nanocomposites and that could only be observed by closed cell *in situ* E-TEM. In fact, while the performance of Pd@CeO_2 was usually attributed to the formation of a core-shell structure, *in situ* calcination at 500°C in a 150 Torr O_2 atmosphere

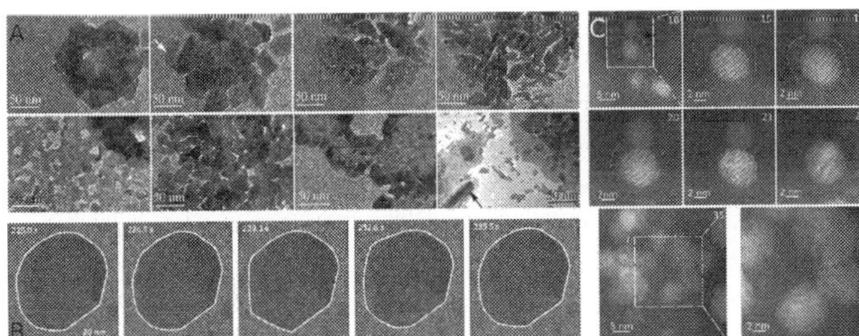

Figure 9. (A) Metallic evolution of Ag NPs in an air-filled cell after exposure. Top, from left to right: after 0 s with 0.18 A cm² current density; after 20 s with 0.18 A cm² current density (a new grain is indicated by a white arrow); after 0 s with 0.44 A cm² current density; after 20 s with 0.44 A cm² current density. Bottom, from left to right: after 0 s with 0.66 A cm² current density; after 20 s with 0.66 A cm² current density; after 20 s with 0.72 A cm² current density; after 20 s with 0.8 A cm² current density (the black arrow marks the vertical alignment of the AgO vapor phase). Adapted with permission from [100]; (B) time-resolved TEM images of a Pt nanoparticle at the gas exit of the reaction zone. The elapsed time is indicated in the top left corner. Adapted with permission from [97]; (C) (top) disappearance of an atom cloud and growth of a nearby particle, with subsequent coalescence at 650°C in 150 Torr O_2. The atomic cloud is delimited by a dashed line; (bottom) appearance of subnanometric features. The elapsed time indicated in minutes is at the top right corner of each image. Adapted with permission from [101].

revealed the dissociation of the Pd and CeO_2 particles to atomic clouds. Subsequent calcinations lead to re-organization of the atomic species into bigger sized ceria and Pd NPs or to the nucleation of a new structure featuring very small, highly dispersed silica, Pd, and ceria (**Figure 9C**). Assuming a conventional catalytic behavior of the bigger particles, the exceptional catalytical performance of Pd@CeO_2 above 800°C was attributed to this new structure.

4. *In situ* (S)TEM imaging of liquid specimens

As mentioned in the Section 1, since the invention of the electron microscope, one of the main scientific challenges has always been the imaging of liquid or wet samples. Even if this is usually possible by using light/fluorescence/confocal microscopy, and even taking into account the most recent and dramatic resolution improvements achieved due to the super-resolution approaches [102], its resolution (currently less than 100 nm) is still far from that of any electron microscope and represents a strong limit for nanoscale analyses of liquid samples. That being said, this section deals with the methods and devices that allow to achieve much higher resolution of liquid/wet samples performing *in situ* TEM instead of light-microscopy-based imaging. Analogously to what described in the previous section, also here the term "E-Cell" will be used to simply define the volume in which the interaction between the liquid environment and the specimen of interest occurs, irrespective of the device used to allow this interaction to happen.

A brief historical overview looks helpful in understanding what were, since the first decades of electron microscopy, the main limitations related to the possibility to perform *in situ* TEM of specimens kept in a liquid environment. First, the liquid has to be isolated from the rest of the column, which being under vacuum conditions would give rise to an almost immediate liquid evaporation, unless it would have a very low vapor pressure, and to the concomitant loss of the high vacuum conditions. The first trials to realize an *in situ* specimen holder capable to locally isolate the liquid in an E-Cell was proposed in the works of Abrams and McBain [103, 104]. It can be basically described as a platinum cell containing the liquid sample and constituted by two parts that were glued together, each provided with a very small hole where a window was attached, the latter thin enough to allow the electron beam to pass through them and the internal part that contained the liquid. This kind of E-Cell was just put in a normal TEM holder instead of the grid. The material used to fabricate the 50 nm thin electron-transparent windows was called collodion, and it was basically constituted by nitrocellulose. In none of the two papers of Abrams and McBain is reported any EM image of a liquid sample (gold particles in water) imaged by using this *in situ* specimen holder, and the authors affirmed that this was due to the particles' too rapid-free Brownian movement in a liquid as fluid as water. It was needed waiting almost 30 years to read a further paper [105] in which Ernest Fullam showed a novel way to fabricate a wet E-Cell for TEM. The design of this latter cell basically followed that of Abrams and McBain, but Fullam indicated that one of the most important failure of the colleagues' previous cell was due to the material chosen to fabricate the thin windows: the nitrocellulose films were invariably broken by the pressure difference between inside and outside the E-Cell, once it was inserted in the TEM column. In fact, the most important novelty shown by Fullam was the manner followed to fabricate the thin and electron-transparent windows, which were now constituted by a triple-layer film: Formvar and nitrocellulose separately cast from fresh solutions and deposited on 400 mesh copper TEM grids. Fullam estimated this configuration having a total thickness ranging from 30 to 40 nm and showed that it did not suddenly ruptured, allowing to image chromium hydroxide spheres in water suspension even if, despite the low magnification (5k) used, still in such a case the loss in spatial resolution due to the water electron scattering as well as once again to the Brownian movement of the particles suspended in water is clearly reported. To limit both these effects, only a possible solution was indicated: making the liquid thickness contained by the E-Cell as low as possible.

A further and strong novelty was then proposed in the work of Parsons [66], who described a new kind of experimental design to perform *in situ* TEM of liquid samples. It is the birth of the so-called environmental TEM (E-TEM) that consisted, and still consists, in a completely different type of E-Cell inside the column, the details of which have been already described in the previous section. Basically, and to remind its working principle, this kind of E-Cell is no longer physically limited by two thin electron-transparent windows and then just inserted into the TEM column as a part of a special specimen holder, but it is itself a part of the TEM column, where the ability to kept it under low gaseous pressure conditions is given by a system constituted by two differentially pumped and apertures' limited volumes put just above and below the E-TEM cell. Parson indicated that in the E-TEM cell he proposed a water vapor pressure of around 0.03 bar was enough to establish and keep a wet compartment. However, by using such a configuration he also mentioned that a maximum resolution of about 4.0 nm in the presence of water vapor pressure could be attainable, with the microscope operating at an acceleration voltage of 200 kV, and showed an E-TEM image of unstained and unfixed coliform bacteria and intracytoplasmic organelles, even if not highly resolved.

A variation of E-TEM is that shown by Gai in [106], where both environmental and high-resolution TEM (EHRTEM) were shown with a double possible configuration: first, a specific microreactor inserted inside the E-Cell of the EHRTEM column enables studies of thermally driven gas-solid reactions, while keeping an atomic resolution. Similarly, a way to develop *in situ* TEM studies of wet samples that could be even heated was indicated. It consisted in inserting a dedicated specimen holder in the E-Cell, capable to inject liquids onto the sample, located in the furnace of the holder.

However, what was mainly developed in the following years and so far to perform *in situ* TEM imaging is mainly based on sample holders with the isolating double thin window systems. This is not only due to the not secondary point that purchasing an E-TEM is obviously much more expensive than acquiring a dedicated sample holder to be inserted into a conventional S(TEM). However, it was actually shown that *in situ* E-TEM imaging of liquid/wet specimen unavoidably brings to work with a liquid film thickness not less than 1 μm and whose fine control is very hardy achievable as a consequence of a high and systematic loss in the final point resolution.

More recently, double window-based E-Cells for liquids were then and again used for *in situ* specimen holders' fabrication, being adaptable to any standard. The sample of interest can be sealed inside the double thin-window E-Cell together with gases or liquids, the latter possibly also let to flow. Although the E-Cell thin windows' presence unavoidably makes a little worse the final resolution, by such a way the sample can be surrounded by a thin gaseous or liquid environmental layer, with a thickness that can be *a priori* chosen within a large range, providing more or less resolution depending on both the chosen TEM imaging geometries and the total thickness of both sealing windows and interposed sample [107, 108].

The major improvement achieved for the development of specimen holders capable to allow *in situ* (S)TEM imaging of liquid samples was the fabrication of devices with robust but thin windows for the relevant E-Cells. These were realized producing thick microchips of amorphous silicon nitride (SiN) with a central zone having a thin or a very thin thickness (20–100 nm). In addition, further specific advances in the microchip fabrication allowed the integration of electrodes and microfluidic circuits to mimic physiological and/or dynamic conditions in the E-Cell for more specific *in situ* (S)TEM imaging such as, just for instance, the one of proteins or NPs interactions with whole cells [107, 109, 110], electrochemical reactions [111, 112], nanocrystals' growth in liquid solution [113, 114].

In the following, we first deal with the equations needed to determine the final point resolution when an *in situ* sample holder containing a water layer of given thickness is used in both TEM and STEM geometries since each of them brings to the optimal way to prepare the specimen that should be looked at. Second, it will continue and end with some examples of *in situ* (S)TEM imaging for both materials and biological science studies.

4.1. The point resolution when using an *in situ* sample holder for liquids

4.1.1. TEM

To determine how to calculate the final point resolution when using an *in situ* sample holder for liquids, a first approximation could be adopted. It consists in considering the material constituting the E-Cell windows with a scattering factor negligible for typical beam energies (200–300 keV), with respect to that of the liquid layer contained in the E-Cell, which is valid for most of the cases. Under such a condition, the main resolution's limiting factor is the chromatic aberration due to the inelastic scattering of electrons by the liquid. Taking then into account the most diffused and simple case of water as liquid medium, the electron energy broadening due to the beam-water inelastic scattering and the fact that chromatic aberration is the one effectively governing the final resolution, then d_{TEM} could be expressed by the following equation:

$$d_{TEM} = 6 \times 10^{12} \left(\frac{\alpha C_c T}{E^2} \right) \qquad (2)$$

where C_C is the spherical aberration of the objective lens, usually ranging between 0.5 and 2 mm depending on the microscope used, α is the objective semi angle, E is the beam energy, and T is the whole thickness of the cell filled by water. The water features (density, mean atomic number, and atomic weight) have been taken into account in the numerical coefficient appearing in the equation. Just for instance, with E = 200 keV, α = 10 mrad, C_C = 2mm, and T =1 µm, a final resolution d_{TEM} of 4 nm is calculated [108].

Thus, taking into account what indicated by Eq. (2) to look at any liquid sample (i.e., at a solid sample surrounded by a liquid environment) by TEM using the parallel electron beam geometry, the highest attainable resolution is obtained for solid objects surrounded by liquid in a configuration where electrons have to pass the lowest possible liquid volume after having crossed the solid objects, in order to minimize the energy broadening that they suffer in scattering with the liquid's molecules. For this reason, the solid part of the sample should be put onto the internal side of that E-Cell windows from which the electrons definitely exit from the E-Cell or, analogously, if the solid objects are homogeneously distributed all over the liquid the best choice to image them by TEM is that to focalize the ones closest to the lower E-Cell's window. In other words, to obtain the best resolution, the objective lens focal plane should be placed as near as possible to the internal side of the lower E-Cell's window. **Figure 10A** shows this experimental configuration.

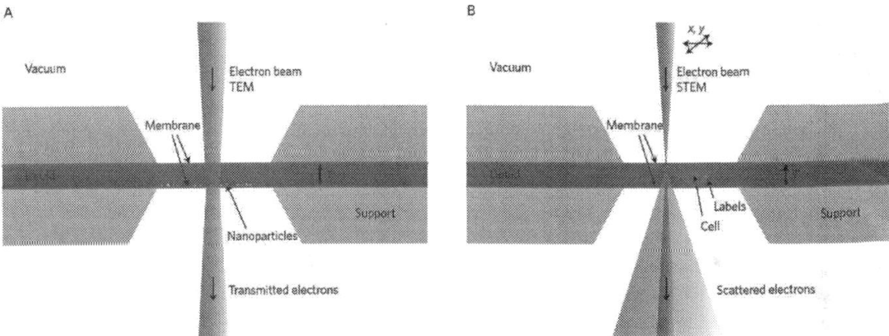

Figure 10. Different possible configurations for electron microscopy in liquid. (A) TEM imaging of nanoparticles in a liquid fully enclosed between two electron-transparent windows. Adapted with permission from [108]. (B) STEM imaging in a fully enclosed light liquid, used to image nanoparticle labels on whole biological cells. Adapted with permission from [108].

4.1.2. STEM

Differently from the TEM geometry, in which a parallel electron beam is used to illuminate whole areas of the sample, in the STEM the imaging is performed by a convergent beam that scans the sample point by point and where the contrast is then obtained by collecting the electron signal produced by each irradiated point with an appropriate detector. Even considering that the transmitted beam is the simplest way to form an image (the so-called Bright field mode), to image the contrast due to small but atomically heavy objects (like NPs) dispersed in a lighter liquid medium, the signal emitted with an high angle of divergence is usually the one collected to form the STEM image, being this STEM detection's geometry called HAADF. In such a case, the contrast is due to the fact that the detected signal's intensity is roughly proportional to Z^2, with Z being the atomic number of the material irradiated by the electron beam. Under these experimental conditions, the STEM resolution (d_{STEM}) achievable for a NPs' sample surrounded by a liquid environment is usually defined as the diameter of the smallest NP visible above the background noise as given by the following equation:

$$d_{STEM} = 5l_{nanop} \sqrt{\frac{T}{N_0 l_{liquid}}} \tag{3}$$

where l_{nanop} and l_{liquid} are the mean free path length for elastic scattering of NP and surrounding liquid, respectively, N_0 is the number of incident electrons, and T is the whole thickness of the liquid cell [107, 108].

Thus, the highest resolution for the STEM imaging of an object within a liquid is achieved when the object is at the entrance side of the sample. Moreover, keeping the specimen holder filled with the same water volume, the NP STEM imaging will be performed with a resolution in principle quite higher than the one obtainable by using a TEM imaging.

According to Eq. (3), a gold NP (marker) with a 1.4 nm in diameter can be resolved on a water layer of $T = 5$ μm. For NPs positioned deeper in the liquid, imaging takes place with a blurred probe, but a resolution better than 10 nm is still achieved in the top 1 μm layer of the liquid. As previously illustrated for the TEM, **Figure 10B** shows the experimental configuration corresponding to the STEM geometry with an *in situ* sealed E-Cell for liquid specimens.

4.2. *In situ* liquid TEM for materials science studies

Just as short overview of the different fields of application of *in situ* (S)TEM of liquid sample and devices, in the following some quite known examples are reported. One of the important papers that started to show the terrific possibilities offered by this technique was that of Williamson et al. [111], while a quite similar study was also published by Radisic et al. [112]. Both works deal with electrochemical processes monitored over time by providing a sealed electrochemical E-Cell well fitted for both vacuum environment and restricted volume available in the TEM column. The liquid E-Cell is filled with a water-based solution (usually electrolytes) and also contains electrodes as shown in **Figure 11A**. Both papers showed time-resolved imaging of electrochemically driven process of nucleation and following growth of individual Cu clusters onto an electron-transparent Au electrode patterned over one of the windows of an E-TEM liquid cell, with simultaneous measurement of voltage and current involved. It should be noticed that in both papers the TEM operating mode and the corresponding beam geometry was chosen, and in fact the Cu nanoclusters' formation is observed when they reach a size of several tens of nanometers (**Figure 11B** and **C**).

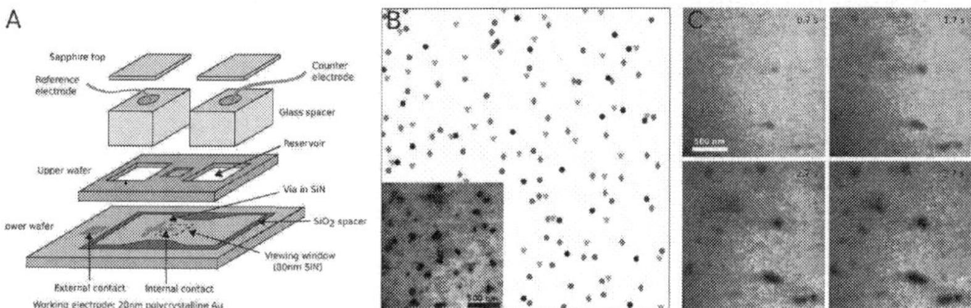

Figure 11. (A) Sketch of the liquid cells' components. Adapted with permission from [111]. (B) The center-of-mass positions of Cu clusters during four deposition experiments (different colors) on the same area of the electrode. Inset is the micrograph that corresponds to the black circles. Adapted with permission from [111]. (C) Small regions extracted from a video recorded during deposition at 5 mA cm⁻². Images taken at the times shown after current flow began. Adapted with permission from [111].

In 2009, Zheng et al. again used TEM geometry and a double SiN window *in situ* E-Cell for liquid specimens to image Pt nanocrystals' growth and coalescence, obtaining a final resolution well lower than 1 nm, mainly due to the low spherical aberration of the high-resolution objective lens but also to the 200–300 kV accelerating voltage used [113]. For the particle synthesis, a stock solution was prepared by dissolving Pt(acetylacetonate)₂ in a mixture of *o*-dichlorobenzene and oleylamine. About 100 nanoliters of such a growth solution were put into

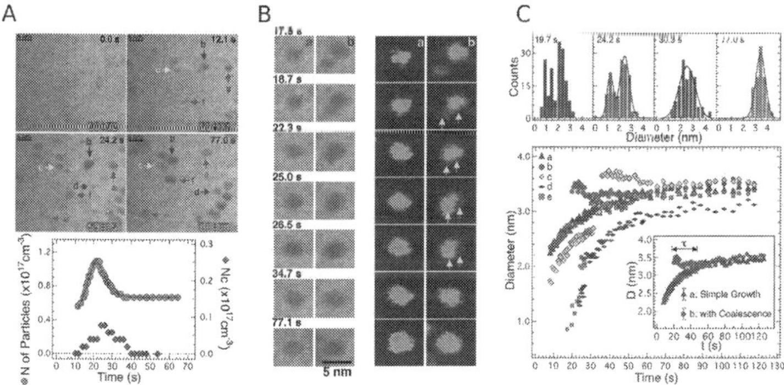

Figure 12. Growth and coalescence of Pt nanocrystals. (A) Top images: video frames acquired at different time points of exposure to the electron beam are reported, with specific particles labeled by arrows. Bottom image: the number of particles (left axis) and the number of coalescence events (Nc, right axis) vs. time, as measured with a time rate of 2.0 s. Adapted with permission from [113]. (B) Video images showing nanoparticle's growth occurred by monomer addition (left column, a) or growth by means of coalescence (left column, b). Particles are selected from the same field of view. Right columns (a) and (b): color images of the left ones, enlarged 1.5 times. The yellow arrows are indicating recrystallization observed in the coalesced particle. Adapted with permission from [113]. (C) Top images: histogram of particle size distribution at different time points, with black curves corresponding to Gaussian fittings. Bottom images: particle size vs. growth time of particles highlighted in panel (A). Inset shows two types of growth trajectories: simple growth and growth with coalescence. A relaxation time (τ) was observed after a coalescence event. Adapted with permission from [113].

the E-Cell, where the SiN windows and the liquid layer had a thickness of 25 and 200 nm, respectively. It is crucial to highlight here that the driving force used to promote the nucleation, growth, and coalescence of the Pt NPs was the sole electron beam irradiation. The main results published in the paper of Zheng et al. are summarized in **Figure 12A–C**. They consisted in showing not only the nucleation and growth of the Pt nanoclusters over time in the sealed E-Cell, but also in showing the fundamental role played by the coalescence of the larger NPs with the smaller ones during the growth process.

Similarly to what cited above, Evans et al. synthesized PbS NPs using as electrolyte a solution containing lead acetate, poly(vinyl alcohol), isopropyl alcohol, and thioacetamide [114]. The selective decomposition of thioacetamide was again promoted using the illumination with high-energy electron beam as driving force, and it gave rise to the production of free sulfur ions in solution. Surrounding lead ions then reacted with the free sulfur ions to allow the PbS NPs to nucleate and grow, while the poly(vinyl alcohol) acted as a stabilizing agent. However, two quite important improvements were shown in the work of Evans et al. and consisted in (a) using a continuous flow of reactants, drawn in and out from the E-Cell by appropriate microfluidic lines and (b) monitoring the processes occurring inside the E-Cell by using a spherical aberration corrected STEM, with higher resolution than the one attainable using a conventional TEM, for the reasons explained above.

4.3. *In situ* liquid TEM for biological sciences studies

For biological purposes, both E-TEM and sealed E-Cells for liquid specimen holders have been used, with a large predominance of the sealed ones. However, a premise is crucial before showing some examples. When using any EM imaging technique, either TEM or SEM based, no living specimen could be directly looked at without destroying it. That is due to the intrinsic conditions needed for any electron microscope to work (high vacuum and radiation damage due to the electrons-matter interactions) that are incompatible with the high water content, low density, and very low hardness of any living organism. To look at any living matter *post mortem*, so far a very large series of experimental protocol have been then developed, which allow making it rigid, dehydrated or frozen, and possibly with enough contrast if seen by a transmission geometry. In any case, all preparation protocols unavoidably require to kill either the cells or the micro-organisms prior to inserting them in any EM. The *in situ* liquid sample holders are then the only devices that are currently able to allow looking at living matter without killing it, and keeping it into a physiological aqueous environment, even if some effects of the radiation damage have to be taken into account. First, the small temperature rise can alter diffusion and reaction rates via convective motions within the liquid volume. Second, nonthermal effects may be more critical and more complicated to deal with: (a) high-energy electrons in water give rise to the pretty long-lived species' production such as OH radicals and hydrated electrons, even if charged species could quench radicals resulting in the concomitant reduction of their reactivity in solution; (b) substrate chemistry also plays a nonnegligible role since reactive species can be generated from interactions between the beam and the substrate, i.e., the thin windows that seal the E-Cell; (c) organic specimens are known to be intensely affected by a small enough electron dose, which means that for pristine biological materials, structural damage begins to happen at the subnanometer scale when the low electron dose of $\sim 10^2$ electrons per nm^2 is exceeded. This effect could be limited: (a) considering that the electron dose usually scales quadratically with the resolution, and then the latter could be appropriately chosen in order to have still good images but low radiation damage; (b) making use of an ultrasensitive CMOS camera, which allows to work using very low electron doses. If higher electron doses are required, the main way to preserve the cell/organism ultrastructure is stabilizing it by a chemical or cryofixation.

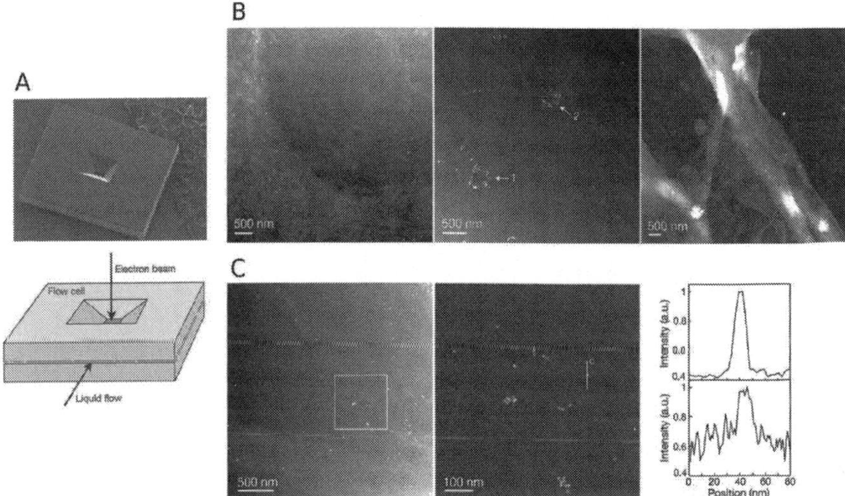

Figure 13. (A) Top image: single Si microchip with an electron-transparent SiN window. The sizes of the microchip are 2.0 × 2.3 × 0.3 117. Bottom image: sketch of a microfluidic chamber constituted by two Si microchips such as the ones presented in the top image and showing the liquid flow direction. Adapted with permission from [107, 108]. (B) Liquid STEM images of COS7 fibroblast cells labeled EGF-Au. Left image: edge of a fixed COS7 cell after 5 min incubation with EGF-Au. The labels are visible as bright spots and the cellular material is shown as light-gray matter on a dark-gray background (pixel size: 5.7 nm). Central image: COS7 cell incubated with EGF-Au for 10 min and then in buffer (without EGF-Au) for additional 15 min. Circular cluster of labeled EGF receptors are visible. Arrow 1 indicates sharp-edged gold labels in the cluster, whereas the labels in the cluster at arrow 2 appear blurred and cannot be distinguished as individual entities, indicating that cluster 2 is out of the vertical plane of focus. The observation of circular clusters of labels at different vertical positions in the cell indicates that the gold nanoparticles labels were internalized (pixel size: 5.7 nm). Right image: sample shown in left image, but imaged after the flow cell opening and sample drying in air. The dried sample displays much stronger contrast on the cellular material: the white large zones in the image correspond to the edge of a dried cell, while the gold nanoparticles labels are still visible as bright spots (pixel size: 8.9 nm). Adapted with permission from [107]. (C) Left image: COS7 cell edge labeled with EGF-Au, where the bright spots indicate the labels (pixel size: 2.9 nm). Central image: magnification of the area defined by the white square in left image showing the individual labels. Right top line profile: HAADF signal as a function of the electron probe position over an individual gold nanoparticle, corresponding to the line labeled "C" in the central image. Right bottom line profile: HAADF signal as detected on a 10 nm gold nanoparticle at the bottom of 1.3 μm thick liquid layer (10% PBS buffer in water) in a test sample without cells (pixel size was 0.91 nm). Adapted with permission from [107].

As first example of study of biological living matter by *in situ* E-Cell for liquid specimens, we report here what Liu et al. showed in their paper [115]. It reports the detailed fabrication using microelectromechanical system (MEMS) techniques to obtain a novel microchip constituted by two SiO_2 films used as a substrate and the 9 nm thin windows for *in situ* imaging of living organisms in aqueous environment using a conventional TEM geometry. Such a kind of microchips—with a gap between them ranging from 2 to 5 μm—allowed the effective TEM observation of living bacteria *Escherichia coli* and the tellurite reduction process in *Klebsiella pneumonia*, the height of which were estimated ranging between 0.5 and 1μm. The most important results of this paper were (a) the capability of bacterium *K. pneumoniae* and the yeast *Saccharomyces cerevisiae* to stay alive onto the microchip substrate under continuous electron beam irradiation needed for the TEM imaging for up to 14 and 42 s, respectively; (b) the possibility to follow over time, even if with not very high resolution as expected from Eq. (2), biological processes such as the tellurite reduction by *K. pneumonia*.

As already mentioned and shown by Eq. (3), the use of STEM allowed to achieve a dramatically increased resolution and to image by using HAADF geometry even very small heavy particles dispersed in a light liquid medium. This was the case of Au-labeled eukaryotic fibroblast cells (COS7) immersed in a several micrometer-thick liquid layer [107, 109]. In both the papers a flow liquid E-Cell was used, allowing the aqueous solution to pour through the E-Cell and around the whole fibroblast cells. The experimental configuration is reported in **Figure 13A**. The cells were grown, then labeled, and finally fixed prior to be inserted in the flow E-Cell, on the internal side of the silicon nitride film that was finally placed to seal the upper part of the E-Cell, in order to maximize the STEM resolution according to Eq. (3). The cells' immunolabeling was realized by using gold NPs sizing 10 nm, in order to target the epidermal growth factor (EGF) molecules bound to the cellular EGF receptors. Thus, using a water layer 10 μm thick, the theoretical resolution limit determined by using Eq. (3) was found to be equal to 1.9 nm. In agreement with that, the 10 nm sized labeling gold NPs were detected with a spatial resolution of about 4 nm, as shown in **Figure 13B** and **C**, using the STEM in HAADF geometries, and performed with a beam accelerated by a voltage of 200 kV and a dwell time of 20 μs. If it is not weird that the electron beam did not perturb the fixed cells, it is however remarkable that for the dose used (7×10^4 electrons per nm^2) and within the flowing liquid no visible damage consequences were observed on the spatial distribution of the labeling NPs on the cells.

As last and most recent example the results achieved by Pohlmann *et al.* are shown [110]. In their paper they showed an *in situ* TEM imaging, time resolved, of the uptake of typical drug delivery vehicles—in such a case gold nanorods (GNR) encapsulated in polyvinylpyridine (PVP)—from glioblastoma stem cells (GSCs), which actually is one of the most aggressive cancers among brain tumors. This kind of study is quite interesting, allowing to image by TEM the evolution over time of NP uptake from living cells, a phenomenon so far investigated just by light/confocal microscopy. To perform it, the authors first prepared a multiwell E-Cell with a maximum height of 2 μm where the GSCs were forced to be tethered on the lower side, thanks to a surface functionalization that involved their NOTCH1 receptors, as shown in **Figure 14A**. Then, the GSCs contained in the multiwell E-Cell were incubated together with the

PVP-GNR NRs for 20 and 60 min, respectively, prior to inserting the sealed E-Cell into the TEM. The evolution of the PVP-GNRs within the GSCs was then followed over a total time of 30 s, using the TEM classical geometry performed at medium magnification, and with a high contrast objective lens and an electron beam acceleration voltage of 120 kV. Even under these conditions, the resolution was enough to study the PVP-GNRs uptake evolution over time. As shown in **Figure 14B**, the movement of the PVP-GNRs inside the GSCs is similar for both the incubation times: a local displacement of the PVP-GNRs and the solution surrounding them were observed. The main difference between the GSCs incubated for 20 and 60 min with the PVP-GNRs consisted in the number of those uptaken from the GSCs: higher the incubation time, higher the amount of PVP-GNRs internalized from the tumor cells.

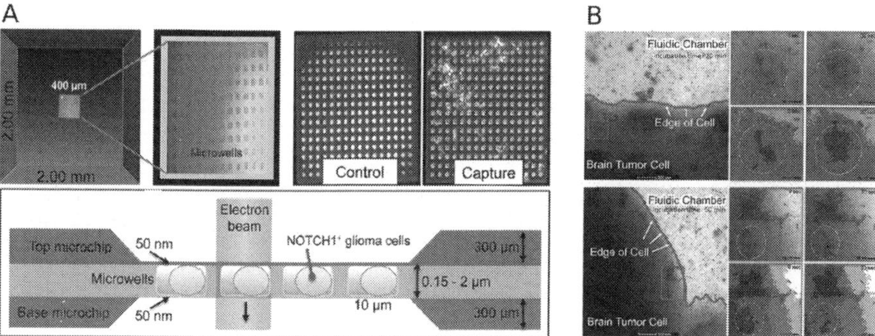

Figure 14. (A) First row, left images: sketches of microchip with integrated microwells (10 µm in diameter) functionalized to specifically capture GS9-6/NOTCH+ glioblastoma stem cells; right images: comparison between the functionalized microwells (called "capture") and the negative control, i.e., identical microwells not treated for functionalization and then not able to tether GSCs. Second row: sketch showing a cross-sectional view of the microfluidic system containing GSC positioned in the TEM column. Adapted with permission from [110]. (B) Gold nanorods penetrated in a GSCs after two different times of incubation. Top images show the capability of unlabeled gold nanorods to perforate the cells within 20 min incubation period; moreover, the liquid dispersed around the nanorods displaces their position within a common region. These movements are represented in a contour map generated at the same threshold level and within the same region of interest at the initial imaging stage (0 s) and final stage (30 s). Regions exhibiting the greatest density (included both nanorods and their surrounding liquid) were contoured at the darker colored threshold while regions showing less density were represented using increasingly lighter gradients of color. Bottom images: cell incubated for 60 min, where more NPs inside the cell than in the solution are observed, if compared to the sample incubated for 20 min. Adapted with permission from [110].

5. Conclusions

In this chapter, some of the most recent developments for the *in situ* TEM and STEM imaging techniques have been shown. The chapter aimed at showing the impact on the *in situ* TEM/STEM imaging of the progress achieved in the last 10–15 years in terms of novel microscope architecture developments, fabrication and commercial availability of novel and much more performing dedicated specimen holders. This in turn has allowed achieving unprecedented results in terms of sample stability and diversity of setups, variation of the local sample environment, and possible external *stimuli*. Due to the vastness of the topic, the three sections of this chapter have been dedicated to describe three particular applications of the *in situ* TEM/STEM that witness how this technique evolved in the last decade: the *in situ* sample heating under high vacuum condition, the *in situ* sample heating in a controlled gaseous atmosphere, and, finally, the *in situ* imaging of samples immersed in a liquid layer of user-controlled thickness. The importance of these improvements lies in the widening of their possible application not only to inorganic but also to biological materials, even in living conditions, which paves the way to novel, so far unattainable studies.

Author details

Alberto Casu, Elisa Sogne, Alessandro Genovese, Cristiano Di Benedetto, Sergio Lentijo Mozo, Efisio Zuddas, Francesca Pagliari and Andrea Falqui*

*Address all correspondence to: andrea.falqui@kaust.edu.sa

King Abdullah University of Science and Technology (KAUST), Biological and Environmental Sciences and Engineering (BESE) Division, Thuwal, Kingdom of Saudi Arabia

References

[1] Freitag B, Knippels G, Kujawa S, Van der Stam M, Hubert D, Tiemeijer PC, Kisielowski C, Denes P, Minor A, Dahmen U. First performance measurements and application results of a new high brightness Schottky field emitter for HR S/TEM at 80–300 kV acceleration voltage. Microsc. Microanal. 2008; 14:1370–1371. DOI: 10.1017/S1431927608087370.

[2] Haider M, Uhlemann S, Schwan E, Rose H, Kabius B, Urban K. Electron microscopy image enhanced. Nature 1998; 392:768–769. DOI: 10.1038/33823.

[3] Krivanek OL, Dellby N, Lupini AR. Towards sub Angstrom electron beams. Ultramicroscopy. 1999; 78:1–11. DOI: 10.1016/S0304 3991(99)00013 3.

[4] Tiemeijer PC, Van Lin JHA, De Long AF. First results of a monochromatized 200 kV TEM. Microsc. Microanal. 2001; 7:234–235.

[5] Kasper E. Field electron emission systems. In: Barer R, Cosslett VE, editors. Advances in Optical and Electron Microscopy. Vol. 8. Academic, London. p. 207; 1982.

[6] Von Harrach HS, Dona P, Freitag B, Soltau H, Niculae A, Rohde M. An integrated multiple silicon drift detector system for transmission electron microscopes. J. Phys. Conf. Ser. 2010; 241:012015. DOI: 10.1088/1742 6596/241/1/012015.

[7] Gubbens A, Barfels M, Trevor C, Twesten R, Mooney P, Thomas P, Menon N, Kraus B, Mao C, McGinn B. The GIF quantum, a next generation post column imaging energy filter. Ultramicroscopy 2010; 110:962–970. DOI: 10.1016/j.ultramic.2010.01.009.

[8] Ruskina RS, Yub Z, Grigorieff N. Quantitative characterization of electron detectors for transmission electron microscopy. J. Struct. Biol. 2013; 184:385–393. DOI: 10.1016/j.jsb. 2013.10.016.

[9] Heinemann K, Poppa H. Direct observation of small cluster mobility and ripening. Thin Solid Films. 1976; 33:237–251. DOI: 10.1016/0040 6090(76)90084 5.

[10] Ruault MO, Chaumont J, Bernas H. Transmission electron microscopy study of ion implantation induced Si amorphization. Nucl. Instrum. Methods Phys. Res. 1983; 209:351–356. DOI: 10.1016/0167 5087(83)90822 0.

[11] Chen SH, Zheng LR, Carter CB, Mayer JW. Transmission electron microscopy studies on the lateral growth of nickel silicides. J. Appl. Phys. 1985; 57:258–263. DOI: 10.1063/1.335482.

[12] Dannenberg R, Stach E, Groza JR, Dresser BJ. TEM annealing study of normal grain growth in silver thin films. Thin Solid Films. 2000; 379:133–138. DOI: 10.1016/S0040 6090(00)01570 4.

[13] Kamino T, Saka H. A newly developed high resolution hot stage and its application to materials characterization. Microsc. Microanal. Microstruct. 1993; 4:127–135. DOI: 10.1051/mmm:0199300402 3012700.

[14] Zhang M, Olson EA, Twesten RD, Wen JG, Allen LH, Robertson IM, Petrov I. In situ transmission electron microscopy studies enabled by microelectromechanical system technology. J. Mater. Res. 2005; 20:1802–1807. DOI: 10.1557/JMR.2005.0225.

[15] Allard LF, Bigelow WC, Jose Yacaman M, Nackashi DP, Damiano J, Mick SE. A new MEMS based system for ultra high resolution imaging at elevated temperatures. Microsc. Res. Tech. 2009; 72:208–215. DOI: 10.1002/jemt.20673.

[16] Sperling RA, Rivera Gil P, Zhang F, Zanella M, Parak WJ. Biological applications of gold nanoparticles. Chem. Soc. Rev. 2008; 37:1896–1908. DOI: 10.1039/b712170a.

[17] DeLong RK, Reynolds CM, Malcolm Y, Schaeffer A, Severs T, Wanekaya A. Function- alized gold nanoparticles for the binding, stabilization, and delivery of therapeutic DNA, RNA, and other biological macromolecules. Nanotechnol. Sci. Appl. 2010; 3:53– 63. DOI: 10.2147/NSA.S8984.

[18] Zhou ZY, Tian N, Li JT, Broadwell I, Sun SG. Nanomaterials of high surface energy with exceptional properties in catalysis and energy storage. Chem. Soc. Rev. 2011; 40:4167–4185. DOI: 10.1039/c0cs00176g.

[19] Li N, Zhao P, Astruc D. Anisotropic gold nanoparticles: synthesis, properties, applications, and toxicity. Angew. Chem. Int. Ed. 2014; 53:1756–1789. DOI: 10.1002/anie.201300441.

[20] Langille MR, Personick ML, Zhang J, Mirkin CA. Defining rules for the shape evolution of gold nanoparticles. J. Am. Chem. Soc. 2012; 134:14542–14554. DOI: 10.1021/ja305245g.

[21] Koga K, Ikeshoji T, Sugawara KI. Size and temperature dependent structural transitions in gold nanoparticles. Phys. Rev. Lett. 2004; 92:115507. DOI: 10.1103/PhysRevLett.92.115507.

[22] Kuo CL, Clancy P. Melting and freezing characteristics and structural properties of supported and unsupported gold nanoclusters. J. Phys. Chem. B. 2005; 109:13743–13754. DOI: 10.1021/jp0518862.

[23] Barnard AS, Young NP, Kirkland AI, van Huis MA, Xu H. Nanogold: a quantitative phase map. ACS Nano. 2009; 3:1431–1436. DOI: 10.1021/nn900220k.

[24] Young NP, van Huis MA, Zandbergen HW, Xu H, Kirkland AI. Transformations of gold nanoparticles investigated using variable temperature high resolution transmission electron microscopy. Ultramicroscopy. 2010; 110:506–516. DOI: 10.1016/j.ultramic.2009.12.010.

[25] Baumgardner WJ, Yu Y, Hovden R, Honrao S, Hennig RG, Abruna HD, Muller D. Nanoparticle metamorphosis: an in situ high temperature transmission electron microscopy study of the structural evolution of heterogeneous Au:Fe2O3 nanoparticles. ACS Nano. 2014; 8:5315–5322. DOI: 10.1021/nn501543d.

[26] Figuerola A, van Huis M, Zanella, M, Genovese A, Marras S, Falqui A, Zandbergen HW, Cingolani R, Manna L. Epitaxial CdSe Au nanocrystal heterostructures by thermal annealing. Nano Lett. 2010; 10:3028–3036. DOI: 10.1021/nl101482q.

[27] Lavieville R, Zhang Y, Casu A, Genovese A, Manna L, Di Frabrizio E, Krahne R. Charge transport in nanoscale "all inorganic" networks of semiconductor nanorods linked by metal domains. ACS Nano. 2012; 6:2940–2947. DOI: 10.1021/nn3006625.

[28] Liu Z, Che R, Elzatahry AA, Zhao D. Direct imaging Au nanoparticle migration inside mesoporous silica channels. ACS Nano. 2014; 8:10455–10460. DOI: 10.1021/nn503794v.

[29] Costi R, Saunders AE, Banin U. Colloidal hybrid nanostructures: a new type of functional materials. Angew. Chem. Int. Ed. 2010; 49:4878–4897. DOI: 10.1002/anie.200906010.

[30] Jun YW, Choi JS, Cheon J. Shape Control of Semiconductor and Metal Oxide Nanocrystals through Nonhydrolytic Colloidal Routes. Angew. Chem. Int. Ed. 2006; 45:3414–3439. DOI: 10.1002/anie.200503821.

[31] Zheng H, Smith RK, Jun YW, Kisielowski C, Dahmen U, Alivisatos AP. Observation of single colloidal platinum nanocrystal growth trajectories. Science. 2009; 324:1309–1312. DOI: 10.1126/science.1172104.

[32] Asoro MA, Desiderio K, Ferreira PJ. In situ transmission electron microscopy observations of sublimation in silver nanoparticles. ACS Nano. 2013; 7:7844–7852. DOI: 10.1021/mn402771j.

[33] Li H, Zanella M, Genovese A, Povia M, Falqui A, Giannini C, Manna L. Sequential cation exchange in nanocrystals: preservation of crystal phase and formation of metastable phases. Nano Lett. 2011; 11:4964–4970. DOI: 10.1021/nl202927a.

[34] Hellebusch DJ, Manthiram K, Beberwyck BJ, Alivisatos AP. In situ transmission electron microscopy of cadmium selenide nanorod sublimation. J. Phys. Chem. Lett. 2015; 6:605–611. DOI: 10.1021/jz502566m.

[35] Hudak BM, Chang YJ, Yu L, Li G, Edwards DN, Guiton BS. Real time observation of the solid liquid vapor dissolution of individual tin(IV) oxide nanowires. ACS Nano. 2014; 8:5441–5448. DOI: 10.1021/nn5007804.

[36] Yalcin AO, Fan Z, Goris B, Li WF, Koster RS, Fang CM, Van Blaaderen A, Casavola M, Tichelaar FD, Bals S, Van Tendeloo G, Vlugt TJH, Vanmaekelbergh D, Zandbergen HW, Van Huis MA. Atomic resolution monitoring of cation exchange in CdSe PbSe heteronanocrystals during epitaxial solid–solid–vapor growth. Nano Lett. 2014; 14:3661–3667. DOI: 10.1021/nl501441w.

[37] De Trizio L, De Donato F, Casu A, Genovese A, Falqui A, Povia M, Liberato M. Colloidal CdSe/Cu^{3}P/CdSe and their evolution upon thermal annealing. ACS Nano. 2013; 7:3997–4005. DOI: 10.1021/nn3060219.

[38] Geim AK, Novoselov KS. The rise of graphene. Nat. Mater. 2007; 6: 183–191. DOI: 10.1038/nmat1849.

[39] Balandin AA, Ghosh S, Bao W, Calizo I, Teweldebrhan D, Miao F, Lau CN. Superior thermal conductivity of single layer graphene. Nano Lett. 2008; 8:902–907. DOI:10.1021/nl0731872.

[40] Soldano C, Mahmood A, Dujardin E. Production, properties and potential of graphene. Carbon. 2010; 48:2127–2150. DOI: 10.1016/j.carbon.2010.01.058.

[41] Novoselov KS, Geim AK, Morozov SV, Jiang D, Zhang Y, Dubonos SV, Grigorieva IV, Firsov AA. Electric Field Effect in Atomically Thin Carbon Films. Science. 2004; 306:666–669. DOI: 10.1126/science.1102896.

[42] Reina A, Jia X, Ho J, Nezich D, Son H, Bulovic V, Dresselhaus MS, Kong J. Large area, few layer graphene films on arbitrary substrates by chemical vapor deposition. Nano Lett. 2009; 9:30–35. DOI: 10.1021/nl801827v.

[43] Yannopoulos SN, Siokou A, Nasikas NK, Dracopoulos V, Ravani F, Papatheodorou GN. CO2-laser-induced growth of epitaxial graphene on 6H-SiC (0001). Adv. Funct. Mater. 2012; 2:113–120. DOI: 10.1002/adfm.201101413.

[44] Park J, Mitchel WC, Grazulis L, Smith HE, Eyink KG, Boeckl JJ, Tomich DH, Pacley SD, Hoelscher JE. Epitaxial graphene growth by carbon molecular beam epitaxy (CMBE). Adv. Mater. 2010; 22:4140–4145. DOI: 10.1002/adma.201000756.

[45] Liu Z, Lin YC, Lu CC, Yeh CH, Chiu PW, Iijima S, Suenaga K. In situ observation of step edge in plane growth of graphene in a STEM. Nat. Commun. 2014; 5:4055. DOI: 10.1038/ncomms5055.

[46] Kotakoski J, Santos Cottin D, Krasheninnikov AV. Stability of graphene edges under electron beam: equilibrium energetics versus dynamic effects. ACS Nano. 2011; 6:671–676. DOI: 10.1021/nn204148h.

[47] He Z, He K, Robertson AW, Kirkland AI, Kim D, Ihm J, Yoon E, Lee GD, Warner JH. Atomic structure and dynamics of metal dopant pairs in graphene. Nano Lett. 2014; 14:3766–3772. DOI: 10.1021/nl500682j.

[48] Pi K, McCreary KM, Bao W, Han W, Chiang YF, Li Y, Tsai SW, Lau CN, Kawakami RK. Electronic doping and scattering by transition metals on graphene. Phys. Rev. B. 2009; 80:075406. DOI: 10.1103/PhysRevB.80.075406.

[49] Kano E, Hashimoto A, Kaneko T, Tajima N, Ohno T, Takeguchi M. Interactions between C and Cu atoms in single layer graphene: direct observation and modelling. Nanoscale. 2016; 8:529–535. DOI: 10.1039/C5NR05913E.

[50] Ma J, Alfe D, Michaelides A, Wang E. Stone Wales defects in graphene and other planar sp2 bonded materials. Phys. Rev. B. 2009; 80:033407. DOI: 10.1103/PhysRevB.80.033407.

[51] Ramasse QM, Zan R, Bangert U, Boukhvalov DW, Son YW, Novoselov, KS. Direct experimental evidence of metal mediated etching of suspended graphene. ACS Nano. 2012; 6:4063–4071. DOI: 10.1021/nn300452y.

[52] Kim MJ, Wanunu M, Bell DC, Meller A. Rapid fabrication of uniformly sized nanopores and nanopore arrays for parallel DNA analysis. Adv. Mater. 2006; 18:3149–3153. DOI: 10.1002/adma.200601191.

[53] Liu K, Feng J, Kis A, Radenovic A. Atomically thin molybdenum disulfide nanopores with high sensitivity for DNA translocation. ACS Nano. 2014; 8:2504–2511. DOI: 10.1021/nn406102h.

[54] Xu T, Xie X, Yin K, Sun J, He L, Sun L. Controllable atomic-scale sculpting and deposition of carbon nanostructures on graphene. Small. 2014; 10:1724–1728. DOI: 10.1002/smll.201303377.

[55] He K, Robertson AW, Gong C, Allen CS, Xu Q, Zandbergen H, et al. Controlled formation of closed edge nanopores in graphene. Nanoscale. 2015; 7:11602–11610. DOI: 10.1039/C5NR02277K.

[56] Asoro MA, Kovar D, Ferreira PJ. Effect of surface carbon coating on sintering of silver nanoparticles: in situ TEM observations. Chem. Commun. 2014; 50:4835–4838. DOI: 10.1039/c4cc01547a.

[57] Chen C, Hu Z, Li Y, Liu L, Mori H, Wang Z. In Situ high resolution transmission electron microscopy investigation of overheating of Cu nanoparticles. Sci. Rep. 2016; 6:19545. DOI: 10.1038/srep19545.

[58] Romankov S, Park YC. In situ high temperature TEM observation of material escape from a surface of CoFeNi/Cu/ZrAlO composite into the amorphous carbon layer. J. Alloys Compd. 2015; 632:408–416. DOI: 10.1016/j.jallcom.2015.01.214.

[59] Tang DM, Liu C, Yu WJ, Zhang LL, Hou PX, Li JC, Li F, Bando Y, Golberg D, Cheng HM. Structural changes in iron oxide and gold catalysts during nucleation of carbon nanotubes studied by in situ transmission electron microscopy. ACS Nano. 2014; 8:292–301. DOI: 10.1021/nn403927y.

[60] Boston R, Schnepp Z, Nemoto Y, Sakka Y, Hall SR. In situ TEM observation of a microcrucible mechanism of nanowire growth. Science. 2014; 344:623–626. DOI: 10.1126/science.1251594.

[61] Casu A, Genovese A, Manna L, Longo P, Buha J, Botton GA, Lazar S, Kahaly MU, Schwingenschloegl U, Prato M, Li H, Ghosh S, Palazon F, De Donato F, Lentijo Mozo S, Zuddas E, Falqui A. Cu2Se and Cu nanocrystals as local sources of copper in thermally activated in situ cation exchange. ACS Nano. 2016; 10:2406–2414. DOI: 10.1021/acsnano.5b07219.

[62] Marton L. Marton L. Electron microscopy of biological objects (La microscopie electronique des objets biologiques). Bull. Classe. Sci. Acad. Roy Belgique. 1935; 21:553–564

[63] Hashimoto H, Naiki T, Eto T, Fujiwara K. High temperature gas reaction specimen chamber for an electron microscope. Jpn. J. Appl. Phys. 1968; 7:946–952. DOI: 10.1143/JJAP.7.946.

[64] Parsons DF. Structure of wet specimens in electron microscopy. Science. 1974; 186:407–414. DOI: 10.1126/science.186.4162.407.

[65] Mehraeen S, McKeown JT, Deshmukh PV, Evans JE, Abellan P, Xu P, Reed BW, Taheri ML, Fischione PE, Browning ND. A (S)TEM gas cell holder with localized laser heating for in situ experiments. Microsc. Microanal. 2013; 19:470–478. DOI: 10.1017/S1431927612014419.

[66] Turner JN, See CW, Ratkowski AJ, Chang BB, Parsons DF. Design and operation of a differentially pumped environmental chamber for the HVEM. Ultramicroscopy. 1981; 6:267–280. DOI: 10.1016/s0304 3991(81)80162 3.

[67] Lee TC, Dewald DK, Eades JA, Robertson IM, Birnbaum HK. An environmental cell transmission electron microscope. Rev. Sci. Instrum. 1991; 62:1438–1444. DOI: 10.1063/1.1142464.

[68] Boyes ED, Gai PL. Environmental high resolution electron microscopy and applications to chemical science. Ultramicroscopy. 1997; 67:219–232. DOI: 10.1016/S0304 3991(96)00099 X.

[69] Sharma R. Design and applications of environmental cell transmission electron microscope for in situ observations of gas–solid reactions. Microsc. Microanal. 2001; 7:494–506. DOI: 10.1007/s10005 001 0015 1.

[70] Hansen TW, Wagner JB. Environmental transmission electron microscopy in an aberration corrected environment. Microsc. Microanal. 2012; 18:684–690. DOI: 10.1017/S1431927612000293.

[71] Boyes ED, Gai PL. Visualising reacting single atoms under controlled conditions: advances in atomic resolution in situ environmental (scanning) transmission electron microscopy (E(S)TEM). C. R. Physique 2014; 15:200–213. DOI: 10.1016/j.crhy.2014.01.002.

[72] Kawasaki T, Ueda K, Ichihashi M, Tanji T. Improvement of windowed type environmental cell transmission electron microscope for in situ observation of gas solid interactions. Rev. Sci. Instrum. 2009; 80:113701. DOI: 10.1063/1.3250862.

[73] Baker RTK, Barber MA, Harris PS, Feates FS, Waite RJ. Nucleation and growth of carbon deposits from the nickel catalyzed decomposition of acetylene. J. Catal. 1972; 26:51–62. DOI: 10.1016/0021 9517(72)90032 2.

[74] Baker RTK, Harris PS. Controlled atmosphere electron microscopy. J. Phys. E. 1972; 5:793–797. DOI: 10.1088/0022 3735/5/8/024.

[75] Parkinson GM. High resolution, in situ controlled atmosphere transmission electron microscopy (CATEM) of heterogeneous catalysts. Catal. Lett. 1989; 2:303–307. DOI: 10.1007/BF00770228.

[76] Giorgio S, Joao SS, Nitsche S, Chaudanson D, Sitja G, Henry CR. Environmental electron microscopy (E TEM) for catalysts with a closed E cell with carbon windows. Ultramicroscopy. 2006; 106:503–507. DOI: 10.1016/j.ultramic.2006.01.006.

[77] de Jonge N, Bigelow WC, Veith GM. Atmospheric pressure scanning transmission electron microscopy. Nano Lett. 2010; 10:1028–1031. DOI: 10.1021/nl904254g.

[78] Petkov N. In situ real time TEM reveals growth, transformation and function in one dimensional nanoscale materials: from a nanotechnology perspective. ISRN Nanotechnology. 2013; 2013:893060. DOI: 10.1155/2013/893060.

[79] Hansen TW, Wagner JB. Controlled Atmosphere Transmission Electron Microscopy. Springer International Publishing; Switzerland, 2016. 329 pp. DOI: 10.1007/978 3 319 22988 1.

[80] Suzuki M, Yaguchi T, Zhang XF. High resolution environmental transmission electron microscopy: modeling and experimental verification. Microscopy. 2013; 62:437–450. DOI: 10.1093/jmicro/dft001.

[81] Wagner JB, Cavalca F, Damsgaard CD, Duchstein LDL, Hansen TW. Exploring the environmental transmission electron microscope. Micron. 2012; 43:1169–1175. DOI: 10.1016/j.micron.2012.02.008.

[82] Xin HL, Niu K, Alsem DH, Zheng H. In situ TEM study of catalytic nanoparticle reactions in atmospheric pressure gas environment. Microscopy Microanal. 2013; 19:1558–1568. DOI: 10.1017/S1431927613013433.

[83] Zhang XF, Kamino T. Imaging gas–solid interactions in an atomic resolution environmental TEM. Micros. Today. 2006; 14:16–18.

[84] Bright AN, Yoshida K, Tanaka N. Influence of total beam current on HRTEM image resolution in differentially pumped E TEM with nitrogen gas. Ultramicroscopy. 2013; 124:46–51. DOI: 10.1016/j.ultramic.2012.08.007.

[85] Takeda S, Kuwauchi Y, Yoshida H. Environmental transmission electron microscopy for catalyst materials using a spherical aberration corrector. Ultramicroscopy. 2015; 151:178–190. DOI:10.1016/j.ultramic.2014.11.017.

[86] Jinschek JR, Helveg S. Image resolution and sensitivity in an environmental transmission electron microscope. Micron. 2012; 43:1156–1168. DOI: 10.1016/j.micron.2012.01.006.

[87] Cavalca F, Laursen, AB, Wagner JB, Damsgaard CD, Chorkendorfflb, Hansen TW. Light induced reduction of cuprous oxide in an environmental transmission electron microscope. Chem. Cat. Chem. 2013; 5:2667–2672. DOI: 10.1002/cctc.201200887.

[88] Jeangros Q, Hansen TW, Wagner JB, Dunin Borkowski RE, Hébert C, Van herle J, Hessler Wyser A. Oxidation mechanism of nickel particles studied in an environmental transmission electron microscope. Acta Mater. 2014; 67:362–372. DOI: 10.1016/j.actamat.2013.12.035.

[89] Xin HL, Pach EA, Diaz RE, Stach EA, Salmeron M, Zheng H. Revealing correlation of valence state with nanoporousstructure in cobalt catalyst nanoparticles by in situ environmental TEM. ACS Nano. 2012; 6:4241–4247. DOI: 10.1021/nn3007652.

[90] Serve A, Epicier T, Aouine M, Cadete Santos Aires FJ, Obeid E, Tsampas M, Pajot K, Vernoux P. Investigations of soot combustion on yttria stabilized zirconia by environ-

mental transmission electron microscopy (E TEM). Appl. Catal. A Gen. 2015; 504:74–80. DOI: 10.1016/j.apcata.2015.02.030.

[91] Chenna S, Crozier PA. Operando transmission electron microscopy: a technique for detection of catalysis using electron energy loss spectroscopy in the transmission electron microscope. ACS Catal. 2012; 2:2395–2402. DOI: 10.1021/cs3004853.

[92] Creemer JF, Helveg S, Hoveling GH, Ullmann S, Molenbroek AM, Sarro PM, Zandbergen HW. Atomic scale electron microscopy at ambient pressure. Ultramicroscopy. 2008; 108:993–998. DOI: 10.1016/j.ultramic.2008.04.014.

[93] Allard LF, Overbury SH, Bigelow WC, Katz MB, Nackashi DP, Damiano J. Novel MEMS based gas cell/heating specimen holder provides advances imaging capabilities for in situ reaction studies. Microsc. Microanal. 2012; 18:656–666. DOI: 10.1017/S1431927612001249.

[94] Yokosawa T, Alan T, Pandraud G, Dam B, Zandbergen H. In situ TEM on (de)hydrogenation of Pd at 0.5–4.5 bar hydrogen pressure and 20–400°C. Ultramicroscopy. 2012; 112:47–52. DOI: 10.1016/j.ultramic.2011.10.010.

[95] Doll T, Hochberg M, Barsic D, Scherer A. Micro machined electron transparent alumina vacuum windows. Sens. Actuat. A Phys. 2000; 87:52–59. DOI: 10.1016/S0924 4247(00)00461 1.

[96] Wu F, Yao N. Advances in windowed gas cells for in situ TEM studies. Nano Energy 2015; 13:735–756. DOI: 10.1016/j.nanoen.2015.03.015.

[97] Vendelbo SB, Elkjær CF, Falsig H, Puspitasari I, Dona P, Mele L, Morana B, Nelissen BJ, van Rijn R, Creemer JF, Kooyman PJ, Helveg S. Visualization of oscillatory behaviour of Pt nanoparticles catalysing CO oxidation. Nat. Mater. 2014; 13:884–890. DOI: 10.1038/nmat4033.

[98] Creemer JF, Helveg S, Kooyman PJ, Molenbroek AM, Zandbergen HW, Sarro PM. A MEMS reactor for atomic scale microscopy of nanomaterials under industrially relevant conditions. J. Microelectromech. Syst. 2010; 19:254–264. DOI: 10.1109/JMEMS.2010.2041190.

[99] Yaguchi T, Suzuki M, Watabe A, Nagakubo Y, Ueda K, Kamino T. Development of a high temperature–atmospheric pressure environmental cell for high resolution TEM. J. Electr. Microsc. 2011; 6:217–225. DOI: 10.1093/jmicro/dfr011.

[100] Sun L, Wook Noh K, Wen JG, Dillon SJ. In situ transmission electron microscopy observation of silver oxidation in ionized/atomic gas. Langmuir 2011; 27:14201–14206. DOI: 10.1021/la202949c.

[101] Zhang S, Chen C, Cargnello M, Fornasiero P, Gorte RJ, Graham GW, Pan X. Dynamic structural evolution of supported palladium ceria core shell catalysts revealed by in situ electron microscopy. Nat. Commun. 2015; 6. DOI: 10.1038/ncomms8778.

[102] Hell SW. Far field optical nanoscopy. Science. 2007; 316(5828):1153–1158. DOI:10.1126/science.1137395.

[103] Abrams IM, McBain JW. A closed cell for electron microscopy. Science. 1944; 100(2595): 273–274. DOI: 10.1126/science.100.2595.273.

[104] Abrams IM, McBain JW. A closed cell for electron microscopy. Appl. Phys. 1944; 15:607–609. DOI: 10.1063/1.1707475.

[105] Fullam EF. A closed wet cell for the Electron Microscope. Rev. Sci. Instrum. 1972; 43:245–247. DOI: 10.1063/1.1685603.

[106] Gai PL. Development of wet environmental TEM (wet E TEM) for in situ studies of liquid catalyst reactions on the nanoscale. Microsc. Microanal. 2002; 8:21–28.

[107] de Jonge N, Peckys DB, Kremers GJ, Piston DW. Electron microscopy of whole cells in liquid with nanometer resolution. PNAS. 2009; 106:2159–2164. DOI: 10.1073/pnas.0809567106.

[108] de Jonge N, Ross FM. Electron microscopy of specimens in liquid. Nat. Nanotech. 2011; 6:695–704. DOI: 10.1038/nnano.2011.161.

[109] Dukes M, Peckys DB, de Jonge N. Correlative fluorescence microscopy and scanning transmission electron microscopy of quantum dot labeled proteins in whole cells in liquid. ACS Nano. 2010; 4(7):4110–4116. DOI: 10.1021/nn1010232.

[110] Pohlmann ES, Patel K, Guo S, Dukes MJ, Sheng Z, Kelly DF. Real time visualization of nanoparticles interacting with glioblastoma stem cells. Nano Lett. 2015; 15:2329–2335. DOI: 10.1021/nl504481k.

[111] Williamson MJ, Tromp RM, Vereecken PM, Hull R, Ross FM. Dynamic microscopy of nanoscale cluster growth at the solid–liquid interface. Nat. Mater. 2003; 2:532–536. DOI: 10.1038/nmat944.

[112] Radisic A, Vereecken PM, Hannon JB, Searson PC, Ross FM. Quantifying electrochemical nucleation and growth of nanoscale clusters using real time kinetic data. Nano Lett. 2006; 6:238–242. DOI: 10.1021/nl052175i.

[113] Zheng H, Smith RK, Jun Y, Kisielowski C, Dahmen U, Alivisatos AP. Observation of single colloidal platinum nanocrystal growth trajectories. Science. 2009; 324:1309–1312. DOI: 10.1126/science.1172104.

[114] Evans JE, Jungjohann KL, Browning ND, Arslan I. Controlled growth of nanoparticles from solution with in situ liquid transmission electron microscopy. Nano Lett. 2011; 11:2809–2813. DOI: 10.1021/nl201166k.

[115] Liu KL, Wu CC, Huang YJ, Peng HL, Chang HY, Chang P, Hsu L, Yew TR. Novel microchip for in situ TEM imaging of living organisms and bio reactions in aqueous conditions. Lab. Chip. 2008; 8:1915–1921. DOI: 10.1039/B804986F.

[116] Kim TH, Bae JH, Lee JW, Shin K, Lee JH, Kim MY, Yang CW. Temperature calibration of a specimen heating holder for transmission electron microscopy. Appl. Microsc. 2015; 45:95–100. DOI: 10.9729/AM.2015.45.2.95.

[117] Tanigaki T, Ito K, Nagakubo Y, Asakawa T, Kanemura T. An in situ heating TEM analysis method for an interface reaction. J. Electron Microsc. (Tokyo). 2009; 58:281–287. DOI: 10.1093/jmicro/dfp020.

Chapter 11

In Situ Transmission Electron Microscopy Studies in Gas/Liquid Environment

Fan Wu and Nan Yao

Additional information is available at the end of the chapter

http://dx.doi.org/10.5772/62551

Abstract

Conventional transmission electron microscopy (TEM) typically operates under high vacuum conditions. However, *in situ* investigation under real-world conditions other than vacuum, such as gaseous or liquidus environment, is essential to obtain practical information for materials including catalysts, fuel cells, biological molecules, lithium ion batteries, etc. Therefore, the ability to study gas/liquid–solid interactions with atomic resolution under ambient conditions in TEM promises new insights into the growth, properties, and functionality of nanomaterials. Different platforms have been developed for *in situ* TEM observations in ambient environment and can be classified into two categories: open-cell configuration and sealed gas/liquid cell configuration. The sealed cell technique has various advantages over the open-cell approach. This chapter serves as a review of windowed gas/liquid cells for *in situ* TEM observations.

Keywords: *In situ* TEM, sealed gas cell, sealed liquid cell, lithium ion battery, open-cell configuration

1. Introduction

Transmission electron microscopy (TEM) is one of the most powerful techniques to characterize structure and chemistry of solids at the atomic scale. The simultaneous acquisition of nanoscale chemical analysis, atomic resolution images, and diffraction patterns provides comprehensive information that other characterization tools cannot compete with. Conventional TEM typically operates under high vacuum conditions ~1.5 × 10⁻⁷ Torr [1]. However, *in situ* investigation under real-world conditions other than vacuum, such as gaseous or liquidus environment, is essential to obtain practical information [2] for materials including

catalysts, fuel cells, biological molecules, lithium ion batteries, etc. Therefore, the ability to study gas/liquid–solid interactions with atomic resolution under ambient conditions in TEM promises new insights into the growth, properties, and functionality of nanomaterials. *In situ* controlled-environment TEM (ETEM) [3, 4] enabling TEM study of specimens in ambient environment is necessary for various nanomaterial-based technologies, such as efficient energy conversion/use/ storage, transportation, food production, and environmental protection [5] etc.

So far, different platforms have been developed for *in situ* ETEM observations in ambient environment and can be classified into two categories: 1) platforms with an open-cell configuration, and 2) platforms with a sealed gas/liquid cell configuration. The sealed-type ETEM using a sealed cell has various advantages over the open-cell approach. First of all, the reaction volume and the specimen are confined by electron-transparent top and bottom "windows," allowing gas/liquid to be introduced and sealed within a tiny space, and separated from the other parts of the TEM column. The resulting electron path length is on the order of a few microns [6, 7], much thinner than the opened-type approach and allowing much better resolution to observe lattice images. This is especially good for gas cell because the acceptable reaction pressures within the gas cell can equal or exceed a full atmosphere [6, 8–10] while maintaining the ability to record atomic resolution images [8–11]. Furthermore, much more rapid thermal response than standard heating holders and more rapid stabilization of specimen drift can be realized by integrating miniaturized, low mass heating devices [12], or laser heating [1] into the sealed cell. Therefore, a better control of the reaction process and the imaging experiments is achieved. An additional advantage is that the sealed-cell approach only modifies a small device on the tip of a TEM sample holder, thus can be used in any normal TEM without modifications to any other parts of a TEM. The cost of performing ETEM studies using the sealed-cell approach is typically a tiny fraction of the cost of a dedicated ETEM using open-type approach, because the latter requires modifications to the whole column. Thereby the sealed-cell approach allows *in situ* ETEM studies to be easily extended to many laboratories in the field. Last but not least, the sealed-cell platforms enable *in situ* ETEM characterization with the introduction of any types of volatile carbon-based electrolytes, which is impossible for open-type approach due to the high vacuum requirement inside TEM chamber.

Due to the various advantages over the open-type approach, sealed-cell approach has become the dominant way to perform ETEM studies under ambient conditions. A fast-growing number of research groups worldwide are conducting researches using this technology. This chapter discusses *in situ* ETEM studies in ambient environment by using sealed gas/liquid cells. Different designs and applications of the sealed cells for *in situ* TEM observations are summarized. Future research directions of the sealed gas/liquid cells are demonstrated for the benign development of this field.

2. Sealed gas cells

TEM is one of the most powerful techniques to characterize structure and chemistry of solids at the atomic scale. The simultaneous acquisition of nanoscale chemical analysis, atomic resolution images, and diffraction patterns provides comprehensive information that other characterization tools cannot compete with. However, information about the structural and chemical changes under ambient conditions, especially under gaseous environment, is usually not available since conventional TEM operates under high vacuum conditions ~1.5 × 10^{-7} Torr [1]. For materials such as catalysts, fuel cells, and biological molecules, *in situ* investigation under real-world conditions other than vacuum is essential to obtain practical information [2]. Therefore, the ability to study gas-solid interactions with atomic resolution at ambient pressures in TEM promises new insights into the growth, properties, and functionality of nanomaterials. Significant improvements in scanning/TEM (S/TEM) technologies containing a gaseous environment have enabled now the atomic scale study during gas-solid interactions [5] with energy resolution in the sub-eV range, and sensitivity to detect single atoms [13]. *In situ* controlled-ETEM [3, 4] enabling TEM study of specimens in ambient environment is necessary for various nanomaterial-based technologies, such as efficient energy conversion/use/storage, transportation, food production, and environmental protection [5], etc.

The original designs for ETEM observations under gaseous conditions have been around for over 70 years [14] and are made available by two main kinds of methods [15]: one is the opened type, which confines the gas near the sample by means of pressure-limiting apertures and maintain the vacuum in the remaining column by a differential pumping scheme [16–19] (e.g. in 1991, Nan Yao et. al. [20] used two pole pieces to confine the gas near the sample region, for studying supported metal catalysts during catalytic process in a TEM column); the other is the sealed type, which uses a sealed gas cell [4, 7, 8, 21–24] to enclose the sample and the high-pressure gas within a tiny space. For the opened type, the pressure-limiting apertures with small holes are positioned in the objective lens in close proximity to the sample, and the differential pumping system is equipped to avoid diffusion of the gas molecules from the chamber toward other parts of TEM, especially the electron gun. Any type of specimen holder can be accepted by the opened-type ETEM. However, differential pumping ETEM has many obvious disadvantages [9], such as a long time needed to ramp up to and down from a selected temperature, difficulty of stabilizing specimen drift due to the large power consumption and heating effects of the heating unit, huge cost needed to modify the TEM column, and the long gas path (on the order of ~1 cm [2, 6, 25]) through which the electron beam must pass that limits reaction pressures to a level of about 15–20 Torr [1].

On the other hand, the sealed-type ETEM using a sealed gas cell has various advantages over the differential pumping approach. First of all, the reaction volume and the specimen are confined by electron-transparent top and bottom "windows," allowing a gas to be introduced and sealed within a tiny space, and separated from the other parts of the TEM column. The resulting gas path length is on the order of a few microns [6, 7], much thinner than the opened-type approach and allowing much better resolution to observe lattice images. Consequently, the acceptable reaction pressures within the gas cell can equal or exceed a full atmosphere [6, 8–10] while maintaining the ability to record atomic resolution images [8–11]. Furthermore, much more rapid thermal response than standard heating holders and more rapid stabilization of specimen drift are realized by integrating miniaturized, low mass heating devices [12] or laser heating [1] into the sealed gas cell. Therefore, a better control of the reaction process and the imaging experiments is achieved. An additional advantage is that the sealed-gas-cell approach only modifies a small device on the tip of a TEM sample holder, thus can be used in any normal TEM without modifications to any other parts of a TEM. The cost of performing ETEM studies using the gas-cell approach is typically a tiny fraction of the cost of a dedicated ETEM using differential-pumping approach, because the latter requires modifications to the whole column. Thereby, the sealed-gas-cell approach allows *in situ* ETEM studies to be easily extended to many laboratories in the field. Due to the various advantages, sealed-gas-cell approach has become the dominant way to perform ETEM studies under gaseous environment. A fast-growing number of research groups worldwide are conducting researches using this technology.

Advances in 0D [26], 1D [27], and 2D [28–43] material fabrication technologies have enabled various forms of nanoscale materials, which increased the needs of *in situ* ETEM studies through the closed-type approach, i.e. sealed gas cells. Sealed gas cells enabled *in situ* TEM observations, thus allowing the evaluation of the effect of external stimuli including mechanical, electrical, and magnetic force on nanomaterials. Some advantages of *in situ* TEM observations with sealed gas cells are listed as follows [44]:

1.   Concurrent observations of structural, morphological, and chemical changes in ambient atmosphere are enabled.

2.   The same area is observed during the whole reaction process in ambient atmosphere, when sample is subjected to external stimuli.

3.   Intermediate steps during reactions in the ambient atmosphere can be identified.

4.   Both thermodynamic and kinetic data leading to nanomaterials synthesis or functioning in ambient atmosphere can be obtained.

5.   Considerable time saving as the synthesis and characterization are performed concurrently in ambient atmosphere.

The sealed gas cells have been applied in a variety of research fields and topics, including dynamic observation of catalytic reactions [2, 8, 21, 23–45], oxidation and reduction of metals [22], interaction between materials and ionized gas [46], de/hydrogenation processes [47], biological studies [4, 7], etc. Some groups just used their newly developed sealed cells to demonstrate their properties and improved technical limits for *in situ* TEM observations [1, 9, 10, 48, 49]. These applications are discussed in details as follows.

2.1. (De)hydrogenation processes

Hydrogen storage materials are needed for hydrogen fuel, particularly in the automotive industry. To enhance the kinetics and modify the thermodynamics of hydrogenation, nanostructured hydrogen storage materials are needed and study of the *in situ* hydrogenation/dehydrogenation process on the atomic scale is essential [47]. Using the MEMS-type sealed gas cell, Tadahiro Yokosawa et al. [47] observed the hydrogenation and dehydrogenation of Pd with a very consistent precision and a nanometer resolution, allowing a distinction between hydrogenation behaviors of individual grains. The electron beam was found to have no disturbing influence on the determination of the (de)hydrogenation temperatures in the case of Pd, under normal working conditions and at pressures of 750 and 2400 Torr. The relationship between (de)hydrogenation and pressure fitted well with bulk experiments in which the pressure was varied. Fast determination of the hydrogenation and dehydrogenation temperatures was allowed by realizing a very fast change in temperature.

2.2. Interactions between materials and gases

In 1976, Hiroshi Fujita et. al. [50] designed and used a sealed gas cell for a 3MV-class electron microscope to observe the reaction between H_2 gas and iron, as shown in **Figure 1(a)**. After electron irradiation damage in vacuum, the secondary defects were preferentially formed around the dislocations, which were linear structures in **Figure 1(a)**. Fern-leaf-like structures were formed around individual dislocation lines when the iron foil was exposed to wet H_2 gas of ~1200 Torr for ~30 min during electron irradiation of 2×10^{19} e/sec.cm^2 in intensity, as seen in micrograph (a′), which is an enlargement of a framed part in **Figure 1(a)**. These fern-leaf-like structures were quite different from those in **Figure 1(b)**, therefore they might be some sort of Fe hydrides that were closely related to the hydrogen embrittlement of iron.

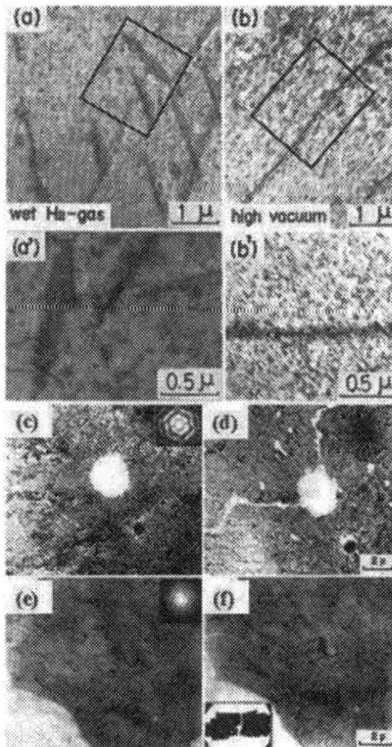

Figure 1. Reaction between wet H_2 gas and an iron foil. Micrographs (a) and (b) were taken after a heavy electron irradiation in wet H_2 gas of about 1200 Torr and a vacuum of 1×10^{-6} Torr, respectively. Micrographs (a′) and (b′) are enlargement of framed parts in (a) and (b), respectively. (c)–(d): Evaporation of 18/8 type stainless steel at 650°C in a vacuum of 1×10^{-6} Torr. (e)–(f): Evaporation of 18/8 type stainless steel at higher than 700°C in Ar-10 vol % H_2 gas of 760 Torr [50].

2.3. Suppression of specimen evaporation

The same sealed gas cell by Hiroshi Fujita et al. [50] was also used for suppression of the evaporation of specimen at high temperatures, thus decreasing the damage made by evaporation when metals and alloys were annealed at considerably high temperatures in vacuum. **Figure 1(c)–(f)** show the suppression of evaporation of 18/8 type stainless steel during annealing. The specimen (**Figure 1(c)**) was partly evaporated in a vacuum of 1×10^{-6} Torr by heating at 650°C, as seen in **Figure 1(d)**. In contrast, the evaporation of specimen was remarkably suppressed in a mixed gas of 760 Torr consisting of commercially pure Ar gas and 10 volume% H_2 gas even when the specimen was heated at high temperatures (more than 700°C). Microstructures in the specimen could be seen clearly even when a gas layer was as thick as ~100 μm, as shown in **Figure 1(e)** and **(f)**.

2.4. Oxidation and reduction of metals

Making use of the heating element and the enclosed gases, *in situ* observations of oxidation and reduction processes can be performed with sealed gas cells. **Figure 2** shows the oxidation process of a Cu thin film in a sealed gas cell designed by M. Komatsu et al. [22] in 2005. Initially, a small amount of Cu oxide formed during evaporation, as shown in the bright field image (**Figure 2(a)**) and the corresponding selected area electron diffraction pattern (**Figure 2(a')**), respectively. Oxygen was then introduced into the cell to 9.75 Torr and the specimen was gradually heated to 470K. Cu oxide was found to preferentially nucleate on the film surface (**Figure 2(b)** and **(b')**). As the temperature increased, Cu was oxidized to very fine oxide particles, as shown in **Figure 2(c)** and **(c')**. After the specimen was heated to 670K, all particles changed to CuO (**Figure 2(d)** and **(d')**). The CuO grains grew larger when the specimen temperature reached 770K (**Figure 2(e)** and **(e')**). The reduction of CuO was also observed *in situ*, as shown in **Figure 2**. The film was gradually reheated in 9.75 Torr of H_2. As the specimen temperature was further increased, CuO was completely reduced to Cu at 670 K (**Figure 2(h)** and **(h')**).

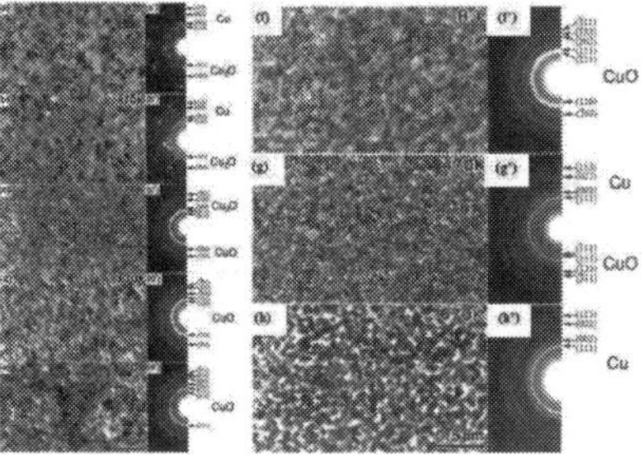

Figure 2. Successive stages of the oxide growth on a 100-nm-thick Cu thin film between room temperature and ~770K under 1.3×10^3 Pa of O_2. (a)–(e): Bright field images. (a')–(e'): The corresponding SAEDs. Successive stages of the reduction of CuO between room temperature and ~670K under 1.3×10^3 Pa of H_2. (f)–(h): Bright field images. (f')–(h'): The corresponding SAEDs [22].

2.5. *In situ* growth of nanostructures

In situ observation of the growth process of CuO whiskers was carried out in the same sealed gas cell by M. Komatsu et al. [22]. A series of electron micrographs show the successive stages of growth of Cu oxide whiskers in 30 Torr of O_2 (**Figure 3**). Initially, a non-uniform oxide film formed on the Cu surface, resulting to a jagged edge (**Figure 3(a)** and **(b)**). After 40s, the oxide layer stabilized with a smoother edge (**Figure 3(c)**). Then whiskers started to grow on the oxide layer gradually (**Figure 3(d)–(f)**).

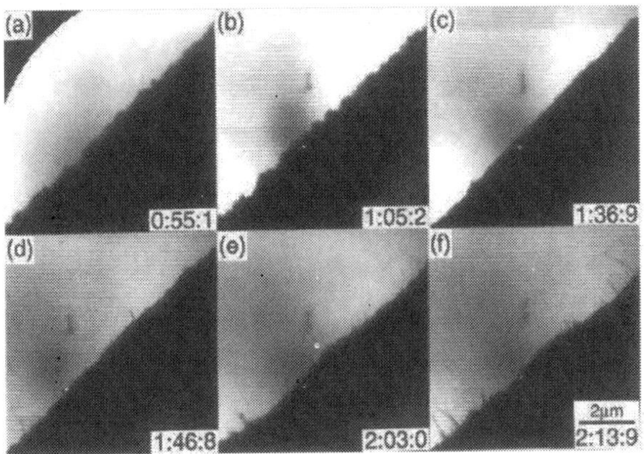

Figure 3. Successive stages of growth of Cu oxide whiskers in 30 Torr of O_2 [22].

2.6. Reactions with atomic/ionized gases

The earliest report of interactions with ionized/atomic gas induced by the electron beam of TEM was made by Li Sun et al. [46] in 2011. The enhanced electron flux could increase the concentrations of both reducing electrons and oxygen ions. Below a certain threshold, oxidization dominated the system response and resulted in accelerated interaction between silver and oxygen ions. The average size of silver grains continued to decrease, as shown in **Figure 4(c)–(f)**. At current densities of 0.44 A cm^{-2}, the silver grains were rod-shaped (**Figure 4(c), (d)**). At current densities greater than 0.65 A cm^{-2}, the silver grains reverted back to a more compact angular morphology (**Figure 4(f)**). Due to the increased oxygen fugacity associated with higher concentrations of ionized and atomic oxygen, all Ag_2O phase was further oxidized to AgO at current densities greater than 0.65 A cm^{-2}. Above 0.75 A cm^{-2}, a significant portion of noncrystalline phase existed (**Figure 4(g)**). Once the electron current density increased beyond 0.77 A cm^{-2}, new grains nucleated (**Figure 4(h)**) out of the vapor phase, exhibiting a twin structure. The reaction between the nanoparticles (NPs) and gas produced a concentration gradient around the particles that was observed as a bright ring around each silver grain. The silver oxide depletion width in the gas phase indicated a strong chemical interaction between the solid and vapor phases. The AgO vapor phase often aligned itself into 2D sheets perpendicular to the beam that subsequently became unstable precipitate clusters of new silver grains or swept through a region randomly (**Figure 4(h)**). The competition between oxidation and electron-beam-induced reduction also provided excess heat. For significantly high fluxes of ionized oxygen, a thermal effect could induce local vaporization at the surface, and sequentially an *in situ* nanoscale reaction ion sputtering. This investigation revealed a variety of microstructural processes associated with the oxidation of Ag by atomic and ionized gas species. The electron beam was demonstrated to be an important source of both oxidation and reduction. The sealed cell approach provided an opportunity to make early observations of real-time nanoscale dynamics associated with oxidation in ionized and atomic gas, the movement of a partial pressure of a gas phase, and interactions between the condensed and vapor phases of a material. The results provided new insights into manipulating nano-structure and chemistry through ionized gas treatment and offered unique access to simulate reactions with atomic and ionized gas.

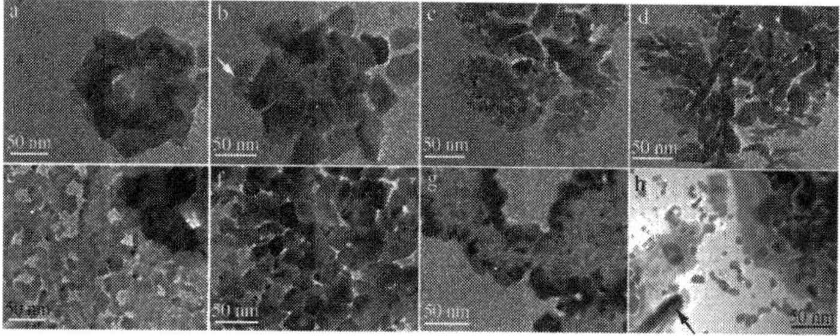

Figure 4. Microstructure of the observation area in an air-filled cell after exposure of (a) 0 s with 0.18 A cm^{-2} current density; (b) 20 s with 0.18 A cm^{-2} current density (the arrow marks a new grain); (c) 0 s with 0.44 A cm^{-2} current density; (d) 20 s with 0.44 A cm^{-2} current density; (e) 0 s with 0.66 A cm^{-2} current density; (f) 20 s with 0.66 A cm^{-2} current density; (g) 20 s with 0.72 A cm^{-2} current density; and (h) 20s with 0.8 A cm^{-2} current density (the arrow marks the vertical alignment of the AgO vapor phase) [46].

2.7. Dynamic observation of catalysts and catalytic reactions

One of the earliest attempts to observe a catalyst in a sealed gas cell was reported by Parkinson et al. [21] in 1989. Using a narrow-gap, sealed gas cell and a 400-kV TEM, images of the crystal lattice of ceria (0.31 nm) were recorded under flowing nitrogen gas at 20 Torr. Structural information of chemical significance became discernible at ~0.3 nm, which offered real hope of carrying out fundamental dynamic studies of the activation, reaction, and passivation of gas/solid systems at close to the atomic level.

Seventeen years later, the atomic-scale *in situ* observations of catalysts were performed by S. Giorgio et al. [23], during a chemical reaction. For the first time, Au and Pd clusters supported on TiO$_2$ and amorphous carbon were observed with a sealed gas cell with the resolution of (111) lattice fringes. Initially, an Au cluster in vacuum was strongly contaminated, but the contamination disappeared while the faceting and the crystalline lattice were visible in the cluster after circulation of H_2 at a pressure of 3 Torr at room temperature. The cluster was completely faceted after annealing until 350°C in the same reducing atmosphere, then cooling down to room temperature.

The resolution of *in situ* observation of catalysts was improved by a novel MEMS-type nanoreactor in 2008 [8]. More importantly, the nanoreactor facilitated the direct observation of Cu nanocrystal growth and mobility on a sub-second time scale at a higher temperature (500°C) and higher gas pressure (900 Torr of H_2). The *in situ* TEM images showed atomic lattice fringes in the Cu nanocrystals with spacing of 0.18 nm, attesting the spatial resolution limit of the system. The system of Cu nanocrystals on a ZnO support is commonly used as catalyst for methanol synthesis and for conversion of hydrocarbons in fuel cells. Also, it is a prototype example of the industrially important group of 3d transition metal catalysts. The catalyst was heated in the H_2 atmosphere to the maximum operation temperature of 500°C. ZnO crystallites with diameters of 20–100 nm appeared in the precursor with facetted, compact shapes (**Figure 5(a)**). The CuO appeared as smaller patches of more irregular shapes at the edges of

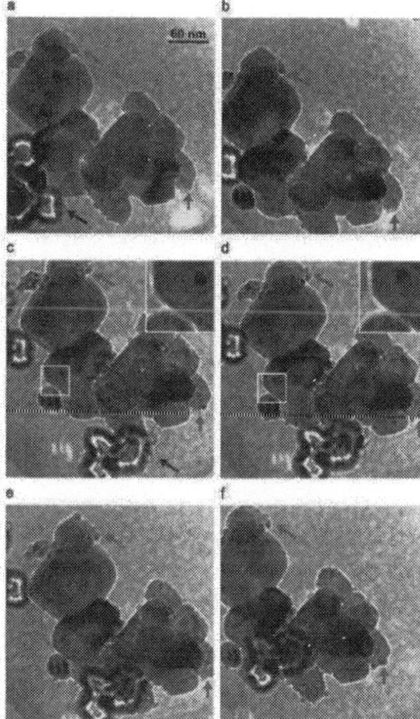

Figure 5. Image sequences of the Cu nanocrystal growth and mobility on ZnO. Nanocrystals (darker contrast) form from CuO precursors (blue arrows) during heating from room temperature to 500°C in 900 Torr H_2. After growth, nanocrystals can exhibit transient mobility (white square). Crystallites on the opposite window are seen out of focus (black arrows in (a) and (c)). The frames are recorded at (a) room temperature, (b) 260°C, (c) 330°C, (d) 365°C, (e) 410°C, and (f) 500°C. All frames are averaged over four consecutive images. The exposure time for each image is 0.145 s (color online) [8].

ZnO. As temperature increased to ~260°C, the CuO patches broke up into several particles with diameters of 5–10 nm (**Figure 5(b)–(e)**). The state of the nanocrystals was inferred from atomic-resolution TEM images during exposure to 900 Torr H_2 at 500°C. Atomic lattice fringes were clear in both the brighter ZnO support crystallites and the darker Cu nanocrystals (**Figure 6(a)**). Lattice fringes with spacings of 0.21 and 0.18 nm could be recorded in the nanocrystals [24], corresponding to (111) and (200) planes of Cu, identified by Fourier transform (**Figure 6(b)**) of the TEM image.

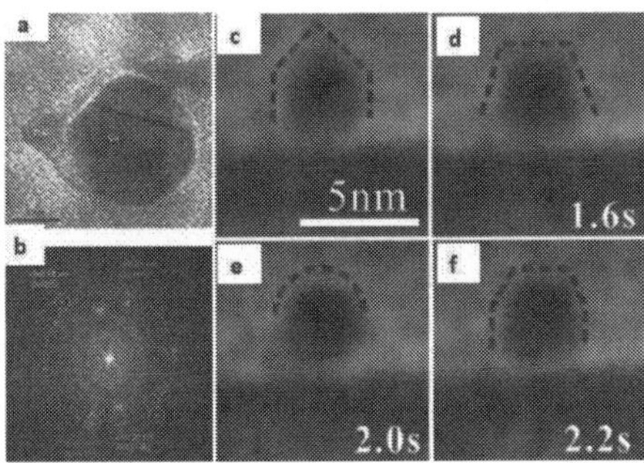

Figure 6. A representative HRTEM image of the Cu/ZnO catalyst during exposure to 900 Torr hydrogen at 500°C. (a) The image displays lattice fringes of a twinned Cu nanocrystal and of the ZnO support. (b) A Fourier transform of (a). The bright dots represent sets of lattice fringes. Their lattice spacing corresponds to the distance to the origin and reveals the crystallographic identity. The large, red circle corresponds a spacing of 0.21 nm. The smallest, resolved lattice spacing is 0.18 nm. (c)–(f) [2]: *In situ* TEM images of the nanoparticulate gold catalyst supported on TiO_2 recorded sequentially. The time shown in the lower right-hand corners of (d)–(f) correspond to intervals measured from the time at which (c) was recorded (color online) [8].

One year later, the MEMS-type sealed gas cell developed by Tadahiro Kawasaki et. al. [2] was applied for *in situ* TEM observations of a gold nanoparticulate catalyst supported on TiO_2. One percent CO in dry air was introduced to react with O_2 to form CO_2 on the catalyst surface. **Figure 6(c)–(f)** show the sequential morphologies of the gold NP. The shape of the gold particle changed markedly over a short period of time, such as the 0.4 s interval between **Figure 6(d)** and **(e)** and the 0.2 s between **(e)** and **(f)**. Various facets of the gold appeared in **Figure 6(c)**, **(d)**, and **(f)**. They sometimes disappeared and the gold particle formed a spherical shape in **Figure 6(e)**. However, the lattice fringes of the gold could not be observed due to electron scattering by the high-pressure gas.

The most recent *in situ* visualization of oscillatory behavior of Pt NPs catalyzing CO oxidation was reported by S. B. Vendelbo et al. [45] in 2014. TEM image series of the Pt NPs were acquired at windows both at the entrance and exit of the reaction zone, at a rate (1–2 frames per second) faster than the rate of the reaction oscillations, to directly visualize the NPs on this timescale. Near the reaction zone entrance, the NPs had a stationary and more spherical morphology during the oscillating reaction. In contrast, near the reaction zone exit, the Pt NPs switched between spherical and facetted morphology (**Figure 7(1)**). As the CO conversion increased rapidly, Pt NPs started a gradual transformation from the more spherical shape towards a more facetted shape. The fully facetted shape was reached within 3 s after the CO peak conversion (**Figure 7(1) III**). On decrease in the CO conversion, the NP transformed back to the more spherical shape (**Figure 7(1) IV**) and retained that shape until the CO conversion rose steeply again. Thus, the individual NPs near the exit from the reaction zone underwent oscillatory and reversible shape changes with a temporal frequency matching the oscillations in reaction power, indicating that the oscillatory CO conversion and the dynamic shape change of the Pt NPs were coupled. To address the mechanism governing the oscillatory reaction, the state of the Pt NPs was examined at the atomic scale (**Figure 7(2) (c)–(e)**). The spacing of crystalline lattice planes and the uniform contrast across the projected image of the NPs were consistent with metallic Pt. The combined high-resolution TEM and DFT analyses indicated that the Pt surfaces remained in the metallic state under the present conditions. Time-resolved series of high-resolution TEM images show that in the more spherical state, the Pt NPs were terminated by close-packed (111) planes, more open (110) planes and step sites (**Figure 7(2) (a), (c), (e)**), while for the more facetted state, the NPs were terminated by extended (111) planes as well as a reduced abundance of higher index terminations and steps (**Figure 7(2) (b), (d)**).

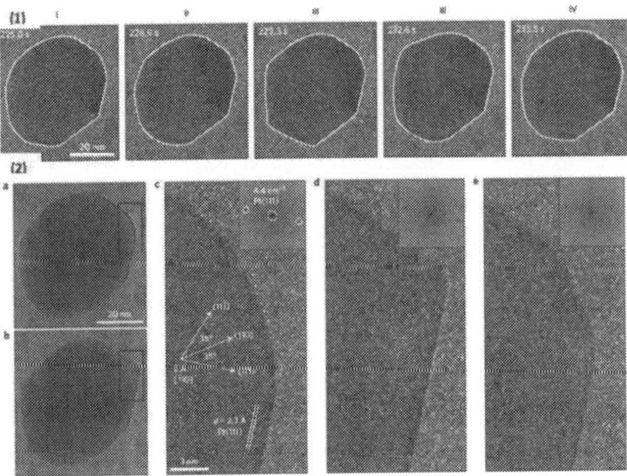

Figure 7. (1) Time-resolved TEM images of a Pt NP at the gas exit of the reaction zone. (2) Atomic-scale visualization of the dynamic refacetting of a Pt NP during the oscillatory CO oxidation. Time-resolved high-resolution TEM images of a Pt NP at the gas exit of the reaction zone. The gas entering the reaction zone is 1.0 bar of CO:O$_2$:He at 4.2%:21.0%: 74.8% and nanoreactor temperature is 727K. (a)–(e): The TEM images showing the more spherical shape (a, c, e) and the more facetted shape (b, d), during the oscillatory reaction. Fast Fourier transforms included as insets in (c)–(e) reveal a lattice spacing corresponding to the Pt(111) lattice planes. The orientation of the observed Pt(111) lattice fringes is consistent with the superimposed crystal lattice vectors and zone axis (color online) [45].

Apart from the homemade sealed gas cells above, commercial MEMS-type sealed gas cells have also been applied to *in situ* studies of catalysts. The membrane-type heating chip manufactured by Hummingbird Scientific (Lacey, WA, USA) provided a temperature controllable reaction platform for oxidation reactions of cobalt NPs with flowing oxygen (0.2 sccm), while ramping temperature from 150 to 250°C and 250 to 350°C at ~5°C/s [25]. **Fig-**

ure 8(a) shows the time-lapse images of three selected Co particles. The metallic cobalt core could shrink with a unidirectional retraction front (**Figure 8(aI)**) and a sweeping retraction front (**Figure 8(aII)**). In projection, the residual metallic puddle was faceted, which was likely shaped by the faceted hollow shell (**Figure 8(aIII)**). The quantification of the volume trajectory of the metallic core (**Figure 8(d)**) shows that the metallic core volume started to rapidly decrease when temperature reached 250°C. After the first volume-decreasing phase, volume shrinkage dwelled for a short period of time at the first plateau (① in **Figure 8(b)**). Then a second rapid decreasing phase initiated with a lower volume shrinkage rate (**Figure 8(b)**). The metallic core was finally eliminated, but at an even slower volume shrinking rate. The particle's oxide shell was in contact with other particles with upper and lower right boundaries open.

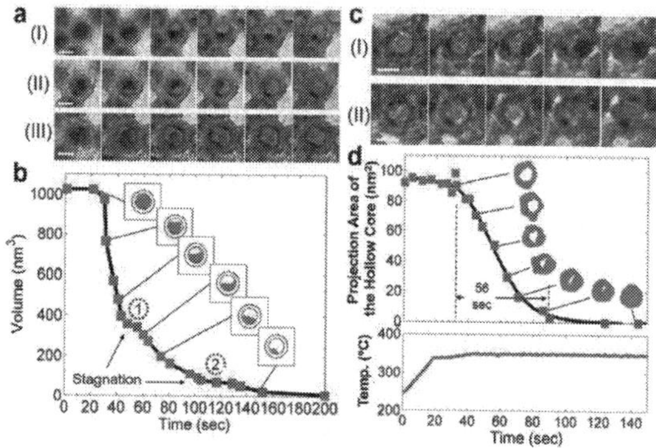

Figure 8. *In situ* heating of cobalt NPs in flowing oxygen. (a) Real-time reaction dynamics of the Kirkendall effect. (b) Metallic core volume trajectory of particle I in (a). ① and ② mark two diffusion stagnation plateaus. (c) Restructuring of the hollow oxide structure at 250–350°C in flowing oxygen. (d): Hollow core volume trajectory of particle I in (c). Scale bar is 10 nm [25] (color online).

2.8. Biological studies

The sealed gas cells encapsulated specimens in a thin gas layer, preventing specimens from destruction as in vacuum. Therefore, the sealed gas cells have also been widely applied into biological studies. This section will discuss the biological applications of the sealed gas cells for *in situ* TEM observation.

The first application of sealed gas cells in biology was reported by H. G. Heide in 1962 [4]. For organic specimens, it is necessary to prevent carbon removal from increasing to a rate higher than the rate of contamination, which would destruct the specimen. A rapid dehydration of the specimen can be prevented if unnecessary heating is avoided even at pressures of 100–200 Torr, which was proved by TEM pictures of small water droplets in air at 100 Torr. It was possible to prevent carbon removal in the specimen if H_2, He, N_2, or Ar instead of air was used at this pressure and the illuminated area was reduced to ~2 μm diameter with the double condenser.

In 2001, an *in situ* sealed gas cell was used to study the reduction of Cr (VI) by bacterium (*Shewanella oneidensis*) by T. L. Daulton et al. [7]. Bacteria from rinsed cultures were placed directly in the gas cell and examined under 97.5 Torr pressure of air saturated with water vapor, showing rod-shaped morphology typical of flagellated and non-spore-forming species (**Figure 9**). Cells remained plump/hydrated while the EPS retained moisture and appeared as a continuous capsule surrounding the cells. However, damage to the cells was observed within minutes of electron-beam exposure, arising from the primary destruction of weak Van der Waals biomolecular bonds. Direct *in situ* TEM imaging revealed two distinct populations of *S. oneidensis* in the cultures: bacteria exhibiting low image contrast (**Figure 9(a), (c)**) and bacteria encrusted/impregnated with electron-dense particles (**Figure 9(b), (d)**). Further examination of the encrusted bacteria showed that their gram-negative, cell envelope was electron dense (**Figure 9(d)**) and appeared darkest along the perimeter where the electron path length was the greatest. The cell envelopes of non-encrusted cells produced very low image contrast as compared to encrusted bacteria. The increase in contrast indicated that the cell envelope was saturated with absorbed elements of heavy mass, such as Cr. The binding of heavy elements in the cell envelope was associated with Cr reduction.

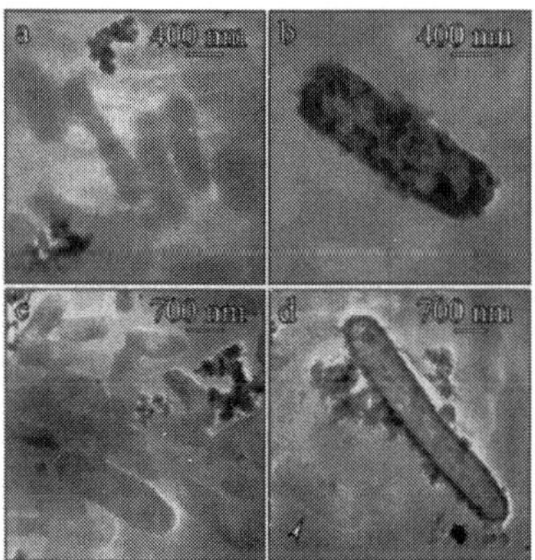

Figure 9. *S. oneidensis* imaged in the environmental cell at 100 Torr: bacteria exhibiting low contrast in bright field EC-TEM imaging (a, c), and bacteria encrusted/impregnated with electron dense particulates (b, d). The arrowhead in panel (d) points to a low contrast bacterium in the same field of view as a bacterium with electron dense particulates, illustrating the dramatic contrast difference. The low-contrast, diffuse background, best seen in panel (a), represents the extracellular polymeric substances that surround the cells [7].

2.9. *In situ* investigations on cladding materials

Cladding is the outer layer of the fuel rods, preventing radioactive fission fragments from escaping the fuel into the coolant and contaminating it. Exposures to irradiation, temperature changes, and stresses may induce microstructural changes, and ultimately result in failure of the cladding. It is thus essential to use *in situ* TEM to observe microstructural changes at the nanoscale dynamically, for predicting the performance of cladding in-service and during storage, understanding the dominant processes related to these changes and their consequences. In 2012, a sealed gas cell developed by K. Hattar et al. [51] was used to investigate the radiation tolerance of potential Generation IV cladding materials and the degradation mechanisms in Zr-based claddings of importance for dry storage. Examination of a Zircaloy foil enclosed by top and bottom windows (**Figure 10(a)**) showed deterioration of resolution due to expected additional scattering of electrons by the 5-μm-thick air, after initial scattering by the foil. Despite loss in resolution, prominent features of the foil that were previously observed under vacuum still remained visible. After annealing at 300°C for over 15 min, negligible changes in the Zircaloy morphology occurred (**Figure 10(b)**). Following the 15-min annealing at 300°C, the temperature of the gas cell was raised to 600°C and a dramatic morphology change within the sample was observed almost instantaneously (**Figure 10(c–d)**).

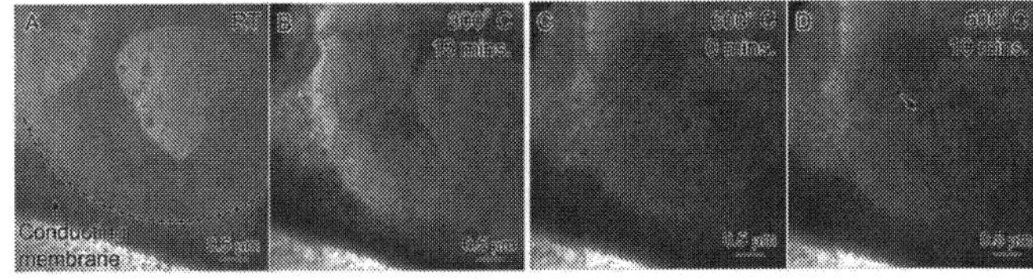

Figure 10. Images of Zircaloy lamella at nominally atmospheric pressure. (a) Initial structure. (b) 15 min at 300°C, (c) 600°C, and (d)600°C after 10 min [51].

The same device was applied to study hydride formation in Zirlo™ cladding material [12] recently. The formation of hydrides, their dissolution, and re-precipitation, is particularly important for long-term dry storage of currently used fuel assemblies, as the size, shape, and orientation of hydrides play a strong role in the mechanical properties of spent claddings. During *in situ* observation, hydrogen was introduced and maintained at a pressure of 330 Torr. The temperature was then increased at a rate of 1°C/s to approximately 400°C, and held for 90 min. **Figure 11(a)** and **(b)** show a comparison between the microstructures of the Zirlo™ prior to annealing and during later stages of annealing. The disappearance of microstructural features (arrow 1) and the formation of a new grain (arrow 2) are evident. The region around the new grain (**Figure 11(c)**) and the analysis of diffraction information (**Figure 11(d)**) indicated the formation of either ε-ZrHx (x > 1.8) or γ-ZrH. These results show that *in situ* environmental heating TEM can be applied to study this mechanism at the nanoscale in order to verify predictive material models.

3. Sealed liquid cells

Conventional TEM is not compatible with studies of electrochemical energy storage processes, but the development of TEM holders and sealed liquid cell (SLC) platforms encapsulating thin liquid layers promise *in situ* imaging and spectroscopy of electrochemical processes [52, 53] (e.g. electrodeposition [54] and dendrite growth [55]) on the nanoscale [56–58], by incorporating electrodes [54, 59] in a liquid environment. One major application of SLCs for *in situ* TEM observation is for lithium ion battery(LIB) research. Unlike ex situ studies, which involve unexpected reactions due to the removal of the particles from their native and reactive environment [60], *in situ* TEM electrochemical characterization will mimic the true environment in a commercial LIB cell. The *in situ* liquid TEM has allowed quantitative analysis of processes (e.g. NP growth from solution [61–63]), and direct observation of beam-sensitive systems (including macromolecular complexes [64, 65], soft materials [66, 67]) and of processes that span from the electrochemical deposition of metals [54, 55], to growth of different nanostructures [61, 62, 68–71]. Now it gains growing attention for LIB research.

Apart from nanostructured anodes/cathodes, the development of platforms enabling *in situ* TEM electrochemical characterization is also required by various other aspects of LIB research. For example, one of the most well-known reactions at the electrode/electrolyte interface is the formation of the solid-electrolyte interphase (SEI), which is a reaction product of mixed composition formed on high-voltage anodes (e.g. Li metal or graphite-lithium intercalation compounds) or cathodes, by electrochemical reduction or oxidation of the electrolyte [72], respectively. The study of the SEI layer requires the use of commonly used LIB electrolytes (volatile carbonate-based solution), the ability to monitor the changes in SEI layer with cycling, time or temperature, and probes having sufficient spatial resolution to detect a reaction product layer of a few nm thick. All of these requirements make ex situ characterization inappropriate for the study of SEI, since the SEI layer is highly sensitive to moisture, air, and other kinds of contaminations [73]. The importance and critical need to develop platforms enabling *in situ* TEM electrochemical characterization of LIBs are thus obvious. Furthermore, the detailed understanding in dendritic growth of lithium metal on the electrode also requires such *in situ* TEM platforms, because the dendritic growth of lithium metal on the electrode can lead to short-circuit and thus battery failure [73].

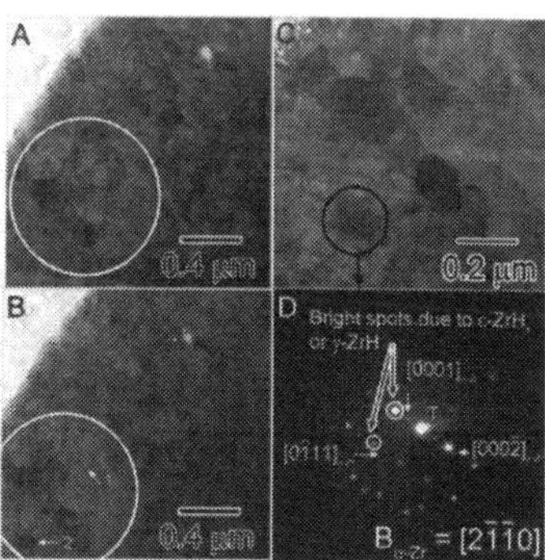

Figure 11. (a) and (b) Overview of images before the start of *in situ* experiment, and after 72 min at 400°C in H_2 atmosphere, (c) zoomed-in image of the region of interest around arrow 2 in (b), and (d) shows the superimposition of two diffraction patterns obtained from the circled dark grain shown in (c), one of the parent α-Zr phase (diffraction spots of weaker intensity) and the other is consistent with a face-centered tetragonal hydride phase [12].

Typical commercially available Li-ion batteries usually use carbonate-based liquids as electrolytes, such as diethyl carbonate (DEC), dimethyl carbonate (DMC) mixed with ethylene carbonate (EC), etc. To enable *in situ* TEM characterization of electrochemical reactions in real lithium ion batteries, sealing those volatile liquids inside a sufficiently narrow channel for electron transmission is a wise option. The SLC platforms enable *in situ* electrochemical characterization with any types of electrolytes with a sealed-cell configuration, thus promoting the potential use of volatile carbon-based electrolytes for LIB research. One of the first TEM SCLs was created by F.M. Ross et al. [54, 74] to study Cu electro-deposition during TEM imaging. This SLC platform sealed the aqueous electrolyte by assembling two silicon chips with thin silicon nitride membranes in a face-to-face configuration. This flip-chip approach allows imaging chemical reactions in liquids with high spatial resolution [54, 57, 58, 62, 75] with different membranes of silicon nitride, silicon dioxide, or polymer, such that it has been adopted in various studies, including cell imaging [76–78] and NP synthesis [58] in solutions. For example, electrochemical deposition of polycrystalline Au [75], anisotropic electrodeposition of nickel nanograins [79], and electrochemical growth of single crystal lead dendrites through nucleation, aggregation, alignment, and attachment of randomly oriented small grains [80] were imaged by using electrochemical SLCs.

To date, large progress has been made on fabrication and testing of the design features (including sealing, assembly, alignment, etc.) of SLCs [81], which opens the opportunity to address key questions on the electrode-electrolyte interfaces in native liquid environments, e.g. Kyong Wook Noh et al. [82] captured cyclic formation and dissolution of solid-electrolyte interphase at a Sn electrode in commercial liquid. Due to the reduced length scale of the electrodes, limited electrolyte volume, low current measurements [83], high vapor pressure of commercial electrolytes, and low contrast of lithium during TEM imaging through the membrane window, the application of SLCs as *in situ* electrochemical TEM cells for LIB research is still a great challenge and very limited. The application of sealed liquid cells for *in situ* TEM electrochemical characterization of lithium ion batteries are discussed as follows.

To track the lithiation process and elucidate the lithiation mechanism, the following techniques can be used: morphological imaging, electron diffraction [84, 85], energy-dispersive X-ray (EDX), and electron energy loss spectroscopy (EELS). Morphological imaging cannot give chemical information [60], and diffraction spots are quickly obscured in thicker liquid films. Moreover, lithium scatters electrons so weakly that elastic imaging is challenging and EDX signal for lithium has a much too low energy for detection. EELS offers chemical fingerprints (core-loss EELS) and electronic structure information (valence EELS), but EELS of Li in liquid will be degraded by multiple scattering events in thick liquids [86] and the lithium K-edge (~54 eV) cannot be distinguished from many transition metal (e.g. Fe [87]) edges and the superimposed bulk plasmon of the thick liquid films, which makes the majority of Li-edge spectroscopy to be ex situ [88] and core-loss EELS of the lithium practically impossible in a liquid cell. Therefore, valence EELS is the best way to interrogate electronic structure and detect the state of lithiation of battery electrodes in SLCs, because valence EELS provides strong signals due to large scattering cross-sections and low background from the liquid (the electronic structure shift usually occurs at energies below ~6–7 eV where the electrolyte is transparent and stable

[60]). The spatial resolution of valence EELS is ultimately limited by the delocalization of the low-energy excitations [89], multiple scattering in the liquid environment, and low-dose imaging conditions, to be on the nanometer scale. Megan E. Holtz et al. [60] successfully observed the lithiation state by valence energy-filtered TEM (EFTEM) in thicker liquid layers than commonly allowed by core-level spectroscopy [86], probing the low-energy regime at ~1–10 eV. They employed ab initio theory to calculate optical gaps of the relevant solvated species, taking solution effects into account with a hybrid function [90] including a nonlinear description of the polarization response of the surrounding liquid. By combining electrochemistry in the TEM with valence spectroscopic imaging and theory, they identified the lithiation state of both the electrode and electrolyte during *in situ* operation. Their work demonstrated the unique ability of an *in situ* TEM SLC to observe the Li de/insertion dynamics and degradation of $LiFePO_4$ cathode in real time. The real-time evolution of individual grains and NPs of $LiFePO_4$ (cathode) [60] was studied in the native environment of a battery in a liquid cell TEM (as mentioned above). Particles (lithium-rich/poor) were observed to delithiate one at a time in a mosaic fashion, with different delithiation mechanisms in neighboring particles. Core-shell structures and anisotropic growth in different particles within the same agglomerate on the electrode were directly imaged along with the phase transformations, thanks to the *in situ* SLC design. Although they used Li_2SO_4 aqueous electrolyte due to its high abundance, less viscosity, low weight, and nontoxicity [91], volatile electrolytes could be used in their SLCs. They imaged at 5 eV with a 5 eV wide energy window [60] to track the state of lithiation (**Figure 12**). There were clear differences between the charged (**Figure 12**, right) and the discharged state (**Figure 12**, left) in both the particles and the solution in the 5 eV spectroscopic

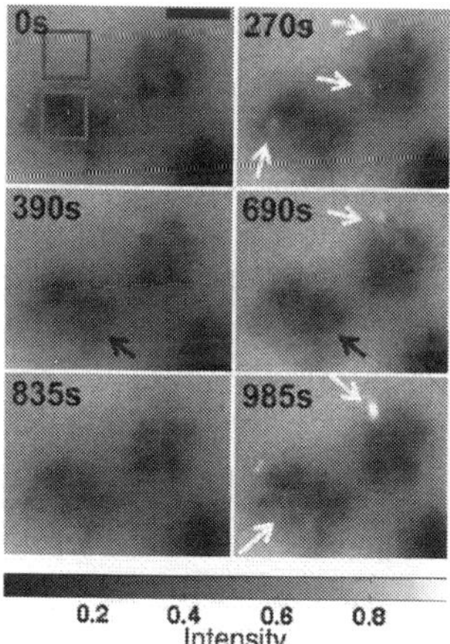

Figure 12. *In situ* charging and discharging of the cathode material $LiFePO_4$ in 0.5 M Li_2SO_4 aqueous electrolyte: the 5 eV spectroscopic EFTEM images of charging and discharging at indicated times. Scale bar is 400 nm. Bright regions are delithiated $FePO_4$ and dark regions are $LiFePO_4$. There are more bright regions of $FePO_4$ at the end of charge cycles and less during the discharges. White arrows point toward "bright" charged particles, and black arrows point toward "dark" discharged particles [60].

images. Particles showed more bright regions (corresponding to delithiated FePO$_4$) in the charged state. The cluster of particles was brighter in the charged image as marked by black arrows, especially around the edges of the cluster. The brightest particles may correspond to completely delithiated FePO$_4$, whereas the overall slight increase in intensity in the particles may indicate partially delithiated particles. On discharge, these bright regions of FePO$_4$ disappeared, transitioning back to LiFePO$_4$.

The electrochemical lithiation of Au electrode, dendritic growth of crystalline lithium, and the subsequent stripping of lithium and thinning of Li-Au layer under the applied cyclic voltammetry was observed by Zeng et al. [73], using commercial LiPF$_6$/EC/DEC electrolyte, which proved that real electrolyte of LIBS can be used in electrochemical SLCs [73]. **Figure 13 (A)–(J)** shows the sequential images representing the early stage of electrolyte decomposition, lithiation of gold electrode, and the subsequent growth and dissolution of lithium dendrites. **Figure 13 (k)** shows the corresponding applied electrical potential and measured electrical current from frame **(A)** to frame **(J)**. The thickness of Li-Au alloy did not change drastically at the later stage, as shown in **Figure 13(L)**. During stripping, the dissolution of plated lithium starts from the tip and the kink points as a reverse process of plating **(Figure 13(M))**. The formation of SEI layer on the other side of the electrode was also captured for better understanding of correlation between cyclic stability and the passivating film formed during the charge-discharge process in real LIBs. The drawback in their design was a lack of lithium metal source inside the SLC to supply the consumed lithium ions, such that the Li ion concentration in the electrolyte changed during the reaction. Adding a lithium metal source and an additional reference electrode into the SLC is necessary for direct comparison between the electrochemical processes inside a TEM column and that in real LIBs.

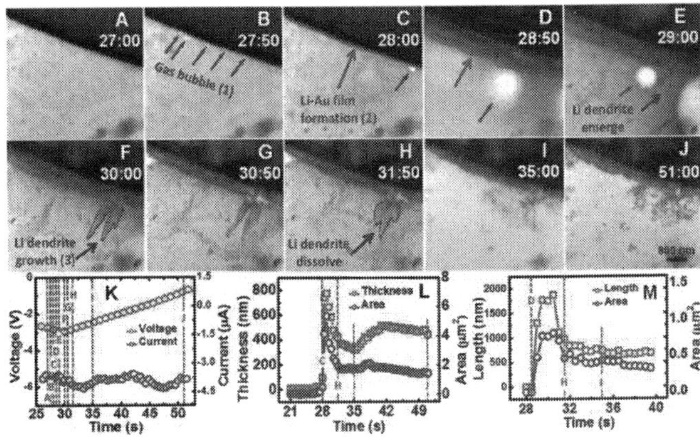

Figure 13. (A–J) Time evolution of the growth and dissolution of Li-Au alloy and lithium dendrite; (K) the corresponding applied electric potential and measured electric current from frame A to frame J; (L) plot of Li-Au layer thickness and area as a function of time; (M) dimension and area evolution of the lithium dendrite tip as a function of time during cyclic voltammetry in the voltage range of 0 to −3 V at scan rate of 0.1 V/s [73].

SLCs have also been used to study the stabilities of different electrolytes commonly used for Li-ion and Li-O$_2$ battery [92, 93]. Five different electrolytes [94], including LiAsF$_6$ salt dissolved in three different organic solvents: (1) 1,3-dioxolane (DOL), (2) DMC, (3) a mixture of DMC and EC and LiTf in dimethyl sulfoxide (DMSO), LiPF$_6$ in EC/DMC were studied. **Figure 14** shows six different time series of bright-field (BF) STEM images corresponding to the five electrolyte solutions and the EC/DMC solvent alone. To ensure that the observed lack of degradation products when imaging the LiTf:DMSO mixture was not a result of improper focus, the edge of the window was recorded as a reference. **Figure 14** shows that apart from LiTf in DMSO, all the other salt-containing solutions tested showed some evidence of degradation. It is worth to note that the degradations of the electrolytes were triggered by the imaging electrons (300 kV), instead of extra electrodes. This work shows that the electron-beam in the STEM can be used as an effective tool for evaluating stability and degradation in battery electrolytes by allowing direct visualization of the reductive decomposition of the electrolyte components, instead of postmortem analysis (chromatography) [95, 96]. This *in situ* approach can potentially be used for more rapid identification of next-generation electrolytes.

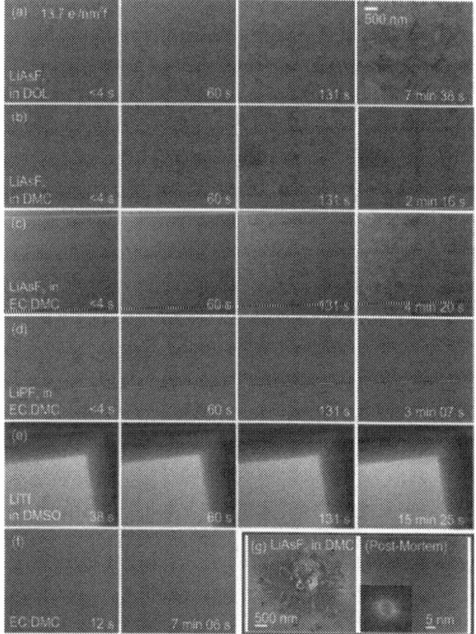

Figure 14. e⁻ beam-induced breakdown of different electrolytes upon irradiation. (a–e) Cropped BF STEM images showing the time evolution of five different electrolytes at serial exposure times. (f) Frames from a data set probing the stability of the solvent EC/DMC for the same dose conditions as above over 7 min of continuous irradiation. (g) TEM images of an irradiated area of the LiAsF6 in DMC mixture after separating and washing the Si chips for performing postmortem analysis. Low-magnification (left) and high-resolution TEM and consequent fast Fourier transform of the irradiated area shows the presence of LiF nanocrystals [94].

The lithiation and delithiation process of fully submerged electrodes is another important application of SLCs. For example, the de/lithiation process of Si NW electrodes during electrochemical testing was observed [88] by using *in situ* SLC platform and real electrolyte (as mentioned previously). The structural evolution of the Si NW upon lithiation is illustrated in **Figure 15(a)–(c)**. The pristine Cu-Si NW has an overall diameter of ~100 nm as revealed by the dark contrast in **Figure 15(a)**. The width of the Cu coating on the Si NW was measured to be ~80 nm. The lithiation of the Si nanowire immersed in the liquid electrolyte progressed in the core-shell fashion. The total diameter of the wire changed from 100 to 298 nm at 1658 s (**Figure 15(b)**) and to 391 nm at 2462 s (**Figure 15(c)**). The diameter as a function of lithiation time is plotted in **Figure 15(d)**. The increase of the diameter was quicker at the beginning of the lithiation and slowed down with the progression of the lithiation process. The lithiation behavior observed by the *in situ* SLC was also compared with that obtained based on the open-cell configuration in their study. For the case of SLC, the Si NW was fully immersed in the liquid electrolyte so that the insertion of lithium ions into Si was from all possible directions at the same time. The lithiation of the single nanowire proceeded in a core-shell mode with a uniform shell thickness along the axial direction of the whole nanowire, providing a global view of the response of the whole single NW with lithium insertion. However, for the open-cell configuration the lithium ion source was only in contact with the end of the Si nanowire, leading to the sequential lithiation process of the nanowire in only one direction.

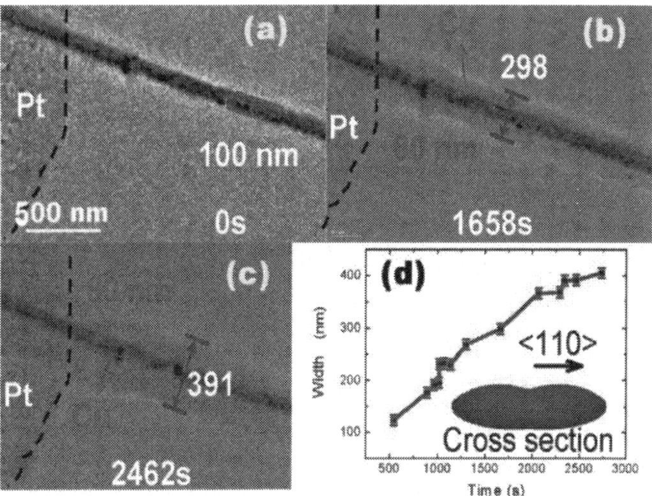

Figure 15. *In situ* liquid-cell TEM observation of the lithiation of the Cu-coated Si (Cu-Si) NW. (a) TEM image showing the pristine state of the Cu-Si NW at 0 s; (b) core-shell formation of the Cu-Si NW during lithiation at 1658 s; (c) TEM image of the Cu-Si NW at 2462 s; (d) plotted width changes of the NW as a function of time. Note that, in all images from a to c, the Pt contact region is labeled by the black lines in the left of the image. The inset in panel c illustrating the cross-sectional image after anisotropic swelling of the Si nanowire upon lithium insertion with maximum volume expansion along the <100> direction [88].

Apart from Si NW electrodes, Si NPs as electrode material were also studied by using G-SLC [56, 97], which showed that the very first lithiation at the Si-electrolyte interface had the strong orientation dependence favoring the <110> directions, but then the Li diffusion occurred isotropically after passing the initial stage regardless of the NP size. This indicated that the rate-limiting diffusion barrier is at Si-electrolyte interfaces instead of within Si or at the interfaces between lithiated and unlithiated regions. The orientation-dependent initial lithiation phenomenon was evidenced by HRTEM images as well as electron diffraction analyses, as shown in **Figure 16**. For the representative three Si NPs (whose original diameters are 34, 83, and 103 nm), their morphological and dimensional changes along <110>, <111>, and <100> directions were monitored. The selected-area electron diffraction patterns (the left ones in **Figure 16 (a-c)**) indicated that all of the three Si NPs were single-crystalline with <110> zone axes. The lithiation progressed predominantly along <110> directions, leading to the aniso-tropic volume expansion along the same crystal orientations, as indicated by the white arrows in **Figure 16 (a–c)**.

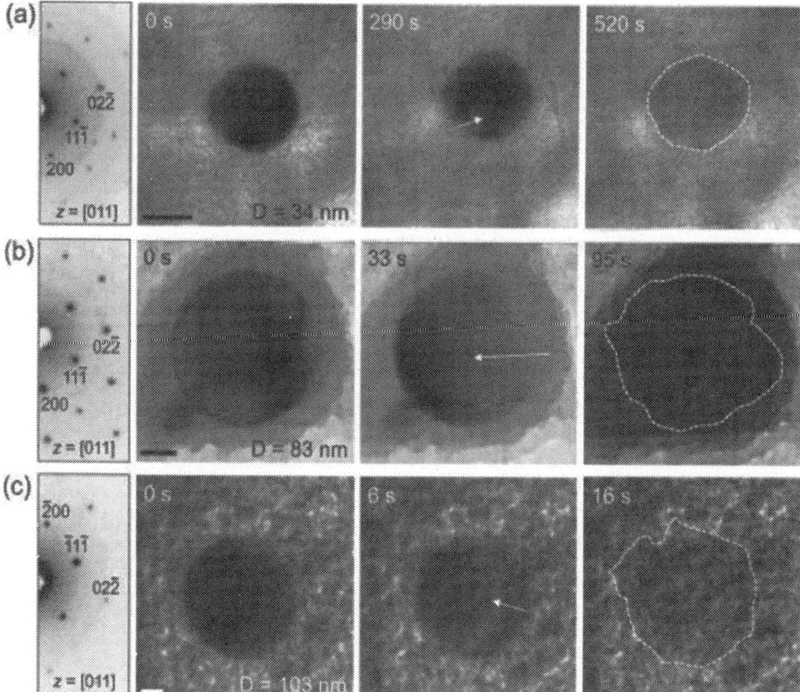

Figure 16. Morphological and dimensional changes of Si NPs analyzed by GLC-TEM during the course of lithiation: (a–c) Time series bright-field TEM images of the Si NPs with initial diameters of 34, 83, and 103 nm, respectively. The white arrows indicate the Si<110> directions. The SA-EDPs in (a–c) indicate crystalline nature of the pristine Si NPs and their crystallographic orientations along <110> zone axes. The scale bars in (a–c) are 20 nm [97].

Based on the above discussions, the major research achievements to date in applications of SLCs for *in situ* TEM electrochemical characterization of LIBs are highlighted as follows:

1. The stabilities and degradation mechanisms of commercial electrolytes commonly used for Li-ion and Li-O$_2$ battery were studied [94], providing reference for future choices on electrolytes.

2. The lithiation and delithiation process of fully submerged electrodes, such as Si NW electrodes [88, 98], Sn electrodes [82], and Si NPs [97], during electrochemical testing were observed by using *in situ* SLC platform and real electrolyte.

3. The electrochemical lithiation process, such as dendritic growth of crystalline lithium [60], stripping of lithium [60], SEI layer formation [99, 100], etc., were observed in real time with nanoscale resolution during electrochemical charge and discharge [60] using commercial electrolyte [73].

4. Summary and future research directions

In sum, sealed gas cells for *in situ* ETEM observation and sealed liquid cells for *in situ* TEM electrochemical characterization of LIBs have been reviewed in this article. Sealed cells have various advantages over the opened-type approach, thus becoming the dominant way to perform ETEM studies under gaseous/liquid environment. However, improvements are still needed in the following aspects for the benign development of this technology.

For sealed gas cells, thermal expansion/contraction and the consequent sample drift need to be minimized during *in situ* ETEM observation. The key to realize this is to improve ways of heating the sample. Localized heating sources such as laser or infrared light can be used for this purpose. Furthermore, the gas path length for electrons to go through the sealed gas cells needs to be decreased, possibly by better designs of the cell configurations. For example, if the heater can be integrated into the window instead of hanging out in the middle of the cell, the thickness of the spacer can be decreased such that a shorter gas path length is accomplished. Moreover, the pressure limit within the sealed gas cell should be increased, so that this technique can be applied to increasingly more fields of study. This can be achieved by various approaches: 1) since the major concern for limiting the pressure is the mechanical strength of the windows, either modifying the manufacturing process of the windows or replacing the current material with new ones may work well; 2) the configurations of the windows can be improved to withstand a higher pressure. For example, a single window with a large surface area is easier to break under high pressure, compared with several smaller windows distributing evenly. Last but not least, the solid-gas reactions need to be controlled and monitored better by developing ways to measure the parameters of the sample inside the cell more precisely. The direct approach to achieve this goal is to integrate measurement devices, such as nanoscale thermometers and pressure gauges into the sealed gas cells. The indirect way is to use chemical analysis techniques to reflect temperature and pressure changes. For example, the drift of EELS peak positions can be used to tell the temperature change.

For sealed liquid cells, TEM characterization of lithium through a liquid is challenging because lithium is a weak elastic scatterer and multiple scattering from the liquid could dominate the signal, resulting to a poor spatial resolution and contrast. Consequently, the spatial resolution and contrast needs to be enhanced. The future research direction is to improve SLC designs for better spatial resolution, which can be realized by decreasing the thickness of the liquid layer, redesigning the electrode configuration, utilizing alternative viewing window materials, employing different electrolytes, controlling electron dose, optimizing spacer thickness, etc. With an improved spatial resolution, various new research frontiers, including defect structure within SEI layer during (de)lithiation process, can be explored by using SLCs in TEM, rendering it the most promising technique for *in situ* electrochemical characterization of LIBs.

Author details

Fan Wu* and Nan Yao

*Address all correspondence to: fanwu@princeton.edu

Princeton Institute for the Science and Technology of Materials (PRISM), Princeton University, USA

References

[1] Mehraeen S, McKeown JT, Deshmukh PV, Evans JE, Abellan P, Xu P, et al. A (S)TEM gas cell holder with localized laser heating for in situ experiments. Microscopy and Microanalysis. 2013;19:470–8.

[2] Kawasaki T, Ueda K, Ichihashi M, Tanji T. Improvement of windowed type environmental-cell transmission electron microscope for in situ observation of gas-solid interactions. Review of Scientific Instruments. 2009;80:113701.

[3] Heide H. Elektronenmikroskopie von Objekten unter Atmosphärendruck oder unter Drucken, welche ihre Austrocknung verhindern. Naturwissenschaften. 1960;47:313–7.

[4] Heide HG. Electron microscopic observation of specimens under controlled gas pressure. The Journal of Cell Biology. 1962;13:147–52.

[5] Jinschek JR. Advances in the environmental transmission electron microscope (ETEM) for nanoscale in situ studies of gas-solid interactions. Chemical Communications. 2014;50:2696–706.

[6] Hansen TW, Wagner JB. Catalysts under controlled atmospheres in the transmission electron microscope. ACS Catalysis. 2014;4:1673–85.

[7] Daulton TL, Little BJ, Lowe K, Jones-Meehan J. In situ environmental cell–transmission electron microscopy study of microbial reduction of chromium(VI) using electron energy loss spectroscopy. Microscopy and Microanalysis. 2001;7:470–85.

[8] Creemer JF, Helveg S, Hoveling GH, Ullmann S, Molenbroek AM, Sarro PM, et al. Atomic-scale electron microscopy at ambient pressure. Ultramicroscopy. 2008;108:993–8.

[9] Allard LF, Overbury SH, Bigelow WC, Katz MB, Nackashi DP, Damiano J. Novel MEMS-based gas-cell/heating specimen holder provides advanced imaging capabilities for in situ reaction studies. Microscopy and Microanalysis. 2012;18:656–66.

[10] Yaguchi T, Suzuki M, Watabe A, Nagakubo Y, Ueda K, Kamino T. Development of a high temperature-atmospheric pressure environmental cell for high-resolution TEM. Journal of Electron Microscopy. 2011;60:217–25.

[11] de Jonge N, Bigelow WC, Veith GM. Atmospheric pressure scanning transmission electron microscopy. Nano Letters. 2010;10:1028–31.

[12] Rajasekhara S, Hattar KM, Tikare V, Dingreville RPM, Clark B. Hydride formation in cladding materials studied via in-situ environmental heating transmission electron microscopy. 2012.

[13] Rose HH. Historical aspects of aberration correction. Journal of Electron Microscopy. 2009;58:77–85.

[14] Marton L. La microscopie electronique des objets biologiques. Bulletin de l'Academie de Belgique Classe des Sciences (5). 1937;28:672–5.

[15] Tanaka N, Usukura J, Kusunoki M, Saito Y, Sasaki K, Tanji T, et al. Development of an environmental high-voltage electron microscope for reaction science. Microscopy. 2013;62:205–15.

[16] Baker RTK, Harris PS. Controlled atmosphere electron microscopy. Journal of Physics E: Scientific Instruments. 1972;5:793.

[17] Butler EP. In situ experiments in the transmission electron microscope. Reports on Progress in Physics. 1979;42:833.

[18] Boyes ED, Gai PL. Environmental high resolution electron microscopy and applications to chemical science. Ultramicroscopy. 1997;67:219–32.

[19] Sharma R. Design and applications of environmental cell transmission electron microscope for in situ observations of gas–solid reactions. Microscopy and Microanalysis. 2001;7:494–506.

[20] N. Yao GES, R. A. Kemp, D. C. Guthrie, R. D. Cates, C. M. Bolinger. Environmental cell TEM studies of catalyst particle behavior. In: Bailey G, editor. 49th Annual Conference of EMSA. San Francisco: San Francisco Press; 1991. p. 1028–9.

[21] Parkinson GM. High resolution, in-situ controlled atmosphere transmission electron microscopy (CATEM) of heterogeneous catalysts. Catal Lett. 1989;2:303–7.

[22] Komatsu M, Mori H. In situ HVEM study on copper oxidation using an improved environmental cell. Journal of Electron Microscopy. 2005;54:99–107.

[23] Giorgio S, Sao Joao S, Nitsche S, Chaudanson D, Sitja G, Henry CR. Environmental electron microscopy (ETEM) for catalysts with a closed E-cell with carbon windows. Ultramicroscopy. 2006;106:503–7.

[24] Creemer JF, Helveg S, Kooyman PJ, Molenbroek AM, Zandbergen HW, Sarro PM. A MEMS reactor for atomic-scale microscopy of nanomaterials under industrially relevant conditions. Microelectromechanical Systems, Journal of. 2010;19:254–64.

[25] Xin HL, Niu K, Alsem DH, Zheng H. In situ TEM study of catalytic nanoparticle reactions in atmospheric pressure gas environment. Microscopy and Microanalysis. 2013;19:1558–68.

[26] Gao M, Chen X, Pan H, Xiang L, Wu F, Liu Y. Ultrafine SnO_2 dispersed carbon matrix composites derived by a sol–gel method as anode materials for lithium ion batteries. Electrochimica Acta. 2010;55:9067–74.

[27] Gbordzoe S, Kotoka R, Craven E, Kumar D, Wu F, Narayan J. Effect of substrate temperature on the microstructural properties of titanium nitride nanowires grown by pulsed laser deposition. Journal of Applied Physics. 2014;116:4310.

[28] Wu F, Rao SS, Prater JT, Zhu YT, Narayan J. Tuning exchange bias in epitaxial Ni/MgO/TiN heterostructures integrated on Si(1 0 0). Current Opinion in Solid State and Materials Science. 2014;18:263

[29] Rao S, Prater J, Wu F, Nori S, Kumar D, Narayan J. Integration of epitaxial permalloy on Si (100) through domain matching epitaxy paradigm. Current Opinion in Solid State and Materials Science. 2013;18:1–5.

[30] Rao SS, Prater J, Wu F, Shelton C, Maria J-P, Narayan J. Interface magnetism in epitaxial $BiFeO_3$-$La_{0.7}Sr_{0.3}MnO_3$ heterostructures integrated on Si (100). Nano letters. 2013;13:5814–21.

[31] Rao SS, Prater JT, Wu F, Nori S, Kumar D, Yue L, et al. Positive exchange bias in epitaxial permalloy/MgO integrated with Si (100). Current Opinion in Solid State and Materials Science. 2014;18:140–6.

[32] Singamaneni SR, Prater J, Wu F, Narayan J. Interface magnetism of two functional epitaxial ferromagnetic oxides integrated with Si (100). APS March Meeting Abstracts. 2014;1:1249

[33] Singamaneni SR, Prater J, Wu F, Nori S, Kumar D, Yue L, et al. Positive exchange bias in epitaxial permalloy/MgO integrated with Si (100). APS March Meeting Abstracts. 2014;1:1254

[34] Wu F. Planar defects in metallic thin film heterostructures: North Carolina State University; 2014.

[35] Bayati M, Molaei R, Wu F, Budai J, Liu Y, Narayan R, et al. Correlation between structure and semiconductor-to-metal transition characteristics of VO_2TiO_2/sapphire thin film heterostructures. Acta Materialia. 2013;61:7805–15.

[36] Wu F, Narayan J. Controlled epitaxial growth of body-centered cubic and face-centered cubic Cu on MgO for integration on Si. Crystal Growth & Design. 2013;13:5018–24.

[37] Molaei R, Bayati R, Wu F, Narayan J. A microstructural approach toward the effect of thickness on semiconductor-to-metal transition characteristics of VO_2 epilayers. Journal of Applied Physics. 2014;115:4311.

[38] Wu F, Zhu YT, Narayan J. Grain size effect on twin density in as-deposited nanocrystalline Cu film. Philosophical Magazine. 2013;93:4355–63.

[39] Lee YF, Wu F, Narayan J, Schwartz J. Oxygen vacancy enhanced room-temperature ferromagnetism in Sr_3SnO/c-YSZ/Si (001) heterostructures. MRS Communications. 2014;4:7–13.

[40] Wu F, Zhu YT, Narayan J. Macroscopic twinning strain in nanocrystalline Cu. Materials Research Letters. 2013;2:63–9.

[41] Lee YF, Wu F, Kumar R, Hunte F, Schwartz J, Narayan J. Epitaxial integration of dilute magnetic semiconductor Sr_3SnO with Si (001). Applied Physics Letters. 2013;103:2101.

[42] Wu F, Wen HM, Lavernia EJ, Narayan J, Zhu YT. Twin intersection mechanisms in nanocrystalline fcc metals. Materials Science and Engineering: A. 2013;585:292–6.

[43] Gupta N, Singh R, Wu F, Narayan J, McMillen C, Alapatt GF, et al. Deposition and characterization of nanostructured Cu2O thin-film for potential photovoltaic applications. Journal of Materials Research. 2013;28:1740–6.

[44] Sharma R. Experimental set up for in situ transmission electron microscopy observations of chemical processes. Micron (Oxford, England : 1993). 2012;43:1147–55.

[45] Vendelbo SB, Elkjær CF, Falsig H, Puspitasari I, Dona P, Mele L, et al. Visualization of oscillatory behaviour of Pt nanoparticles catalysing CO oxidation. Nat Mater. 2014;13:884–90.

[46] Sun L, Noh KW, Wen J-G, Dillon SJ. In situ Transmission Electron Microscopy Observation of Silver Oxidation in Ionized/Atomic Gas. Langmuir. 2011;27:14201–6.

[47] Yokosawa T, Alan T, Pandraud G, Dam B, Zandbergen H. In-situ TEM on (de)hydrogenation of Pd at 0.5–4.5 bar hydrogen pressure and 20–400°C. Ultramicroscopy. 2012;112:47–52.

[48] Vendelbo SB, Kooyman PJ, Creemer JF, Morana B, Mele L, Dona P, et al. Method for local temperature measurement in a nanoreactor for in situ high-resolution electron microscopy. Ultramicroscopy. 2013;133:72–9.

[49] Alan T, Yokosawa T, Gaspar J, Pandraud G, Paul O, Creemer F, et al. Micro-fabricated channel with ultra-thin yet ultra-strong windows enables electron microscopy under 4-bar pressure. Applied Physics Letters. 2012;100:–.

[50] Hiroshi F, Masao K, Isao I. A universal environmental cell for a 3MV-class electron microscope and its applications to metallurgical subjects. Japanese Journal of Applied Physics. 1976;15:2221.

[51] Hattar K, Rajasekhara S, Clark BG. In situ TEM ion irradiation and atmospheric heating of cladding materials. MRS Online Proceedings Library. 2012;1383:null-null.

[52] Liu XH, Zhang LQ, Zhong L, Liu Y, Zheng H, Wang JW, et al. Ultrafast Electrochemical Lithiation of Individual Si Nanowire Anodes. Nano Letters. 2011;11:2251–8.

[53] Huang JY, Zhong L, Wang CM, Sullivan JP, Xu W, Zhang LQ, et al. In situ observation of the electrochemical lithiation of a single SnO2 nanowire electrode. Science. 2010;330:1515–20.

[54] Williamson MJ, Tromp RM, Vereecken PM, Hull R, Ross FM. Dynamic microscopy of nanoscale cluster growth at the solid–liquid interface. Nat Mater. 2003;2:532–6.

[55] White ER, Singer SB, Augustyn V, Hubbard WA, Mecklenburg M, Dunn B, et al. In situ transmission electron microscopy of lead dendrites and lead ions in aqueous solution. ACS Nano. 2012;6:6308–17.

[56] Yuk JM, Park J, Ercius P, Kim K, Hellebusch DJ, Crommie MF, et al. High-resolution EM of colloidal nanocrystal growth using graphene liquid cells. Science. 2012;336:61–4.

[57] de Jonge N, Ross FM. Electron microscopy of specimens in liquid. Nat Nano. 2011;6:695–704.

[58] Zheng H, Smith RK, Jun Y-w, Kisielowski C, Dahmen U, Alivisatos AP. Observation of single colloidal platinum nanocrystal growth trajectories. Science. 2009;324:1309–12.

[59] Grogan JM, Bau HH. The nanoaquarium: a platform for in situ transmission electron microscopy in liquid media. Journal of Microelectromechanical Systems. 2010;19:885–94.

[60] Holtz ME, Yu Y, Gunceler D, Gao J, Sundararaman R, Schwarz KA, et al. Nanoscale imaging of lithium ion distribution during in situ operation of battery electrode and electrolyte. Nano Letters. 2014;14:1453–9.

[61] Woehl TJ, Evans JE, Arslan L, Ristenpart WD, Browning ND. Direct in situ determination of the mechanisms controlling nanoparticle nucleation and growth. Acs Nano. 2012;6:8599–610.

[62] Liao H-G, Cui L, Whitelam S, Zheng H. Real-Time Imaging of Pt3Fe Nanorod Growth in Solution. Science. 2012;336:1011–4.

[63] Woehl TJ, Park C, Evans JE, Arslan I, Ristenpart WD, Browning ND. Direct observation of aggregative nanoparticle growth: kinetic modeling of the size distribution and growth rate. Nano Lett. 2014;14:373–8.

[64] Evans JE, Jungjohann KL, Wong PCK, Chiu P-L, Dutrow GH, Arslan I, et al. Visualizing macromolecular complexes with in situ liquid scanning transmission electron microscopy. Micron (Oxford, England : 1993). 2012;43:1085–90.

[65] Mirsaidov UM, Zheng H, Casana Y, Matsudaira P. Imaging protein structure in water at 2.7 nm resolution by transmission electron microscopy. Biophysical Journal. 2012;102:L15-L7.

[66] Huang T-W, Liu S-Y, Chuang Y-J, Hsieh H-Y, Tsai C-Y, Wu W-J, et al. Dynamics of hydrogen nanobubbles in KLH protein solution studied with in situ wet-TEM. Soft Matter. 2013;9:8856–61.

[67] Proetto MT, Rush AM, Chien M-P, Abellan Baeza P, Patterson JP, Thompson MP, et al. Dynamics of soft nanomaterials captured by transmission electron microscopy in liquid water. Journal of the American Chemical Society. 2014;136:1162–5.

[68] Evans JE, Jungjohann KL, Browning ND, Arslan I. Controlled growth of nanoparticles from solution with in situ liquid transmission electron microscopy. Nano Letters. 2011;11:2809–13.

[69] Zheng H, Claridge SA, Minor AM, Alivisatos AP, Dahmen U. Nanocrystal diffusion in a liquid thin film observed by in situ transmission electron microscopy. Nano Lett. 2009;9:2460–5.

[70] Jungjohann KL, Bliznakov S, Sutter PW, Stach EA, Sutter EA. In situ liquid cell electron microscopy of the solution growth of Au–Pd core-shell nanostructures. Nano Letters. 2013;13:2964–70.

[71] Parent LR, Robinson DB, Cappillino PJ, Hartnett RJ, Abellan P, Evans JE, et al. In situ observation of directed nanoparticle aggregation during the synthesis of ordered nanoporous metal in soft templates. chemistry of materials. 2014;26:1426–33.

[72] Aurbach D, Markovsky B, Levi MD, Levi E, Schechter A, Moshkovich M, et al. New insights into the interactions between electrode materials and electrolyte solutions for advanced nonaqueous batteries. Journal of Power Sources. 1999;81–82:95–111.

[73] Zeng Z, Liang W-I, Liao H-G, Xin HL, Chu Y-H, Zheng H. Visualization of electrode-electrolyte interfaces in LiPF6/EC/DEC electrolyte for lithium ion batteries via in situ TEM. Nano Letters. 2014;14:1745–50.

[74] Ross FM. Growth processes and phase transformations studied by in situ transmission electron microscopy. IBM J Res Dev. 2000;44:489–501.

[75] Radisic A, Vereecken PM, Hannon JB, Searson PC, Ross FM. Quantifying electrochemical nucleation and growth of nanoscale clusters using real-time kinetic data. Nano Letters. 2006;6:238–42.

[76] Thiberge S, Nechushtan A, Sprinzak D, Gileadi O, Behar V, Zik O, et al. Scanning electron microscopy of cells and tissues under fully hydrated conditions. Proceedings of the National Academy of Sciences of the United States of America. 2004;101:3346–51.

[77] Liu K-L, Wu C-C, Huang Y-J, Peng H-L, Chang H-Y, Chang P, et al. Novel microchip for in situ TEM imaging of living organisms and bio-reactions in aqueous conditions. Lab on a Chip. 2008;8:1915–21.

[78] de Jonge N, Peckys DB, Kremers GJ, Piston DW. Electron microscopy of whole cells in liquid with nanometer resolution. Proceedings of the National Academy of Sciences of the United States of America. 2009;106:2159–64.

[79] Chen X, Noh KW, Wen JG, Dillon SJ. In situ electrochemical wet cell transmission electron microscopy characterization of solid–liquid interactions between Ni and aqueous $NiCl_2$. Acta Materialia. 2012;60:192–8.

[80] Sun M, Liao H-G, Niu K, Zheng H. Structural and morphological evolution of lead dendrites during electrochemical migration. Sci Rep. 2013;3.

[81] Sullivan JP, Huang J, Shaw MJ, Subramanian A, Hudak N, Zhan Y, et al. Energy harvesting and storage: materials, devices, and applications In: Understanding Li-ion battery processes at the atomic-to nano-scale; Dhar NK, Wijewarnasuriya PS, Dutta AK, editors. 2010

[82] Noh KW, Dillon SJ. Morphological changes in and around Sn electrodes during Li ion cycling characterized by in situ environmental TEM. Scripta Materialia. 2013;69:658–61.

[83] Unocic RR, Sacci RL, Brown GM, Veith GM, Dudney NJ, More KL, et al. Quantitative electrochemical measurements using in situ ec-S/TEM devices. Microscopy and Microanalysis : The Official Journal of Microscopy Society of America, Microbeam Analysis Society, Microscopical Society of Canada. 2014;20:452–61.

[84] Brunetti G, Robert D, Bayle-Guillemaud P, Rouviere JL, Rauch EF, Martin JF, et al. Confirmation of the domino-cascade model by LiFePO4/FePO4 Precession Electron Diffraction. Chemistry of Materials. 2011;23:4515–24.

[85] Wang C-M, Xu W, Liu J, Zhang J-G, Saraf LV, Arey BW, et al. In situ transmission electron microscopy observation of microstructure and phase evolution in a SnO2 nanowire during lithium intercalation. Nano Letters. 2011;11:1874–80.

[86] Holtz ME, Yu Y, Gao J, Abruna HD, Muller DA. In situ electron energy-loss spectroscopy in liquids. Microscopy and microanalysis : the official journal of Microscopy

Society of America, Microbeam Analysis Society, Microscopical Society of Canada. 2013;19:1027–35.

[87] Moreau P, Boucher F. Revisiting lithium K and iron M(2),(3) edge superimposition: the case of lithium battery material LiFePO(4). Micron (Oxford, England : 1993). 2012;43:16–21.

[88] Gu M, Parent LR, Mehdi BL, Unocic RR, McDowell MT, Sacci RL, et al. Demonstration of an electrochemical liquid cell for operando transmission electron microscopy observation of the lithiation/delithiation behavior of Si nanowire battery anodes. Nano Letters. 2013;13:6106–12.

[89] Muller DA, Silcox J. Delocalization in inelastic scattering. Ultramicroscopy. 1995;59:195–213.

[90] Ernzerhof M, Scuseria GE. Assessment of the Perdew–Burke–Ernzerhof exchange-correlation functional. The Journal of Chemical Physics. 1999;110:5029–36.

[91] Wang Y, Yi J, Xia Y. Recent progress in aqueous lithium-ion batteries. Advanced Energy Materials. 2012;2:830–40.

[92] Xu K. Nonaqueous liquid electrolytes for lithium-based rechargeable batteries. Chemical Reviews. 2004;104:4303–418.

[93] Nasybulin E, Xu W, Engelhard MH, Nie Z, Burton SD, Cosimbescu L, et al. Effects of electrolyte salts on the performance of Li–O2 batteries. The Journal of Physical Chemistry C. 2013;117:2635–45.

[94] Abellan P, Mehdi BL, Parent LR, Gu M, Park C, Xu W, et al. Probing the degradation mechanisms in electrolyte solutions for Li-ion batteries by in situ transmission electron microscopy. Nano Letters. 2014;14:1293–9.

[95] Gachot G, Grugeon S, Armand M, Pilard S, Guenot P, Tarascon J-M, et al. Deciphering the multi-step degradation mechanisms of carbonate-based electrolyte in Li batteries. Journal of Power Sources. 2008;178:409–21.

[96] Gachot Gg, Ribière P, Mathiron D, Grugeon S, Armand M, Leriche J-B, et al. Gas chromatography/mass spectrometry as a suitable tool for the Li-ion battery electrolyte degradation mechanisms study. Analytical Chemistry. 2010;83:478–85.

[97] Yuk JM, Seo HK, Choi JW, Lee JY. Anisotropic lithiation onset in silicon nanoparticle anode revealed by in situ graphene liquid cell electron microscopy. ACS Nano. 2014;8:7478–85.

[98] Layla Mehdi B, Gu M, Parent LR, Xu W, Nasybulin EN, Chen X, et al. In-situ electro-chemical transmission electron microscopy for battery research. Microscopy and Microanalysis. 2014;20:484–92.

[99] Unocic RR, Sun X-G, Sacci RL, Adamczyk LA, Alsem DH, Dai S, et al. Direct visuali-zation of solid electrolyte interphase formation in lithium-ion batteries with in situ

electrochemical transmission electron microscopy. microscopy and microanalysis. 2014;20:1029–37.

[100] Sacci RL, Dudney NJ, More KL, Parent LR, Arslan I, Browning ND, et al. Direct visualization of initial SEI morphology and growth kinetics during lithium deposition by in situ electrochemical transmission electron microscopy. Chemical Communications. 2014;50:2104–7.

Chapter 12

Advanced Scanning Tunneling Microscopy for Nanoscale Analysis of Semiconductor Devices

Leonid Bolotov and Toshihiko Kanayama

Additional information is available at the end of the chapter

http://dx.doi.org/10.5772/62552

Abstract

Significant attention has been addressed to high-spatial resolution analysis of modern sub-100-nm electronic devices to achieve new functions and energy-efficient operations. The chapter presents a review of ongoing research on charge carrier distribution analysis in nanoscale Si devices by using scanning tunneling microscopy (STM) employing advanced operation modes: a gap-modulation method, a molecule-assisted probing method, and a dual-imaging method. The described methods rely on detection and analysis of tunneling current, which is strongly localized within an atomic dimension. Representative examples of applications to nanoscale analysis of Si device cross-sections and nanowires are given. Advantages, difficulties, and limitations of the advanced STM methods are discussed in comparison with other techniques used in a field of device metrology.

Keywords: scanning tunneling microscopy, semiconductor devices, charge carrier distribution, resonant electron tunneling, silicon-on-insulator, photocarrier profiling, fullerene molecule

1. Introduction

Since invention of solid-state electric junctions, charge carrier distribution has become the primary requirement of electronic device design to achieve desirable device performance. Typically, a spatial distribution of charge carriers in semiconductor devices is created by introduction of electronic impurity atoms with particular electron configuration allowing to donate a free electron to the host semiconductor (donor impurity) or to trap a valence electron (acceptor impurity) from the host material. Thus, the host semiconductor with donor impuri-

ty atoms has become a negative-charge (electrons) conductor and is called *n-type*. The host semiconductor with acceptor impurity atoms has become a positive-charge (holes) conductor and is called *p-type*. Typical semiconductor devices have concentration of impurity atoms in a range of 10^{15}–10^{21}/cm^3, which is less than 1 % of total number of atoms. Defects and atom vacancy often behave like impurity atoms.

Early days, charge carrier distribution was derived from spatial distributions of impurity atoms in semiconductor materials. Secondary ion mass spectrometry (SIMS) has been used to obtain a depth distribution profile of impurity atoms in semiconductor materials by sputtering with high-energy ions. As modern high-performance Si devices such as complementary metal-oxide-semiconductor (CMOS) transistors are less than 100 nm in size, and have complex material structures, the 1D SIMS profiling becomes inadequate. **Figure 1** shows a typical structure of a metal-oxide-semiconductor field effect transistor (MOSFET) consisting of gate, channel, and source/drain regions with high impurity concentrations.

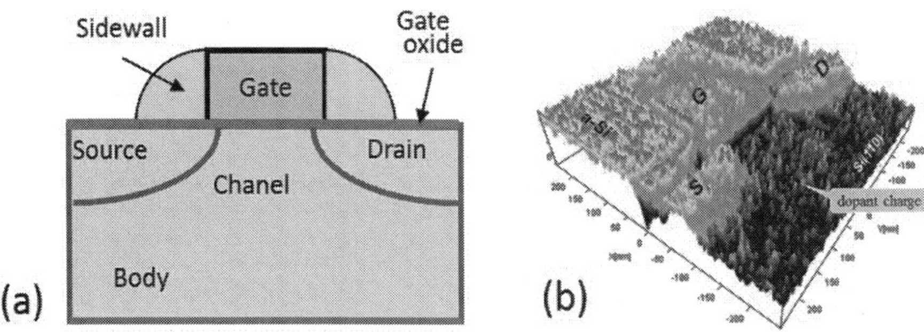

Figure 1. (a) A sketch of a MOSFET device in a cross-section. (b) 3D view of a charge distribution in a MOSFET measured by the STM gap modulation method. Charge concentration is emphasized by color: blue color for low charge concentration in p-Si channel and red color for high charge concentration in source (S), drain (D), and gate (G).

Recently, a new technique of tree-dimensional (3D) atom mapping, which is called atom probe tomography, was introduced based on counting of atom ions ejected from a needle-like device specimen [1–6]. Aside from complexity of the sample preparation and 3D data reconstruction of the atom probe technique, the charge carrier distribution is assumed to be equal to that of impurity atoms. However, the carrier distribution deviates significantly from the impurity atom distribution as a result of internal electric field at material interfaces, trapped charges in oxide, and fractional activation of impurity atoms in areas of high impurity concentration. Therefore, techniques allowing to measure local distribution of charge carriers within the electronic device interior have been a focus of attention from scientific and practical points of view.

Significant attention has been addressed to high-spatial resolution analysis of modern sub-100-nm electronic devices, nanowire devices which meet miniaturization to less than 10 nm in order to achieve new functions and energy-efficient operation. Last decade, various techniques have been developed for charge carrier mapping. A common high-resolution imaging technique, scanning electron microscopy (SEM), has been upgraded with an energy-filtering option, allowing us to obtain the image contrast as a function of the surface electrostatic potential [7–10].

Scanning probe techniques are an important tool for local probing of electric properties and have played important roles in scientific research on electronic materials and in evaluations of device structures in fabrication processes. Scanning probe microscopy (SPM) techniques are based on the ability to position a sharp probe electrode in very close proximity with high precision to the sample surface under investigation [11]. Different physical quantities can be measured by the probe including electric tunneling current, atomic and electrostatic forces, or other types of probe-sample interactions. By moving the probe laterally over the sample surface and performing measurements at different locations, two-dimensional distributions of surface atomic structure, electric current, electrostatic potential, or other properties can be obtained.

SPM techniques employed in local electrical measurements are atomic force microscopy with a conductive probe (c-AFM) [12], scanning spreading resistance microscopy (SSRM) [13], scanning Kelvin probe microscopy (SKPM) [14], and scanning tunneling microscopy (STM) [15]. These scanning probe techniques create two-dimensional (2D) maps of variations in the surface electric potential or electric current density along a cross-section of a semiconductor device, when the surface states, defects, adsorbates, and foreign particles on the cross-sectional surface do not affect the initial charge carrier distribution. In majority of cases, certain surface treatments of the cross-sectional surface are applied prior to measurements to eliminate undesirable surface effects. Quantitative impurity profiles by SSRM and SKPM have been demonstrated for high impurity concentrations, where a spatial resolution on the order of the probe tip radius (~5 nm) was obtained under optimum conditions [16–19].

STM has been used for impurity distribution measurements in Si devices by analyzing current-voltage spectra [20–23]. To derive quantitatively variation in the charge carrier distribution from STM measurements, one must analyze complex dependence of the tunneling current on the bias voltage, the tunneling gap, and the band-bending potential beneath the STM probe tip on a semiconductor surface. Thus, simulations of STM operation are an essential part of the data analysis.

In this chapter, we focus on advanced STM-based spectroscopy techniques as nanoscale methods for two-dimensional (2D) charge carrier analysis. It represents original development of scanning probe microscopy methods for Si device metrology with ultimate spatial resolution. We describe the principles of the advanced STM methods and give representative examples of applications to nanoscale analysis of Si CMOS devices and nanowires. Advantages, difficulties, and limitations of the advanced STM modes will be discussed in comparison with other techniques used in a field of device metrology.

The chapter begins with description of device cross-section preparation methods and essential features of STM measurements on a semiconductor surface. Measurement principles of original STM-based techniques and application examples will be given. Current development in STM simulations will be outlined. Prospects toward research in new 2D materials will be elaborated.

2. Preparation of Si device cross-sections

Figure 2 shows a common way for making solid crystal cross-sections. The process includes a number of steps. (1) Cleavage and/or *dicing* of a thin crystal wafer are used to define a desired location of the cross-sectional plane. (2) *Chemical-mechanical planarization-polishing* (CMP) and focus ion beam (FIB) techniques are applied to tune location of the cross-sectional plane with a sub-micrometer accuracy. (3) Chemical and electric *passivation* of the cross-sectional Si surface by hydrogenation or thin oxide is carried out to prevent distortion of original charge carrier distributions by surface states and contamination.

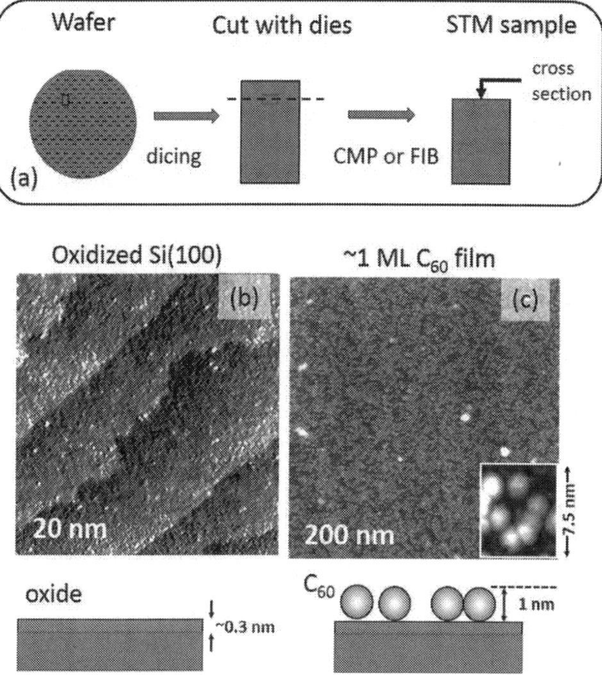

Figure 2. (a) Fabrication of a device cross-section for STM measurements. (b)–(c) STM images (a set-point: 300 pA, 2.0 V) of an oxide-passivated Si surface before (b) and after (c) 1 ML C$_{60}$ film formation. Color scale is 0.8 nm (b) and 2.5 nm (c). Insert shows an image of 10 C$_{60}$ molecules.

Chemical and electric passivation of solid surfaces is the subject of extended research in catalysis to control on charge transfer process and chemical reactions in solid-liquid and solid-gas interfaces [24]. Moreover, chemical and electric passivation of semiconductor surfaces are a basic process in fabrication of modern Si devices, enabling to reduce off-state leakage current and photocarrier losses in solar cells [25]. Without passivation treatment, silicon surfaces have pronounced bands of surface states, which dominate the contrast of the STM images, so that it becomes difficult to characterize the underlying electrical interfaces. Therefore, passivation of Si surfaces by hydrogenation or oxidation has been employed in order to reproducibly prepare uniform surfaces of device cross-sections and to obtain very low density of surface states.

2.1. Passivation by hydrogenation

Hydrogenation of Si surfaces is achieved by etching in fluoric acid solutions. Etching removes the native Si oxide and terminates the Si dangling bonds with hydrogen atoms making a stable, passivated surface with very low density of surface states in the Si band gap [26]. A number of investigations have confirmed that tunneling spectra of such stabilized Si surfaces show variation with dopant type and concentration due to passivation of dangling bond states and the suppression of surface states [27–32].

Si(111) surfaces can be atomically flattened by wet treatment in NH_4F aqueous solutions [33]. In the procedure, the samples were dipped in a 5% HF solution to remove the residual oxide layer, then immersed in a 40% NH_4F solution at room temperature, and rinsed in ultrapure water for 1 min. This treatment renders the Si surface mono-hydride, well suited for STM analysis. In this treatment, hydrogen also reacts with near-surface impurity atoms forming electrically inactive complexes, thus, changing the initial charge distribution. To reactivate the impurity atoms, heating of the samples around 200–250°C is necessary [33, 34].

To prepare atomically flat Si(001) surfaces, a combined process is adopted, which consists of wet treatment using a fluoric acid solution and subsequent annealing in H_2 atmosphere at ~600°C and a pressure of ~2 × 10^3 Pa [35]. The authors showed the formation of an atomically flat Si(001) surface that have well-ordered step-terrace structures in the active device area. The flattening was attributed to the enhanced migration of Si atoms when anisotropic etching was suppressed.

2.2. Passivation by an ultrathin oxide

Hydrogenation of Si surfaces may not always be compatible with processing steps in a particular application, as Si surface etching usually introduces topographic contrast due to etching rate dependence on doping concentration, crystal orientation, and material composition. An alternative way to passivate Si surface is oxidation. The passivation of Si surfaces by controlled growth of ultra-thin oxide layer relies on the layer-by-layer oxidation kinetics at low oxygen pressure [36–38]. We adopted the preparation of cross-sectional surfaces of Si devices as follows [39, 40]. First, dicing and ultra-fine polishing are used to expose either (100) or (110) surfaces of the device. The polished surfaces are cleaned by few cycles of etching in dilute fluoric acid solution and wet-oxidation in H_2SO_4:H_2O_2 (3:1) solution to remove a damage layer. Finally, ultra-thin (~0.3 nm) oxide layer is grown at ~600°C under an O_2 pressure of 3 ×

10^{-3} Pa following etch-cleaning in HF:HCl (1:19). This procedure left a flat surface without any ordered structure as seen in **Figure 2(b)**, where the atomically flat terraces are separated by atomic steps of 0.24–0.27 nm in height. The oxide thickness was 0.32–0.35 nm as determined by x-ray photoelectron spectroscopy, and by scanning reflection electron microscopy (SREM). The low-pressure oxidation process results in a residual density of surface state traps of ~10^{12} cm^{-2} for Si(100) surfaces [24, 41, 42], which is suitable for STM spectroscopy analysis.

2.3. Formation of C$_{60}$ monolayer films

When a well-defined mono-molecular layer is prepared on a passivated surface, its molecular level can be utilized to quantitatively analyze the electrical properties of the underlying substrate. We call this method as a molecule-assisted spectroscopy. For this purpose, mono-molecular thick films of C$_{60}$ (fullerene) were formed by vapor sublimation of C$_{60}$ to the oxidized Si surfaces to a thickness of 3–5 molecular layers. The excess of C$_{60}$ layers was removed by sample heating at 170–190°C for 10 min. Because electrostatic interaction between the molecule and the underlying Si is stronger than the Van der Waals interaction between molecules within the film, a C$_{60}$ molecules adjacent to the Si surface remain at high coverage (~80%) as seen in **Figure 2(c)** [41].

3. Tunneling microscopy: basics

The STM operation principle is based on quantum mechanical phenomenon—electron tunneling through a potential barrier formed by a gap between the outermost atoms on the metal tip and the sample. When the gap is about 1 nm or less, electrons from the STM tip can penetrate into the sample with certain probability owing to the wave nature of the quantum particle.

Under external electric field, electron tunneling creates a measurable electric current, the tunneling current. In the single particle approximation, the tunneling current density is given by a difference in the particle flow across the gap from the STM tip and that from the semi-conductor and is expressed as an integral over particle's energy

$$J_{tun} = J_{\rightarrow} - J_{\leftarrow} = \int_0^{\infty} \rho_{tip}(E) \cdot f(E) \cdot \rho_{sample}\left(qV_{gap} - E\right) \cdot \left[1 - f\left(qV_{gap} - E\right)\right]$$
$$\cdot T\left(Z, E, V_{gap}\right) dE, \tag{1}$$

where $T(Z, E, V)$, the transmission factor, is a function of gap width (Z), electron energy (E), and external gap voltage (V_{gap}). $\rho_{tip}(E)$ and $\rho_{sample}(E)$ are the density of electron states at the surface of the STM tip and the sample, respectively. $f(E)$ is the Fermi function describing which energy states are occupied with electrons.

Here, we outline the important features of the STM technique essential for analysis of charge carrier distribution in semiconductors. They are

- tunneling barrier shape,

- sharing of applied voltage between the tunnel gap and a surface band-bending region, and

- surface charge density in the semiconductor beneath the STM probe electrode.

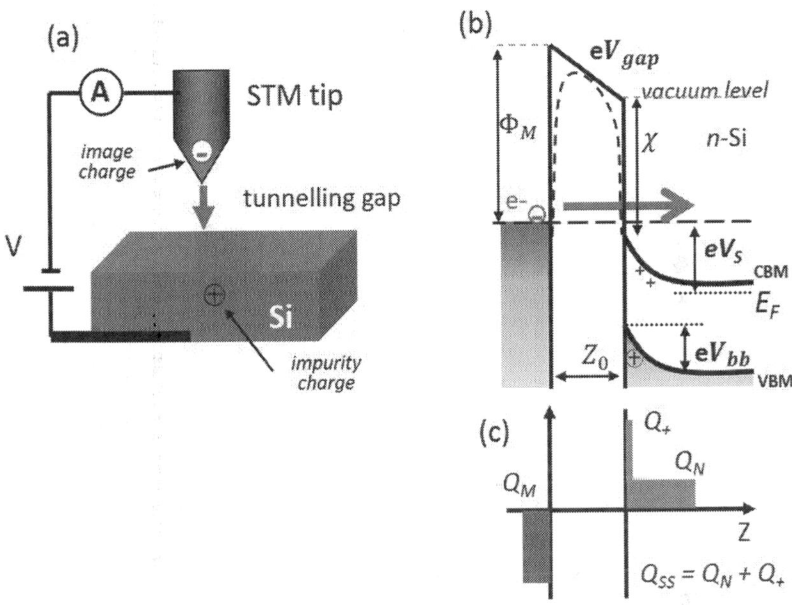

Figure 3. The principle of scanning tunneling microscopy of a semiconductor. (a) An STM setup, (b) an energy band diagram of a tunnel junction, and (c) a charge balance diagram.

The tunneling barrier shape determines the electron transmission factor and the value of the tunneling current. **Figure 3** shows an STM measurements setup and an energy band diagram of an ideal STM junction for n-type Si. Rectangular shape of the tunneling barrier is used in simple STM models. The actual potential barrier profile is different because of image potential lowering ($\Delta\phi$) owing to strong Coulomb interaction between charge and image charge in conductive materials [43–45]. Also, the tunneling gap may include an insulating layer such as ultrathin oxide and a molecular film with different dielectric properties. Therefore, the tunneling electrons experience an effective potential barrier of a barrier height (BH) given by

$$BH = \left(\Phi_M + E_F - qV_{gap}\right)/2 - \Delta\phi,\tag{2}$$

where Φ_M is the work function of the metal tip and E_F is the Fermi energy of the semiconductor, q is the elementary charge. For electron energy smaller than BH, the transmission factor is approximated by [46].

$$T(E,V) = \exp\left(-\alpha \cdot \sqrt{BH} \cdot Z\right),\tag{3}$$

The tunneling constant $\alpha = 10.2$ when the gap width is in units of nanometer and BH – in eV.

Because electric charge density in semiconductors is lower than that in metals, applied electric field penetrates deep beneath the semiconductor surface. To maintain the charge neutrality, a band-bending region is created beneath the STM probe. The applied voltage V_S is shared between the gap and the band-bending region and is given by

$$V_S = V_{gap} + V_{bb} + \phi_{MS},\tag{4}$$

where last term $\phi_{MS} = (\Phi_M - E_F)$ is an electrostatic potential difference between the work function of the STM tip and the semiconductor Fermi energy, and χ is the electron affinity of the semiconductor. In thermal equilibrium and $V_S = 0$, the charge neutrality is conserved, and the electric charge in the STM tip (Q_M) is equal to the local electric charge at the semiconductor surface beneath the STM tip. At $V_S = 0$, the band-bending region is created owing to the electrostatic potential difference ϕ_{MS}. **Figure 3** illustrates the case when an electron depletion region is formed for n-type Si under an external positive bias voltage $V_S > 0$ to the sample. For n-Si, the surface charge (Q_{SS}) includes positive charge of impurity atoms (Q_N) and mobile carriers (holes) (Q_+):

$$Q_M = Q_{SS} = Q_+ + Q_N,\tag{5}$$

According to the Gauss law [43, 47, 48], the voltage across the gap is given by

$$V_{gap} = \frac{|Q_{SS}|}{\epsilon_0} \cdot Z_0,\tag{6}$$

where ϵ_0, and ϵ_{Si} are the permittivity of the vacuum gap and Si, respectively. The depth of the band-bending region (w) depends on the electric field screening by the electric charge in the semiconductor and is given by

$$w \cong \sqrt{\frac{2\,\epsilon_{Si}\,V_{bb}}{Q_{SS}}},\tag{7}$$

It is straightforward that the tunneling current strongly depends on the local electric charge at the semiconductor surface. When there were surface states and interface traps, these trapped charges would alter the initial charge carrier distribution, and great care must be taken to prepare clean, well-defined cross-sectional surfaces. In fact, conventional furnace oxidation produces a gap-state density of about 10^{10} cm^{-2} for Si(100) and less than 10^{12} cm^{-2} for Si(111) surfaces [47]. Low-pressure oxidation below 600°C results in a density of $\sim 10^{12}$ cm^{-2} for Si(100) surfaces [26, 41, 42]. The surface oxidation effectively reduces density of surface states on Si surfaces, making that the current behavior becomes dependent on charge carrier concentration in the Si bulk beneath the STM probe.

Topographic STM images of a sample surface are formed when the STM probe is moved along the surface while keeping pre-determined tunneling current value (I_{tun}) at an applied voltage (V_S) by adjusting the gap width with a piezoelectric scanning unit. The STM technique offers ultimate spatial resolution down to a sub-nanometer range because tunneling current is strongly localized around the outermost atom of the STM tip owing to exponential current decay with the tip-sample distance. Three advanced STM-based modes discussed below rely on measurements and analysis of the tunneling current and, thus, offer high spatial resolution. Details of the SPM system construction and operation have been reviewed in original papers and textbooks [11].

4. Advanced STM modes

To study charge carrier distribution in semiconductor devices, we describe three STM-based techniques: a vacuum gap modulation method, a molecule-assisted probing method, and a dual-imaging method.

4.1. Vacuum gap modulation method

A vibrating electrode technique was used to measure the surface potential on solid surfaces by using the Kelvin method [49]. Present-day noncontact atomic force microscopy (nc-AFM) uses vibrating probes for detecting atomic, electrostatic and magnetic forces [50]. In metals, mechanical modulation of the tunnel barrier has been applied as a method to evaluate local work function of the sample [46, 51–54]. In semiconductors, a model of STM junction considering both transparency of the tunnel barrier and the band-bending potential was elaborated [22, 23].

When the STM probe vibrates normal to the sample surface, the gap width changes as

$$Z = Z_0 - dz \cdot \sin\left(\omega t\right), \tag{8}$$

where $\omega = 2\pi \bullet f$ is the angular frequency, dz is an amplitude of the vibration. For small vibration amplitude, $dz \ll Z_0$, the transmission factor periodically changes with the time-dependent change of both the gap width and the gap voltage. When the STM probe approaches toward the surface, V_{gap} is reduced while increasing the surface potential (V_{bb}). A change of the gap voltage V_{gap} is related to the mean charge Q_{SS} at the surface by the Gauss law [43, 47, 48] and is expressed as

$$dV_{gap} = -d\psi = -\frac{|Q_{SS}|}{\epsilon_0} \cdot dz, \tag{9}$$

where $d\psi$ is a change of the band-bending potential.

To determine the tunneling current response (dI) to a time-dependent variation of the gap width, the tunneling current is expressed as

$$I_{tun}(\omega t) = I_0 + dI_1 \cdot \sin(\omega t), \tag{10}$$

where I_0 is the mean tunneling current. In the linear approximation [46], the current response is dominated by variation of the mean transparency of the vacuum gap. Thus, in-phase amplitude of the tunneling current response is given as

$$dI_1 \propto I_0 \cdot \frac{|Q_{SS}|}{\epsilon_0} \cdot dz. \tag{11}$$

In our experiments, the mean tunneling current I_0 is held constant; thus, the quantity (dI_1/dz) is proportional to the local charge density at the surface beneath the STM tip under the bias voltage. There is a 90°-phase-shifted current component representing a displacement current owing to change in the STM junction capacitance as discussed in details in Reference [55]. We used the capacitive signal for fine-tuning of the signal phase in the measurements of in-phase current by a lock-in technique.

In the model above, terms due to the shape of the tunnel barrier and capacitance effects associated with modulation of the band-bending region beneath the STM probe are neglected, albeit the effects are essential at high frequency and low impurity concentration [55].

When the modulation of band-bending region is taken into account, the tunneling current response is given by two terms (Appendix A)

$$\left(\frac{dI}{dz}\right) \propto I_0 \cdot K_3; \quad K_3 = \alpha\sqrt{BH} - \left(\frac{\alpha Z_0}{4\sqrt{BH}} + \beta\right) \cdot \frac{|Q_{SS}|}{\epsilon_0}, \tag{12}$$

The first term represents the contribution of the gap width modulation, and the second term accounts for variations of V_{gap} and V_{bb} .

It is constructive to take a look at origin of charge Q_{SS} for n-type and p-type Si under positive bias voltage. In n-Si in **Figure 3**, the electric field from the STM probe repels mobile electrons deep into the bulk creating a surface depletion region, and $Q_{SS} = Q_N + Q_+ \approx Q_N > 0$. The larger the bias voltage, the larger the amount of positive charge accumulated beneath the STM probe. As a consequence, the amplitude of the current response (dI) depends predominantly on density of accumulated positive charge. On the contrary, in p-type Si under the same polarity bias, the electric field attracts mobile majority carriers (holes) to the surface reducing amount of negative charge of acceptor impurities (Q_P) beneath the STM probe. As a consequence, the amplitude of the current response (dI) depends predominantly on small amount of accumulated positive charge, and $Q_{SS} = Q_P + Q_+ \approx Q_+$. At the position of electrical p-n junction, the balance of positive and negative charges exists, and $Q_{SS} \approx 0$. Thus, we are able to derive position of electrical p-n junction through analysis of the (dI/dz) profiles. In addition, detection of charge centres near the Si surface at a depth of ~1 nm has been reported for epitaxial Si layers [56].

Experimentally, differential tunneling current (dI/dZ) maps were obtained by vibrating the STM probe normal to the sample surface. The STM probe-sample gap was vibrated at a frequency of 12–50 kHz and an amplitude of 20–50 pm while keeping the vacuum gap at constant mean tunneling current I_0 (the constant current mode). In-phase current response dI was measured with a lock-in amplifier at each point in the topographical image. The vibration frequency was selected sufficiently larger than the feedback circuit bandwidth (~10 kHz) and away from the electromechanical resonances of the STM measurement system.

4.2. Molecule-assisted probing method

The ability of specific molecules to selective reactions on the surface is well known in catalysis. Recently, functionalization of SPM probes by attaching functional groups to achieve the chemical selectivity in recognition of DNA sequences and biological molecules has been performed, for example, see [57–59].

The method described here is different. A molecule-assisted probing method makes use of a discrete energy level of an adsorbed molecule as a *marker* of the local Fermi energy. It takes advantage of resonant electron tunneling (RET) to monitor the energy level of the marker molecule, such as fullerene C_{60}, introduced into a tunneling barrier between the STM probe and the oxidized Si surface. The fact that the C_{60}-derived conductance peaks shift in energy depending on dopant concentration in the underlying substrate makes this technique usable as a probing method of the charge carrier profiling on semiconductors [39, 41, 60]. The C_{60} molecule was selected as it satisfies the selection criteria: small size, chemical stability, and an energy position of molecular orbital outside of the Si energy band gap.

A model of a double-barrier junction (DBJ) was elaborated based on the theory of planar resonant tunnel diodes [61] and alignment of molecular states [62]. **Figure 4(a)** and **Figure 4(b)** show the experimental setup and an energy band diagram of an ideal DBJ consisting of the vacuum gap (B1), the C_{60} layer and the thin oxide (B2) under a resonant injection bias V_{RET}. E_A is the electron affinity of the C_{60} layer, and E_i is the Fermi energy for intrinsic Si. At the resonance condition, the Fermi energy of the STM tip aligns with the lowest unoccupied molecular orbital (LUMO), and thus, the strength of electric field in the vacuum gap is given by $F = (\Phi_M - E_A)/Z_0$. For an ideal oxide and neutrality of C_{60}, continuity of the electric displacement is preserved across the DBJ, and the RET voltage is given by

$$V_{RET} = \epsilon_{Si} F \cdot \left(Z_0 + \frac{d_{C_{60}}}{\epsilon_{C_{60}}} + \frac{d_{ox}}{\epsilon_{ox}} \right) + V_{bb},\tag{13}$$

where d_{60} and d_{ox} are the thickness of C_{60} molecule and the oxide, respectively. $\epsilon_{C_{60}}$ and ϵ_{ox} are the permittivity of C_{60} and oxide, respectively. V_{bb} voltage is obtained as a function of the electric field F at the Si surface by solving the 3D Poisson equation at quasi-equilibrium.

To measure the RET voltage, mono-molecular fullerene films were prepared by vapor sublimation of C_{60} to the oxidized Si surfaces at room temperature followed by re-evaporation of excess molecules as described in Section 2.3. Differential conductance $(dI/dV) - V$ spectra in **Figure 4(c)** were obtained at a constant probe-sample gap by using a lock-in technique where a small *ac* voltage (20 mV$_{pp}$, 50 kHz) was superimposed on the sample bias voltage. The initial tunneling conditions were set with a tunneling current of 200 pA at a set-point voltage of 2.5 V. Each $(dI/dV) - V$ spectrum was fitted to Lorentzian function to determine a voltage of the C_{60}-derived conductance peak, the RET voltage [41, 64]. For high conductance of the tunnel gap, the STM tip is close to the molecule layer, and another transport mechanism, the single electron tunneling [66], becomes apparent and hinders the RET voltage detection. Thus, optimization of the gap width is required.

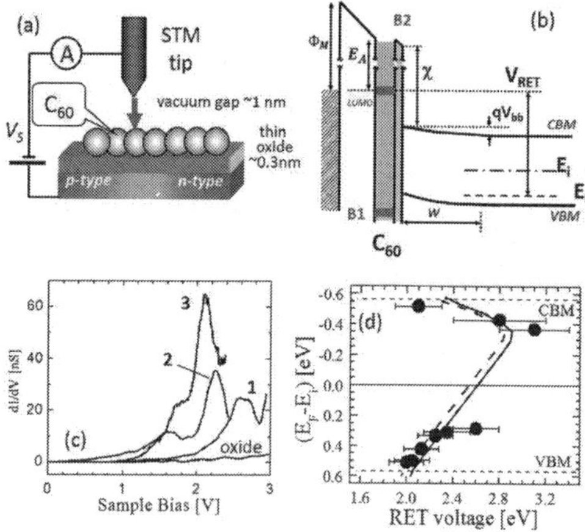

Figure 4. Molecule-assisted probing method. (a) A setup. (b) An energy band diagram of a double-barrier junction under the resonance conditions. B1 is the tunneling gap, and B2 is thin oxide. (c) (dI/dV) spectra of C_{60} on p-type Si substrates with a boron concentration of 8×10^{14} cm^{-3} (curve 1), 4×10^{15} cm^{-3} (curve 2), 3×10^{18} cm^{-3} (curve 3), and without C_{60}. (d) RET voltage as a function of the Si Fermi energy $(E_F - E_i)$ from measurements (symbols) and 3D numerical calculations for oxide thickness of 0.3 nm (broken line) and 0.7 nm (solid line) according to Eq. (13) and Reference [41].

The measured RET voltage obtained for uniformly doped Si wafers with different dopant concentrations is shown in **Figure 4(d)**. The data are well reproduced by the numerical calculations according to Eq. (13) where STM probe emitter was modeled as a cone with a hemispherical end and a radius of curvature of 10 nm, and $Z_0 = 1$ nm, $d_{C60} = 1$ nm , and $d_{ox} = 0.3$ nm, $\Phi_M = 4.5$ eV for W(111) probes and $E_A = 2.6$ eV. The good agreement between the calculated RET voltage and the experimental data for uniform-doped wafers verifies the calibration relationship for Si [41, 63].

The spatial resolution of the method is restricted to the size of the *marker* molecule and to the electric field penetration length. It has been demonstrated by the (dI/dV) mapping that the RET peaks are localized within the C_{60} core (~1 nm) due to their origin in resonant tunneling mediated by one lowest unoccupied molecular orbital (LUMO+1) of C_{60} [41]. Since the LUMO +1 was localized at the pentagonal rings [65] and C_{60} molecule rotates at room temperature, the observed peak intensity represents the orientation-averaged orbital conductance of C_{60}. The estimate of the penetration depth is a Debye length of ~1.5 nm for *p*-Si under large positive bias, though the length depends on the dopant concentration for n-Si [41, 63].

4.3. A dual-imaging method

STM technique is limited to conductive surfaces and is inapplicable to the imaging of novel device structures, including insulator surfaces such as silicon-on-insulator (SOI) devices. Strong interest to such measurements is stimulated by the fact that discrete dopant distribution enables attractive applications such as quantum computing [67] and single-electron devices [68]. Therefore, a dual-imaging method was developed to enable simultaneous measurements of electric current and interaction force acting on the scanning probe. It was achieved by attaching an STM metal tip to a special force sensor [67–76].

Figure 5 shows the experimental setup for the simultaneous measurement of tunneling current (I_{tun}) and force between the metal probe tip and the Si surface. In our technique, the interaction force gradient between the metal probe tip and the surface was detected as a shift in the resonance frequency (Δf) of a quartz length extension resonator (qLER) which vibrated at ~1 MHz (Q factor ~50,000) with an amplitude of 0.05–0.3 nm [67–70]. The probe tips were made of a tungsten wire with a diameter of 10 μm. The wire was attached to the quartz resonator and sharpened by the focused ion beam technique (FIB). Typically, the probe tips had a diameter of Ø30 nm and the aspect ratio of more than 10, resulting in small stray capacitance. Detection of the frequency shift by electric means makes such sensors suitable for measurements in ultra-high vacuum environment and at different temperature, which are often required in nanomaterial and nanoscale device research.

The advantages of our multimode scanning probe microscopy (MSPM) system are

- tunneling current and forces acting on the probe tip are measured *simultaneously* at a mean probe-sample gap of about 1 nm in constant current (CC) or constant force (CF) operation modes;

- small vibration amplitude (0.1–0.2 nm) enables us to drastically reduce the probe-sample gap, leading to better *spatial resolution*;

- the *sensitivity* to electrostatic forces is increased at an optimal gap;

- the force detection is performed in a *noncontact* manner, which is suitable for measurements of solid crystals and thin films.

In the CC mode, a force gradient map is measured while the mean gap (Z_0) maintains a set-point tunneling current. Typically, the measurement condition corresponds to a gap of approximately 1 nm, as estimated from the distance dependence of the tunneling current [72]. The spatial variation of the frequency shift (Δf) reflects variations in the interaction force caused by charge carriers, impurity charges, and surface imperfections as illustrated in **Figure 5(b)**. When a donor is present in proximity to the STM tip, the attractive force acting on the tip increases owing to Coulomb interaction between the donor charge and the image charge induced in the STM tip, leading to measurable change in the Δf value [75, 76]. The interaction strength depends on the depth of the donor location and the electrostatic screening by mobile carriers. Experimentally, lateral extent of 5–10 nm and a detection depth of ~1 nm have been reported for phosphorus and boron atoms in Si [32, 33, 76]. Change in the interaction force on grains with different work function was employed for recognizing crystal orientation of sub-10-nm-size grains in nano-crystalline TiN films [77].

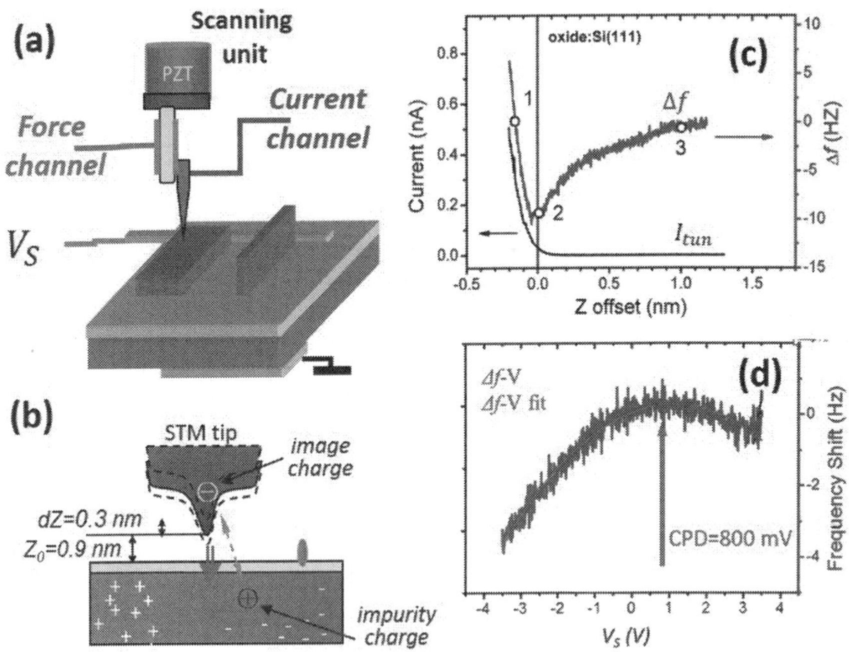

Figure 5. Dual-imaging method. (a) A measurement setup. (b) A sketch of interaction force acting on a vibrating STM probe. (c) (I_{tun}-Z) and (Δf-Z) spectra showing ranges of repulsive interaction (1–2) and attractive Coulomb interaction (2–3) for an oxide-passivated Si(111) surface (a set-point: 30 pA, 2.0 V). (d) A measured (Δf-V_S) spectrum at position 3(blue curve), and a result of fitting to Eq. (14) (red curve).

In the CF mode, a tunneling current (I_{tun}) map is measured while the mean gap (Z_0) is maintained at a constant frequency shift. There are two ranges in distance dependences of I_{tun} and Δf as indicated in **Figure 5(c)** for an oxide-passivated Si(111) surface. At short distances (range 1–2), repulsive interaction dominates, and current exponentially grows when the STM tip approaches the surface. At longer distances (range 2–3), the electrostatic Coulomb interaction dominates. There is an optimal distance indicated as position 2 in **Figure 5(c)** where the sensitivity to electrostatic force is maximum [72]. At this distance, the ($\Delta f - V_S$) spectrum has the largest curvature.

Under the applied voltage V_S, the electrostatic force gradient between the probe tip and the sample is expressed according to the theory in References [73, 78] for small vibration amplitude

$$\Delta f \propto \frac{\partial F}{\partial z} = -\frac{1}{2}(V_S - CPD)^2 \cdot \frac{\partial^2 C}{\partial z^2}, \tag{14}$$

where C is the effective tip-sample capacitance. CPD, the contact potential difference, refers to the difference between the work function of the metal probe (Φ_M) and the Fermi energy of the underlying Si (E_F), and is given by

$$CPD = \frac{1}{q}(E_F - \Phi_M), \tag{15}$$

where q is the elementary charge. A local value of the CPD voltage, which is determined by local charge concentration in the underlying Si, can be obtained by fitting of the spectrum to Eq. (14). In the example in **Figure 5(d)**, a CPD voltage of +0.8 V was obtained for an oxidized p-Si(111) surface. The CPD voltage mapping was employed in 2D analysis of the built-in potential in small Si MOSFET devices [79] and p-n junctions [72] showing the attainable spatial resolution better than 3 nm. Particular applications of the CF mode also include analysis of impurity distribution profiles from I_{tun} maps measured at different bias voltage [80], non-uniform distribution of photocarrier in Si stripes [81], and nanoscale conductance switching in phase-change GeSbTe thin films [82].

5. Application examples

5.1. Channel length in small MOSFET

For STM measurements, cross-sections of Si MOSFETs were prepared by ultra-fine polishing to expose (110) surfaces and were passivated by ultra-thin oxide layer as described in Section 2.2. Si n-type MOSFET with nominal gate lengths (L_G) in the range of 20–150 nm were fabricated according to a process described in Reference [83]. The measurements were done with W(111) crystal probes in an ultrahigh vacuum (~4 × 10^{-9} Pa) at room temperature.

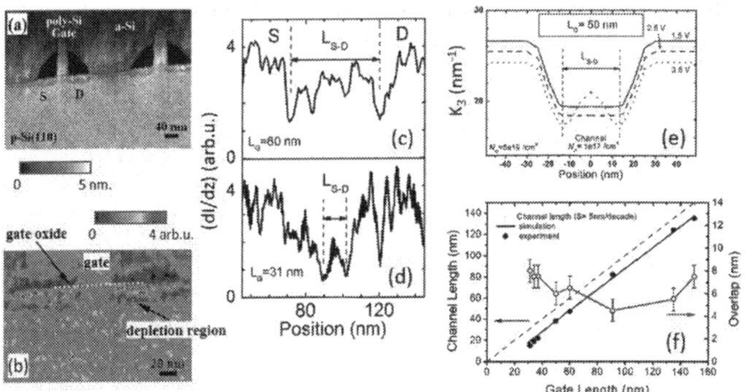

Figure 6. (a) Topographic image of a cross-section containing two small Si MOSFET devices. (b) A (dI/dZ) map of a device with a gate length of 31 nm (a set-point: 230 pA, 3.4 V, dz = 20 pm). (c)–(d) Line profiles measured at 12 nm depth beneath the gate electrodes showing the electric channel length (L_{S-D}). (e) Profiles calculated by Eq. (12) for expected impurity distribution. (f) Measured electric channel length (symbols) as a function of gate length. Line is the calculation result.

Topographic image of two small MOSFET is shown in **Figure 6(a)**, where the gate electrodes are surrounded by two black cavities produced by sidewall oxide etching during the surface preparation. The source/drain (S/D) extensions on the left- and right-hand sides of the gate electrode are seen as bright stripes in the (dI/dZ) map in **Figure 6(b)**. Depletion regions separate the S/D extensions from the p-type channel beneath the gate electrode and the Si bulk. The extension depth is ~18 nm as measured from the gate oxide. The electric channel length (L_{S-D}) was determined as the distance between 2 minima in (dI/dZ) line profiles measured at a depth of 12 nm beneath the gate oxide as indicated in **Figure 6(c, d)**. Calculated profiles of the K_3 factor in **Figure 6(e)** reproduce the measured (dI/dZ) profiles, confirming that each minimum in (dI/dZ) signal represents the position of the electric p-n junction. L_G was determined from STM topographs. Results summarized in **Figure 6(f)** give an overlap value of 6 ± 1 nm, which is in excellent agreement with a transverse straggle of 7 nm for an implanted ion energy of 25 keV. An accuracy of the channel measurements was about 1 nm at 3.4 V, while the measurements were affected by random positions of individual ionized dopant atoms in the extension regions.

5.2. Super-junction devices fabricated by the channeling ion implantation

The C_{60}-assisted probing technique has been actually applied to quantitative analysis of charge carrier profiles on cross-sections of power MOSFET, where the precise control over the doping profile is essential to obtain low ON-state resistance and high breakdown voltage [39, 40]. **Figure 7(a)** depicts a schematic structure of a super-junction power MOSFET. Two p-type islands were formed by multiple boron ion implantations into the low-doped n-type epitaxial layer with a carrier density of ~1 × 10^{16} cm^3. In **Figure 7(b)**, we clearly see that two p-type islands are separately formed with the same peak concentrations, confirming the anticipated dopant concentration. Moreover, the experimental data revealed an extension of island 1 beyond the expected depth, which is attributed to a scatter-less travel of boron ions through Si crystal at high implantation energy, the ion channeling effect[84].

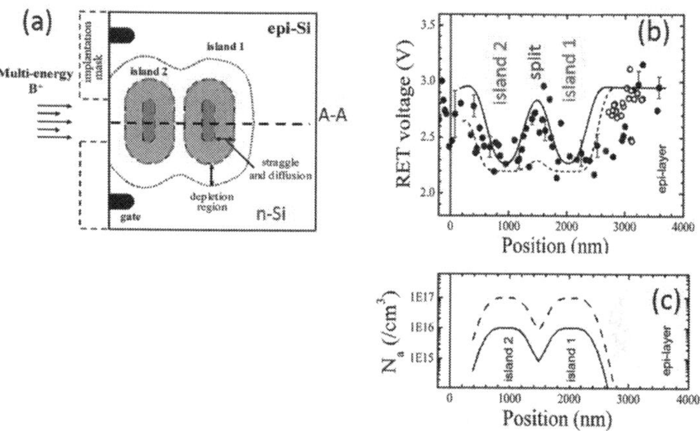

Figure 7. (a) Schematic structure of a super-junction device showing two p-Si islands made by boron ion implantation. (b) Depth profiles of the RET voltage along center of the device: measured data (symbols) taken with 20-nm steps. Profiles (lines) were calculated for the two boron density profiles shown in (c). Reproduced with the permission from Reference [39].

5.3. Length-dependent resistivity of Si nanowires

The ability of the dual-imaging method for characterization of modern silicon-on-insulator (SOI) devices is illustrated by analysis of the structure and electric conductance of SOI nanowires (NW) with different surface passivation. Note that the NW is the promising structure for sub-10-nm MOSFETs and for such functional devices as chemical sensors. **Figure 8** shows high-resolution measurements of a Si NW with a cross-section area of 20×20 nm^2 acquired at a set point of $\Delta f = 0.6$ Hz, $dz = 95$ pm, $V_S = -1.5$ V. We see in **Figure 8(c)** the current gradually decreases in the NW interior with the distance from the Si pad owing to the dependence of the NW resistivity on its length. We note that an apparent NW width in the current map is about 2-fold of that in the topograph. As the NW is protruded above the buried oxide (BOX) by 20 nm, a side surface of the sharp tip touches the NW as illustrated in the insert of **Figure 8(c)**, and this results in a so-called "sidewall" current outside the Si NW body. The current value and fluctuations were reduced for the NW passivated with an ultrathin oxide layer compared to the hydrogen passivation. The tunneling current decreased within a distance of ~300 nm from the Si pad electrode for both types of surface termination. At the negative voltage, the tunneling current is defined by electrons traveling from large Si pad through the SOI nanowire, and the current value is determined by resistivity of the NW volume and the surface conduction. The macroscopic conduction model including the conductance contributions of the nanowire volume and the surface states confirmed the length-dependent conductance of thin Si nanowires [85].

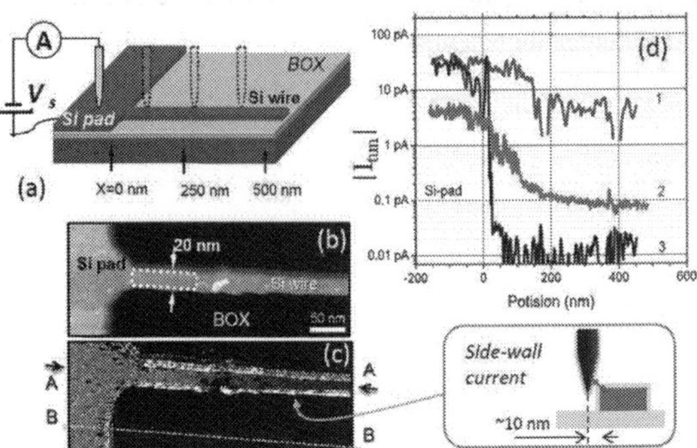

Figure 8. (a) An experiment setup. (b) Topographic image of silicon-on-insulator nanowire with a cross-section of 20 × 20 nm², and (c) corresponding current map acquired at -1.5 V and Δf = 0.6 Hz, dz = 95 pm. (d) Current profiles along A-A line for Si nanowires after hydrogen-passivation (curve 1), oxide passivation (curve 2), and along B-B line (curve 3). Adopted from Reference [85] (Copyright 2013 Trans. Mat. Res. Soc. Japan).

5.4. Wavelength-dependent photocarrier distribution across strained Si stripes

Photo-carrier generation in semiconductors is a fundamental process utilized in solar cells and photo-detectors. For reduced size of modern detectors, the role of structural elements in carrier accumulation and transport has been increasing [86]. In particular, photocarrier distribution on textured surfaces of Si can be a factor to improve the efficiency of solar cells. Analysis of spatial distribution of photocurrent (PC) in strained Si stripes under tilted illumination gives an insight into photocarrier behavior near the stripe edges with an effective spatial resolution of ~10 nm [81].

Figure 9 shows the sample structure and the measurement setup, where inhomogeneous light intensity profile was created under tilted (50° off-normal) illumination and different light wavelength (λ). Strained Si stripes of 50–1000 nm in width and 300 nm in height were fabricated on Si(001) wafer, and separated by SiO_2. The stripe surface was passivated by an ultrathin oxide as described in Section 2.2. The light intensity was mechanically modulated at frequency of ~3 kHz, and the PC signal was measured by a lock-in unit. Topographs and PC maps were measured by the dual-imaging method where the tip-sample gap was set by a set-point of Δf = 1.2 Hz, dz = 130 pm, and V_S = − 0.8 V, using the CF mode.

Topographic image in **Figure 9(b)** shows uniform surface of the Si stripe. The PC signal was not uniform, and large at a distance of ~50 nm from the stripe edge on the light illumination side, when stripes were illuminated with laser light and an intensity of 12 mW/cm² as seen in **Figure 9(c)**. Large PC signal at stripe edges was observed irrespective of the scanning directions, when light with λ = 405 and 364 nm was used as seen in line profiles in **Figure 9(d, e)**. In contrast, illumination with red light (λ = 675 nm) produced uniform PC distribution. As the absorption depth in Si is ~11 nm for λ = 364 nm, ~130 nm for λ = 405 nm, and ~4000 nm for λ = 675 nm [87], the respective illumination produces different light intensity profiles. Calculated PC profiles in **Figure 9(f)** reproduced the observed PC distributions when a rectangular bar geometry, non-coherent light, and a photocarrier diffusion length of 100 nm were used [81].

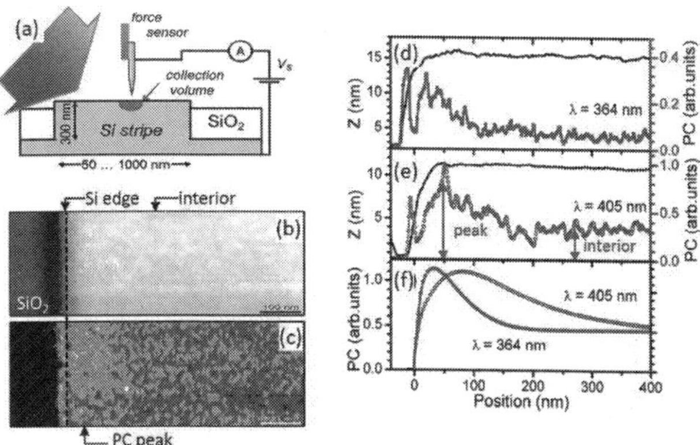

Figure 9. (a) Photocurrent (PC) measurement setup. (b) A topograph and (c) corresponding PC map of a Si stripe under illumination with λ = 405 nm. (d)–(e) Measured line profiles of height (black lines) and PC (dotted lines) across the stripe edge under tilted illumination for two wavelengths (λ). (a set-point: Δf = 1.2 Hz, dZ = 130 pm, V_s = − 0.8 V). (f) PC line profiles calculated for a rectangular bar exposed to light at top and side surfaces. Adopted from Reference [81] (Copyright 2012 The Japan Society of Applied Physics).

We note that the relative intensity of a PC peak at a position of ~30 nm for λ = 364 nm is ~3.2-fold the signal in the stripe interior. Enhancement of light intensity by ~3.5-fold at strained Si stripe edges has been reported for λ = 364 nm [88, 89]. The enhancement mechanism may be related to increased photocarrier generation owing to interference of coherent laser light [81], narrowing of the Si energy gap under stress [90] or increase in the tunneling probability through electromagnetic field coupling to the sharp STM tip [91].

6. Simulations of tunneling current spectra

STM has the capability to 2D impurity profiling by employing advanced STM methods as shown above. Although, accurate analysis of charge carrier distributions in actual 2D and 3D device structures has been a substantial challenge. STM tunneling current is a complex function of structural, material, and electronic parameters of the system consisting of a 3D probe tip and a semiconductor. On the basis of fundamental theory, there have been theoretical discussions of 1D and 2D treatments for the STM junction geometry. A 3D numerical simulator has been reported that solves the 3D potential distribution of the sample STM probe system and calculates the tunneling current, so-called *the potential-based model* [23, 92, 93]. However, to describe the precise physics of STM measurements, the charge carrier flow in the sample must be included, as evidenced by the NW measurements in **Figure 8**. Recently, new model evolves solving the charge carrier transport between a probe tip and a sample consistently with the current continuity equation, so-called *the current-continuity model*. The current-continuity model accounts for charge carrier transport between states in an STM probe and the conduction and the valence band of Si and was implemented on the basis of a technology computer-aided design (TCAD) semiconductor device simulator code [94]. It is a significant advancement in the field.

An analysis based on the current-continuity model has been applied successfully to extracting impurity distribution profiles in a MOSFET from experimental current maps measured by the dual-imaging method [80], and for evaluating photocarrier dynamics in Si nanowires with a cross-section of 10×10 nm^2 [95].

The remaining challenge is to include the effect of single impurity scattering on charge carrier transport in nanoscale devices. The impurity scattering for a thin semiconductor wire has been solved using the 3D Green function approach and the numerical Monte-Carlo method [96]. An atomistic view into an impurity atom appearance in STM images has been elaborated within the framework of a self-consistent-charge density functional tight-binding method (SCCDFTB), for example, see [97, 98].

7. Conclusion

Advanced STM-based methods for 2D analysis of charge carrier distributions in semiconductor devices with high spatial resolution represent the substantial development of scanning probe microscopy. The described methods rely on detection and analysis of tunneling current which is strongly localized within an atomic dimension. This leads to significant improvement in the sensitivity and spatial resolution for measuring local electric characteristics of Si devices and nanowires, when effects of surface states are suppressed by adequate surface treatment.

The gap modulation method can attain an ultimate spatial resolution comparable to that of STM topographic images in p-n junction regions, and can detect individual charged impurity atoms along the surface at a depth of few nanometers. Quantitative evaluation of charge distributions can be derived by comparing experimental data and simulations of the underlying charge concentration. The accuracy relies on the ability of the simulation to account for quantum phenomena, and further development of simulations based on the current-continuity model will be essential.

The capability of the molecule-assisted probing method has been demonstrated with the use of C_{60} molecules. A spatial resolution of ~1 nm is determined by the size of the molecule. However, the C_{60} film on oxidized Si surfaces leaves ~20% uncovered areas. The coverage can be increased by the use of chemically modified C_{60} or other small molecules those formed a monomolecular-thick film on SiO_2 surface. For high conductance of the tunnel gap, another transport mechanism, the single electron tunneling [66], becomes dominant and obscures the RET voltage measurements. Thus, optimization of the gap width is required.

The presented methods can be used for measuring on rough surfaces, but careful data analysis should be performed to discard "artifacts." In the gap modulation method, the tip vibration amplitude (dz) varies with tilt angle of the underlying surface, causing changes in the (dI/dZ) signal. In the dual-imaging method, large "sidewall" current such as shown in **Figure 8** must be considered in data analysis. Also, atomically ordered surfaces can be obtained by cleavage, yet, to attain ultimate spatial resolution, STM measurements in well-controlled environment such as in an ultrahigh vacuum are necessary, where we can avoid undesirable effects caused by absorption of charged particles and molecules from air.

To summarize, specific features of the presented 2D STM-based methods are (a) noncontact, stress-free measurements allowing analysis of delicate sample structures; (b) high spatial sensitivity to electrostatic field, which is substantial advancement in comparison with scanning Kelvin probe microscopy; (c) the ability to study nanoscale structures with a lateral size of 20 nm and below, which are inaccessible by other techniques.

Further applications of the advanced STM methods will contribute to high-spatial resolution analysis of modern sub-100-nm electronic devices, functional nanowire devices, and novel devices incorporating two-dimensional materials such as graphene and topological superlattices. It will advance our understanding of charge carrier transport at nanoscale and encourage inventing novel energy-efficient devices.

8. Appendix A

The tunneling current is described as a periodic function as

$$I_{tun}\left(t\right) = I_0 + dI \cdot \sin \omega t = I_0 \cdot \left(1 + K_3 \cdot dz \cdot \sin \omega t\right), \tag{A1}$$

The mean tunneling current is given in terms of the thermionic emission approximation including the vacuum tunneling term according to Reference [99] as

$$I_0 = A_e^* T^2 \cdot \exp\left(-\alpha\sqrt{BH} \cdot Z_0\right) \cdot \exp\left(-\beta V_{bb}\right) \cdot \left[\exp\left(\beta V_{gap}\right) - 1\right], \tag{A2}$$

For $dz \ll Z_0$, the factor K_3 is derived considering only linear terms of dz and $d\psi$, and is given by

$$K_3 = \alpha\sqrt{BH} - \left(\frac{\alpha Z_0}{4\sqrt{BH}} + \beta\right) \cdot \frac{|Q_{SS}|}{\epsilon_0}, \tag{A3}$$

The area charge concentration at the Si surface (Q_{SS}) is obtained by solving the Poisson equation. An analytic solution for a 1D abrupt junction is given by [47]

$$Q_{SS} = -\frac{\epsilon_{Si}}{\beta \cdot \Lambda} \cdot G(V_{bb}), \tag{A4}$$

$$G(y) = \left[\left(e^{-\beta y} + \beta y - 1 \right) + \frac{n}{p_0} \left(e^{\beta y} - \beta y - 1 \right) \right]^{1/2} \geq 0, \tag{A5}$$

$$\Lambda = \sqrt{\frac{\epsilon_{Si}}{\beta Q_N}}, \tag{A6}$$

Λ is the extrinsic Debye length, and volume densities of positive (p_0) and negative (n) charge are in the Si bulk. The factor $\beta = 1/k_B T$, and k_B is the Boltzman constant, T is temperature.

For 3D structures, a charge concentration at the semiconductor surface (Q_{SS}) is obtained by numerically solving the Poisson equation.

Acknowledgements

The authors would like to thank colleagues of Nanoelectronics Research Institute (AIST, Japan) for valuable discussions and constructive comments motivating the research works.

Author details

Leonid Bolotov* and Toshihiko Kanayama

*Address all correspondence to: bolotov.leonid@aist.go.jp

National Institute of Advanced Industrial Science and Technology (AIST), Tsukuba, Ibaraki, Japan

References

[1] Müller E. W, Panitz J. A., and McLane S. B.: The atom probe field ion microscope. Review of Scientific Instruments. 1968; 39: 83–86. DOI:10.1063/1.1683116

[2] Kelly T. F. and Miller M. K.: Atom probe tomography. Review of Scientific Instruments. 2007; 78: 031101. DOI:10.1063/1.2709758

[3] Miller M. K., Cerezo A., Hetherington M. G., and Smith G. D. W. Atom Probe Field Ion Microscopy. New York: Oxford University Press; 1996. 509 p. ISBN-13: 978–0198513872.

[4] Han B., Takamizawa H., Shimizu Y., Inoue K., Nagai Y., Yano F., Kunimune Y., Inoue M., and Nishida A.: Phosphorus and boron diffusion paths in the polycrystalline silicon gate of a trench-type three-dimensional metal-oxide-semiconductor field-effect transistor investigated by atom probe tomography. Applied Physics Letters. 2015; 107: 023506. DOI:10.1063/1.4926970.

[5] Kambham A. K., Mody J., Gilberta M., Koelling S., and Vandervorst W.: Atom-probe for FinFET dopant characterization. Ultramicroscopy. 2011; 111: 535–539. DOI:10.1016/j.ultramic.2011.01.017

[6] CAMECA: Science & Metrology Solutions [Internet]. 2016. Available from: http://www.cameca.com/instruments-for-research/atom-probe.aspx [Accessed 2016-02-08].

[7] Kazemian P., Mentink S. A. M., Rodenburg C., and Humphreys C. J.: Quantitative secondary electron energy filtering in a scanning electron microscope and its applications. Ultramicroscopy. 2007; 107(2–3): 140–150. DOI:10.1016/j.ultramic.2006.06.003

[8] Venables D., Jain H., and Collins D. C.: Secondary electron imaging as a two-dimensional dopant profiling technique: Review and update. Journal of Vacuum Science and Technology B. 1998; 16(1): 362–366. DOI:10.1116/1.589811

[9] Kazemian P., Mentink S. A. M., Rodenburg C., and Humphreys C. J.: High resolution quantitative two-dimensional dopant mapping using energy-filtered secondary electron imaging. J. Applied Physics. 2006; 100: 054901. DOI:10.1063/1.2335980

[10] Masters R. C., Pearson A. J., Glen T. S., Sasam F.-C., Li L., Dapor M., Donald A. M., Lidzey D. G., and Rodenburg C.: Sub-nanometre resolution imaging of polymer–fullerene photovoltaic blends using energy-filtered scanning electron microscopy. Nature Communications. 2015; 6: 6928. (9 pages) DOI:10.1038/ncomms7928

[11] Weisendanger R. Scanning Probe Microscopy and Spectroscopy: Methods and Applications. Cambridge: Cambridge University Press; 1994. 637p. DOI:10.1017/CBO9780511524356

[12] Zavyalov V. V., McMurray J. S., and Williams C. C.: Scanning capacitance microscope methodology for quantitative analysis of p-n junctions. Journal of Applied Physics. 1999; 85: 7774–7783. DOI:10.1063/1.370584

[13] DeWolf P., Vandervorst W., Smith H., and Khalil N.: Comparison of two-dimensional carrier profiles in metal-oxide-semiconductor field-effect transistor structures obtained with scanning spreading resistance microscopy and inverse modeling. Journal of Vacuum Science and Technology B. 2000; 18: 540–544. DOI:10.1116/1.591228

[14] Tabe M., Moraru D., Ligowski M., Anwar M., Yokoi K., Jablonski R., and Mizuno T.: Observation of discrete dopant potential and its application to Si single-electron devices. Thin Solid Films. 2010; 518: S38–S42. doi:10.1016/j.tsf.2009.10.051

[15] Binnig G., Rohrer H., Gerber C., and Weibel E.: Surface studies by scanning tunneling microscopy. Physical Review Letters. 1982; 49: 57–82. DOI:10.1103/PhysRevLett.49.57

[16] Alvarez D., Hartwich J., Fouchier M., Eyben P., and Vandervorst W.: Sub-5-nm-spatial resolution in scanning spreading resistance microscopy using full-diamond tips. Applied Physics Letters. 2003; 82: 1724–1726. DOI:10.1063/1.1559931

[17] Zhang L., Ohuchi K., Adachi K., Ishimaru K., Takayanagi M., and Nishiyama A.: High-resolution characterization of ultrashallow junctions by measuring in vacuum with scanning spreading resistance microscopy. Applied Physics Letters. 2007; 90: 192103. DOI:10.1063/1.2736206

[18] Moraru D., Ligowski M., Yokoi K., Mizuno T., and Tabe M.: Single-electron transfer by inter-dopant coupling tuning in doped nanowire silicon-on-insulator field-effect transistors. Applied Physics Express. 2009; 2: 071201. DOI:10.1143/APEX.2.071201

[19] DeWolf P., Stephenson R., Trenkler T., Clarysse T., Hantschel T., and Vandervorst W.: Status and review of two-dimensional carrier and dopant profiling using scanning probe microscopy. Journal of Vacuum Science and Technology B. 2000; 18: 361–368. DOI:10.1116/1.591198

[20] Jager N. D., Marso M., Salmeron M., Weber E. R., Urban K., and Ebert P.: Physics of imaging p–n junctions by scanning tunneling microscopy and spectroscopy. Physical Reviews B. 2003; 67: 165307. DOI:10.1103/PhysRevB.67.165307

[21] Fukutome H., Arimoto H., Hasegawa S., and Nakashima H.: Two-dimensional characterization of carrier concentration in metal-oxide-semiconductor field-effect transistors with the use of scanning tunneling microscopy. Journal of Vacuum Science and Technology B. 2004; 22: 358–363. DOI:10.1116/1.1627792

[22] Weimer M., Kramar J., and Baldeschwieler J. D.: Band bending and the apparent barrier height in scanning tunneling microscopy. Physical Reviews B. 1989; 39: 5572–5577. DOI: 10.1103/PhysRevB.39.5572

[23] Feenstra R. M.: Electrostatic potential for a hyperbolic probe tip near a semiconductor. Journal of Vacuum Science and Technology B. 2003; 21: 2080–2088. DOI: 10.1116/1.1606466

[24] Morrison S. R. The chemical physics of surfaces. New York: Plenum Press; 1977. 415p. DOI:10.1007/978-1-4615-8007-2

[25] Mariani G., Scofield A. C., Hung C.-H., and Huffaker D. L.: Gas nanopillar-array solar cells employing in situ surface passivation. Nature Communications. 2013; 4: 1497. DOI:10.1038/ncomms2509

[26] Yablonovitch E., Allara D. L., Chang C. C., Gmitter T., and Bright T. B.: Unusually low surface-recombination velocity on silicon and germanium surfaces. Physical Review Letters. 1986; 57: 249–252. DOI:10.1103/PhysRevLett.57.249

[27] Liu L., Yu J. and Lyding J. W.: Atom-resolved three-dimensional mapping of boron dopants in Si(100) by scanning tunneling microscopy. Applied Physics Letters. 2001; 78: 386–388. DOI:10.1063/1.1339260

[28] Liu F. Y., Griffin P. B., Plummer J. D., Lyding J. W., Moran J. M., Richards J. F., and Kulig L.: Carrier profiling via scanning tunneling spectroscopy: comparison with scanning capacitance microscopy. Journal of Vacuum Science and Technology B. 2004; 22: 422–426. DOI:10.1116/1.1643054

[29] Bell L. D., Kaiser W. J., Hecht M. H., and Grunthaner F. J.: Direct control and characterization of a Schottky barrier by scanning tunneling microscopy. Applied Physics Letters. 1988; 52: 278–280. DOI:10.1063/1.99493

[30] Okui T., Hasagawa S., Nakashima H., Fukutome H., and Arimoto H.: Visualization of 0.1-μm-metal-oxide-semiconductor field-effect transistors by cross-sectional scanning tunneling microscopy. Applied Physics Letters. 2002; 81: 2475–2477. DOI: 10.1063/1.1509118

[31] Lequn L., Yu J., and Lyding J. W.: Subsurface dopant-induced features on the Si(100)2×1:H surface: fundamental study and applications. IEEE Transactions on. Nanotechnology. 2002; 1: 176–183. DOI:10.1109/TNANO.2002.807391

[32] Nishizawa M., Bolotov L., Tada T., and Kanayama T.: Scanning tunneling microscopy detection of individual dopant atoms on wet-prepared Si(111):H surfaces. Journal of Vacuum Science and Technology B. 2006; 24: 365–369. DOI:10.1116/1.2162564

[33] Nishizawa M., Bolotov L., and Kanayama T.: Scanning tunneling microscopy observation of individual boron dopant atoms beneath Si(001)-2x1 surface. Japanese Journal of Applied Physics. 2005; 44: L1436–L1438. DOI:10.1143/JJAP.44.L1436

[34] Nishizawa M., Bolotov L., and Kanayama T.: Simultaneous measurement of potential and dopant atom distributions on wet-prepared Si(111):H surfaces by scanning tunneling microscopy. Applied Physics Letters. 2007; 90: 122118. DOI: 10.1063/1.2716837

[35] Morita Y. and Nishizawa M.: Surface preparation of Si(001) substrate using low-pH HF solution. Applied Physics Letters. 2005; 86: 171907. DOI:10.1063/1.1915515

[36] Miyata N., Watanabe S., and Ichikawa M.: Nanometer-scale Si-selective epitaxial growth using an ultrathin SiO$_2$ mask. Journal of Vacuum Science and Technology B. 1999; 17: 978–982. DOI:10.1116/1.590679

[37] Cai Q., Hu Y. F., Hu S. T., and Wang X.: Scanning tunneling microscopy studies of ultrathin gate oxide films grown on highly B-doped Si(100) substrates. Journal of Vacuum Science and Technology B. 2000; 18: 2384–2387. DOI:10.1116/1.1289927

[38] Yasuda T., Nishizawa M., Kumagai N., Yamasaki S., Oheda H., and Yamabe K.: Atomic-layer resolved monitoring of thermal oxidation of Si(001) by reflectance difference

oscillation technique. Thin Solid Films. 2004; 455–456: 759–763. DOI:10.1016/j.tsf. 2003.11.262

[39] Bolotov L., Nishizawa M., Miura Y., and Kanayama T.: Carrier concentration profiling on oxidized surfaces of Si device cross sections by resonant electron tunneling scanning probe spectroscopy. Journal of Vacuum Science and Technology B. 2008; 26: 415–419. DOI:10.1116/1.2802103.

[40] Kanayama T., Nishizawa M., and Bolotov L.: Dopant and carrier concentration profiling with atomic resolution by scanning tunneling microscopy. ECS Transactions. 2009; 19(1): 117–124. DOI:10.1149/1.3118937

[41] Bolotov L., Uchida N., and Kanayama T.: Scanning tunneling spectroscopy of atomic clusters deposited on oxidized silicon surfaces: induced surface dipole and resonant electron injection. Journal of Physics: Condensed Matter. 2003; 15: S3065–S3081. DOI: 10.1088/0953-8984/15/42/006.

[42] Bitzer T., Rada T., Richardson N. V., Dittrich T., and Koch F.: Gap state formation during the initial oxidation of Si(100)-2×1. Applied Physics Letters. 2000; 77: 3779–3783. DOI: 10.1063/1.1330222

[43] Simmons J. G., Hsueh F. L., and Faraone L.: Two-carrier conduction in MOS tunnel-oxides II-theory. Solid State Electronics. 1984; 27: 1131–1139. DOI: 10.1016/0038-1101(84)90055-8

[44] Simmons J. G.: Generalized formula for the electric tunnel effect between similar electrodes separated by a thin insulating film. Journal of Applied Physics. 1963; 34: 1793–1803. DOI:10.1063/1.1702682

[45] Simmons J. G.: Electric tunnel effect between dissimilar electrodes separated by a thin insulating film. Journal of Applied Physics. 1963; 34: 2581–2590. DOI:10.1063/1.1729774

[46] Tersoff J. and Hamann D. R.: Theory and application for the scanning tunneling microscope. Physical Review Letters. 1983; 50: 1998–2001. DOI:10.1103/PhysRevLett. 50.1998

[47] Sze S. M. Physics of Semiconductor Devices. New York: Wiley; 1981. 868 p. ISBN: 0-471-05661-8.

[48] Filip V., Wong H., and Nicolaescu D.: Quantum charge transportation in metal-oxide-Si structures with ultrathin oxide. Journal of Vacuum Science and Technology B. 2006; 24: 38–45. DOI:10.1116/1.2138720

[49] The Kelvin Probe Principles – KP Technology Ltd. [Internet]. Available from: http://www.kelvinprobe.info/technique-theory.htm [Accessed 2016-02-10].

[50] Morita S., Wiesendanger R., and Meyer E., editors. Noncontact Atomic Force Microscopy. Berlin: Springer; 2002. 440 p. DOI:10.1007/978-3-642-56019-4

[51] Binnig G. and Roher H.: Scanning tunneling microscopy. Surface Science. 1983; 126: 236–244. DOI:10.1016/0039-6028(83)90716-1

[52] Saida M., Horikawa K., Sato T., Yamomoto A., Sasaki M.: Local tunneling barrier height observations of NiAl(1 1 0). Surface Science. 2006; 600: L139–L141. DOI:10.1016/j.susc. 2006.03.023

[53] Jia J. F., Inoue K., Hasegawa Y., Yang W. S., and Sakurai T.: Variation of the local work function at steps on metal surfaces studied with STM. Physical Reviews B. 1998; 58: 1193–1197. DOI:10.1103/PhysRevB.58.1193

[54] Arai T. and Tomitori M.: Bias dependence of Si(111)7×7 images observed by noncontact atomic force microscopy. Applied Surface Science. 2000; 157: 207–211. DOI:10.1016/ S0169-4332(99)00527-9

[55] Majima Y., Oyama Y., and Iwamoto M.: Measurement of semiconductor local carrier concentration from displacement current–voltage curves with a scanning vibrating probe. Physical Reviews B. 2000; 62: 1971–1977. DOI:10.1103/PhysRevB.62.1971

[56] Bolotov L., Nishizawa M., and Kanayama T.: Surface potential mapping and charge center detection on oxidized silicon surfaces by vacuum-gap modulation scanning tunneling spectroscopy. AIP Conference Proceedings. 2007; CP931: 535–538. DOI: 10.1063/1.2799431

[57] Bruker Nano Surfaces Business. Application Note #130 [Internet]. Common Approaches to Tip Functionalization for AFM-Based Molecular Recognition Measurements. http://www.bruker.co.jp/axs/nano/imgs/pdf/AN130.pdf [Accessed 2016-02-10].

[58] Benoit M., Gabriel D., Gerisch G., and Gaub H. E.: Discrete interactions in cell adhesion measured by single-molecule force spectroscopy. Nature Cell Biology. 2000; 2(6): 313–317. DOI:10.1038/35014000

[59] Sonny C. H., Ailey K. C., Wilbur A. L., Carolyn R. B., Daniel A. F., and Matthew B. F.: DNA-coated AFM cantilevers for the investigation of cell adhesion and the patterning of live cells. Angewandte Chemie International Edition. 2008; 47(44): 8473–8477. DOI: 10.1002/anie.200802525

[60] Bolotov L., Okui T., and Kanayama T.: Scanning resonant tunneling spectroscopy of fullerene molecules on Si surfaces for carrier density profiling across p-n junctions. Applied Physics Letters. 2005; 87: 133107. DOI:10.1063/1.2058221

[61] Mizuta H. and Tanoue T. The Physics and Applications of Resonant Tunneling Diodes. Cambridge: Cambridge University Press; 1995. 239 p. DOI:10.1017/CBO9780511629013

[62] Mönch W. Semiconductor Surfaces and Interfaces. 3rd ed. Berlin: Springer; 2001. 548 p. ISBN: 978-3-540-679028.

[63] Bolotov L., Nishizawa M., and Kanayama T.: Two dimensional dopant profiling by scanning tunneling microscopy. Journal of Vacuum Society of Japan. 2011; 54: 412–419.

[64] Mazur U. and Hipps K. W.: Resonant tunneling in metal phthalocyanines. Journal of Physical Chemistry. 1994; 98: 8169–8172. DOI:10.1021/j100084a040

[65] Grobis M., Wachowiak A., Yamachika R., and Crommie M. F.: Tuning negative differential resistance in a molecular film. Applied Physics Letters. 2005; 86: 204102. DOI:10.1063/1.1931822

[66] Grabert H. and Devoret M., editors. Single Charge Tunneling. New York: Springer; 1992. 335 p. DOI:10.1007/978-1-4757-2166-9

[67] Kane B. E.: A silicon-based nuclear spin quantum computer. Nature. 1998; 393: 133–137. DOI:10.1038/30156

[68] Likharev K. K.: Dynamics of some single flux quantum devices: I. Parametric quantron. IEEE Transactions on Magnetics. 1977; 13(1): 242–244. DOI:10.1109/TMAG.1977.1059351

[69] Heike S. and Hashizume T.: Atomic resolution noncontact atomic force/scanning tunneling microscopy using a 1 MHz quartz resonator. Applied Physics Letters. 2003; 83: 3620–3623. DOI:10.1063/1.1623012

[70] An T., Nishio T., Eguchi T., Ono M., Nomura A., Akiyama K., and Hasegawa Y.: Atomically resolved imaging by low-temperature frequency-modulation atomic force microscopy using a quartz length-extension resonator. Reviews Scientific Instruments. 2008; 79: 033703–033707. DOI:10.1063/1.2830937

[71] An T., Eguchi T., Akiyama K., and Hasegawa Y. Atomically-resolved imaging by frequency-modulation atomic force microscopy using a quartz length-extension resonator. Applied Physics Letters. 2005; 87: 133114. DOI:10.1063/1.2061850

[72] Bolotov L., Tada T., Iitake M., Nishizawa M., and Kanayama T.: Measurements of electrostatic potential across p–n junctions on oxidized Si surfaces by scanning multimode tunneling spectroscopy. Japanese Journal of Applied Physics. 2011; 50: 04DA04. DOI:10.1143/JJAP.50.04DA04.

[73] Giessible F. J. Forces and frequency shifts in atomic-resolution dynamic-force microscopy. Physical Reviews B. 1997; 56: 16010–16015. DOI:10.1103/PhysRevB.56.16010

[74] Giessible F. J. A direct method to calculate tip–sample forces from frequency shifts in frequency-modulation atomic force microscopy. Applied Physics Letters. 2001; 78: 123–125. DOI:10.1063/1.1335546

[75] Giessible F. J. Advances in atomic force microscopy. Reviews Modern Physics. 2003;75: 949–983. DOI:10.1103/RevModPhys.75.949

[76] Bolotov L., Tada T., Iitake M., Nishizawa M., and Kanayama T.: Electrostatic potential fluctuations on passivated Si surfaces measured by integrated AFM-STM. e-J. Surface Science and Nanotechnology. 2011; 9: 117–120. DOI:10.1380/ejssnt.2011.117.

[77] Bolotov L., Fukuda K., Tada T., Matsukawa T., and Masahara M.: Spatial variation of the work function in nano-crystalline TiN films measured by dual-mode scanning

tunneling microscopy. Japanese Journal of Applied Physics. 2015; 54: 04DA03. DOI: 10.7567/JJAP.54.04DA03

[78] Hasegawa Y. and Eguchi T.: Potential profile around step edges of Si surface measured by nc-AFM. Applied Surface Science. 2002; 188: 386–390. DOI:10.1016/S0169-4332(01)00955-2

[79] Bolotov L., Tada T., Arimoto H., Fukuda K., Nishizawa M., and Kanayama T.: Built-in potential mapping of silicon field effect transistor cross sections by multimode scanning probe microscopy. Transactions of MRS of Japan. 2013; 38: 257–260.

[80] Bolotov L., Fukuda K., Arimoto H., Tada T., and Kanayama T.: Quantitative evaluation of dopant concentration in shallow silicon p-n junctions by tunneling current mapping with multimode scanning probe microscopy. Japanese Journal of Applied Physics. 2013; 52: 04CA04. DOI:10.7567/JJAP.52.04CA04

[81] Bolotov L., Tada T., Poborchii V., Fukuda K., and Kanayama T.: Spatial distribution of photocurrent in Si stripes under tilted illumination measured by multimode scanning probe microscopy. Japanese Journal of Applied Physics. 2012; 51: 088005. DOI:10.1143/JJAP.51.088005

[82] Bolotov L., Tada T., Saito Y., and Tominaga J.: Changes in morphology and local conductance of GeTe-Sb$_2$Te$_3$ superlattice films on silicon made by scanning probe microscopy in a lithography mode. *Japanese Journal of Applied Physics*. 2016; 55:04EK02. DOI:10.7567/JJAP.55.04EK02

[83] Fukutome H., Arimoto H., Hasegawa S., and Nakashima H.: Two-dimensional characterization of carrier concentration in metal oxide-semiconductor field-effect transistors with the use of scanning tunneling microscopy. Journal of Vacuum Science and Technology B. 2004; 22: 358–361. DOI:10.1116/1.1627792

[84] Breese M. B. H., King. J. C.P, Grime G. W., Smulders P. J. M., Seiberling L. E., and Boshart M. A.: Observation of planar oscillations of MeV protons in silicon using ion channeling patterns. Physical Reviews B. 1996; 53: 8267–8276. DOI:10.1103/PhysRevB.53.8267

[85] Bolotov L., Tada T., Morita Y., Poborchii V., and Kanayama T.: Nanoscale characterization of silicon-on-insulator nanowires by multimode scanning probe microscopy. Transactions of MRS of Japan. 2013; 38: 265–268. DOI: 10.14723/tmrsj.38.265

[86] Sato S., Li W., Kakushima K., Ohmori K., Natori K., Yamada K., and Iwai H.: Extraction of additional interfacial states of silicon nanowire field-effect transistors. Applied Physics Letters. 2011; 98: 233506. DOI:10.1063/1.3598402

[87] Aspnes D. E., Studna A. A., and Kinsbron E.: Dielectric properties of heavily doped crystalline and amorphous silicon from 1.5 to 6.0 eV. Physical Reviews B. 1984; 29: 768–778. DOI:10.1103/PhysRevB.29.768

[88] Poborchii V., Tada T., and Kanayama T.: High-spatial-resolution Raman microscopy of stress in shallow-trench-isolated Si structures. Applied Physics Letters. 2006; 89: 233505. DOI:10.1063/1.2400057

[89] Poborchii V., Tada T., and Kanayama T.: Edge-enhanced Raman scattering in Si nanostripes. Applied Physics Letters. 2009; 94: 131907. DOI:10.1063/1.3110964

[90] Furuhashi M. and Taniguchi K.: Additional stress-induced band gap narrowing in a silicon die. Journal of Applied Physics. 2008; 103: 026103. DOI:10.1063/1.2833435

[91] Houard J., Vella A., Vurpillot F., and Deconihout B.: Optical near-field absorption at a metal tip far from plasmonic resonance. Physical Reviews B. 2010; 81: 125411. DOI: 10.1103/PhysRevB.81.125411

[92] Bardeen J.: Tunnelling from a many-particle point of view. Physical Review Letters. 1961; 6(2): 57–59. DOI:10.1103/PhysRevLett.6.57

[93] Bono J. and Good Jr. R. H.: Theoretical discussion of the scanning tunneling microscope applied to a semiconductor surface. Surface Science. 1986; 175: 415–420. DOI:DOI: 10.1016/0039-6028(86)90243-8

[94] Fukuda K., Nishizawa M., Tada T., Bolotov L., Suzuki K., Satoh S., Arimoto H., and Kanayama T. Three-dimensional simulation of scanning tunneling microscopy for semiconductor carrier and impurity profiling. Japanese Journal of Applied Physics. 2014; 116: 023701. DOI:10.1063/1.4884876

[95] Fukuda K., Nishizawa M., Tada T., Bolotov L., Suzuki K., Sato S., Arimoto H., and Kanayama T.: Simulation of light-illuminated STM measurements. In: Proceedings of the International Conference on Simulation of Semiconductor processes and Devices (SISPAD 2014); 9–11 September 2014; Yokohama, Japan; 2014. pp. 129–132

[96] Sano N. Impurity-limited resistance and phase interference of localized impurities under quasi-one dimensional nano-structures. Journal of Applied Physics. 2015; 118: 244302. DOI:10.1063/1.4938392

[97] Advanced Algorithm and Systems. [Internet]. 2016. Available from: https://www.aasri.jp/pub/spm/pdf/spm_concept_eng.pdf [Accessed 2016-02-10].

[98] Frauenheim T., Seifert G., Elsterner M., Hajnal Z., Jungnickel G., Porezag D., Suhai S., and Scholz R.: A self-consistent charge density-functional based tight-binding method for predictive materials simulations in physics, chemistry and biology. Physica Status Solidi (b). 2000; 217(1): 41–62. DOI:10.1002/(SICI)1521-3951(200001)217:1<41::AID-PSSB41>3.0.CO;2-V

[99] Card H. C. and Rhoderick E. H. Studies of tunnel MOS diodes I. Interface effects in silicon Schottky diodes. Journal of Physics D. 1971; 4: 1589–1605. DOI: 10.1088/0022-3727/4/10/319

Chapter 13

Electron Orbital Contribution in Distance-Dependent STM Experiments

Alexander N. Chaika

Additional information is available at the end of the chapter

http://dx.doi.org/10.5772/63270

Abstract

Scanning tunneling microscopy (STM) is one of the most powerful techniques for the analysis of surface reconstructions at the atomic scale. It utilizes a sharp tip, which is brought close to the surface with a bias voltage applied between the tip and the sample. The value of the tunneling current, flowing between the tip and the sample, is determined by the structure of the surface and the tip, the bias voltage, and the tip-sample distance. By scanning the tip over the surface, a tunneling current map is produced, which reflects the local atomic and electronic structures. This chapter focuses on the role of the tip-surface distance in ultrahigh vacuum STM experiments with atomic and subatomic resolution. At small distances, i.e., comparable with interatomic distances in solids, the interaction between the tip and the surface atoms can modify their electronic structure changing the symmetry of the atomically resolved STM images and producing unusual features at the subatomic scale. These features are related to changes of the relative contribution of different electron orbitals of the tip and the surface atoms at varying distances.

Keywords: atomic resolution, density of states, electron orbital, gap resistance, scanning tunneling microscopy

1. Introduction

The invention of scanning probe microscopy (SPM) [1–3] allowed studying the surface structures with extremely high spatial resolution. The SPM methods use sharp tips, which are ultimately ended with a single atom at the apex, for surface imaging [4–7], fabricating low-dimensional structures from individual atoms and molecules [8–14], studying the physical properties of the nanoobjects [15, 16], and getting the information about the chemical [17, 18] and magnetic

order on surfaces [19–21]. SPM methods are utilized in the fields of physics, chemistry and biology for precise studies of organic and inorganic nanoobjects. The subatomic spatial resolution [22–33] has been achieved during the last decade in the atomic force microscopy (AFM) and scanning tunneling microscopy (STM) experiments.

STM is based on quantum tunneling of electrons from atoms present at the surface of a sample to the front atom of an STM tip (or vice versa) through the vacuum gap. Because of the exponential distance dependence [34], the tunneling current drops approximately by one order of magnitude with every 1 Å increase in the tip-sample separation. Because of the strong distance dependence, more than 90% of the current can be localized on only two, closest to each other, tip and surface atoms. This provides unique spatial resolution of STM, reaching the picometer scale [15, 22–24] and allowing even direct imaging individual electron orbitals at certain tunneling parameters [22–28]. However, the ultimate orbital resolution could rarely be achieved in experiments because of simultaneous contribution of different electron states of the tip and the surface atoms and nonideal geometries of the tips used. Probing particular atomic orbitals with STM is further impeded by possible modifications of the tip and surface electronic structure at very small tunneling gaps (2.0–5.0 Å), generally required for atomically resolved STM imaging.

The role of different electron states and tip-surface distance in high-resolution STM imaging has been discussed since 1980s. Apparently, the first anomalous distance dependence of the STM contrast was reported in reference [35]. Afterward, the distance dependence has been studied in a number of works [22, 23, 27, 28, 36–46]. However, because of complexity of the theoretical calculations for realistic tip-surface systems and experimental studies with precise control of the tunneling gap and the tip state, there is still no detailed knowledge about the role of particular electron orbitals in the distance-dependent STM experiments. This chapter is focused on the experimental and theoretical studies of the role of the tip-sample distance in STM experiments with atomic and subatomic spatial resolution which can pave the way to selective probing particular electron orbitals, controllable chemical analysis at the atomic scale, and surface imaging with picometer lateral resolution.

2. Role of the electron orbitals and tip-surface distance in STM experiments: theory

2.1. Spatial resolution with different atomic orbitals at the tip apex

One of the critical issues of high-resolution STM studies is the lack of reliable information about the tip apex structure, which complicates theoretical calculations and comparison with experimental results. To simplify the calculations, Tersoff and Hamann considered a spherically symmetric s-wave STM tip [47, 48]. In this case, the tunneling current is proportional to the local density of states (DOS) of the surface at the position of the tip, integrated in the energy range defined by the bias voltage. The STM images, calculated using the Tersoff-Hamann model, reflect the surface DOS distribution. This theory has provided a good agreement between the simulated and experimental images for numerous atomically resolved STM studies.

The ultimate lateral resolution R within the Tersoff-Hamann approach [47] is defined by the formula $R = \sqrt{d \cdot 2k^{-1}}$, where d is the distance between the interacting tip and surface atoms and $k^{-1} \approx 1$ Å. The limit of the lateral resolution for a tip-sample distance of 4.5 Å should be about 3 Å. The theory [47] predicts an enhancement of the spatial resolution with decreasing tip-sample distance but cannot explain the large atomic corrugations in STM images of metal surfaces [49–51] and the experimentally observed sub-Ångström lateral resolution [7, 22–24], even at tunneling gaps in the range of 2.0–2.5 Å. The former can be explained either by the tip and surface atom relaxations [37] or by decisive contribution of higher momentum (l) electron states with nonzero momentum projections on the z-axis ($m \neq 0$ states) [52–56]. For example, calculations by Tersoff and Lang for different tip atoms (Mo, Na, Ca, Si, and C) suggest that atomic corrugations on the same surface can vary by one order of magnitude depending on the relative contribution of the electron states with different values of l and m [52].

To get more general description of the tip structure and to clarify the role of particular electron orbitals in STM imaging, Chen introduced the so-called derivative rule [54], where the tunneling matrix elements, corresponding to individual orbital contributions, are proportional to the z derivatives of the surface atom wave functions at the center of the tip apex atom. The total tunneling current can be calculated by summing up all individual contributions of the electron orbitals with different values of l and m [54]. According to the theory [54, 55], d_{z}^{2} and p_{z} tip electron states can be responsible for enhanced atomic resolution and large corrugations on metallic surfaces. For the tips having $m - 0$ electron states at the apex, the simulated STM images should correspond to the surface DOS distribution, similar to the s-wave tip. This assumes that the best STM tips should possess either d_{z}^{2} or p_{z} electron states at the apex. At the same time, the $m \neq 0$ tip states (e.g., d_{xz}, d_{yz}, d_{xy}, and $d_{x^2-y^2}$) can provide enhanced atomic corrugations and intra-atomic effects at small distances [53–56]. For example, d_{xz}, d_{yz}, d_{xy}, and $d_{x^2-y^2}$ electron orbitals at the apex can produce twofold and fourfold split subatomic features, respectively, instead of a single atom at very small tunneling gaps [56]. The dependence of atomic corrugations on the tunneling gap resistance, calculated for different electron orbitals by Sacks [56], suggested that d_{z}^{2} and $m \neq 0$ d-orbitals can produce enhanced corrugations in comparison to the s-state. Furthermore, the corrugations can substantially increase with decreasing gap resistance.

The improvement of the lateral resolution of the SPM to the picometer scale [7, 22–24, 57–60], observation of orbital channels in tunneling conductance [61], and subatomic contrast in SPM experiments [22–32] led to the development of new theories accounting for the distance dependence of the electron orbital contribution [62, 63], energy-dependent combinations of the tip orbitals [64–68], and effects of inter- and intra-atomic interference of the electron orbitals in the tunneling junction [69–73]. Recently, the revised Chen's derivative rule, accounting for the orbital interference effects, has been proposed by Mandi and Palotas [74], who considered the realistic electronic structure and arbitrary spatial orientation of the tip minimizing the computational efforts. Their results showed that the electronic interference effects have a considerable effect on the STM images. As an example, in certain cases a tip with a mixture of s and p_z electron states can provide even higher spatial resolution than pure p_z-orbital tip [74].

2.2. Electronic structure of realistic tips at small tunneling gaps

Although the developed theories can provide a correlation between experimental and theoretical images in certain cases, they rarely take into account possible interaction-induced changes of the tip orbital structure at very small tunneling gaps. It has been demonstrated recently [62–64, 75] that electronic structure of the transition metal tips and relative weights of different d-orbitals in the tip's DOS near E_F strongly depend on the tip element, tip cluster size, and local environment around the tip. In particular, it has been shown that d_{z^2}-orbital is more sensitive to the local environment and the tip cluster size than $m \neq 0$ d-orbitals (d_{xz}, d_{yz}, d_{xy}, and $d_{x^2-y^2}$) [63]. The electronic structure of different transition metal tips can nonequally depend on the tip-surface distance [63, 75]; therefore, determination of the most suitable tip for particular surface can be crucial for reaching highest spatial resolution in the experiments. Stimulated by the subatomic resolution AFM experiments [33], Wright and Solares calculated

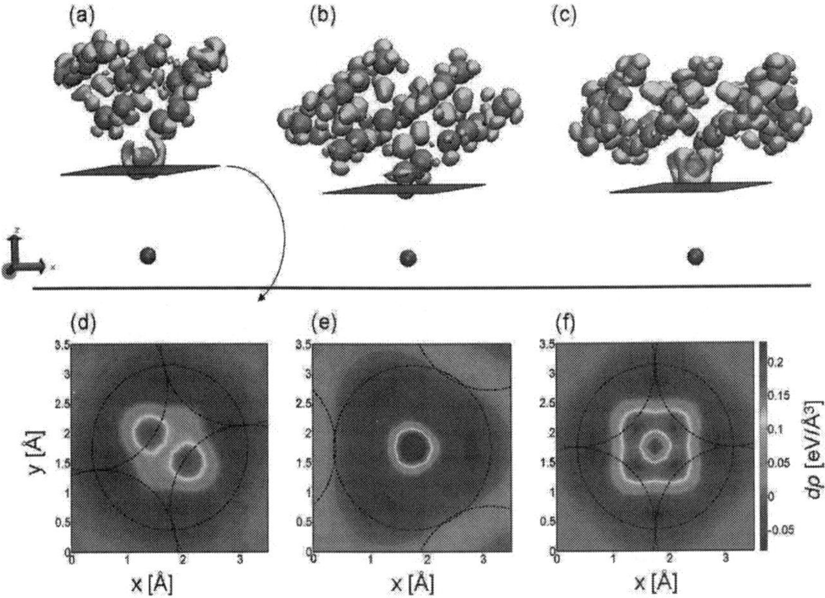

Figure 1. (a–c) The $d\rho = 0.08$ eV/Å^3 isosurface of the change in electron density for the (a) He–W[011], (b) He–W[111], and (c) He–W[001] systems. The helium atom is positioned directly below the front atom of the tip, with a gap of 4.0 Å for the W[011] tip and 3.5 Å for the W[111] and W[001] tips. (d–f) Constant-height slices through the electron density distribution. The relative sizes and positions of the first two layers of the tips' tungsten atoms are drawn as black circles. Reproduced from Reference [75] with permission of IOP.

the electronic structure of the tungsten tips with different crystallographic orientations in proximity to different surface clusters [75–77]. The calculations showed that the experimental SPM images with subatomic resolution [22–24, 33] were, most probably, related to the tip electronic structure, which could become asymmetric at small tip-sample distances because of the tip-sample interaction. The stability of the tungsten tips was found to be dependent on the crystallographic orientation of the apex. In particular, W[001] tip did not substantially relax even at the tip-sample distance of 1.50 Å, while [011]- and [111]-oriented tungsten tips were stable only at gaps above 2.25 and 2.50 Å, respectively. **Figure 1(a)–(c)** shows the isosurfaces of the change in the electron density calculated for the [011]-, [111]-, and [001]-oriented tungsten tips interacting with the helium atom positioned directly below the front atom of the tips [75]. **Figure 1(d)–(f)** illustrates that essentially asymmetric charge density distribution is observed near the front atom of the W[011] and W[001] tips. These areas of increased charge density (dark red color) reveal the twofold and fourfold symmetry representatives of each tip's crystallographic orientation. However, for the W[111] tip, the areas of increased density do not reveal the threefold symmetry proposed from the crystal symmetry consideration [33]. These results suggest that the highest spatial resolution can be achieved in STM experiments using W[111] tips which have symmetrical charge density distribution along the tip axis even at relatively small distances from the surface. In contrast, W[001] and W[011] tips can produce subatomic features at small tip-sample distances [22–24] but can be more suitable for scanning tunneling spectroscopy (STS) experiments demanding high tip stability.

2.3. Role of atomic relaxations at small tip-surface distances

An enhancement of the lateral resolution and atomic corrugations with decreasing tip-surface distance was anticipated from the theories [47, 56] omitting possible relaxations of the interacting tip and surface atoms. Additionally, the topographic contrast in STM experiments can be enhanced at small distances because of the tip and surface atom relaxations reported for the first time almost three decades ago [49]. The role of elastic effects in high-resolution STM experiments on close-packed metal surfaces was thoroughly studied in Ref. [37]. To evaluate possible corrugation enhancement related to atomic relaxations, the simulations were performed on Cu(111) and Al(111) surfaces at different tip configurations and tunneling gaps [37]. The relaxations were found to be substantially stronger for Al(111) because of the higher elasticity of the aluminum surface. As a result, the atomic corrugations in the simulated images of Al(111) increased almost by one order of magnitude (from 10 to 70 pm) with decreasing distance from 600 to 400 pm. The calculations [37] explain the anomalously high corrugation amplitudes observed in numerous STM experiments on metal surfaces. The large outward relaxations of the surface atoms under typical tunneling conditions can lead to atomic corrugations which can hardly be explained from the electron DOS and simplified theoretical considerations [47, 56]. The corrugation enhancement was proved on different metal surfaces; therefore, one can anticipate the validity of the work [37] for other metals, although the effect cannot be so well pronounced as calculated for aluminum surface scanned by an aluminum-terminated tip.

2.4. Reduction in tunneling current channels with decreasing tip-surface distance

First observations of the tunneling current channels associated with different electron orbitals were reported almost two decades ago [61]. However, their role in STM imaging at different gaps is still not clear. Recent theoretical studies showed that relative contributions of different electron orbitals can be substantially modified because of tip-sample interaction changing the tip and surface DOS near E_F [62–64, 75, 78–80]. For example, **Figure 1** demonstrates the distinct asymmetry of the [001]- and [011]-oriented tungsten probes appearing only at small gaps when the overlap of the orbitals of the interacting tungsten and helium atoms takes place [75]. **Figure 2** shows the theoretical calculations demonstrating a drastic reduction of the DOS near E_F associated with the Si(111)7×7 surface atom interacting with the [001]-oriented tungsten tip [79]. The interaction modifies the electronic structure of the surface atom at distances below 4.75 Å, as shown in **Figure 2(d)**. The calculations predict a nonmonotonous dependence of the interaction forces [**Figure 2(b)**] and the tunneling current [**Figure 2(c)**] at these distances. The calculations reveal the exponential current–distance dependence at large distances and drop of tunneling conductance at a distance of 4.75 Å [**Figure 2(c)**]. **Figure 2(b)** and **(c)** demonstrates a correlation between the upward adatom displacement, the decrease of the conductance, and the formation of the chemical bond between the apex atom and the surface adatom. At large tip-sample distances, there is no significant vertical displacement of the adatom [**Figure 2(c)**]. Large upward relaxations of the adatom are observed at the distances of around 5 Å. At the same distances, the short-range chemical force between the tip and the sample changes rapidly

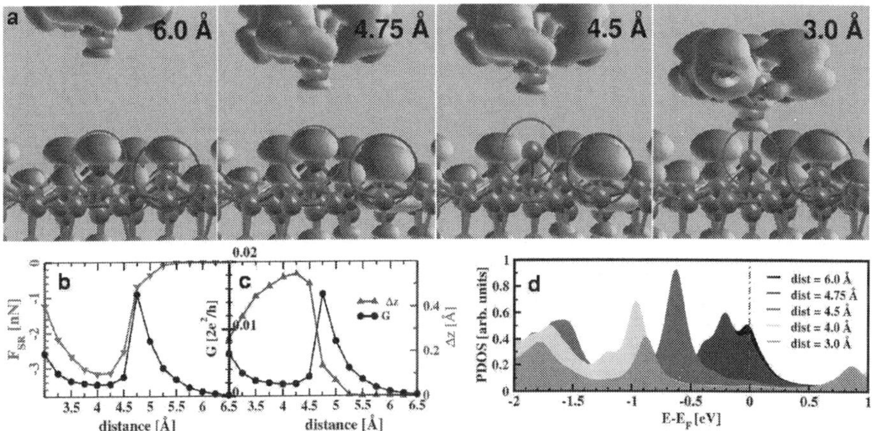

Figure 2. (a) Isosurfaces of electron charge density (the isovalue is 0.08 $e/Å^3$) integrated in the energy range from E_F to $E_F - 0.4$ eV, for different tunneling gaps. The probe is placed over the corner adatom in the Si(111)7×7 faulted unit cell. (b) Evolution of the quantum conductance G (right axis) and the atomic force (left axis) as a function of the tip-sample distance. (c) The quantum conductance (left axis) and the vertical adatom displacement (right axis) as a function of the distance. (d) PDOS of the silicon corner adatom as a function of the tip-sample distance. Reproduced from Reference [78] with permission of APS.

[**Figure 2(b)**]. Then, the vertical adatom displacement decreases until it reaches its initial value at a tip-sample distance of 3 Å. The electron DOS isosurfaces calculated for different tip-sample distances [**Figure 2(a)**] show the typical charge density distribution of the Si(111)7×7 surface at 6.0 Å, a charge transfer from the adatom to neighboring surface atoms at 4.75 Å, the suppression of the adatom dangling bond at 4.5 Å, and the chemical bond formation at 3.0 Å that prevents the electron transport between the tip and the surface at small distances. The W[001] tip electronic structure remained practically unchanged in this tip-sample distance range. Note that calculations of Jelinek et al. [78] were carried out at low voltages, corresponding to probing the p_z-states dominating in the Si(111)7×7 surface DOS near E_F. Therefore, it can be assumed that the overlap of the silicon p_z-orbital with the tungsten tip orbitals is responsible for the strong modification of the tunneling current as shown in **Figure 2(d)**. Furthermore, one can expect similar phenomena in other systems where the electron transport occurs through p_z electron orbitals.

Recent theoretical calculations of the tunneling current between an STM tip and a single Cu/Co atom adsorbed on a Cu(001) surface [62] have shown that conductance in these systems can be decomposed into several orbital contributions. The tunneling probabilities of these individual channels can provide the information about the modifications of the adatom's DOS caused by the tip-adatom interaction. The calculations revealed that the d_z^2-orbitals of the interacting atoms in such systems are especially sensitive to the tip-adatom distance because they start to overlap at larger distances. The DOS associated with the $d_{x^2-y^2}$-orbitals starts to decrease at smaller tunneling gaps and the total reduction at distances between 4.0 and 2.5 Å for this orbital is smaller in comparison with that of the d_{xz}-, d_{yz}-, and d_z^2-orbitals. At tip-sample distances exceeding 3.0 Å, the channels contributing to the most of the tunneling current were generally related to hybrids of s- and d_z^2-orbitals. The contribution of $m \neq 0$ electron states increased with decreasing distances.

3. Electron orbital resolution in distance-dependent STM experiments

The first STM images with the atomic orbital resolution were obtained at the beginning of 1980s. For example, typical images of the Si(111)7×7 surface [5] correspond to direct visualization of the p_z surface orbitals. Similar selective visualization of the atomic orbitals at certain tunneling conditions was achieved on other semiconducting surfaces [17]. The control of the orbital contribution on multicomponent metal surfaces is more difficult. Up to date, several STM studies demonstrating subatomic, electron orbital contrast have been published. In several cases, the observed features were related to direct visualization of the electronic structure of the tip atom by more localized surface atomic orbitals [22–28]. However, some STM studies revealed asymmetric subatomic features associated with the $m \neq 0$ electron orbitals of the surface atoms [30, 81]. The STM studies presented in this section can be important for optimizing the tunneling parameters for high-resolution STM imaging and shed light on the gap resistance dependence of atomically resolved STM images and chemical contrast observed on metallic and semiconducting surfaces.

3.1. Tip orbitals resolved using p_z states of the Si(111)7×7 surface atoms

SPM imaging with subatomic resolution was first claimed by Giessibl et al. [31] who reported the AFM images of the Si(111)7×7 reconstruction demonstrating a regular twofold splitting of the surface atomic features. This was explained by direct visualization of the two atomic orbitals of an Si[001] tip atom by the p_z-orbitals of the surface atoms. This is schematically shown in **Figure 3(b)**. Later studies [28] showed that qualitatively the same asymmetric features can be resolved with STM using a silicon-terminated tungsten tip. **Figure 3(a)** demonstrates the change of the contrast in the STM experiments [28] with decreasing tunneling gap resistance. Note that precise structure of the tip apex, responsible for the subatomic contrast in both SPM experiments [28, 31], was unknown because of the tip preparation procedure. Therefore, the origin of the subatomic features in the Si(111)7×7 SPM images was disputed in a number of papers [82–87]. Alternative explanations based on the feedback loop

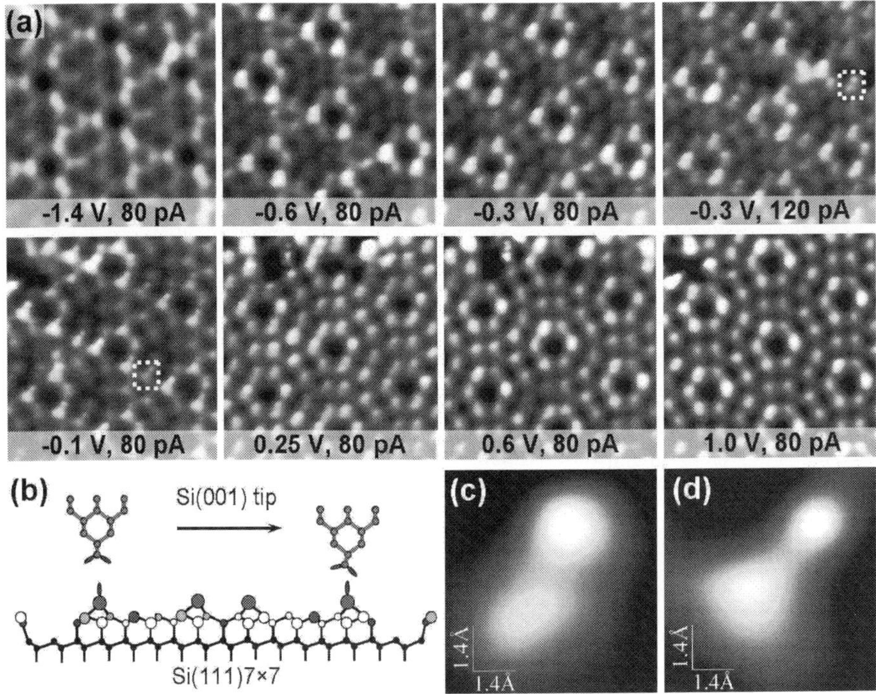

Figure 3. (a) 7.1×7.1 nm² STM images of the Si(111)7×7 surface measured at different bias voltages and tunneling currents (shown on each frame) with the silicon-terminated tungsten tip [28]. The fast scanning direction was from left to right. (b) Side view of a [001]-oriented silicon tip over the Si(111)7×7 surface; right panel depicts a tip bending opposite to the scan direction. (c,d) Zooms of the double features indicated by squares on images measured at $U = -0.3$ V, $I = 80$ pA (c) and $U = -0.1$ V, $I = 80$ pA (d).

artifacts [83], presence of a carbon atom at the apex [86], and visualization of the surface atoms' backbonds [87] were proposed. Nevertheless, independent theoretical calculations [31, 32, 82, 84] support the possibility of direct visualization of the asymmetric charge distribution around the [001]-oriented Si tip atom [**Figure 3(b)**] at small (2.5–4.0 Å) tip-surface separations. This is in agreement with the gap-resistance–dependent STM experiments [28]. **Figure 3(a)** demonstrates that the twofold splitting of the adatoms becomes discernible only at small bias voltages (small distances), while at large negative voltages a (7×7) pattern with with well-resolved adatoms, rest atoms, corner holes, and single-atom defects are observed, which is only possible with extremely sharp single atom–terminated tips [88]. The effect can hardly be explained by the formation of a two-atom terminated apex and visualization of the surface atom backbonds because deep corner holes, rest atoms, and single atom defects on the surface were simultaneously resolved even at low gap resistances. The double features become sharper and change their appearance from symmetric to asymmetric with decreasing gap resistance, as shown in **Figure 3 (c)** and **(d)**. The shape of the double features could be changed both by increasing tunneling current at fixed voltage and decreasing voltage at the same current [28]. The asymmetry of the double features at small gaps can be explained by the relaxation of the apex atom, as shown in **Figure 3(b)**. The distance between the two maxima of the double features decreased from 2.7 to 2.2 Å with decreasing gap resistance from 7.5 to 2.5 GΩ and then increased to 2.75 Å with further decreasing gap resistance from 2.5 to 1.25 GΩ. The decrease of the distance between the subatomic maxima is in line with the theoretical spatial distribution of the two sp^3 dangling bonds at the [001]-oriented silicon apex [32]. The increase of the splitting at smaller distances can be related to tip-sample interaction modifying the electronic structure of the apex atom and inducing lateral relaxations of the tip and surface atoms in near-to-contact regime.

Herz et al. applied a Co_6Fe_3Sm tip for high-resolution STM experiments on the Si(111)7×7 surface [25]. They utilized a dynamic-STM mode with an oscillating probe to reduce the lateral forces in tip sample contact and increase the tip stability at extremely small tunneling gaps. At some tunneling parameters, the adatoms were resolved as extremely sharp spherically symmetric features surrounded by lower lying crescents. These asymmetric features, observed at small gaps, were explained by a convolution of the p_z-orbitals of the surface atoms and the f_z^3-orbital of an Sm atom at the apex tilted to the surface normal by 37°. The validity of this interpretation was supported by the theoretical calculations in an assumption of the tilted Sm f_z^3 tip orbital [25].

3.2. d_{yz} electron orbital of a MnNi tip resolved in STM experiments on the Cu(014)–O surface

Figure 4(a) shows the STM image of the Cu(014)–O surface measured using a polycrystalline MnNi tip [26, 27]. The image demonstrates regular twofold splitting of the copper atomic features along the [1–10] direction. The experimental image displays the 7.2±0.2 Å wide terraces, step edges along the [100] direction and an additional fine structure within the terraces. The image reveals a single atom defect proving the sharpness of the MnNi tip. The doubling of atomic features was observed in rare experiments at small negative bias voltages between -30 and -50 mV in a very narrow range of the tunneling currents. That was explained by the distance dependence of the Cu(014)–O STM images and strong dependence of the tip's electronic structure on the crystallographic orientations of the apex [26, 27].

Figure 4(b) shows the partial density of electron states (PDOS) associated with the Ni and Mn atoms at the apexes of the [001]- and [111]-oriented MnNi tips. The tight binding (TB) calculations demonstrate domination of the d_{yz}-orbital near the Fermi level only for the [111]-oriented MnNi tip terminated by an Mn atom [**Figure 4(b)**]. The density functional theory (DFT) calculations shown in **Figure 4(c)** demonstrate that only certain configurations of the Mn-terminated MnNi[111] tips could produce regular doubling with almost symmetric twofold split features [27]. The regular pattern of the double features reproducing the shape of the Mn atom d_{yz} electron orbital could be resolved when the Cu(014)–O surface d_{z^2}-orbitals and the tip d_{yz}-orbital provided major contribution to the tunneling current. This is schematically shown in **Figure 4(a)**.

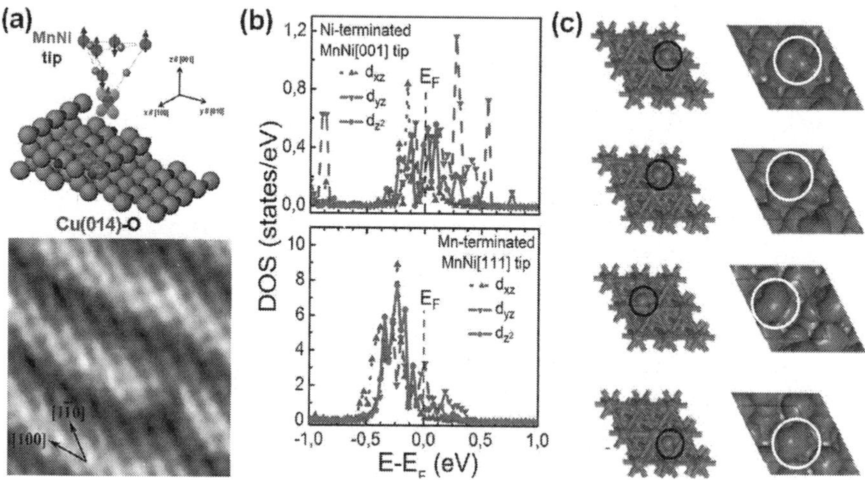

Figure 4. (a) Regular doubling of atomic features in STM images of the Cu(014)–O surface measured with a MnNi tip (bottom) and schematic model of the d_{yz} tip orbital scanning over d_{z^2} surface orbitals (top). Eight instead of four features are resolved along the [1–10] direction within each terrace. The image was taken at U = -30 mV and I = 80 pA. (b) PDOS associated with the d-orbitals of the Ni-terminated MnNi[001] tip (top) and Mn-terminated MnNi[111] tip (bottom). (c) Models (left) and calculated electron density isosurfaces (right) for different MnNi[111] tip configurations. The isosurfaces display the electron density at 2.4 × 10⁻³ electrons per Å³ in the E_F ± 0.22 eV energy range. The apex atom is indicated by circles. Reproduced from Reference [27] with permission of APS.

3.3. Distance dependence of the W[001] tip orbital contribution in STM experiments on graphite

The selection of the tip orbital responsible for high-resolution STM imaging has been demonstrated in references [22–24]. The electron orbitals of the graphite surface atoms were used to study the relative contribution of the [001]-oriented single crystalline tungsten tip orbitals at different distances (**Figure 5**). To avoid apex contamination, W[001] tips were cleaned by flash heating and ion sputtering in ultra-high vacuum (UHV) before the distance-dependent experiments. The transmission electron microscopy (TEM) studies proved the fabrication of the [001]-oriented nanopyramids with well-defined structure at the apexes [24].

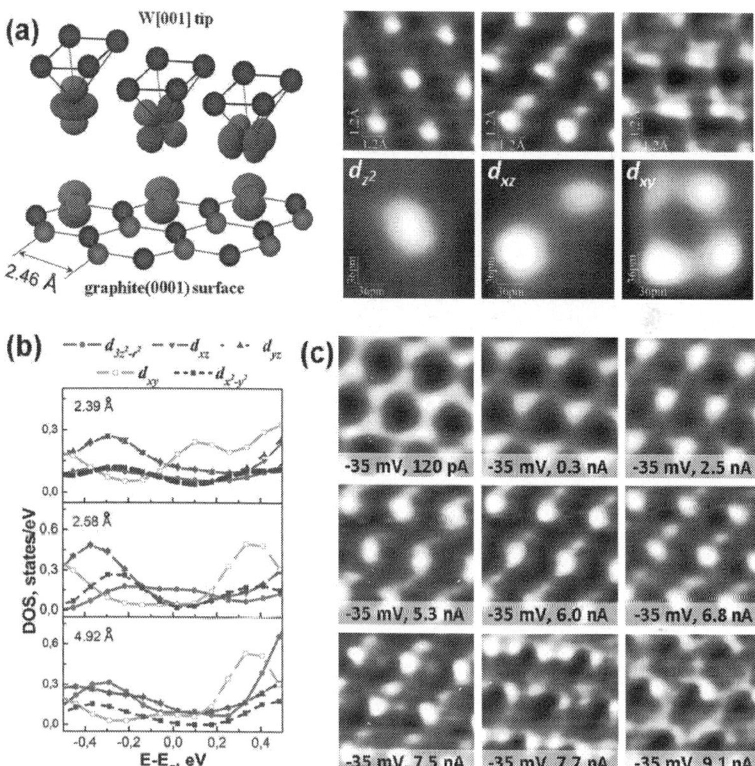

Figure 5. (a) Left: schematic model of a [001]-oriented W tip over a graphite (0001) surface. Right: 6 × 6 Å² (top) and 1.8 × 1.8 Å² (bottom) STM images measured with the W[001] tips at U = 23 mV and I = 2.7 nA (left panels), U = –35 mV and I = 7.4 nA (central panels), U = –100 mV and I = 1.7 nA (right panels). (b) PDOS associated with different d-orbitals of the W[001] tip atom interacting with the graphite (0001) surface. The distances between interacting tip and surface atom nuclei are indicated on each panel. (c) Gap resistance dependence of graphite (0001) STM images measured using a W[001] tip at U = –35 mV. Tunneling currents are indicated on each frame. Panel (b) is reproduced from reference [23] with permission of Elsevier.

Figure 5(c) shows a gap resistance dependence measured with a W[001] tip at a bias voltage of –35 mV. At smaller currents (larger distances) a hexagonal pattern is observed. The atomic features become sharper with increasing tunneling current (decreasing distance). With further increase in the tunneling current, the symmetric features are transformed into two, three, and fourfold split subatomic features at currents between 5.3 and 9.1 nA. Similar subatomic features were earlier observed in AFM experiments with polycrystalline tungsten probes [33] and ascribed to the orbital structure of the tungsten tips with three different crystallographic orientations. The gap resistance dependence measured with the unchanged W[001] tip [**Figure 5(c)**] demonstrates that these images reproduce the electronic structure of the same tip atom modified with decreasing tip-sample distance. The two and fourfold split features in **Figure 5** cannot correspond to the threefold symmetrical graphite surface. At the same time, the observed two and threefold split features cannot be related to the tip atom backbonding because of the symmetry of the W[001] tip used in experiments. The subatomic features can only be explained by a direct visualization of the tip atom's electronic structure modified by the tip-sample interaction at small tunneling gaps. According to the tunneling currents, the observed transformation of the asymmetric features took place in the narrow range of tip-sample distances of the order of 0.1 Å. The actual distance could be slightly above this value because of the tip and surface atom relaxations at small distances [37].

At certain tunneling parameters (voltages and gap resistances), the shapes of the subatomic features in the STM experiments reproduced the electron density distribution associated with the d_z^2-, d_{xz}-, and d_{xy}-orbitals in the tungsten atom. This can be explained by a major contribution of particular electron orbitals of the tip atom at certain tip-sample distances [**Figure 5(a)**]. According to **Figure 5(c)**, the relative contribution of the d_{xz}- and d_{xy}-orbitals increases with decreasing gap resistance. PDOS calculations for the fully relaxed "W[001] tip–graphite" system [**Figure 5(b)**] demonstrate a suppression of the tip d_z^2 electron orbital near E_F at distances below 2.5 Å because of the overlap with the carbon orbitals. The domination of the tip d_{xy} electron orbital near E_F is observed in a narrow range of the tip-surface distances between 2.2 and 2.5 Å, which is in agreement with the current values in **Figure 5(c)**. The surface imaging by the tungsten d_z^2-orbital can be realized at tunneling gaps between 2.5 and 4.0 Å, while d_{xy} electron orbital can yield a maximum contribution to the tunneling current at distances between 2.2 and 2.5 Å. The appearance of the asymmetric features is related to the interaction-induced modification of the PDOS associated with different d-orbitals of the tungsten tip atom.

The DFT calculations [**Figure 5(b)**] and distance-dependent STM experiments [**Figure 5(c)**] show the correlation between the spatial distribution of the atomic orbitals (in particular, extension in the z-direction) and the order of their suppression with decreasing tip-sample distance. The suppression of the further protruded in the z-direction electron orbitals with m = 0 can increase the contribution of the $m \neq 0$ electron states at small gaps. **Figure 5** demonstrates that picoscale spatial resolution and even direct imaging of the transition metal d-orbitals using the p-orbitals of light elements can experimentally be achieved.

3.4. d_{xz}-orbitals of the surface atoms resolved using tungsten tips

STM images shown in **Figures 3–5** correspond to probing the tip electronic structure by the surface atomic orbitals. Similar STM experiments revealing the tip atom's subatomic structure with decreasing tunneling gap have recently been reported for several tip-sample systems [89–92]. **Figure 6(a)** shows the STM image of the Cu(014)–O surface measured with a polycrystalline tungsten tip [30]. This image reveals one bright atomic row within the four-atom-wide terraces [see the model in **Figure 4(a)**]. The atomic features within the well-resolved copper row have two maxima separated by approximately 1 Å [**Figure 6(b)**] reproducing the electron density distribution in the Cu d_{xz} atomic orbital. According to the TB calculations [**Figure 6(c)**], the DOS associated with the d_{xz}-orbitals of the surface copper atoms is maximal for the fourth (down-step) copper row of each terrace suggesting that visualization of one atomic row within the terraces can be achieved if the surface d_{xz}-orbitals yield the largest contribution to the tunneling current. Similar to the STM experiments with the W[001] tips [22–24], this can be reached at small bias voltages and small tip-sample distances, where the d_{xz}-orbital channel dominates in the tunneling current. The STM image with one well-resolved row with double features (**Figure 6(a)**) can be explained by imaging the d_{xz}-orbitals of the Cu(014)–O surface using the d_{z^2}-orbital of a tungsten tip atom.

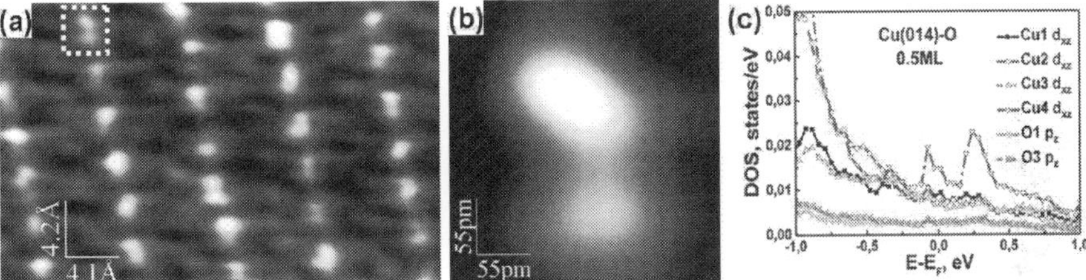

Figure 6. (a) A 3.4×2.1 nm² STM image of the Cu(014)–O surface illustrating the asymmetry of the atomic features associated with the down-step copper row resolved using a polycrystalline W tip at U = –5 mV and I = 0.1 nA [30]. (b) A 2.8×2.8 Å² STM image of a subatomic feature reproducing the shape of the Cu d_{xz} electron orbital. (c) PDOS associated with the Cu d_{xz}-orbitals and O p$_z$-orbitals of the Cu(014)–O surface. The model of the surface is shown in **Figure 4(a)**.

3.5. STM imaging of graphite (0001) using a [111]-oriented diamond tip

Conductive diamond tips, generally considered as AFM probes, can provide carbon atomic orbitals for imaging with picometer lateral resolution (**Figure 5**) and high apex stability at small tunneling gaps. It has been demonstrated that high spatial resolution can be achieved in STM experiments with the boron-doped single crystal diamond tips [93]. **Figure 7(a)–(d)** displays a comparison of the spatial resolution achieved with the [111]-oriented diamond and [001]-oriented tungsten probes. Note that image in **Figure 7(b)** was measured with the W[001] tip used for the high-resolution experiments on graphite (**Figure 5**) and surfaces with nontrivial atomic and electronic structures [94, 95]. The images taken with the diamond and W[001] probes [**Figure 7(a)** and **(b)**] reveal two sublattices corresponding to nonequivalent α and β atoms in the honeycomb lattice. However, the hollow sites are substantially deeper and individual surface atoms are better resolved in the image measured with the diamond probe, as the cross-sections in **Figure 7(c)** and **(d)** illustrate. DFT calculations [93] showed that the DOS at E_F is larger by approximately 25% for β atoms [**Figure 7(f)**] and the difference decreases for the DOS integrated over a wider range of the electron energies. That explains the two slightly nonequivalent sublattices in **Figure 7(a)** and **(b)**. The smaller height difference between the features corresponding to α and β atoms for the image measured at $U = -0.4$ V [**Figure 7(b)**] is in agreement with the decreasing difference in the DOS at larger energies.

Figure 7(g) and **(h)** shows the experimental STM image resolved with the diamond tip and the calculated charge density map corresponding to the surface electron states near E_F, respectively. Both images demonstrate different contrast on α and β atoms. The DOS shown in **Figure 7(e, f)** and the charge density map presented in **Figure 7(g)** were calculated at a tip-sample distance of 4.5 Å where the atomic and electronic structures of the tip and surface atoms are not substantially modified by the interaction [93]. The excellent agreement between experimental and theoretical images suggests that the highest resolution was achieved with the diamond probe at tunneling gaps of 3.5–4.5 Å where the tip did not strongly interact with the surface. This result shows the advantages of the oriented single crystal diamond probes: their structure is stable and well defined while high lateral and vertical resolution can be achieved at larger tunneling gaps comparing to typical tip-sample distances used in experiments with transition metal tips.

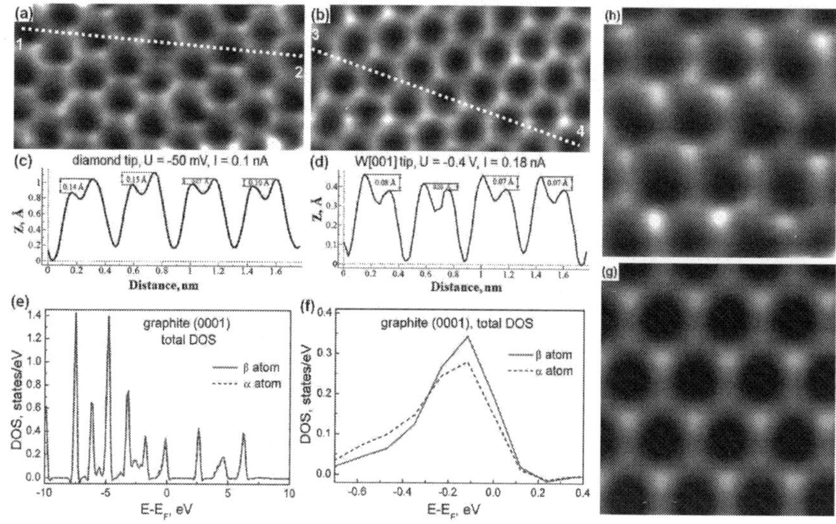

Figure 7. (a, b) 18 × 9 Å² atomically resolved STM images of graphite (0001) measured with a diamond probe at $U = -50$ mV and $I = 0.1$ nA (a) and a W[001] probe at $U = -0.4$ V and $I = 0.18$ nA (b). (c, d) Cross sections 1–2 (c) and 3–4 (d) of the images in panels (a) and (b), respectively. (e, f) Total DOS associated with the α and β atoms of a graphite (0001) surface. (g, h) Comparison of the 9 × 9 Å² calculated electron density distribution map in the energy range from $E_F - 0.2$ eV to E_F (g) and the experimental STM image measured with the diamond probe at $U = -50$ mV and $I = 0.8$ nA (h). Both images show nonequivalence of the α and β atoms in accordance with the DOS shown in panels (e) and (f). Reproduced from Reference [93] with permission of IOP.

Calculations of the partial DOS associated with the electron orbitals of the diamond probe and graphite surface atoms at different tunneling gaps (**Figure 8**) show that electronic structure of the interacting atoms is modified at tip-sample distances below 3.0 Å. The overlap of the tip and surface atomic orbitals leads to decrease in the PDOS associated with the p_z-orbital of the graphite surface atoms when the diamond tip atom is positioned directly above the surface atom [**Figure 7**(**a**, right panel)]. The suppression of the surface atom's p_z-orbital at small tunneling gaps ($d < 2.5$ Å) is in qualitative agreement with the results of the theoretical calculations for the Si(111)7×7 surface interacting with the W[001] tip shown in **Figure 2** [78]. If the diamond tip atom is positioned above the hollow site (left panels in **Figure 8**), the overlap of the front tip atom and the surface orbitals does not take place even at small distances. Therefore, the change of the PDOS of the tip and surface atoms is minor even at the tunneling gaps of 1.5 Å. According to the calculations [93], the electronic structure of the diamond probe is defined by a mixture of the carbon s and p-states with domination of the p_x- and p_y-orbitals, responsible for the STM imaging (**Figure 7**).

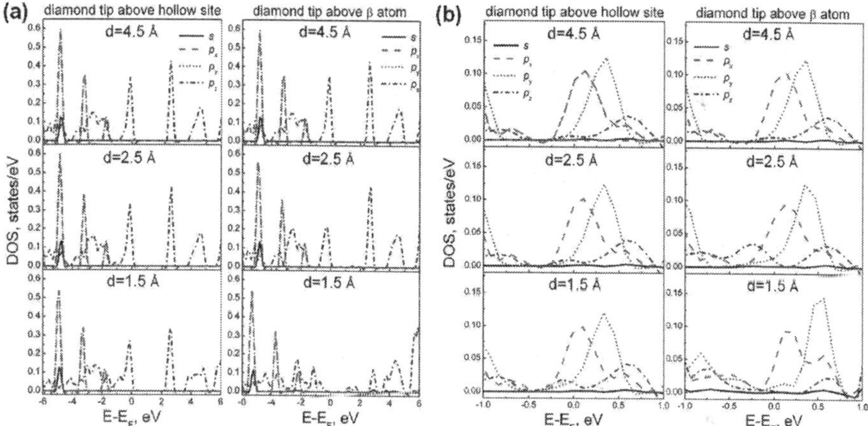

Figure 8. Partial DOS of the β atom of a graphite (0001) surface closest to the diamond tip (a) and the tip apex atom (b) at different tunneling gaps and lateral positions of the tip. Reproduced from Reference [93] with permission of IOP.

3.6. STM experiments with functionalized tips terminated by a light element atom

The experiments with the diamond tip (**Figure 7**) show that an enhancement of the spatial resolution can be achieved using conductive tips having molecules or light element atom at the apex. During the recent years, a number of high-resolution STM studies performed with the conductive tips functionalized by different light elements have been reported [57–59, 89, 96–103]. These studies generally demonstrated an enhanced contrast in the STM experiments with the light element–terminated probes, especially, at small tip-surface distances. For example, the lateral resolution at the level of intramolecular chemical bonds has been reached in STM experiments using molecule-terminated probes [58, 59]. In the recent theoretical work [101], the high resolution in experiments with the functionalized probes was explained by significant apex atom relaxations toward the local minima of the interaction potential at small distances. However, the tip functionalization procedures are generally not capable of producing stable apexes with well-defined structure. These tips could produce substantial noise [58] and asymmetric subatomic features [89] at small tunneling gaps. Besides, STM images measured with the molecule-terminated tips can strongly depend on the precise orientation of the molecule at the apex and the bias voltage [102, 103] that can complicate the interpretation of the atomically resolved STM data.

3.7. STM imaging of the random bond length distortions in graphene using a W[111] tip

Figure 9 demonstrates the picometer lateral resolution achieved in high-resolution STM studies of graphene/SiC(001) system [7] using a [111]-oriented single crystalline tungsten tip. Theoretical calculations shown in **Figure 1** suggest that the front atom of the W[111] tip should possess symmetric charge distribution along the tip axis even at small tip-sample distances. Although the W[111] tips were found to be the least stable from three possible low-index crystallographic orientations [75], they can produce higher spatial resolution in STM experiments without possible tip structure effects which can be observed with the W[001] tips (**Figure 5**). Indeed, although high spatial resolution was achieved on the quasi-freestanding graphene grown on SiC(001) with the stable W[001] tips [24], the best resolution was obtained with the W[111] tip [7]. **Figure 9(a)** demonstrates the rippled morphology of graphene on SiC(001) with the lateral and vertical dimensions of the ripples of about 30–50 and 1 Å, respectively. **Figure 9(b)** and **(c)** demonstrates random picoscale distortions of the carbon bond lengths, which are very close to the values calculated for the freestanding graphene monolayer [104]. The contrast in **Figure 9(b)** and **(c)** was adjusted to enhance the picometer scale bond length distribution. The picometer lateral resolution obtained in the RT STM experiments (**Figure 9**) corresponds to the best standards of the SPM and can be compared with the resolution achieved in the recent AFM experiments [96].

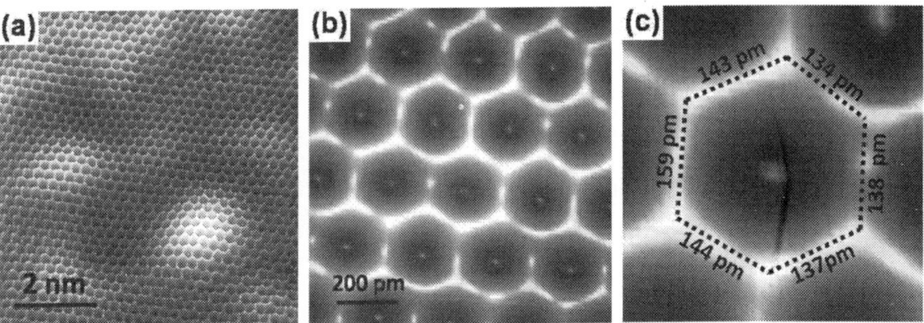

Figure 9. STM images of the graphene synthesized on SiC(001) demonstrating atomic scale rippling (a) and random picoscale distortions of the carbon bond lengths in the graphene lattice (b, c). A distorted hexagon is overlaid on the image on panel (c); the size of the hexagon sides is indicated for clarity. The STM images were measured at U = 22 mV and I = 70 pA (a); U = 22 mV and I=65 pA (b, c).

4. Conclusions

The spatial resolution on the level of individual electron orbitals corresponds to the ultimate resolution of the SPM. During the last decade, a number of SPM studies demonstrating selective visualization of individual electron orbitals and subatomic contrast have been published. The experimental and theoretical works conducted during the recent years demonstrate that selective imaging of the surface electron orbitals can only be achieved at finely adjusted bias voltages and tip-sample distances. Therefore, for the development of the electron orbital imaging capability, the stability of the tip-sample separation should be maintained at the level of 1 pm or below. Distance-dependent STM imaging with electron orbital resolution can lead to further improvement of the lateral resolution down to the picometer scale and development of the chemical-selective imaging of complex multicomponent surfaces.

Acknowledgements

This work was supported by the Russian Academy of Sciences, Russian Foundation for Basic Research (grantNos. 14-02-01234 and 14-02-00949), and Marie Curie International Incoming Fellowship project within the Seventh European Community Framework Programme. The author is grateful to S. N. Molotkov, S. I. Bozhko, S. S. Nazin, V. N. Semenov, A. M. Ionov, V. Yu. Aristov, M. G. Lazarev, N. N. Orlova, A. N. Myagkov, K. N. Eltsov, A. N. Klimov, V. M. Shevlyuga, I. V. Shvets, S. A. Krasnikov, O. Lübben, and B. E. Murphy for help and fruitful discussions.

Author details

Alexander N. Chaika

Address all correspondence to: chaika@issp.ac.ru

Institute of Solid State Physics, Russian Academy of Sciences, Chernogolovka, Russia

References

[1]   Binnig G, Rohrer H, Gerber C, Weibel E. Tunneling through a controllable vacuum gap. Appl. Phys. Lett. 1982;40:178–180. doi:10.1063/1.92999.

[2]   Binnig G, Rohrer H. Scanning tunnelling microscopy. Helv. Phys. Acta 1982;55:726–735. doi:10.5169seals-115309.

[3]   Binnig G, Quate CF, Gerber C. Atomic force microscope. Phys. Rev. Lett. 1986;56:930–933. doi:10.1103/PhysRevLett.56.930.

[4]   Binnig G, Rohrer H, Gerber C, Weibel E. Surface studies by scanning tunneling microscopy. Phys. Rev. Lett. 1982;49: 57–61. doi:10.1103/PhysRevLett.49.57.

[5]   Binnig G, Rohrer H, Gerber C, Weibel E. (7×7) reconstruction on Si(111) resolved in real space. Phys. Rev. Lett. 1983;50:120–123. doi:10.1103/PhysRevLett.50.120.

[6]   Binnig G, Rohrer H, Gerber C, Weibel E. (111) facets as the origin of reconstructed Au(110) surfaces, Surf. Sci. 1983;131:L379–L384. doi:10.1016/0039-6028(83)90112-7.

[7]   Chaika AN, Molodtsova OV, Zakharov AA, Marchenko D, Sanchez-Barriga J, Varykhalov A, et al. Continuous wafer-scale graphene on cubic-SiC(001). Nano Res. 2013;6:562–570. doi:10.1007/s12274-013-0331-9.

[8]   Eigler DM, Schweizer EK. Positioning single atoms with a scanning tunneling microscope. Nature 1990;344:524–526. doi:10.1038/344524a0.

[9]   Crommie MF, Lutz CP, Eigler DM. Confinement of electrons to quantum corrals on a metal surface. Science 1993;262:218–220. doi:10.1126/science.262.5131.218.

[10]   Stroscio JA, Celotta RJ. Controlling the dynamics of a single atom in lateral atom manipulation. Science 2004;306:242–247. doi:10.1126/science.1102370.

[11]   Walsh MA, Hersam MC. Atomic-scale templates patterned by ultrahigh vacuum scanning tunnelling microscopy on silicon. Annu. Rev. Phys. Chem. 2009;60:193–216. doi:10.1146/annurev.physchem.040808.090314.

[12]   Khajetoorians AA, Wiebe J, Chilian B, Wiesendanger R. Realizing all-spin–based logic operations atom by atom. Science 2012;332:1062–1064. doi:10.1126/science.

[13]   Khajetoorians AA, Wiebe J, Chilian B, Lounis S, Blügel S, Wiesendanger R. Atom-by-atom engineering and magnetometry of tailored nanomagnets. Nat. Phys. 2012;8:497–503. doi:10.1038/nphys2299.

[14]   Krasnikov SA, Lübben O, Murphy BE, Bozhko SI, Chaika AN, Sergeeva NN, et al. Writing with atoms: oxygen adatoms on the MoO_2/Mo(110) surface. Nano Res. 2013;6:929–937. doi:10.1007/s12274-013-0370-2.

[15]   Gawronski H, Mehlhorn M, Morgenstern K. Imaging phonon excitation with atomic resolution. Science 2008;319:930–933. doi:10.1126/science.1152473.

[16]   Stipe BC, Rezaei MA, Ho W. Single-molecule vibrational spectroscopy and microscopy. Science 1998;280:1732–1735. doi:10.1126/science.280.5370.1732.

[17]   Feenstra RM, Stroscio JA, Tersoff J, Fein AP. Atom-selective imaging of the GaAs(110) surface. Phys. Rev. Lett. 1987;58:1192–1195. doi:10.1103/PhysRevLett.58.1192.

[18] Schmid M, Stadler H, Varga P. Direct observation of surface chemical order by scanning tunneling microscopy. Phys. Rev. Lett. 1993;70:1441–1444. doi:10.1103/PhysRevLett. 70.1441.

[19] Wiesendanger R, Güntherodt HJ, Güntherodt G, Gambino RJ, Ruf R. Observation of vacuum tunneling of spin-polarized electrons with the scanning tunneling microscope. Phys. Rev. Lett. 1990;65:247–250. doi:10.1103/PhysRevLett.65.247.

[20] Wiesendanger R, Shvets IV, Bürgler D, Tarrach G, Güntherodt HJ, Coey JMD, Gräser S. Topographic and magnetic-sensitive scanning tunneling microscope study of magnetite. Science 1992;255:583–586. doi:10.1126/science.255.5044.583.

[21] Wiesendanger R. Spin mapping at the nanoscale and atomic scale. Rev. Mod. Phys. 2009;81:1495–1550. doi:10.1103/RevModPhys.81.1495.

[22] Chaika AN, Nazin SS, Semenov VN, Bozhko SI, Lübben O, Krasnikov SA, et al. Selecting the tip electron orbital for scanning tunneling microscopy imaging with sub-Ångström lateral resolution. EPL 2010;92:46003. doi:10.1209/0295-5075/92/46003.

[23] Chaika AN, Nazin SS, Semenov VN, Orlova NN, Bozhko SI, Lübben O, et al. High resolution STM imaging with oriented single crystalline tips. Appl. Surf. Sci. 2013;267:219–223. doi:10.1016/j.apsusc.2012.10.171.

[24] Chaika AN, Orlova NN, Semenov VN, Postnova EYu, Krasnikov SA, Lazarev MG, et al. Fabrication of [001]-oriented tungsten tips for high resolution scanning tunneling microscopy. Sci. Rep. 2014;4:3742. doi:10.1038/srep03742.

[25] Herz M, Giessibl FJ, Mannhart J. Probing the shape of atoms in real space. Phys. Rev. B 2003;68:045301. doi:10.1103/PhysRevB.68.045301.

[26] Chaika AN, Semenov VN, Nazin SS, Bozhko SI, Murphy S, Radican K, Shvets IV. Atomic row doubling in the STM images of Cu(014)-O obtained with MnNi tips. Phys. Rev. Lett. 2007;98:206101. doi:10.1103/PhysRevLett.98.206101.

[27] Murphy S, Radican K, Shvets IV, Chaika AN, Semenov VN, Nazin SS, Bozhko SI. Asymmetry effects in atomically resolved STM images of Cu(014)-O and W(100)-O surfaces measured with MnNi tips. Phys. Rev. B 2007;76:245423. doi:10.1103/PhysRevB. 76.245423.

[28] Chaika AN, Myagkov AN. Imaging atomic orbitals in STM experiments on a Si(111)-(7×7) surface. Chem. Phys. Lett. 2008;453:217–221. doi:10.1016/j.cplett.2008.01.025.

[29] Garleff JK, Wenderoth M, Sauthoff K, Ulbrich RG, Rohlfing M. 2×1 reconstructed Si(111) surface: STM experiments versus ab initio calculations. Phys. Rev. B 2004;70:245424. doi:10.1103/PhysRevB.70.245424.

[30] Chaika AN, Nazin SS, Bozhko SI. Selective STM imaging of oxygen-induced Cu(115) surface reconstructions with tungsten probes. Surf. Sci. 2008;602:2078–2088. doi: 10.1016/j.susc.2008.04.014.

[31] Giessibl FJ, Hembacher S, Bielefeldt H, Mannhart J. Subatomic features on the silicon (111)-(7×7) surface observed by atomic force microscopy. Science 2000;289:422–425. doi: 10.1126/science.289.5478.422.

[32] Giessibl FJ, Bielefeldt H, Hembacher S, Mannhart J. Imaging of atomic orbitals with the atomic force microscope—experiments and simulations. Ann. Phys. 2001;10:887–910. doi:10.1002/1521-3889(200111)10:11/12<887::AID-ANDP887=3.0.CO;2-B.

[33] Hembacher S, Giessibl FJ, Mannhart J. Force microscopy with light-atom probes. Science 2004;305:380–383. doi:10.1126/science.1099730.

[34] Fowler RH, Nordheim L. Electron Emission in Intense Electric Fields. Proc. R. Soc. Lond. A 1928;119:173–181. doi:10.1098/rspa.1928.0091.

[35] Bryant A, Smith DPE, Binnig G, Harrison WA, Quate CF. Anomalous distance dependence in scanning tunneling microscopy. Appl. Phys. Lett. 1986;49:936–938. doi: 10.1063/1.97489.

[36] Jurczyszyn L, Mingo N, Flores F. Influence of the atomic and electronic structure of the tip on STM images and STS spectra. Surf. Sci. 1998;402–404:459–463. doi:10.1016/S0039-6028(97)00971-0.

[37] Hofer WA, Garcia-Lekue A, Brune H. The role of surface elasticity in giant corrugations observed by scanning tunneling microscopes. Chem. Phys. Lett. 2004;397:354–359. doi: 10.1016/j.cplett.2004.08.110.

[38] Bode M, Pascal R, Wiesendanger R. Distance-dependent STM-study of the W(110)/C-R(15×3) surface. Z. Phys. B 1996;101:103–107. doi:10.1007/s002570050187.

[39] Wiesendanger R, Bode M, Pascal R, Allers W, Schwarz UD. Issues of atomic-resolution structure and chemical analysis by scanning probe microscopy and spectroscopy. J. Vac. Sci. Technol. A 1996;14:1161–1167. doi:10.1116/1.580259.

[40] Klijn J, Sacharow L, Meyer C, Blugel S, Morgenstern M, Wiesendanger R. STM measurements on the InAs(110) surface directly compared with surface electronic structure calculations. Phys. Rev. B 2003;68:205327. doi:10.1103/PhysRevB.68.205327.

[41] Calleja F, Arnau A, Hinarejos JJ, Vazquez de Parga AL, Hofer WA, Echenique PM, Miranda R. Contrast reversal and shape changes of atomic adsorbates measured with scanning tunneling microscopy. Phys. Rev. Lett. 2004;92:206101. doi:10.1103/PhysRevLett.92.206101.

[42] Blanco JM, González C, Jelínek P, Ortega J, Flores F, Pérez R, et al. Origin of contrast in STM images of oxygen on Pd(111) and its dependence on tip structure and tunneling parameters. Phys. Rev. B 2005;71:113402. doi:10.1103/PhysRevB.71.113402.

[43] Woolcot T, Teobaldi G, Pang CL, Beglitis NS, Fisher AJ, Hofer WA, Thornton G. Scanning tunneling microscopy contrast mechanisms for TiO_2. Phys. Rev. Lett. 2012;109:156105. doi:10.1103/PhysRevLett.109.156105.

[44] Mönig H, Todorovic M, Baykara MZ, Schwendemann TC, Rodrigo L, Altman EI, et al. Understanding scanning tunneling microscopy contrast mechanisms on metal oxides: a case study. ACS Nano 2013;7:10233–10244. doi:10.1021/nn4045358.

[45] Ondracek M, Pou P, Rozsıval V, Gonzalez C, Jelınek P, Perez R, Forces and currents in carbon nanostructures: are we imaging atoms? Phys. Rev. Lett. 2011;106:176101. doi: 10.1103/PhysRevLett.106.176101.

[46] Ondracek M, Gonzalez C, Jelınek P. Reversal of atomic contrast in scanning probe microscopy on (111) metal surfaces. J. Phys.: Condens. Matter 2012;24:084003. doi: 10.1088/0953-8984/24/8/084003.

[47] Tersoff J, Hamann DR. Theory and application for the scanning tunneling microscope. Phys. Rev. Lett. 1983;50:1998–2001. doi:10.1103/PhysRevLett.50.1998.

[48] Tersoff J, Hamann DR. Theory of the scanning tunneling microscope. Phys. Rev. B 1985;31:805–813. doi:10.1103/PhysRevB.31.805.

[49] Wintterlin J, Wiechers J, Brune H, Gritsch T, Hofer H, Behm RJ. Atomic-resolution imaging of close-packed metal surfaces by scanning tunneling microscopy. Phys. Rev. Lett. 1989;62:59–62. doi:10.1103/PhysRevLett.62.59.

[50] Hallmark V, Chiang S, Rabalt J, Swalen J, Wilson R. Observation of atomic corrugation on Au(111) by scanning tunneling microscopy. Phys. Rev. Lett. 1987;59:2879–2882. doi: 10.1103/PhysRevLett.59.2879.

[51] Wintterlin J, Brune H, Hofer H, Behm R. Atomic scale characterization of oxygen adsorbates on Al(111) by scanning tunneling microscopy. Appl. Phys. A 1988;47:99–102. doi:10.1007/BF00619706.

[52] Tersoff J, Lang ND. Tip-dependent corrugation of graphite in scanning tunneling microscopy. Phys. Rev. Lett. 1990;65:1132–1135. doi:10.1103/PhysRevLett.65.1132.

[53] Chen CJ. Origin of atomic resolution on metal surfaces in scanning tunneling microscopy. Phys. Rev. Lett. 1990;65:448–451. doi:10.1103/PhysRevLett.65.448.

[54] Chen CJ. Tunneling matrix elements in three-dimensional space: the derivative rule and the sum rule. Phys. Rev. B 1990;42:8841–8857. doi:10.1103/PhysRevB.42.8841.

[55] Chen CJ. Effects of $m{\neq}0$ tip states in scanning tunneling microscopy: the explanation of corrugation reversal. Phys. Rev. Lett. 1992;69:1656–1659. doi:10.1103/PhysRevLett. 69.1656.

[56] Sacks W. Tip orbitals and the atomic corrugation of metal surfaces in scanning tunneling microscopy. Phys. Rev. B 2000;61:7656–7668. doi:10.1103/PhysRevB.61.7656.

[57] Temirov R, Soubatch S, Neucheva O, Lassise AC, Tautz FS. A novel method achieving ultra-high geometrical resolution in scanning tunnelling microscopy. New J. Phys. 2008;10:053012. doi:10.1088/1367-2630/10/5/053012.

[58] Weiss C, Wagner C, Kleimann C, Rohlfing M, Tautz FS, Temirov R. Imaging Pauli repulsion in scanning tunneling microscopy. Phys. Rev. Lett. 2010;105:086103. doi: 10.1103/PhysRevLett.105.086103.

[59] Gross L, Moll N, Mohn F, Curioni A, Meyer G, Hanke F, Persson M. High-resolution molecular orbital imaging using a p-wave STM tip. Phys. Rev. Lett. 2011;107:086101. doi:10.1103/PhysRevLett.107.086101.

[60] Gross L, Mohn F, Moll N, Schuler B, Criado A, Guitian E, Pena D, Gourdon A, Meyer G. Bond-order discrimination by atomic force microscopy. Science 2012;337:1326–1329. doi:10.1126/science.1225621.

[61] Scheer E, Agrait N, Cuevas JC, Yeyati AL, Ludoph B, Martin-Rodero A, et al. The signature of chemical valence in the electrical conduction through a single-atom contact. Nature 1998;394:154–157. doi:10.1038/28112.

[62] Polok M, Fedorov DV, Bagrets A, Zahn P, Mertig I. Evaluation of conduction eigen-channels of an adatom probed by an STM tip. Phys. Rev. B 2011;83:245426. doi:10.1103/PhysRevB.83.245426.

[63] Choi H, Longo RC, Huang M, Randall JN, Wallace RM, Cho K. A density-functional theory study of tip electronic structures in scanning tunneling microscopy. Nanotechnology 2013;24:105201. doi:10.1088/0957-4484/24/10/105201.

[64] Suominen I, Nieminen J, Markiewicz RS, Bansil A. Effect of orbital symmetry of the tip on scanning tunneling spectra of $Bi_2Sr_2CaCu_2O_{8+\delta}$. Phys. Rev. B 2011;84:014528. doi: 10.1103/PhysRevB.84.014528.

[65] Palotas K, Mandi G, Hofer WA. Three-dimensional Wentzel-Kramers-Brillouin approach for the simulation of scanning tunneling microscopy and spectroscopy. Front. Phys. 2014;9:711–747. doi:10.1007/s11467-013-0354-4.

[66] Palotas K, Mandi G, Szunyogh L. Orbital-dependent electron tunneling within the atom superposition approach: theory and application to W(110). Phys. Rev. B 2012;86:235415. doi:10.1103/PhysRevB.86.235415

[67] Mandi G, Palotas K. STM contrast inversion of the Fe(110) surface. Appl. Surf. Sci. 2014;304:65–72. doi:0.1016/j.apsusc.2014.02.143.

[68] da Silva Neto EH, Aynajian P, Baumbach RE, Bauer ED, Mydosh J, Ono S, Yazdani A. Detection of electronic nematicity using scanning tunneling microscopy. Phys. Rev. B 2013;87:161117. doi:10.1103/PhysRevB.87.161117.

[69] Jurczyszyn L, Stankiewicz B. Interorbital interference in STM tip during electron tunneling in tip–sample system: influence on STM images. Prog. Surf. Sci. 2003;74:185–200. doi:10.1016/j.progsurf.2003.08.014.

[70] Jurczyszyn L, Stankiewicz B. The role of interorbital interference in the formation of STS spectra. Appl. Surf. Sci. 2005;242:70–81. doi:10.1016/j.apsusc.2004.07.066.

[71] Hagelaar JHA, Flipse CFJ, Cerda JI. Modeling realistic tip structures: scanning tunneling microscopy of NO adsorption on Rh(111). Phys. Rev. B 2008;78:161405. doi:10.1103/PhysRevB.78.161405.

[72] Telychko M, Mutombo P, Ondracek M, Hapala P, Bocquet FC, Kolorenc J, et al. Achieving high-quality single-atom nitrogen doping of Graphene/SiC(0001) by ion implantation and subsequent thermal stabilization. ACS Nano 2014;8:7318–7324. doi: 10.1021/nn502438k.

[73] Sachse T, Neel N, Meierott S, Berndt R, Hofer WA, Kröger J. Electronic and magnetic states of Mn2 and Mn2H on Ag(111). New J. Phys. 2014;16:063021. doi:10.1088/1367-2630/16/6/063021.

[74] Mandi G, Palotas K. Chen's derivative rule revisited: role of tip-orbital interference in STM. Phys. Rev. B 2015;91:165406. doi:10.103/PhysRevB.91.165406.

[75] Wright CA, Solares SD. Computational study of tip apex symmetry characterization in high-resolution atomic force microscopy. J. Phys. D: Appl. Phys. 2013;46:155307. doi: 10.1088/0022-3727/46/15/155307

[76] Wright CA, Solares SD. On mapping sub-Ångström electron clouds with force microscopy. Nano Lett. 2011;11:5026–5033. doi:10.1021/nl2030773.

[77] Wright CA, Solares SD. Imaging of subatomic electron cloud interactions: effect of higher harmonics processing in noncontact atomic force microscopy. Appl. Phys. Lett. 2012;100:163104. doi:10.1063/1.3703767.

[78] Jelinek P, Shvec M, Pou P, Perez R, Chab V. Tip-induced reduction of the resonant tunneling current on semiconductor surfaces. Phys. Rev. Lett. 2008;101:176101. doi: 10.1103/PhysRevLett.101.176101

[79] Ternes M, Gonzalez C, Lutz CP, Hapala P, Giessibl FJ, Jelinek P, Heinrich AJ. Interplay of conductance, force, and structural change in metallic point contacts. Phys. Rev. Lett. 2011;106:016802. doi:10.1103/PhysRevLett.106.016802

[80] Neel N, Kröger J, Limot L, Palotas K, Hofer WA, Berndt R. Conductance and Kondo effect in a controlled single-atom contact. Phys. Rev. Lett. 2007;98:016801. doi:10.1103/PhysRevLett.98.016801.

[81] Serrate D, Ferriani P, Yoshida Y, Hla SW, Menzel M, von Bergmann K, et al. Imaging and manipulating the spin direction of individual atoms. Nature Nanotech. 2010;5:350–353. doi:10.1038/nnano.2010.64.

[82] Zotti LA, Hofer WA, Giessibl FJ. Electron scattering in scanning probe microscopy experiments. Chem. Phys. Lett. 2006;420:177–182. doi:10.1016/j.cplett.2005.12.065.

[83] Hug HJ, Lantz MA, Abdurixit A, van Schendel PJA, Hoffmann R, Kappenberger P, Baratoff A. Subatomic features in atomic force microscopy images. Science 2001;291:2509a. doi:10.1126/science.291.5513.2509a.

[84] Huang M, Cuma M, Liu F. Seeing the atomic orbital: first-principles study of the effect of tip termination on atomic force microscopy. Phys. Rev. Lett. 2003;90:256101. doi: 10.1103/PhysRevLett.90.256101.

[85] Chen CJ. Possibility of imaging lateral profiles of individual tetrahedral hybrid orbitals in real space. Nanotechnology 2006;17: S195–S200. doi:10.1088/0957-4484/17/7/S16.

[86] Campbellova A, Ondracek M, Pou P, Perez R, Klapetek P, Jelınek P. "Sub-atomic" resolution of non-contact atomic force microscope images induced by a heterogeneous tip structure: a density functional theory study. Nanotechnology 2011;22:295710. doi: 10.1088/0957-4484/22/29/295710.

[87] Sweetman A, Rahe P, Moriarty P. Unique determination of "subatomic" contrast by imaging covalent backbonding. Nano Lett. 2014;14:2265–2270. doi:10.1021/nl4041803.

[88] Wang YL, Gao HJ, Guo HM, Liu HW, Batyrev IG, McMahon WE, Zhang SB. Tip size effect on the appearance of a STM image for complex surfaces: theory versus experiment for Si(111)-(7×7). Phys. Rev. B 2004;70:073312. doi:10.1103/PhysRevB.70.073312.

[89] Li Z, Schouteden K, Iancu V, Janssens E, Lievens P, van Haesendonck C, Cerdá JI. Chemicaly modified STM tips for atomic-resolution imaging of ultrathin NaCl films. Nano Res. 2015;8:2223–2230. doi:10.1007/s12274-015-0733-y.

[90] Emmrich M, Huber F, Pielmeier F, Welker J, Hofmann T, Schneiderbauer M, et al. Subatomic resolution force microscopy reveals internal structure and adsorption sites of small iron clusters. Science 2015;348:308–311. doi:10.1126/science.aaa5329.

[91] Hofmann T, Pielmeier F, Giessibl FJ. Chemical and crystallographic characterization of the tip apex in scanning probe microscopy. Phys. Rev. Lett. 2014;112:066101. doi: 10.1103/PhysRevLett.112.066101.

[92] Hofmann T, Pielmeier F, Giessibl FJ. Erratum: Chemical and crystallographic characterization of the tip apex in scanning probe microscopy [Phys. Rev. Lett. 112, 066101 (2014)]. Phys. Rev. Lett. 2015;115:109901. doi:10.1103/PhysRevLett.115.109901.

[93] Grushko V, Lubben O, Chaika AN, Novikov N, Mitskevich E, Chepugov A, et al. Atomically resolved STM imaging with a diamond tip: simulation and experiment. Nanotechnology 2014;25:025706. doi:10.1088/09574484/25/2/025706

[94] Chaika AN, Semenov VN, Glebovskiy VG, Bozhko SI. Scanning tunneling microscopy with single crystal W[001] tips: high resolution studies of Si(557)5×5 surface. Appl. Phys. Lett. 2009;95:173107. doi:10.1063/1.3254240.

[95] Yashina LV, Püttner R, Volykhov AA, Stojanov P, Riley J, Vassiliev SY, et al. Atomic geometry and electron structure of the GaTe(1 0 -2) surface. Phys. Rev. B 2012;85:075409. doi:10.1103/PhysRevB.85.075409.

[96]   Gross L, Mohn F, Moll N, Schuler B, Criado A, Guitian E, et al. Bond-order discrimination by atomic force microscopy. Science 2012;337:1326–1329. doi:10.1126/science.1225621.

[97]   Deng ZT, Lin H, Ji W, Gao L, Lin X, Cheng ZH, et al. Selective analysis of molecular states by functionalized scanning tunneling microscopy tips. Phys. Rev. Lett. 2006;96:156102. doi:10.1103/PhysRevLett.96.156102.

[98]   Cheng Z, Du S, Guo W, Gao L, Deng Z, Jiang N, et al. Direct imaging of molecular orbitals of metal phthalocyanines on metal surfaces with an O_2-functionalized tip of a scanning tunneling microscope. Nano Res. 2011;4:523–530. doi:10.1007/s12274-011-0108-y.

[99]   Mohn F, Gross L, Moll N, Meyer G. Imaging the charge distribution within a single molecule. Nature Nanotech. 2012;7:227–231. doi:10.1038/nnano.2012.20.

[100]   Mohn F, Schuler B, Gross L, Meyer G. Different tips for high-resolution atomic force microscopy and scanning tunneling microscopy of single molecules. Appl. Phys. Lett. 2013;102:073109. doi:10.1063/1.4793200.

[101]   Hapala P, Kichin G, Wagner C, Tautz FS, Temirov R, Jelınek P. Mechanism of high-resolution STM/AFM imaging with functionalized tips. Phys. Rev. B 2014;90:085421. doi:10.1103/PhysRevB.90.085421.

[102]   Chiutu C, Sweetman AM, Lakin AJ, Stannard A, Jarvis S, Kantorovich L, et al. Precise orientation of a single C60 molecule on the tip of a scanning probe microscope, Phys. Rev. Lett. 2012;108:268302.doi:10.1103/PhysRevLett.108.268302.

[103]   Lakin AJ, Chiutu C, Sweetman AM, Moriarty P, Dunn JL. Recovering molecular orientation from convoluted orbitals. Phys. Rev. B 2013;88:035447. doi:10.1103/PhysRevB.88.035447.

[104]   Fasolino A, Los JH, Katsnelson MI. Intrinsic ripples in graphene. Nature Mater. 2007;6:858–861. doi:10.1038/nmat2011.

Chapter 14

Wavefunction Analysis of STM Image: Surface Reconstruction of Organic Charge Transfer Salts

Hirokazu Sakamoto, Eiichi Mori, Hideyuki Arimoto,
Keiichiro Namai, Hiroyuki Tahara, Toshio Naito,
Taka-aki Hiramatsu, Hideki Yamochi and
Kenji Mizoguchi

Additional information is available at the end of the chapter

http://dx.doi.org/10.5772/63406

Abstract

In this chapter, the wavefunction analysis is demonstrated, applied to the organic charge transfer salts composed of electron donor and electron acceptor molecules. Scanning tunneling microscopy (STM) images of the surface donor layers in the three charge transfer salts, α-(BEDT-TTF)$_2$I$_3$, β-(BEDT-TTF)$_2$I$_3$, and (EDO-TTF)$_2$PF$_6$, are analyzed with the atomic π electron orbitals of sulfur, oxygen, and carbon atoms. We have deduced three different kinds of surface molecular reconstructions as follows: (1) charge redistribution in α-(BEDT-TTF)$_2$I$_3$, (2) translational reconstruction up to 0.1 nm in β-(BEDT-TTF)$_2$I$_3$, and (3) rotational reconstruction transforming the 1D axis from the a axis to the b axis in (EDO-TTF)$_2$PF$_6$. Finally, it is concluded that the surface reconstruction is ascribed to the additional gain of the cohesive energy of the π electron system, provoked by the reduced steric hindrance with the anions of the missing outside double layer. The investigations of the surface states provide not only interesting behaviors of the surface cation layer, but also important insights into the electronic states of a lot of similar charge transfer crystals, as demonstrated in α-(BEDT-TTF)$_2$I$_3$.

Keywords: STM, surface reconstruction, charge reorientation, translational reconstruction, rotational reconstruction, organic charge transfer salt, wavefunction analysis, topography

1. Introduction

It is well known that scanning tunneling microscopy (STM) is a powerful tool to disclose the atomic images, and is applicable to the conducting materials, such as metals and semiconductors. However, since STM images are constructed with the electron tunneling probability between the wavefunctions of an STM probe tip and those of a sample surface, it is required for us to analyze the wavefunctions of the atoms and the molecules in the surface layer to extract structural information. In this chapter, some examples of the wavefunction analysis [1] are demonstrated in the organic charge transfer salts composed of electron donor and electron acceptor molecules, in which the van der Waals interaction and the π electron transfer integrals between like molecules, and the Coulomb attractive interaction between unlike molecules govern the formation of the crystals. Since the van der Waals interaction between the like molecules is relatively weak, the analysis of atomic π orbitals of sulfur, oxygen, and carbon atoms of the donor molecules would be a good approximation for the wavefunction analysis of STM images.

Organic charge transfer salts have been intensively investigated for a long period more than 40 years (for example, see [2]). The electronic states are mainly governed by the π electron network of donor and acceptor molecules and the ratio of the number of acceptor molecules to that of the donor molecules, which dominates the filling of a donor π electron band. The first candidate of organic metals is a charge transfer complex, TTF-TCNQ (tetrathiafulvalene-tetracyanoquinodimethane), composed of one-dimensional (1D) independent columns of donor TTF and acceptor TCNQ molecules with the fractional charge transfer of $\delta \approx 0.59$ electron between them. TTF-TCNQ shows a sharp and remarkably large conductivity maxima up to $\sigma \approx 3 \times 10^5$ S/cm around 54 K, which is not superconducting fluctuation, as is initially conjectured [3], but is Peierls transition to an insulating state [4]. The first category of organic superconductors is quasi-one dimensional electronic systems, $(\text{TMTSF})_2\text{X}$ (TMTSF = tetramethyltetraselenafulvalene, X = PF_6, AsF_6, ClO_4, etc.) [5] and $(\text{TMTTF})_2\text{Br}$ (TMTTF = tetramethyltetrathiafulvalene) with $T_C \approx 1$ K, mostly under pressure. The organic superconductors with the higher transition temperatures up to $\approx$14 K have been realized with BEDT-TTF donor molecules [bis(ethylenedithio)tetrathiafulvalene, commonly abbreviated with "ET"], which form two dimensional π electron networks [6]. In addition to the superconductivity, the charge transfer salts provide a variety of exotic physical properties by adjusting chemical and physical parameters like chemical modifications of donor and acceptor molecules including molecular symmetry and many crystal phases with the same composition including segregated donor and acceptor stacks and alternately mixed stacks of donor and acceptor molecules. Thus, we are possible to continuously control the structure of the π electron networks from 1D to 2D, the filling of the π electron bands and electron–electron correlation of π bands, resulting in not only non-BCS superconductor, but also metals with variety of ground states like antiferromagnetic (AF), charge density wave (CDW), spin density wave (SDW), spin Peierls (SP), charge ordering (CO) states, and their combinations with exotic magnetic structures, for example, magnetic-field induced superconductivity [7]. Then, STM investigation of these organic systems is useful to collect local information of the π electron systems of the crystal surface layer [1, 8–18].

In this chapter, the STM images in the surface donor layers of the three charge transfer salts, α-(BEDT-TTF)$_2$I$_3$, known as a bulk Dirac Fermion system under pressure [19], β-(BEDT-TTF)$_2$I$_3$ with superconducting transition [20] and (EDO-TTF)$_2$PF$_6$ (EDO–TTF= ethylene dioxy tetrathiafulvalene) with multi-instability around room temperature (RT) [21, 22] are analyzed with the atomic π electron orbitals of sulfur, oxygen and carbon atoms, which are the main carrier of the π electrons. These salts have segregated 2D layers of the BEDT-TTF or EDO-TTF donor molecules and I$_3^-$ or PF$_6^-$ anion molecules. Two (four) donor molecules construct a unit cell with one (two) I$_3^-$ or PF$_6^-$ anion molecule(s); thus, two donor molecules have one electron hole in all the salts [23, 24, 21].

We have deduced three different kinds of surface molecular reconstructions by the electron wavefunction analysis of STM images, as follows.

1.   *Charge* redistribution in α-(BEDT-TTF)$_2$I$_3$

2.   *Translational* reconstruction up to 0.1 nm in β-(BEDT-TTF)$_2$I$_3$

3.   *Rotational* reconstruction transforming the 1D axis from a to b in (EDO-TTF)$_2$PF$_6$

These surface reconstructions stabilize the π electron system with the additional gain of the cohesive energy in the surface donor layer caused by the removed steric hindrance with the anion molecules of the missing outside double layer.

In α-(BEDT-TTF)$_2$I$_3$, it was found that the electronic states of the surface layer are CO state without definite displacement of the molecules, but with small molecular rotation <1°. This surface state is similar to the CO ground state of the bulk system below 135 K, which would be caused by calming down of the thermal vibration in the end ethylene group of BEDT-TTF molecules. It is suggested that the missing steric hindrance of the surface BEDT-TTF molecules with I$_3^-$ ions of the missing outside double layer stabilizes the ground CO state at 300 K even with thermal vibration of the ethylene groups. In β-(BEDT-TTF)$_2$I$_3$, the translational reconstruction up to ≈0.1 nm along the a axis was found in the surface BEDT-TTF layer. This reconstruction removes a staggered structure of the (BEDT-TTF)$_2$ molecular units, resulting in the increase of the cohesive energy between the surface BEDT-TTF molecules. In (EDO-TTF)$_2$PF$_6$, asymmetric EDO-TTF molecules stack with head-to-tail type arrangement, but TTF groups as the main carrier of the π electrons stack without distinctive staggering. The STM image suggests large rotational modifications of the surface EDO-TTF molecules, which drastically modify the electronic structure of the surface layer. The π band of the bulk crystal is one dimensional with a side-by-side single sulfur network along the a axis, but that of the surface EDO-TTF layer is along the stacking b axis with a face-to-face sulfur pair network, which enhances the cohesive energy and stabilizes the π electron system of the surface EDO-TTF layer.

1.1. Surface states of charge transfer salts

When we proceed with the wavefunction analysis of STM images, it is useful to consider what is the same as and what can be different from that of the interior layers. The charge neutrality must be kept not only in the whole crystal, but also in each local double layer formed by donor (cation) and acceptor (anion) layers in the segregated layer salts, as in the present salts.

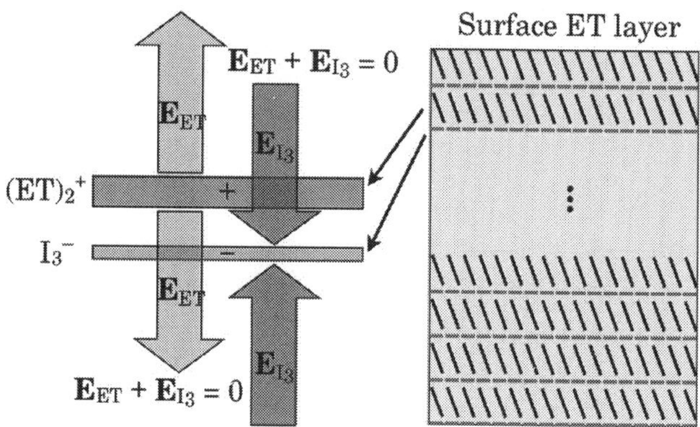

Figure 1. Schematic picture of the electric field produced by a ET_2^+ and I_3^- double layer with the same total charges of each layer in α-(BEDT-TTF)$_2$I$_3$. The electric field of the double layer is confined inside the double layer. That is, outside the double layer, the electric field cancels out. However, the leakage electric field near the double layer forms the electric dipolar field, which binds the neighboring double layers. Thus, the surface ET layer approximately feels only the electric field of the I_3^- layer within the double layer (reused from Ref. [1]).

Figure 1 schematically describes the two-dimensionally stacked structure of the present salts like α-(BEDT-TTF)$_2$I$_3$, in which the total number of (BEDT-TTF)$_2^+$ layers must be equal to that of I_3^- layers to keep the charge neutrality of the crystal. A set of (BEDT-TTF)$_2^+$ and I_3^- layers forms an electric double layer like a capacitor, and then, the whole electric field produced by the double layer is confined inside the double layer, as well known in the elementary electrostatics of a capacitor. Then, the electric field outside the double layer is approximately zero and the neighboring double layers are only weakly bound each other by the short-range electric dipolar field leaked locally from the double layers. This fact agrees with the reported experimental finding that the STM probe tip can peel off only the pairs of cation and anion molecules [11]. As in the present salts with the stable anions such as I_3^- and PF_6^-, the total number of the holes in a cation layer must be equal to that of the anions in a double layer independent of the surface layer or the interior layers, which provides the local charge neutrality.

Thus, the reconstructions of a surface layer can happen as charge redistributions and/or structural redistributions, keeping the charge neutrality of each double layer. Next, we focus on the Coulomb forces exerted on a cation layer by anion layers hereafter.

On the Coulomb forces exerted to the surface cation layer by the interior layers, it is useful to separate the cases:

1. Inside double layers, and

2. The partner anion layer of the surface double layer.

On the first point, since the electric field is absent outside the double layer, the long-range Coulomb force of the inside double layers does not exist, as mentioned above. On the second point, the partner anion layer of the double layer generally forms a flat sheet and generates approximately uniform electric field normal to the anion layer [1]. The uniform electric field attracts uniformly the cation layer sheet, in which the cation molecules are tightly bound each other by the cohesive interaction with the other cation molecules, forming a π band, under constraint of the steric hindrance with the cation molecules and the partner anion molecules in the double layer. This situation is the same as that for all the double layers and there is no special situation even in the surface layer. Only the difference of the surface double layer from the inside is that the short-range dipolar electric field by a missing double layer outside is absent, which affects only limitedly to the attraction between the members of the double layer because of its short-range nature of the dipolar field.

Thus, the Coulomb interaction between the cation and the anion layer of the electric double layer does not the main origin of the structural reconstruction in the surface ET layer, but the unique origin of the structural reconstruction can be the missing steric hindrance with the anion layer of the missing outside double layer, which competes with the cohesive interaction within the surface cation layer.

2. Wavefunction analysis

The analysis of STM images with atomic wavefunctions is described for α-(BEDT-TTF)$_2$I$_3$, as an example of the organic charge transfer salts, where the cohesion is dominated by the van der Waals interaction and the π electron transfer integrals. **Figure 2** shows a schematic picture of the surface layer of the flat donor molecules with the π wavefunctions of the end sulfur atoms, and the STM probe tip with the s wavefunction. An STM topography $\Delta h(x)$ in a constant current mode is obtained by controlling the height of the probe tip to keep the tunneling current constant with scanning the probe tip. The higher the probe tip, the brighter the topography appears. The topography is simulated by a constant amplitude contour of the π wavefunctions, $\Psi(\Delta h, x)$ along the scan direction x parallel to the plane of the donor layer.

The unit cell of α-(BEDT-TTF)$_2$I$_3$ contains four BEDT-TTF molecules (express by ET, hereafter), labeled by A, A$'$, B, and C with different π charge numbers by charge disproportionation (CD) caused by the anisotropic transfer integrals between the molecules in the unit cell. The wavefunction Ψ^i_{S3p} for the 3p orbital of the relevant sulfur atom in the ET(i) molecule (i = A, A$'$, B and C) is expressed as

$$\Psi^i_{S3p} = \sqrt{\rho_i f_S}\,\Psi_{S3p}, \tag{1}$$

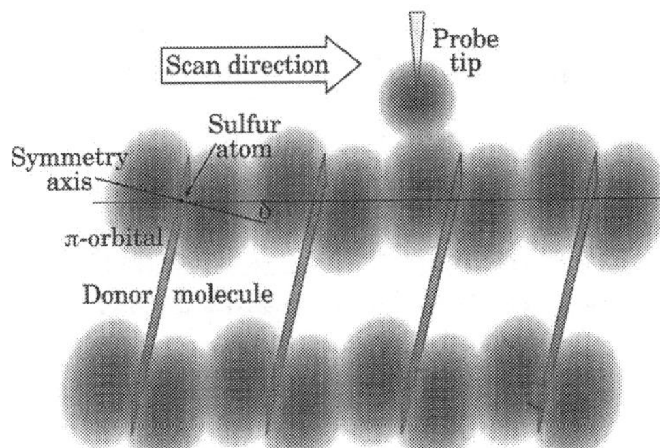

Figure 2. A schematic picture of a donor layer with a flat molecular plane. Electron wavefunctions are described by blue ellipses for π electrons of the donor molecules and a blue circle for s electron of the probe tip. The symmetry axis of the π orbital is perpendicular to the molecular plane, which leans by δ to the plane of the donor layer expressed by the thin solid horizontal line through the sulfur atom.

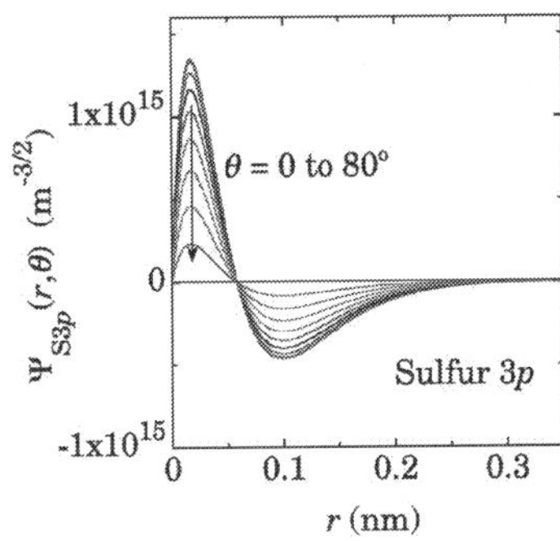

Figure 3. 3p wavefunction of a sulfur atom with the effective nuclear charge of Z_{eff} = 5.48 for every 10° of θ from 0 to 80, which is defined in **Figure 4.** r is the distance from the sulfur nucleus. STM usually observes the region much farther than 0.1 nm (reused from Ref. [1]).

where ρ_i is the number of π charges in the highest occupied molecular orbital (HOMO) band of each ET(i) molecule, f_S is the fraction of π charges at the sulfur atom in each ET(i) molecule, and Ψ_{S3p}, as shown in **Figure 3**, is the wavefunction of the sulfur $3p_\pi$ orbital with the atomic number Z = 16, expressed as

$$\Psi_{S3p} = \sqrt{\frac{2}{\pi}} \left(\frac{Z_{eff}}{a_0}\right)^5 \frac{r}{81} \left(6 - \frac{Z_{eff}r}{a_0}\right) \exp\left(-\frac{Z_{eff}r}{3a_0}\right) \cos\theta, \qquad (2)$$

and for $2p_\pi$ orbitals,

$$\Psi_{2p} = \sqrt{\frac{1}{2\pi}\left(\frac{Z_{eff}}{a_0}\right)^5}\,\frac{r}{4}\exp\left(-\frac{Z_{eff}\,r}{2a_0}\right)\cos\theta, \qquad (3)$$

where $a_0 = 5.29 \times 10^{-11}$ m is the Bohr radius, and $Z_{eff} = 5.48$, 4.45, and 3.14 is the effective nuclear charge for the sulfur, oxygen, and carbon, respectively, in which the screening effect of the inner core electrons is taken into account [25]. The representative contour of the constant Ψ_{S3p} is shown in **Figure 4**. The tunnel current I in the limits of small voltage and low temperature is expressed as [26]

$$I_{tunnel} = \left(\frac{2\pi}{\hbar}\right)e^2 V_a \Sigma_i \left|M_{\mu,i}\right|^2 \delta(E_i - E_F)\delta(E_\mu - E_F), \qquad (4)$$

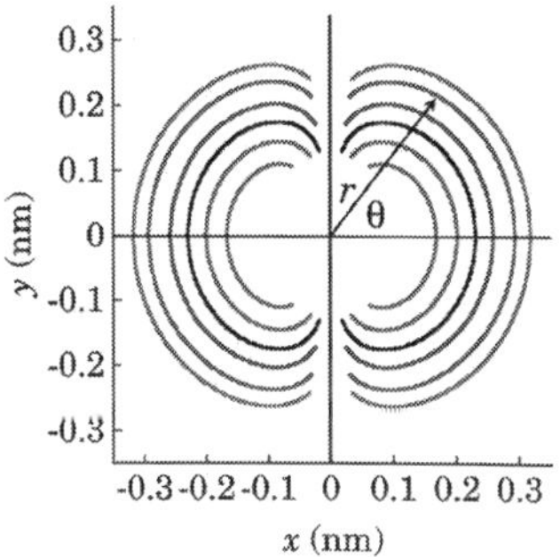

Figure 4. Constant amplitude contour of $3p$ wavefunction of a sulfur atom deduced from **Figure 3** for the STM topography analysis in α-(ET)$_2$I$_3$. The abscissa corresponds to the symmetry axis of the $3p$ orbital of the sulfur atom at the origin, perpendicular to the molecular plane corresponding to the vertical line. Each constant amplitude of $|\Psi_{S3p}|$ is 5×10^{12}, 1×10^{13}, 5×10^{13}, 1×10^{14}, and $2\times10^{14}\,m^{-3/2}$ from the outer to the inner (reused from Ref. [1]).

where V_a is the applied voltage; $M_{\mu,i}$ is the tunneling matrix element between the Ψ_μ of the probe tip and Ψ^i_{S3p} of the relevant sulfur atom; and E_i, E_μ, and E_F are the energies of the states Ψ^i_{S3p} and Ψ_μ in the absence of tunneling, and the Fermi energy of the tip, respectively. The matrix element is expressed as

$$M_{\mu,i} = -\left(\frac{\hbar^2}{2m}\right)\int dS \cdot (\Psi^*_\mu \nabla \Psi^i_{S3p} - \Psi^i_{S3p}\nabla\Psi^*_\mu), \qquad (5)$$

where the integral is over any surface lying entirely within the barrier region. The probe wavefunction Ψ_μ is taken to be independent of ET(i). In the constant current mode, the difference of Ψ^i_{S3p} produces the probe height change Δh, which depends on the local density of states proportional to $\rho_i f_s$ under the constant tunneling current condition.

Thus, the observed topography can be simulated with the adjustable parameter of ρ_i providing f_S is independent of the sites. The relative fraction of ρ_i on the ET(i) site is related to the amplitude of the sulfur $3p$ wavefunction $\Psi_{S3p}(r_i^{max})$, where r_i^{max} is the radial distance from the relevant sulfur nucleus of ET(i), where the tip height is maximum. The condition for providing the same tip current at each maximum tip height is expressed as

$$\Psi^i_{S3p}(r_i^{max}) = \Psi^j_{S3p}(r_j^{max}),$$
$$\frac{\Psi^i_{S3p}(r_i^{max})}{\Psi^j_{S3p}(r_j^{max})} = \sqrt{\frac{\rho_i}{\rho_j}}\,\frac{\Psi_{S3p}(r_i^{max})}{\Psi_{S3p}(r_j^{max})} = 1. \tag{6}$$

Thus, the ratio of the charge number in ET(i) to ET(j) can be described in terms of the amplitude of Ψ_{S3p} as

$$\frac{\rho_i}{\rho_j} = \left(\frac{\Psi_{S3p}(r_j^{max})}{\Psi_{S3p}(r_i^{max})}\right)^2. \tag{7}$$

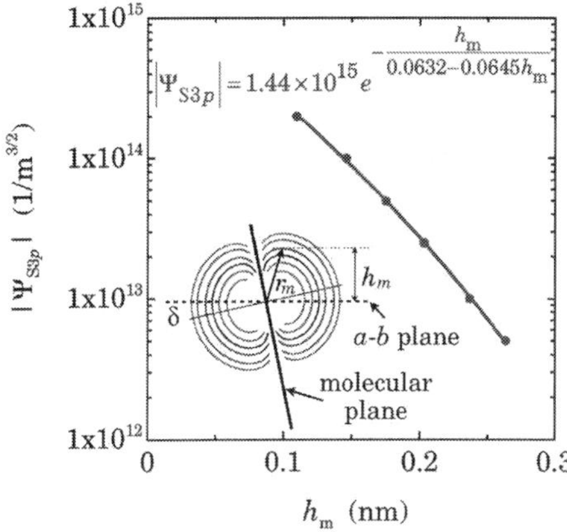

Figure 5. Maximum height h_m estimated from each Ψ_{S3p} wavefunction contour in the inset, where the symmetry axis (thin solid line) of the $3p$ wavefunction tilts by $\delta \approx 12$ deg against the a–b plane described by the dashed horizontal line through the sulfur atom. The analytical expression approximately describes $\left|\Psi_{S3p}\right|$ as a function of h_m, which is applicable to the STM topography analysis in the a–b plane of α-(ET)$_2$I$_3$ (reused from Ref. [1]).

Figure 5 shows a more realistic configuration of h_m with a tilting angle of $\delta = 12°$ for $\Psi_{S3p}(h_m)$ in α-(ET)$_2$I$_3$, where h_m is the maximum height of the contour curve measured from the relevant sulfur atom, as shown in the inset. Since $\Psi_{S3p}(h_m)$ decays exponentially, the following phenomenological formula is derived to reproduce the data:

$$\left|\Psi_{S3p}\right| = a\exp\left(-\frac{h_m}{h_0}\right) = a\exp\left(-\frac{h_m}{b-ch_m}\right). \tag{8}$$

Figure 5 demonstrates that the data points around $h_m \approx 0.2$ nm are represented well by the parameters $a = 1.44 \times 10^{15}$ m$^{-\frac{3}{2}}$, and the correlation length $h_0 = b - ch_m$ nm, where $b = 0.0632$ nm and $c = 0.0645$. Thus, the molecular charge ratio of ET(i) to ET(B) is described with the relative difference $\Delta h_{m,i}$ by

$$\begin{aligned}
\frac{\rho_i}{\rho_B} &= \left(\frac{\Psi_{S3p}(h_{m,B})}{\Psi_{S3p}(h_{m,B}+\Delta h_{m,i})}\right)^2 \\
&= \exp\left\{-2\left(\frac{h_{m,B}}{b-ch_{m,B}} - \frac{h_{m,B}+\Delta h_{m,i}}{b-c(h_{m,B}+\Delta h_{m,i})}\right)\right\},
\end{aligned} \tag{9}$$

where $h_{m,B}$ is the maximum height for ET(B) and $h_{m,i} = h_{m,B} + \Delta h_{m,i}$ for ET(i). $\Delta h_{m,i}$ can be directly measured as Δh_i for each ET(i); the tip height difference from ET(B) with some corrections is described in **Table 1** in the next section.

	Δh_i	$h_{m,i}$	$\Delta_{S,i}$	δ_i(deg)	$\Delta_{\delta,i}$	Case (a)		Case (b)	
						Δ_{CD_i}	$\Delta h_{ap,i}$	$\Delta h_{m,i}$	ρ_i/ρ_B
B	0	0.23	0	10.4	0	0	0	0	1
C	−0.025(7)	0.21	−0.003	12.9	−0.003	−0.007	−0.012	−0.019	0.36
A	−0.012(7)	0.22	0.000	11.3	−0.003	−0.003	−0.006	−0.009	0.61
A′	−0.022(7)	0.21	−0.002	11.2	−0.002	−0.003	−0.015	−0.018	0.38

Table 1. Tip height difference Δh_i measured by STM topography relative to ET(B) from **Figure 10** for α-(BEDT-TTF)$_2$I$_3$. $h_{m,i}$ is estimated from the simulation of the topographies and is utilized to estimate $\Delta_{\delta,i}$. $\Delta_{S,i}$ is the relative height difference of the relevant sulfur atom to ET(B) measured from the a–b plane, extracted from the structural data [23]. δ_i represents the angle of $3p$ orbital axis against the $a - b$ plane and $\Delta_{\delta,i}$ is the relative height change caused by δ_i, which is proportional to $h_{m,i}$. Δ_{CD_i} is the expected tip height due to the charge distribution caused by the CD state at RT and $\Delta h_{m,i} = \Delta h_i - \Delta_{S,i} - \Delta_{\delta,i}$ is the apparent change caused by the surface reconstruction over the CD state in the case (a). $\Delta h_{m,i} = \Delta h_i - \Delta_{S,i} - \Delta_{\delta,i}$ is the experimental relative height in the case (b), which gives the ratio of molecular charge with respect to ET(B), that is, ρ_i / ρ_B. All the length scales are in nm (reused from Ref. [1]).

3. Surface reconstruction in charge transfer salts

3.1. Charge redistribution in α-(BEDT-TTF)$_2$I$_3$

Figure 6 shows the crystal structure of α-(ET)$_2$I$_3$ in the a–b plane and relative locations of I$_3^-$ ions in (B). The electrical property of the bulk crystal is conductive above 135 K and the charge ordered insulating state below 135 K. The unit cell contains four ET molecules and three non-equivalent sites (A, A'), B and C at RT with different charges of the CD state. **Figure 7** shows the molecular structure of ET with HOMO molecular orbitals calculated by MOPAC. The π electrons are mainly located on the TTF group and partially on the "dithio" sulfur atoms, but not on the ethylene group. As a result, the π charge fraction on the end "dithio" sulfur atom(s) is observed by STM in the a–b plane.

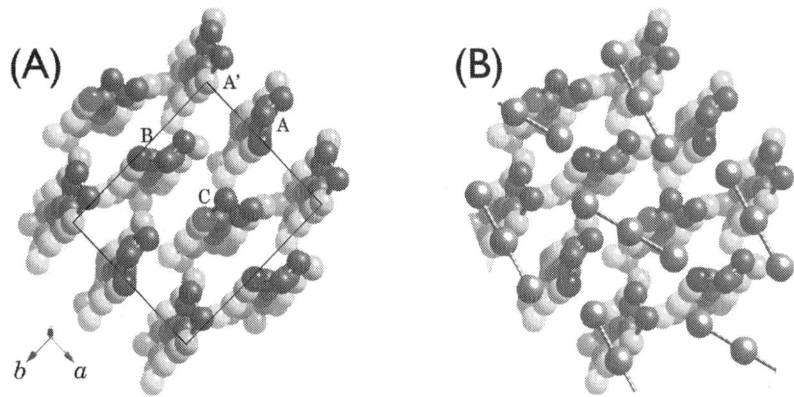

Figure 6. Unit cell structure of α-(ET)$_2$I$_3$ determined by the X-ray analysis at 300 K [23]. Blue and yellow balls indicate carbon and sulfur atoms, respectively. Hydrogen atoms are not indicated. The labels A, A', B, and C show the four molecules in a unit cell. The sulfur atoms observed by STM are marked by the special colors of purple, yellow, red and green for A, A', B, and C molecules, respectively. a = 0.9187 nm, b = 1.0793 nm, c = 1.7400 nm, α = 96.957°, β = 97.911°, and γ = 90.795°. (B) shows relative location of I$_3^-$ ions.

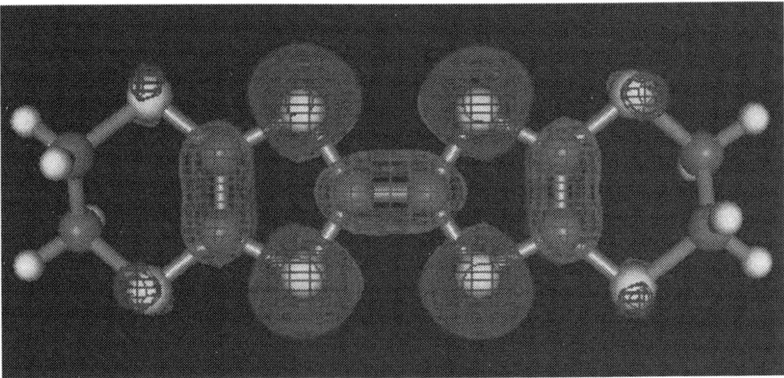

Figure 7. Molecular structure of BEDT-TTF with HOMO molecular orbitals of π electrons calculated by MOPAC. The size of the orbitals represents the relative fraction of HOMO electrons (reused from Ref. [1]).

STM was carried out at RT with easyScan 2 (NanoSurf®). A mechanically sharpened $Pt_{0.8}Ir_{0.2}$ wire, the constant tunneling current (setpoint) of 1 nA and tip potential of 10 mV were used. The α-(ET)$_2$I$_3$ samples typically with 5 × 2 × 0.05 mm^3 were prepared following a previously reported procedure [27]. The SPIP image processing software is utilized to eliminate instrumental drift of STM on the basis of the reported lattice parameters [23].

3.1.1. Site assignment

The charge distribution in the surface (ET)$_2$ layer is analyzed with the wavefunction analysis described in the previous section [1]. **Figures 8** and **9** show the observed STM images in the a–b plane, where the higher the probe tip, the brighter the topography appears. There are two characteristic points in these images: absence of a noticeable long-range modulation and the periodic characteristic structure made of four types of brightness and shapes. These are helpful features to assign the crystal structure. The a axis along ET(A)–ET(A′) and ET(B)–ET(C) is assigned to run from the top left to the bottom right. The brightest areas are assigned to the ET(B) because of the largest molecular charge at RT, as cited in **Table 2** [23]. Then, the less bright areas in the same arrays as ET(B) are assigned to ET(C). The other arrays with small and less bright areas in between the ET(B)–ET(C) arrays correspond to the ET(A)–ET(A′) arrays. Although the charge numbers of ET(A) and ET(A′) are equivalent to each other at RT in the bulk crystal [23], the observed brightness of the ET(A)–ET(A′) arrays of the surface layer suggests some non-equivalence of the alternate brightness.

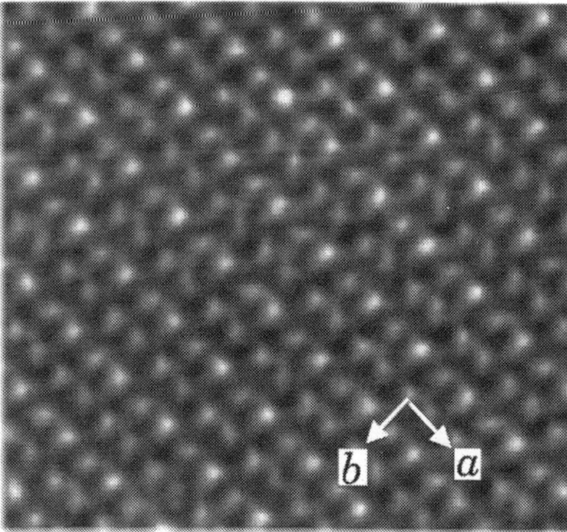

Figure 8. STM image of a thin plate-like single crystal of α-(BEDT-TTF)$_2$I$_3$ in ≈ 7.0 × 6.5 nm^2, where the thermal drift was corrected with the reported lattice parameters [23]. The assigned a and b axes are indicated by the arrows (reused from Ref. [1]).

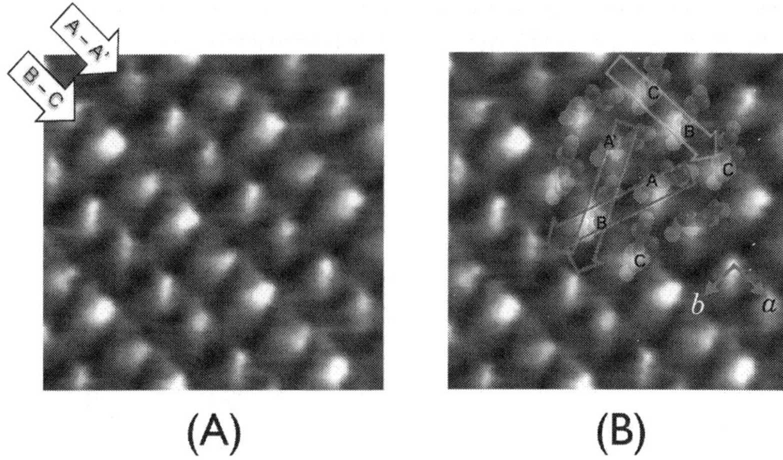

Figure 9. STM image of α-(BEDT-TTF)$_2$I$_3$ in 2.8 × 2.6.nm^2. Four BEDT-TTF molecules are contained in the unit cell. (A) Molecular arrays, A–A' and B–C, run from the top left to the right bottom. (B) Crystal structure assigned to the STM image of α-(BEDT-TTF)$_2$I$_3$. The balls with A', A, B, and C represent the relevant sulfur atoms. The three large arrows represent the direction of the topography shown in **Figure 10** (reused from Ref. [1]).

Site	Present results	X-ray results	
	RT	RT	20 K
B	0.42(8)	0.29(2)	0.35(4)
C	0.16(2)	0.21(3)	0.12(5)
A	0.26(5)	0.25(2)	0.39(5)
A'	0.16(2)	0.25(2)	0.14(5)

Table 2. Relative molecular charge $\rho_i / \sum_i \rho_i$ for α-(BEDT-TTF)$_2$I$_3$ in the case (b), which is compared with that estimated by X-ray analysis [23]. Note the broken inversion symmetry between ET(A) and ET(A'), which suggests the rich charge stripes of the B-A-B type. The parentheses show uncertainty in the last digit (reused from Ref. [1]).

Figure 9B shows the STM image with the structure determined by X-ray analysis [23], where the position and direction of the bright areas agree with the 3p orbitals of the sulfur atoms without recognizable reconstructions. Here, note that the halves of the sulfur 3p orbitals on the single side of the molecular plane are observed in the image because of the tilted symmetry axis of the sulfur 3p orbital by $\approx 12°$ out of the a–b plane.

Figure 10 shows the several superposed topographies along with the three arrow directions of **Figure 9B**, which enables us to average the random error out graphically to estimate the probe height difference Δh_i relative to ET(B). Several parameters derived from Δh_i are summarized in **Table 1**. The characteristic features of the topographies are as follows.

- $\Delta h_{\mathrm{A}}'$ is almost the same as Δh_{C}, and Δh_{A} appears in between Δh_{B} and $\Delta h_{\mathrm{A}}'$ or Δh_{C}, suggesting symmetry breaking between ET(A) and ET(A') in the surface ET layer.

- Both steep changes related to the nodes of $3p$ wavefunctions and long tails caused by overlapping of the wavefunctions are found.

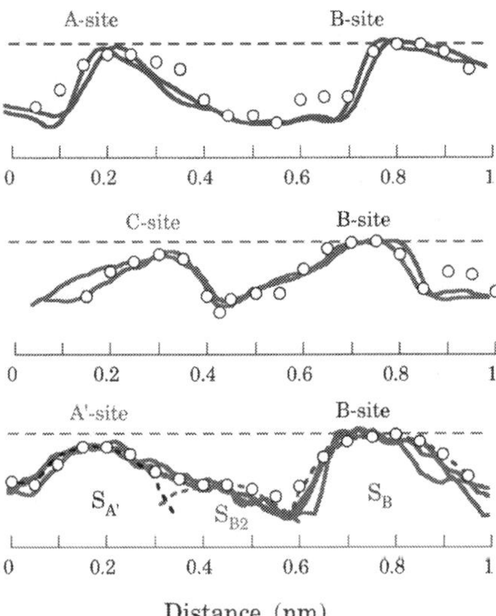

Figure 10. STM topographies of α-(BEDT-TTF)$_2$I$_3$ (solid curves) and the simulations (open circles) with $\Sigma_i \, |\Psi^i_{\mathrm{S}3p}|^2$, where the molecular charge ratio $\rho_i / \rho_{\mathrm{B}}$ in **Table 1** was taken into account. The topographies were measured along the arrows in **Figure 9(B)**. The horizontal dashed straight lines are visual guides at the top of the S_{B} site. The broken curves for the bottom A–B scan show individual contributions of each sulfur atom at S_{B} and S_{A}, and S_{B2}. From S_{B}, note that S_{B2} is 0.1 nm inside. The vertical scale is the same as the horizontal one (reused from Ref. [1]).

Several interpretations on the surface states in comparison with the bulk states are possible:

1. Molecular reconstruction [case (a)]

2. Molecular charge redistribution [case (b)]

3. Molecular reconstruction with charge redistribution [case (c)].

In the following sections, the possibilities of the structural and the charge redistribution are considered.

3.1.2. Structural reconstruction

It is informative to know how the structural parameters change across the phase transition at 135 K from the CD state to the CO state in the bulk system. A remarkable molecular charge redistribution has occurred across the phase transition from nearly equivalent charges within the unit cell at RT to the rich and the poor charge stripes below 135 K. It is a crucial point that the displacement perpendicular to the a–b plane is remarkably small around 0.001 nm or less with the molecular rotation as small as 0.75°, which corresponds to a 0.002 nm change in the position of the relevant sulfur atom perpendicular to the a–b plane [23]. A possibility of the small molecular rotation of the order of 0.75° is not rejected, but the effect to the brightness and the shapes of the STM image would be negligibly small in the present wavefunction analysis. Thus, these facts and the consideration in Section 1.1 strongly suggest that the possibility of the structural reconstruction perpendicular to the a–b plane, which is sensitive to the estimation of the charge redistribution, would not be realistic in α-$(ET)_2I_3$.

3.1.3. Charge redistribution

In this section, the most probable case (b) is discussed; the observed Δh_i's are caused by the charge redistribution in the unit cell. In **Figure 10**, the calculated contour profiles of $\Sigma_i \mid \Psi^i_{S3p} \mid^2$ are shown for the tunneling current. Here, it is demonstrated that the sulfur $3p$ wavefunctions reproduce the characteristic features of the topographies well. The nodes of the $3p$ wavefunctions produce the steep changes near the sulfur atoms. In contrast, overlap between the neighboring S_{3p} wavefunctions reproduces the longer tails of the topographies. Particularly, the presence of the second sulfur atom ET(B2) below 0.1 nm from ET(B) is essential along the A′–B direction.

Table 1 shows the ratios of the molecular charges ρ_i/ρ_B in the case (b). **Table 2** demonstrates the fraction of the molecular charge $\rho_i/\Sigma_i\rho_i$ in the unit cell, which can be compared with the reported results [23] derived with an empirical method [28] both at RT and 20 K for the crystal of α-$(ET)_2I_3$. The charge number of ET(A) is equal to that of ET(A′) in the CD state of the bulk crystal at RT, but the charge equivalence of ET(A) and ET(A′) is completely missing in the surface layer; the fraction in ET(A′) is nearly equal to ET(C), which is similar to the CO state at 20 K in the bulk crystal. Thus, it is demonstrated that the CD state at RT is unstable, but the charge redistribution similar to the CO ground state below 135 K is stabilized in the surface ET layer. This difference from the electronic states of the bulk system can be ascribed to a small-angle molecular rotation up to only 0.75° even at RT in the surface layer, where there is no steric hindrance with the missing outside double layer.

In conclusion, the charge redistribution in the surface ET layer, which is similar to the most stable ground state of the CO state below 135 K in the bulk α-$(ET)_2I_3$, is realized by the absence of the constrained steric interaction between the thermal vibration of the ethylene groups of ET molecules and the I_3^- anion layer in the missing outside double layer. This conclusion helps to interpret the mechanism of the CO phase transition at 135 K in the bulk α-$(ET)_2I_3$ crystal as the thermal vibration of ethylene groups, which prevents the inside $(ET)_2$ layers from forming the ground state molecular conformation at RT. Since the thermal vibration ceases around 135 K, the phase transition from the metallic CD state at RT to the insulating CO state in α-$(ET)_2I_3$ crystals is attained. This would be found in the other organic layered systems, such as β-$(ET)_2PF_6$ [8] and θ-$(ET)_2RbZn(SCN)$ [18].

3.2. Translational reconstruction in β-(BEDT-TTF)$_2$I$_3$

The crystal structure of β-(ET)$_2$I$_3$ is triclinic with (ET)$_2^+$ and I$_3^-$ in the unit cell, as shown in **Figure 11**, with a nearly isotropic two-dimensional Fermi surface. The electrical property is metallic with the conductivity of $\approx$ 30 S/cm at RT [24] and is superconducting at ambient pressure below 1.5 K [20] and 7.4 K under 1.3 kbar [29]. On the basis of the analyzed result of α-(ET)$_2$I$_3$ in the previous section, the missing steric hindrance of the surface (ET)$_2$ layer with the I$_3^-$ layer is also expected to have some effects on the electronic states of the surface (ET)$_2$ layer in β-(ET)$_2$I$_3$. Fortunately, the obtained STM image of β-(ET)$_2$I$_3$ with the setpoint of 1 nA and the tip potential of 10 mV makes us possible to analyze qualitatively the surface reconstruction, as shown in **Figure 12**.

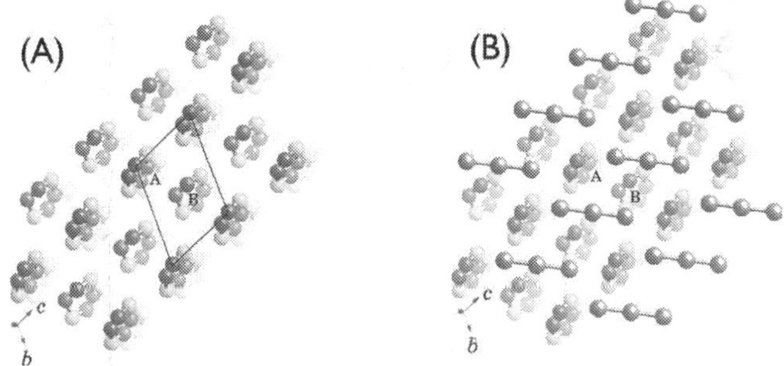

Figure 11. Crystal structure of β-(ET)$_2$I$_3$ in the b–c plane at RT [24]. Crystal data are triclinic with a = 15.243, b = 9.070, c = 6.597 Å, α = 109.73, β = 95.56, γ = 94.32°. Blue and yellow balls indicate carbon and sulfur atoms, respectively. Hydrogen atoms are not indicated. The unit cell contains two ET molecules, ET(A) and ET(B), and one I$_3^-$ ion. (B) shows relative location of I$_3^-$ ions against ET molecules. Note that the steric hindrance of the I$_3^-$ ions with ET(B) is stronger than with ET(A) in this surface and vice versa in the next layer of I$_3^-$ ions.

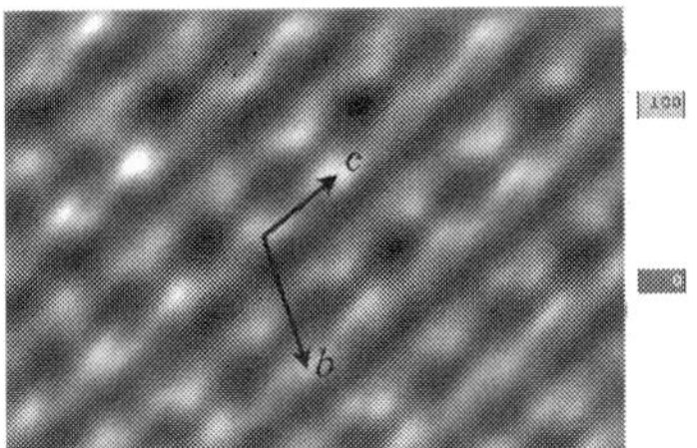

Figure 12. STM image of a needle-like single crystal (0.1 × 0.1 × 3 mm^3) of β-(ET)$_2$I$_3$ in 4.2×2.9 nm^2. Thermal drift was corrected with reference to the reported crystal structure [24], and FFT (fast Fourier transformation) filter was applied. The assigned b and c axes are indicated by the arrows. Note that the two molecules in a unit cell show similar brightness. The brightness reference is shown in the right-hand side (100–0 pm reference).

370 Microscopy and Analysis

Figure 13 shows the crystal structure of β-(ET)$_2$I$_3$ projected along the c axis, which demonstrates the definite difference between ET(A) and ET(B) in the vertical location from the b–c plane. If the surface (ET)$_2$ layer takes the same arrangement as that in the crystal, the remarkable difference of the brightness corresponding to ET(A) and ET(B) is expected in the STM image, but it is not the case. The brightness of the two arrays along the c axis looks almost the same as each other within the uncertainty. Thus, it is concluded that the surface reconstruction is occurred in the (ET)$_2$ surface layer of β-(ET)$_2$I$_3$.

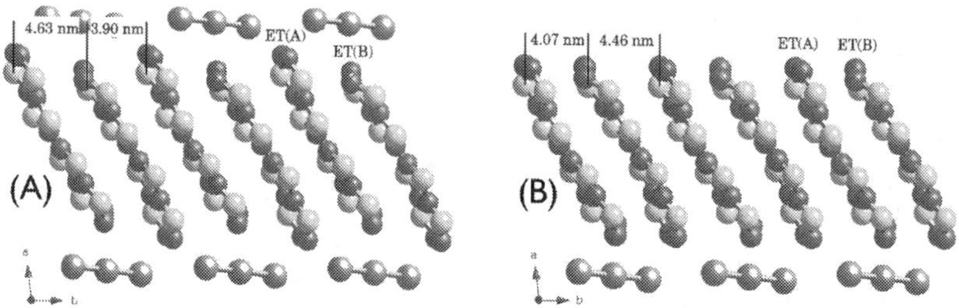

Figure 13. (A) Crystal structure of β-(ET)$_2$I$_3$ projected along the c axis with three unit cells. ET(A) is located nearer to the b–c surface by ≈ 0.1 nm than ET(B). (B) A model for the surface layer of β-(ET)$_2$I$_3$. ET(B)s are slid to align TTF double bonds to the same height as that of ET(A), which increases the cohesive energy in the surface ET layer. The separations between ET(A) and ET(B) are contrasted to (A). The topmost sulfur atoms of ET(B) are slightly deeper by 0.01 nm than ET(A).

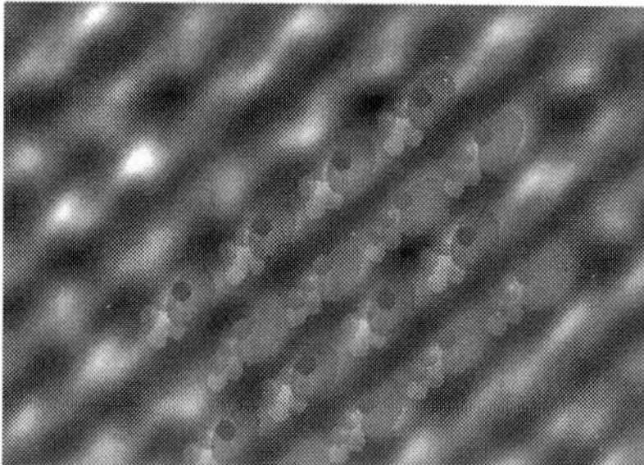

Figure 14. STM image in the b–c plane overlaid by the reconstructed structure with the π clouds. Balls and ellipses indicate the topmost sulfur atoms and the schematic π clouds, respectively. Pink corresponds to ET(A) and blue does ET(B). The difference in the relative location of the π clouds to the sulfur atoms is caused by the difference of the lean angle δ in **Figure 2**.

The origin of the reconstruction is ascribed to the presence of the open space for the surface ET(B) molecules by the missing outside double layer. Thus, the ET molecules in the surface layer are naturally reconstructed to gain additional cohesive energy by relieving the staggered arrangement of ET(A) and ET(B) in the unit cell. In contrast, it is not realistic to ascribe the origin only to some charge redistribution, which must reach three times larger π electron density in ET(B) compared with that in ET(A) to compensate the difference of 0.1 nm in depth from **Figure 5**. Thus, the most probable origin is the structural reconstruction of the two molecules in the unit cell to align to the same height as each other, which gives a similar brightness of ET(B) to ET(A) in the surface layer of β-(ET)$_2$I$_3$. Such a structural reconstruction increases the cohesive energy of the surface ET molecular layer with larger π band width and electrical conductivity. The short intermolecular sulfur-to-sulfur contacts are decreased down to 3.09, 3.43 Å in the surface layer in **Figure 13B** from 3.57, 3.58 Å in the bulk crystal [24] in **Figure 13A**. However, since the shortest 3.09 Å in **Figure 13B** looks unrealistically short, some optimization of the molecular arrangement would be required. The simulated surface molecular arrangement in the b–c plane is overlaid to the STM image in **Figure 14** and reasonably reproduces it, which supports the model of the structural reconstruction in the surface ET layer of β-(ET)$_2$I$_3$.

3.3. Rotational reconstruction in (EDO-TTF)$_2$PF$_6$

The electronic states of (EDO-TTF)$_2$PF$_6$ are of 1D metal along the a axis at RT with internal multi-instability and are transformed to charge ordered insulating states below 280 K associated with distinctive molecular deformations [22]. The mechanism of the metal-insulator transition is interpreted as the cooperation of Peierls instability, charge ordering, and the order–disorder transition of the countercomponent. The crystal structure projected in the a–b plane is shown in **Figure 15**. The unit cell of (EDO-TTF)$_2$PF$_6$ contains (EDO-TTF)$_2$ and PF$_6$. The structure of EDO-TTF molecule is shown in **Figure 16** together with the molecular orbital of π electrons. EDO-TTF molecules stack along the b axis with head-to-tail type arrangement, as described in **Figure 17**.

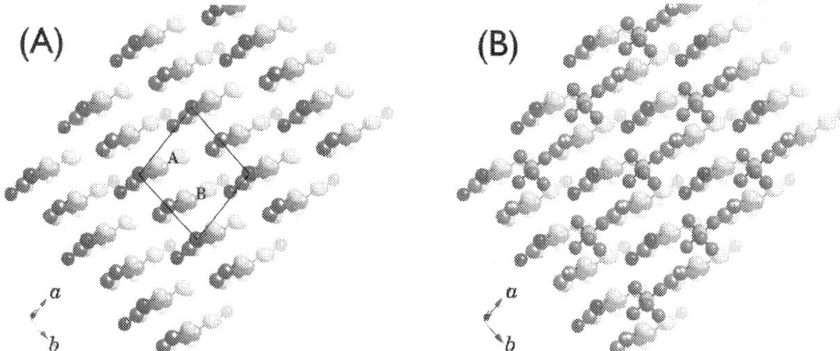

Figure 15. Structure of (EDO-TTF)$_2$PF$_6$ at RT [21]. Crystal data are triclinic with a = 7.197, b = 7.343, c = 11.948 Å, α = 93.454, β = 75.158, γ = 97.405°. Blue, yellow, and red balls indicate carbon, sulfur, and oxygen atoms, respectively. Hydrogen atoms are not indicated. The unit cell contains two EDO-TTF molecules and one PF$_6^-$ ion. (B) shows relative location of PF$_6^-$ ions to EDO-TTF molecules.

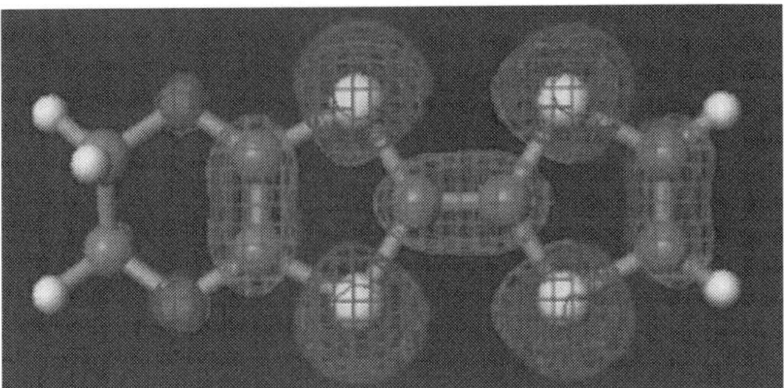

Figure 16. Molecular structure of EDO-TTF with HOMO molecular orbitals of π electrons calculated by MOPAC. The size of the orbitals represents the relative fraction of HOMO electrons. The oxygen sites contain a limited density of π electrons compared with that in the carbon sites.

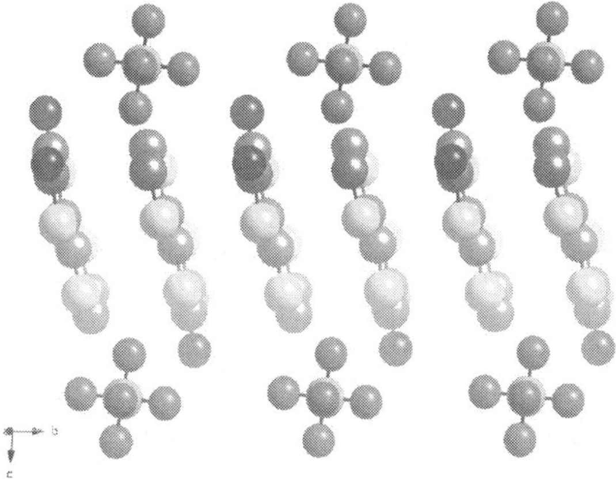

Figure 17. Crystal structure of $(EDO\text{-}TTF)_2PF_6$ along the direction perpendicular to the $b\text{--}c$ plane. PF_6 has no serious steric hindrance with the EDO-TTF layers.

The single crystals of $(EDO\text{-}TTF)_2PF_6$ with needle like thin plate ($1.0 \times 3.0 \times 0.5$ mm⁶) are used for STM study and are too brittle to pass through the transition temperature at 280 K. Thus, STM data were obtained only at RT. The image of $(EDO\text{-}TTF)_2PF_6$ in **Figure 18** was taken in the $a\text{--}b$ plane with the setpoint of $I = 6.1$ pA and the tip potential of $V_{tip} = 200$ mV. The whole image is periodically filled by the characteristic structure with different brightness and sized areas in each unit cell. The a axis was assigned as the direction, in which the profiles without distinctive structure are arrayed. In contrast, some structure caused by the head-to-tail array is expected along the b axis. No superstructure over multiple unit cells is found.

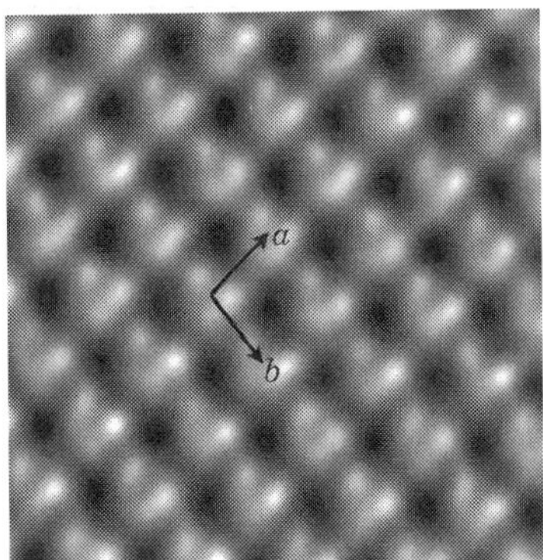

Figure 18. STM image of a single crystal of $(EDO\text{-}TTF)_2PF_6$ in 4.7×4.7 nm^2 with the setpoint of 6.1 pA and the tip potential of 200 mV. Thermal drift was corrected with reference to the reported crystal structure [21] and FFT filter was applied. The assigned a and b axes are shown by the arrows.

Although the EDO-TTF molecules stack with the head-to-tail type arrangement, but the main π electron carriers of TTF groups stack without distinctive staggering. As a result, there would be no motive force for the surface EDO-TTF molecules to adjust their height by sliding along the molecular plane even with the open space of the missing outside double layer, as in the case of β-$(ET)_2I_3$ in Section 3.2.

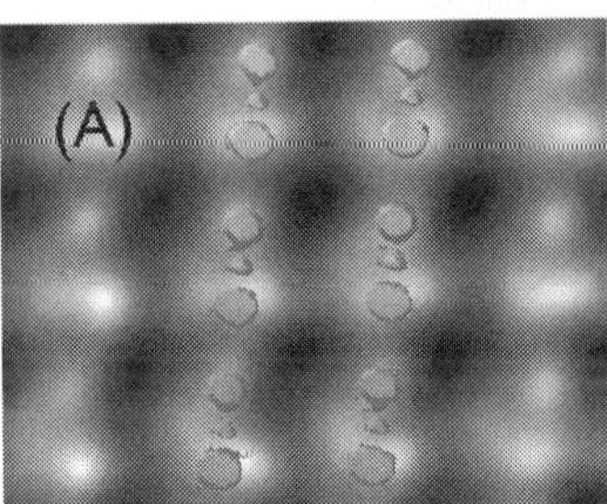

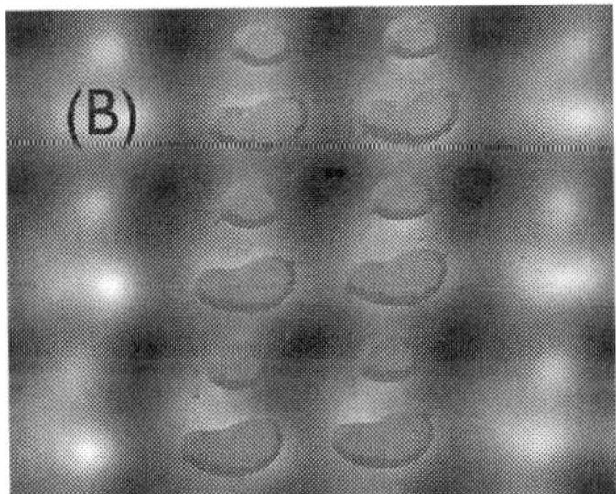

Figure 19. STM image of $(EDO\text{-}TTF)_2PF_6$ with (A) the simulated constant charge density contour of 3×10^{-8} nm^{-3} at 2 Å from the oxygen atom. The second oxygen and the second carbon contribute only a little. The small spots are assigned to the second half of O_{2p} orbital and with (B) the simulation by the rotational reconstruction model with the condition of $\theta_a = 18°$, $\theta_b = 29°$ as described in **Figure 20**, which successfully reproduces the overall profile of the observed image.

The analysis with O_{2p} and C_{2p} wavefunctions is applied to $(EDO\text{-}TTF)_2PF_6$ system. With the ratio of the π charges of the oxygen site to the carbon site, 0.149:0.358 estimated by MOPAC, the constant amplitude contour of the π charge density of 3×10^{-8} nm^{-3} at 2Å from the oxygen atom is compared with the STM image in **Figure 19(A)**. The large and wide bright areas are assigned to the oxygen atoms of EDO groups and the smaller ones are attributed to the carbon atoms of TTF groups. If the simulated contours of the carbon atoms in the TTF end are located on the center of the small areas in **Figure 19(A)**, the simulated contours of the oxygen atoms in the EDO groups show only rough agreement with the STM image, but clearly deviates from the brightest centers at the right-hand side of the large areas in **Figure 19(A)**. The second half of O_{2p} orbital in the simulation is also missing in the STM image. These deviations suggest the presence of some modification in the arrangement of EDO-TTF molecules in the surface layer.

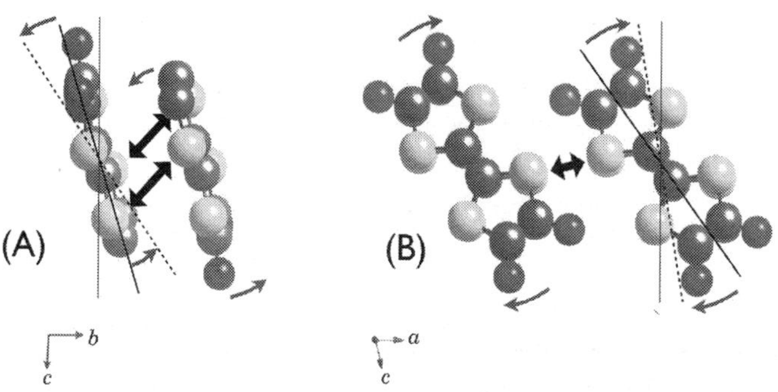

Figure 20. Possible reorientation in the surface EDO-TTF layer. (A) Rotation about the axis parallel with the molecular plane and the a–b plane to lay them down in the a–b plane. For simplicity, the a axis is used. (B) Rotation about the normal to the molecular plane to raise them. In actual, the b axis is used. The black arrows indicate the π electron network (A) along the b axis with the double pathways and (B) along the a axis with the single pathway.

The possible reconstruction consistent with the STM image is rotational reconstruction as follows and as described in **Figure 20**.

1.   Lay the molecules down in the a–b plane about the a axis by θ_a.

2.   Raise EDO-TTF molecules by rotating them about the b axis by θ_b.

The first operation erases the second half of the O_{2p} orbital, which is not observed in the STM image. The second one makes the second oxygen of the EDO group observable, which reproduces the widely spread area corresponding to the oxygen pair.

Figure 19B estimated by the rotational model successfully reproduces the overall profile of the STM image. **Figure 21** demonstrates the topographies (A) along the direction connecting the two oxygen atoms of EDO group and (B) along the direction connecting the oxygen and carbon atoms of EDO-TTF molecules stacking along the b axis. The simulated profiles of the π charge density successfully reproduce the characteristic profiles of the observed STM image, supporting the above reorientational model. Some dip is found between the two peaks in **Figure 21B**, which is caused by the node of the O_{2p} orbital. This kind of sharp variation would not be reproduced by STM because of the limited resolution of STM caused by the $6s$ orbital of PtIr alloy with the large diameter around 0.3–0.4 nm. Here, θ_b is estimated as $\approx 29°$ from $\approx 38°$ in the bulk and 9° observed in the topography of **Figure 21A**. It is, however, difficult to estimate θ_a uniquely from the topography in **Figure 21B**.

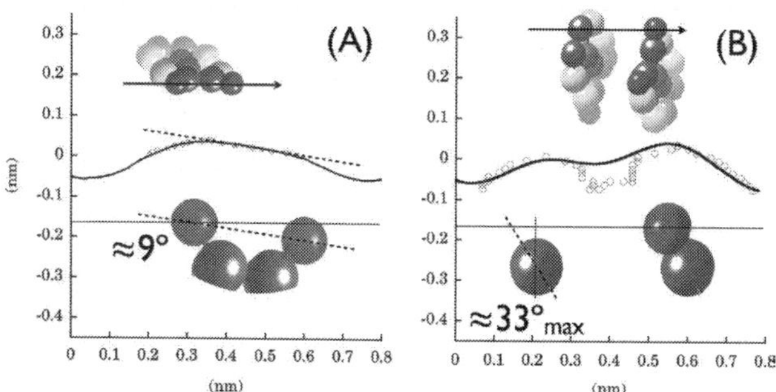

Figure 21. The topographies (A) along the oxygen pair of EDO group and (B) along the oxygen and the carbon atoms of the EDO-TTF molecules. The circles represent the simulated profile at the π charge density of 1×10^{-5} nm^{-3} with $\theta_a =$ 18°, $\theta_b = 29°$. The horizontal line indicates the relative position of the oxygen atom.

What these reorientations modify in the π electron system? The θ_b operation cuts the side-by-side 1D network of the sulfur π electrons along the a axis. The θ_a one forms one-dimensional π electron band along the stacking b axis with the face-to-face pathways of the sulfur pairs in the TTF groups, as shown in **Figure 20**. Furthermore, the $3p$ wavefunction spreads wider along its symmetry axis by <10% than the perpendicular direction, as shown in **Figure 4**. The upper limit of θ_a is estimated as $\approx 18°$ ($\approx 33°$ from the normal of the a–b plane) from the shortest distance of ≈ 3.4 Å between the sulfur atoms. Thus, the rotational reconstruction of the EDO-TTF molecules in the surface layer dramatically change the 1D conductive direction from the a axis to the b axis and can enhance the 1D band width of the π electrons, which stabilizes the surface π electron system and is motivated by the removal of the steric hindrance with the PF$_6^-$ ions of the missing outside double layer.

4. Conclusion

With the wavefunctions of p orbitals, we analyzed the STM images of the three charge transfer salts, α-(BEDT-TTF)$_2$I$_3$, β-(BEDT-TTF)$_2$I$_3$, and (EDO-TTF)$_2$PF$_6$ and found three different kinds of reconstructions: (1) charge redistribution, (2) translational reconstruction and (3) rotational reconstruction.

In α-(BEDT-TTF)$_2$I$_3$, the four characteristic protrusions of the STM image are assigned to the end sulfur atoms of the BEDT-TTF molecules in the unit cell and were compared with the simulation based on the $3p$ orbitals of the sulfur atoms. The obtained STM tip heights are analyzed with the scenarios of the structural reconstruction and the charge redistribution. Finally, the charge redistribution similar to the charge ordered state below 135 K in the bulk crystal is found, which is different from the charge disproportionation of the bulk crystal at RT. The origin is ascribed to the gain of the additional cohesive energy in the ground CO state provoked by the missing steric hindrance between the end ethylene groups and the I$_3^-$ ions of the missing outside double layer, which provides us the insight on the 135 K phase transition of the bulk crystal. This bulk transition from the charge disproportionation to the charge ordered state would be caused by the calming down of the thermal vibration of the end ethylene groups, which assists the BEDT-TTF layers in forming the charge ordered ground state below 135 K.

In β-(BEDT-TTF)$_2$I$_3$, two BEDT-TTF molecules in the unit cell stack with a staggered fashion, in which I$_3^-$ ion is located. Then, the bulk structure suggests the distinctive tip height difference in the unit cell. In contrast, the STM image shows almost the same tip heights in the unit cell, suggesting translational reconstruction up to 0.1 nm to remove the stagger in the unit cell. As a result, the separation between the sulfur atoms of the neighboring BEDT-TTF molecules sizably decreased, which enhances the transfer energy and the cohesive energy of the π electron system. Thus, the missing steric hindrance with I$_3^-$ ions of the missing outside double layer assists the increase of the cohesive energy in the surface BEDT-TTF layer.

In (EDO-TTF)$_2$PF$_6$, the asymmetric molecules of EDO-TTF stack along the b axis with head-to-tail type arrangement, but the TTF groups stack with only tiny stagger, in which the π electrons mainly reside. Thus, any sliding reconstruction, as was found in β-(BEDT-TTF)$_2$I$_3$, would not be important even without the steric hindrance with PF$_6^-$ ions. The comparison between the STM image and the simulated topography suggests large rotational reconstruction about both of the a axis and the b axis. Such a rotational reconstruction drastically changes the direction of the 1D π band from the a axis to the b axis and largely enhances the π electron band width.

Finally, it is concluded that the surface reconstruction is ascribed to the additional gain of the cohesive energy of the π electron system, provoked by the reduced steric hindrance with the anions of the missing outside double layer. The investigations of the surface states provide not only interesting behaviors of the surface cation layer, but also important insights into the electronic states of a lot of similar charge transfer crystals, as demonstrated in α-(BEDT-TTF)$_2$I$_3$.

Acknowledgements

H. Sakamoto and K. Mizoguchi would like to commemorate Hideyuki Arimoto, who passed away by an obstinate disease in June 2014 and to express our sincere thanks to his contribution to the experiment and analysis of β-(ET)$_2$I$_3$.

Author details

Hirokazu Sakamoto[1*], Eiichi Mori[1,4], Hideyuki Arimoto[1], Keiichiro Namai[1], Hiroyuki Tahara[1], Toshio Naito[2], Taka-aki Hiramatsu[3,5], Hideki Yamochi[3] and Kenji Mizoguchi[1]

*Address all correspondence to: sakamoto@phys.se.tmu.ac.jp

1 Department of Physics, Tokyo Metropolitan University, Hachioji, Tokyo, Japan

2 Department of Chemistry and Biology, Graduate School of Science and Engineering, Ehime University, Matsuyama, Ehime, Japan

3 Research Center for Low Temperature and Materials Sciences, Kyoto University, Sakyo-ku, Kyoto, Japan

4 Present adderss: Functional Material Development Center, Imaging Engine Development Division, Ricoh Co., Ltd, Ebina, Kanagawa, Japan

5 Present adderss: Faculty of Agriculture, Meijo University, Nagoya, Japan

References

[1] E. Mori, H. Usui, H. Sakamoto, K. Mizoguchi, and T. Naito. Charge distribution in the surface BEDT-TTF layer of α-(BEDT-TTF)$_2$I$_3$ at room temperature with scanning tunneling microscopy. *J. Phys. Soc. Jpn.*, 81(1):014707 (1–7), 2011.

[2] A series of the proceedings of International Conference on Science and Technology of Synthetic Metals (ICSM). *Synth. Met.*, Elsevier.

[3] L. B. Coleman, M. J. Cohen, D. J. Sandman, F. G. Yamagishi, A. F. Garito, and A. J Heeger. Superconducting fluctuations and the Peierls instability in an organic solid. *Solid State. Commun.*, 12(11):1125–1132, 1973.

[4] J. C. Scott, A. F. Garito, and A. J. Heeger. Magnetic susceptibility studies of tetrathio-fulvalene-tetracyanoquinodimethan (TTF) (TCNQ) and related organic metals. *Phys. Rev. B*, 10(8):3131–3139, 1974.

[5] D. Jérome, A. Mazaud, M. Ribault, and K. Bechgaard. Superconductivity in a synthetic organic conductor (TMTSF)$_2$PF$_6$. *J. Phys. Lett. (Paris)*, 41(4):95–98, 1980.

[6] T. Ishiguro, K. Yamaji, and G. Saito. *Organic superconductors.* Springer, Berlin; New York, 2nd edition, 1998.

[7] H. Kobayashi, H. Cui, and A. Kobayashi. Organic metals and superconductors based on BETS (BETS = Bis(ethylenedithio)tetraselenafulvalene). *Chem. Rev. (Special issue of Molecular Conductors)*, 104(11):5265–5288, 2004.

[8] M. Yoshimura, H. Shigekawa, H. Yamochi, G. Saito, Y. Saito, and A. Kawazu. Surface structure of the organic conductor β-(BEDT-TTF)$_2$I$_3$ observed by scanning tunneling microscopy. *Phys. Rev. B*, 44(4):1970–1972, 1991.

[9] S. N. Magonov, G. Bar, E. Keller, E. B. Yagubskii, and H.-J. Cantow. Atomic scale surface studies of conductive organic compounds 3. Scanning tunneling microscopy studies of

monocrystals of bis(ethylenedithio)tetrathiafulvalene with triiodide, β-ET$_2$I$_3$. *Synth. Met.*, 40:247–256, 1991.

[10] S. N. Magonov, G. Bar, H.-J. Cantow, J. Paradis, J. Ren, M.-H. Whangbo, and E. B. Yagubskii. Scanning tunneling and atomic microscopy images of organic salt conductor (BEDT-TTF)$_2$TlHg(SCN)$_4$. *J. Phys. Chem.*, 97:9170–9176, 1993.

[11] S. N. Magonov, G. Bar, H.-J. Cantow, J. Paradis, J. Ren, and M.-H. Whangbo. Characterization of the surfaces of the conducting salts α-(BEDT-TTF)$_2$X (X = I$_3$, IBr$_2$) by scanning tunneling and atomic force microscopy. *Synth. Met.*, 62:83–89, 1994.

[12] N. Ara-Kato, K. Yase, H. Shigekawa, M. Yoshimura, and A. Kawazu. Scanning tunneling microscopy of TTF-TCNQ single crystal and thin film. *Synth. Met.*, 70(1–3): 1245–1246, 1995.

[13] S. N. Magonov and M.-H. Whangbo. Organic conducting salts. In: *Surface Analysis with STM and AFM*, pp. 189–218. Wiley-VCH Verlag GmbH, Weinheim, 1995.

[14] M. Ishida, K. Hata, T. Mori, and H. Shigekawa. Surface reconstruction formed by ordered missing molecular rows observed on the quasi-one-dimensional organic conductor β-(BEDT-TTF)$_2$PF$_6$. *Phys. Rev. B*, 55(11):6773–6777, 1996.

[15] M. Ishida, T. Mori, and H. Shigekawa. Surface charge-density wave on the one-dimensional organic conductor β-(BEDT-TTF)$_2$PF$_6$. *Phys. Rev. Lett.*, 83(3):596–599, 1999.

[16] K. Hashimoto, T. Nakayama, N. Yoshimoto, M. Yoshizawa, M. Aono, and I. Yamaguchi. Three distinct terraces on a β-(ET)$_2$I$_3$ surface studied by scanning tunneling microscopy. *Jpn. J. Appl. Phys.*, 38:L464–L466, 1999.

[17] M. Ishida, O. Takeuchi, T. Mori, and H. Shigekawa. Comparison of the symmetry breaking in the surface molecular structures of one- and two-dimensional bis(ethylenedithio)tetrathiafulvalene compounds. *Jpn. J. Appl. Phys.*, 39(6B-1):3823–2826, 2000.

[18] N. Yoneyama, T. Sasaki, T. Nishizaki, A. M. Troyanovskiy, and N. Kobayashi. Scanning tunneling microscopy study of the anomalous metallic phases in θ-(BEDT-TTF)$_2$MZn(SCN)$_4$ (M = Rb, Cs). *J. Low Temp. Phys.*, 142:159–162, 2006.

[19] R. Kondo, S. Kagoshima, N. Tajima, and R. Kato. Crystal and electronic structures of the quasi-two-dimensional organic conductor α-(BEDT-TTF)$_2$I$_3$ and its selenium analogue α-(BEDT-TSeF)$_2$I$_3$ under hydrostatic pressure at room temperature. *J. Phys. Soc. Jpn.*, 78(11):114714, 2009.

[20] E. B. Yagubskii, I. F. Shchegolev, V. N. Laukhin, P. A. Kononovich, M. V. Kartsovnik, A. V. Zvarykina, and L. I. Buravov. Normal-pressure superconductivity in an organic metal (BEDT-TTF)$_2$I$_3$ [bis (ethylene dithiolo) tetrathio fulvalene triiodide]. *JETP Lett.*, 39:12, 1984.

[21] A. Ota, H. Yamochi, and G. Saito. A novel metal-insulator phase transition observed in (EDO-TTF)$_2$PF$_6$. *J. Mater. Chem.*, 12(9):2600–2602, 2002.

[22] H. Yamochi and S.-Y. Koshihara. Organic metal (EDO-TTF)$_2$PF$_6$. *Sci. Technol. Adv. Mater.*, 10:024305 (1–6), 2009.

[23] T. Kakiuchi, Y. Wakabayashi, H. Sawa, T. Takahashi, and T. Nakamura. Charge ordering in α-(BEDT-TTF)$_2$I$_3$ by synchrotron X-ray diffraction. *J. Phys. Soc. Jpn.*, 76(11): 113702, 2007.

[24] T. Mori, A. Kobayashi, Y. Sasaki, H. Kobayashi, G. Saito, and H. Inokuchi. Band structures of two types of (BEDT-TTF)$_2$I$_3$. *Chem. Lett.*, 13(6):957–960, 1984.

[25] E. Clementi and D. L. Raimondi. Atomic screening constants from SCF functions. *J. Chem. Phys.*, 38(11):2686–2689, 1963.

[26] J. Tersoff and D. R. Hamann. Theory and application for the scanning tunneling microscope. *Phys. Rev. Lett.*, 50:1998–2001, 1983.

[27] K. Bender, I. Hennig, D. Schweitzer, K. Dietz, H. Endres, and H. J. Keller. Synthesis, structure and physical properties of a two-dimensional organic metal, di[bis(ethylene-dithiolo)tetrathiofulvalene] triiodide, (BEDT-TTF)$_2$I$_3$. *Mol. Cryst. Liq. Cryst.*, 108(3–4): 359–371, 1984.

[28] P. Guionneau, C. J. Kepert, G. Bravic, D. Chasseau, M. R. Truter, M. Kurmoo, and P. Day. Determining the charge distribution in BEDT-TTF salts. *Synth. Met.*, 86:1973–1974, 1997.

[29] K. Murata, M. Tokumoto, H. Anzai, H. Bando, G. Saito, K. Kajimura, and T. Ishiguro. Pressure phase diagram of the organic superconductor β-(BEDT-TTF)$_2$I$_3$. *J. Phys. Soc. Jpn.*, 54(6):2084–2087, 1985.

Chapter 15

Application of Scanning Acoustic Microscopy to Pathological Diagnosis

Katsutoshi Miura

Additional information is available at the end of the chapter

http://dx.doi.org/10.5772/63405

Abstract

Scanning acoustic microscopy (SAM) can obtain high-quality microscopic images of tissues and cells that are comparable with light microscopic images without staining and within only a few minutes. The speed of sound through tissues and cells is correlated with elasticity, thereby indicating their biomechanical properties. The elasticity varies according to the contents, such as collagen or elastic fibers, blood, colloids, mucin, ground substances, and cytoskeleton; therefore, SAM can follow changes in the composition of tissues and cells to determine their functions. Chemical modifications such as fixation, periodic acid-Schiff reaction, and enzymatic digestion may influence acoustic properties, and SAM can follow these changes over time in the same section to facilitate statistical comparisons based on digital values. Digital imaging using SAM is superior to analog methods for modifying images to discriminate changes, such as malignant and benign cell types. The observation ranges are shown in a colored column, and they can be manually adjusted. Thus, precise differences in acoustic properties are readily distinguished by narrowing the range. The resolution of SAM is determined by the wavelength, and it can theoretically exceed that of visible light. Combining these distinct techniques may help to elucidate the structural and functional characteristics of tissues and cells.

Keywords: elasticity, histochemistry, imaging analysis, protease, scanning acoustic microscopy

1. Introduction

Sound travels through different human body tissues at specific speed of sound (SOS) values. Harder materials have greater SOS values; thus, the SOS through each tissue can provide

information regarding its elasticity. Scanning acoustic microscopy (SAM) can be used to calculate the SOS, the attenuation of sound (AOS), and the thickness of tissues, before plotting the data on the screen to form images (**Figure 1**). SAM was first developed in the early 1970s by Lemons and Quate at Stanford University (CA, USA) [1], and the basic design is now used in the biomedical area. In SAM, the image contrast depends on the biomechanical properties of the tissues and the frequency and focusing conditions [2]. In general, sections that contain little structural protein possess SOS and AOS values that are similar to water in the coupling medium. By contrast, sections containing structural proteins such as collagens have significantly greater SOS and AOS values than water.

Figure 1. Appearance of scanning acoustic microscope systems. 1. Signal processor, 2. mechanical scanner with transducer, 3. system control PC, and 4. display.

In previous studies, we have observed various organs, such as the lung [3], gastrointestinal tract [4], thyroid [5], liver [6], heart [6], blood vessels [6], skin [6], and lymph nodes [7], where we used tissue sections and cytological specimens [8]. In further applications, we followed histological changes over time after chemical modifications such as fixation, staining, and digestion [6], where we compared the fresh unfixed sections with formalin-fixed sections. We employed SAM to observe the histological changes following the periodic acid-Schiff (PAS) reaction. After protease digestion, the SAM images were followed over time to examine the sensitivity or resistance to enzymatic treatment.

In this mini review, we explain the SAM system in the following order: principle of SAM, preparation of sample materials, observation procedure, application of SAM to tissue and cytology diagnosis, differentiation between malignant and benign, fixation effects, effects of PAS reaction, effects of collagenase, statistical analysis, and conclusion.

2. Principles of acoustic microscopy

The basic design of a reflective scanning acoustic microscope is shown in **Figure 2**.

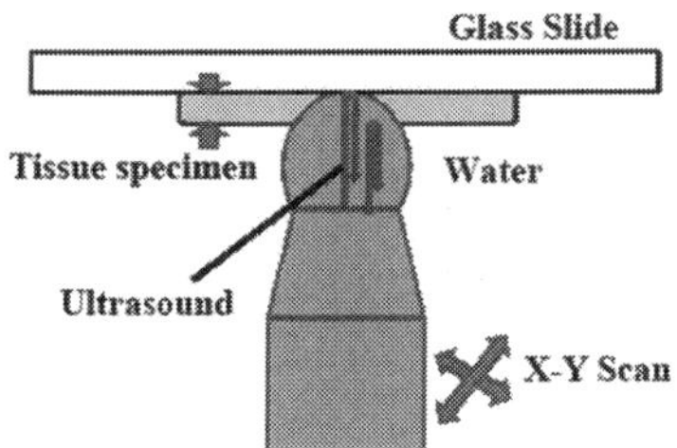

Figure 2. Principle of SAMs. Ultrasonic waves from the transducer reflect off both the glass slide and the sample section before returning to the transducer. The waves pass through 10-μm sample sections with different ultrasonic properties. The transducer automatically scans the section to calculate the speed of sound (SOS) through each area. The section is placed upside down on the transducer, and distilled water is applied between the transducer and the section as a coupling fluid. The control SOS through water is 1495 m/s.

A piezoelectric transducer converts a radio frequency signal into an acoustic wave, which is then focused by a sapphire lens onto the glass slide where the histology specimens are mounted. A coupling liquid (usually water) carries the sound wave between the lens and the specimen. The beam is focused on the slide at a fixed distance from the lens. Components of the acoustic wave may be reflected, scattered, absorbed, or transmitted, which is determined primarily by the elastic properties of the tissue. The waves reflected from both the glass slide and the specimens are then collected by the same lens. The intensity of the resulting acoustic signal is then converted into brightness for display purposes. The specimen is scanned in the horizontal plane in a raster fashion to allow point-by-point analysis of the elastic properties of a cross-section of tissue [9].

The SAM can be used as basic data to assess the biomechanics of tissues and cells. The relationship between the SOS and the elastic bulk modulus of a liquid-like medium is given by the Newton-Laplace equation as follows:

$$c = \sqrt{K / \rho},\tag{1}$$

where c is the SOS, K is the elastic bulk modulus, and ρ is the density.

Thus, the SOS increases with the stiffness (the resistance of an elastic body to deformation by an applied force) of the material, whereas it decreases with the density. Excluding bones (1.750 g/cm^3) and fats (0.9094 g/cm^3), the average soft tissue density is almost 1 g/cm^3, which is the same as water (http://www.scrollseek.com/training/densitiesofdifferentbodymatter.html), and the SOS through soft tissues is strongly correlated with their hardness.

Recent biomechanical studies suggest that the mechanical properties of tissues might not be sufficiently similar to liquids, and thus they should be treated as soft solid materials. The sound waves that generate volumetric deformations (compression) and shear deformations (shearing) are called pressure waves (longitudinal waves) and shear waves (transverse waves), respectively.

The SOS of pressure waves in solid materials, which are calculated by our SAM system, can be described by the following equation if the material is assumed to be isotrophic:

$$c = \sqrt{E(1-v)\,/\,\rho(1+v)(1-2v)}, \qquad (2)$$

where E is Young's modulus or the elastic modulus, ρ is the density, and v is Poisson's ratio. This relationship shows that Young's modulus (the elastic modulus) of the tissue and the SOS are closely related [10].

3. Preparation of sample materials

Fresh frozen or formalin-fixed, paraffin-embedded sections are cut flat to a thickness of 10 μm and used to obtain observations. Excessively thin samples are not suitable for acquiring correct measurements. Excessively thick samples (>20 μm) are not suitable for detecting the returned waves due to the loss of energy. In addition, samples with irregular surfaces will scatter sound and hinder observations.

A cytology specimen is prepared as a single cell layer on a slide using a liquid-based cytology method (BD CytoRich™; Franklin Lakes, NJ, USA). This method can be used to collect cells on the slide by ionic binding. Because the cells are negatively charged, they spontaneously bind to the slide, which has a positive charge [8].

4. Observation procedure of SAM

The SAM (AMS-50SI) system employed in our studies was manufactured by Honda Electronics (Toyohashi, Aichi, Japan), and it was equipped with a 120-MHz (resolution = approximately 13 μm) or 320-MHz (resolution = 4.7 μm) transducer. The SAM images of an object were obtained by plotting the SOS through the sections on the screen.

To perform SAM imaging, the slide sections were placed upside down on the stage above the transducer, and distilled water was applied between the transducer and the section as a coupling fluid because air interferes with sound transmission. After mechanical X–Y scanning, the SOS was calculated from each point on the section and plotted on the screen to create two-dimensional, color-coded images. The vertical bar on the left and the horizontal bar at the bottom of each figure indicate the distance (mm) on the slide. The vertical colored column on the right side of the figure indicates the average SOS of each square area on the section. The color range of observation can be changed manually. The region of interest (0.3, 0.6, 1.2, 2.4, or 4.8 mm^2) employed for acoustic microscopy was determined based on light microscopic (LM) images. The SOS values were calculated at 300 × 300 points and plotted on the screen to create the images, and sound data from 64 cross points on the lattice screen were used for statistical analysis. Other data such as the thickness of the section and AOS were also obtained from each point and shown on the screen.

5. Application of SAM to tissue and cytology diagnosis

In general, pathologists use LM images of sections for histological diagnosis. However, SAM can provide additional or novel information about the same section without staining and within a few minutes. Several examples are described as follows:

5.1. Skin

Changes associated with aging and environmental factors are apparent in the skin [11, 12]. These characteristic changes include thinning of the epidermis with a flattened dermal-epidermal junction, decreased amounts of collagen, elastin, ground substance, and eccrine/apocrine glands in the dermis, and the presence of solar elastosis in the upper dermis of sun-damaged skin.

Figure 3 compares SOS images of the facial skin in elderly and juvenile subjects. The SOS values for the epidermis are greater in the elderly subject than in the juvenile subject, which may indicate the loss of water in the dry skin of the elderly subject compared with the moist skin

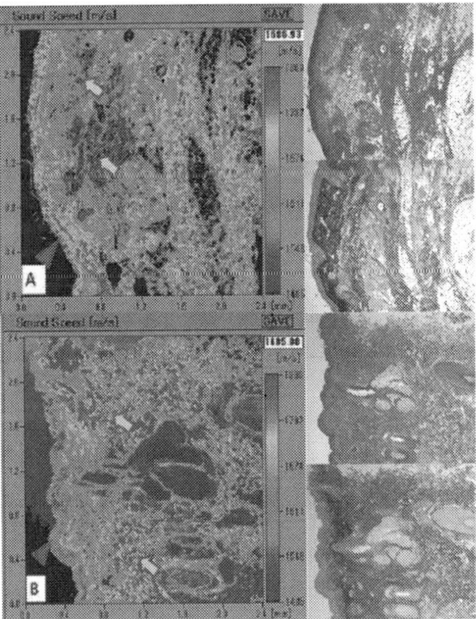

Figure 3. Comparative skin images of SOS in juvenile and elderly subjects. The skin of the elderly subject (**A**) has higher and lower SOS values in the epidermis (arrow head) and superficial dermis (arrows), respectively. By contrast, the skin of the juvenile subject (**B**) has lower and higher SOS values in the epidermis (arrow heads) and dermis (arrows), respectively. In the skin of the elderly subject, elastic fibers are diffusely deposited under the epidermis due to solar elastosis, whereas thick collagen fibers are distributed between the skin adnexa of the superficial dermis in the skin of the juvenile subject. The corresponding LM images with hematoxylin and eosin (right upper) and elastica-van-Gieson (right lower) staining are shown.

of the juvenile subject. The dermal low SOS areas in the elderly subject correspond to solar elastosis, which can be seen as black fibrillary materials by elastic-van-Gieson staining, whereas the dermal high SOS areas in the juvenile subject are consistent with collagen-rich contents. Thus, SAM can provide structural and biochemical information.

5.2. Artery

Narrowing (stenosis) and weakening are the two principle mechanisms of vascular disease [13]. The renal arteries (RAs) of elderly (101-year-old female) and young (30-year-old male) subjects are compared in **Figure 4**. The young subject's RA has a three-layer structure, that is, intima, media, and externa, with uniform thickness and SOS values, respectively. By contrast,

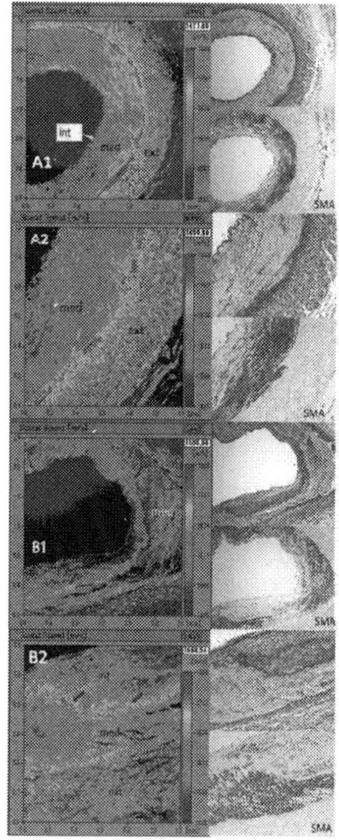

Figure 4. SOS images of the renal arteries in (RAs) young (**A**) and elderly (**B**) subjects. The RA has a three-layered structure, that is, intima, media, and externa with elastic laminae between each layer. The RA of a young subject (A1, A2) has a regular-layered structure, whereas that of an elderly subject (B1, B2) has an irregular shape and thickness. The thick intima with atheroma deposition has a low SOS value. The broad internal elastic lamina (arrows) has higher SOS values than those of the intima and media. The tunica media with variable thickness comprises smooth muscle fibers, which partly disappear with age. The externa comprises thick collagen fibers, which are lower in number in elderly subjects compared with young subjects. The corresponding LM images with elastica-van-Gieson staining and anti-smooth muscle immunostaining (SMA) are shown on the right. ext, externa; int, intima; med, media.

the elderly subject's RA has lost the layered structure with an irregular inner lumen. A thick intima with atheroma (lipids) deposition yields low SOS values. A broad intimal elastic lamina produces higher SOS values than the intima and media. The media with variable thickness comprises smooth muscle fibers that have partly disappeared. The externa has a decreased number of thick collagen fibers, which are clearly visible in the SOS image.

Therefore, SAM performs better than LM by simultaneously displaying structural changes and functional changes indicative of weakening.

5.3. Cardiac valve

Aortic valves (AVs) have a three-layered structure, that is, fibrosa, spongiosa, and ventricularis, as shown in **Figure 5**. The AVs of a young patient possess this layered structure with uniform high, middle, and lowest SOS values for the fibrosa, ventricularis, and spongiosa layers, respectively. By contrast, the elderly patient exhibits changes due to aging such as

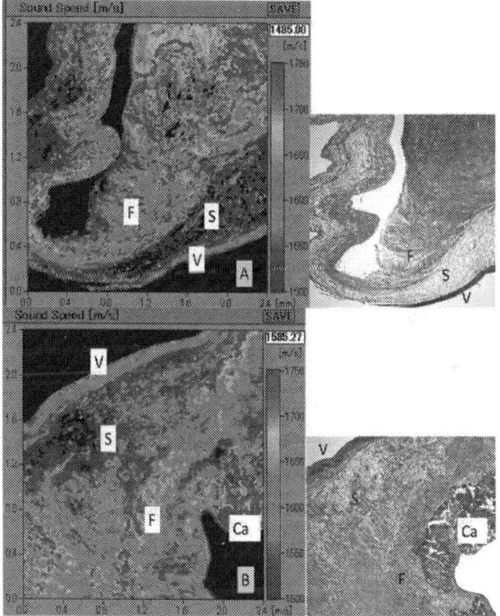

Figure 5. SOS images of aortic valves (AVs) in young (**A**) and elderly (**B**) patients. AVs have a three-layered structure, that is, fibrosa, spongiosa, and ventricularis. The AV of young patients has this layered structure, where the fibrosa, ventricularis, and spongiosa have high, middle, and low SOS values, respectively. The elderly patient exhibits changes in the AV due to aging such as nodular calcification. The SOS image shows that the highest values indicate calcification. The fibrosa has variable thickness and SOS values. The collagen bundles crumble in some places with age. Some collagen fibers drop into the spongiosa, and their interface is vague. The corresponding LM image with elastica-Masson staining is shown on the right. F, fibrosa; Ca; calcification; S, spongiosa; V, ventricularis.

nodular calcification. The SOS image assigns the greatest values to calcification. The fibrosa has a variable thickness and SOS values. The collagen bundles have crumbled in places to form vague interfaces.

Figure 6 shows the AV of a subject with Marfan's syndrome, who needs replacement therapy due to weakening, where the ventricularis is thickened, and fibronectin has accumulated. The SAM image contains variable SOS values at the ventricularis and uneven high SOS values at the fibrosa where disorganized fiber arrays can be seen.

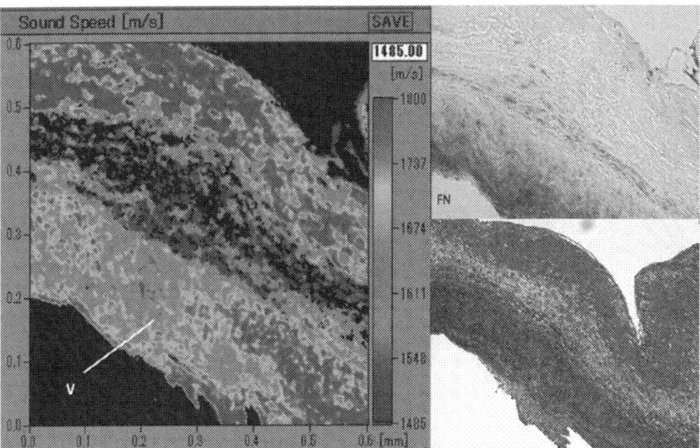

Figure 6. SOS image of the aortic valve in a subject with Marfan's syndrome. The ventricular layer is rich in thick elastic fibers with irregular SOS values. The fibrosa layer is rich in collagen fibers with uneven SOS values and disorganized fiber arrays. The corresponding LM images with elastica-Masson staining and anti-fibronectin immunostaining are shown on the right. V, ventricularis.

The SAM image can visualize collagen and elastin fiber abnormalities both morphologically as well as functionally according to the SOS values.

5.4. Gastrointestinal tract

The gastrointestinal tract basically comprises five layered structures, that is, mucosa, submucosa, muscularis propria, subserosa, and serosa. SAM images can be used to discriminate these layered structures (**Figure 7A**). In cancer, these regular structures are disturbed by cancer invasion (**Figure 7B**). The invaded areas can be detected based on changes in both the morphology and the SOS values.

A carcinoma and a benign adenoma in colon polypectomy are shown in **Figure 8**, where there is an irregular tubular structure with variable high SOS values in the cancer compared with the adenoma.

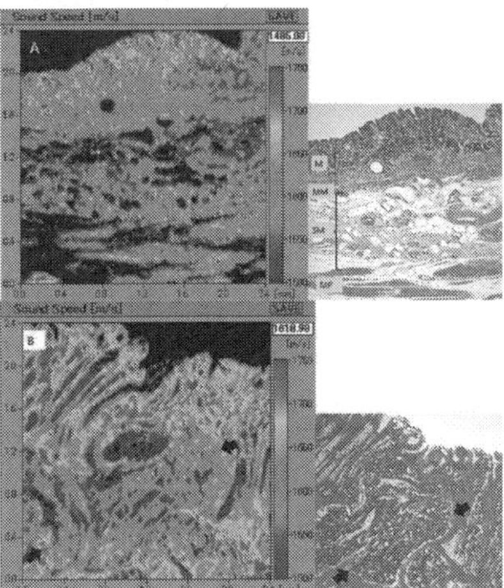

Figure 7. SAM image of normal stomach and gastric adenocarcinoma. From top to bottom, the normal gastric wall (**A**) comprising the mucosa, muscularis mucosae (MM), submucosa (SM), and muscularis propria (MP). A gastric tubular carcinoma (**B**) can be seen invading the normal mucosa (arrows). The stroma in the vicinity of the cancer nest has a higher SOS value than the cancer cells. The corresponding LM images with hematoxylin and eosin staining are shown on the right.

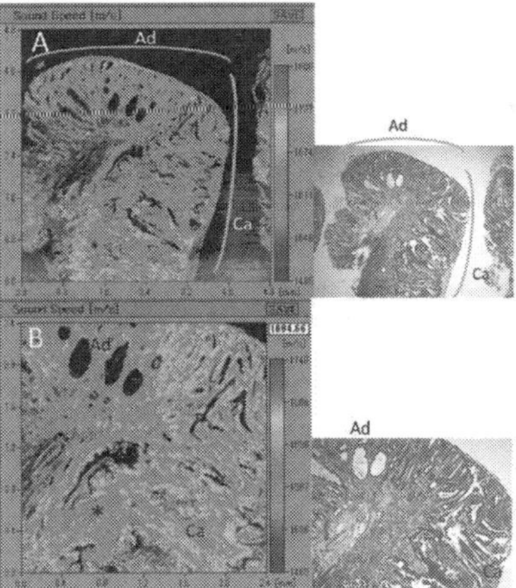

Figure 8. SOS images of adenocarcinoma in tubular adenoma. A colon polypectomy specimen is shown, which contains an adenocarcinoma region in a tubular adenoma (**A**, low magnification; **B**, high magnification). The adenoma comprises regular tubular structures, whereas the carcinoma is made of irregular branched tubules. The SOS image of carcinoma has irregular high values, and the shapes of the tubules are more variable compared with the adenoma. The stroma of the carcinoma at the invasive site (*) has a higher SOS value than the adenoma. The corresponding LM images with hematoxylin and eosin staining are shown on the right. Ad, adenoma; Ca, carcinoma.

5.5. Placenta

A placenta is rich in blood vessels where the exchange of gases and nutrients occurs between the maternal and fetal circulations. Chorionic villi form a delicate mesh in the central stroma surrounded by trophoblasts, which can be visualized well by SAM (**Figure 9**). During inflammation, the villi have increased numbers of blood vessels and inflammatory cells, which can be clearly discriminated using SAM imaging.

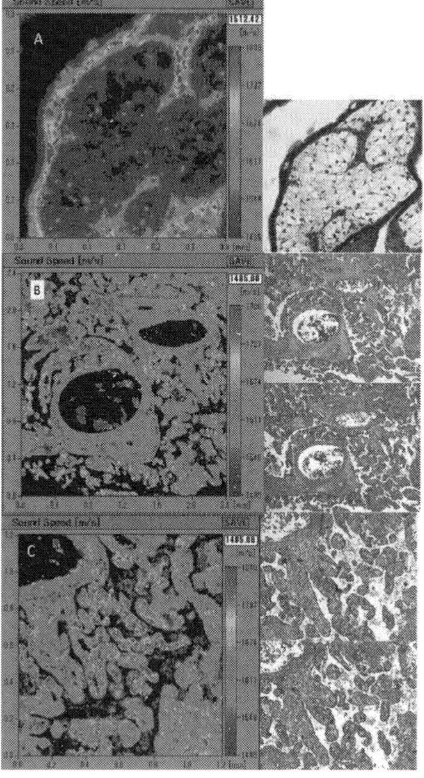

Figure 9. SOS image of placenta. Chorionic villi in the first trimester (**A**), where the central edematous stroma surrounded by trophoblasts is clearly visible. Erythroblasts in the vessels and trophoblasts have high SOS values. The nuclei of macrophages in the central stroma are visible as dot-like higher SOS spots. The third-trimester chorionic villi have chronic villitis (**B**, low magnification; **C**, high magnification) with hypervascular stroma and lymphocytic infiltration. The congested villi with anastomotic capillaries have high SOS values. The corresponding LM images with hematoxylin and eosin staining and Masson trichrome staining are shown on the right.

The blood contents produce very high SOS values, so it is easy to detect changes in the number and size of blood vessels.

5.6. Liver and lung

According to SAM observations, fibrosis comprises collagen fibers with high SOS values. Liver cirrhosis (**Figure 10A**) and pulmonary fibrosis (**Figure 10B**) are representative lesions due to fibrosis. More severe fibrosis yields greater SOS values, and thus the degree of fibrosis can be compared among lesions.

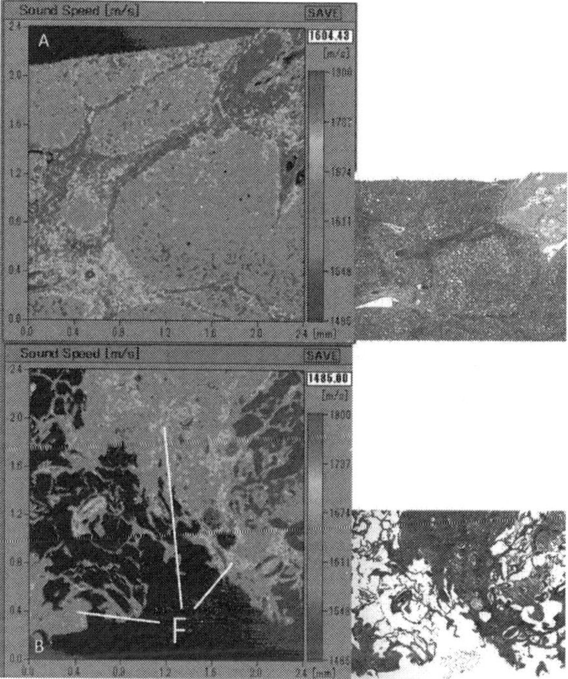

Figure 10. SOS images of liver cirrhosis and pulmonary fibrosis. Liver cirrhosis (**A**) and pulmonary fibrosis (**B**) are representative lesions with fibrosis. In liver cirrhosis, fibrosis with high SOS values surrounds pseudolobules. In pulmonary fibrosis, areas with interstitial fibrosis have high SOS values in the lung. The corresponding LM images with Masson trichrome staining are shown on the right. F, fibrosis.

5.7. Kidney

In diabetes mellitus, the kidney exhibits nodular glomerular sclerosis, and the tubules have thickened basement membranes (**Figure 11**). According to SOS imaging, a sclerotic portion of the glomerulus and tubular basement membranes surrounding the tubular lumen has greater SOS values compared with other glomerular and tubular areas. The SAM imaging system provides digital SOS data, which facilitates comparisons of the severity of sclerosis and elasticity among lesions and cases.

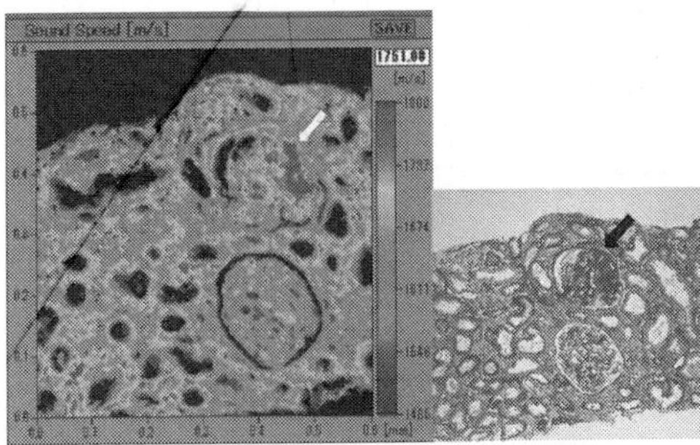

Figure 11. SOS image of diabetes mellitus kidney patient. The kidney of a diabetes mellitus patient has nodular sclerosis of the glomerulus and thickening of the basement membrane. According to SOS imaging, a sclerotic region of the glomerulus (arrow) has a higher SOS value compared with other glomerular areas. Tubular basement membranes surrounding the tubular lumen also have higher SOS values. The corresponding LM image with periodic acid-Schiff staining is shown on the right.

5.8. Heart

Acute myocardial infarction is sometimes difficult to detect in the areas involved. However, SOS images of coagulative necrosis indicate crumbled cardiac muscles with greater SOS values compared with the intact muscles (**Figure 12**). The elasticity of the necrotic area changes during the time course of infarction, and the SOS images may change accordingly.

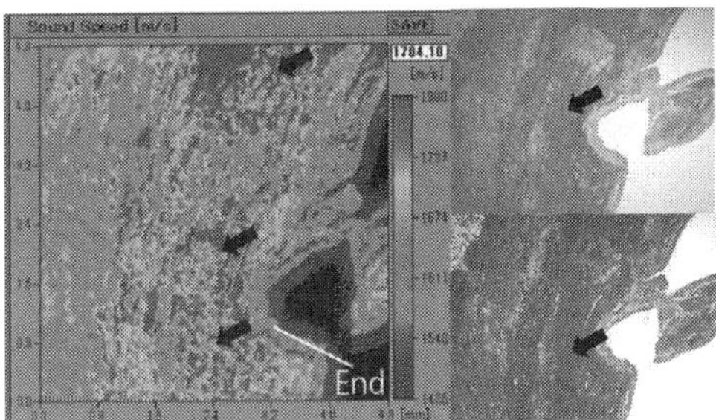

Figure 12. SOS image of acute myocardial infarction. An area with acute infarction and coagulative necrosis contains crumbled cardiac muscles (arrows). The necrotic areas have higher SOS values than intact cardiac muscles on the more endocardial side. The corresponding LM images with hematoxylin and eosin and Masson trichrome staining are shown on the right. End, endocardium.

5.9. Thyroid

Papillary carcinomas are the most common carcinoma in the thyroid. Their typical histology comprises papillary structures with thin elongating and branching cores. The SOS images (**Figure 13**) are comparable with optical images. Compressed follicles and fibrous stroma in the invasive sites have very high SOS values, where any abnormalities make it easy to detect cancer.

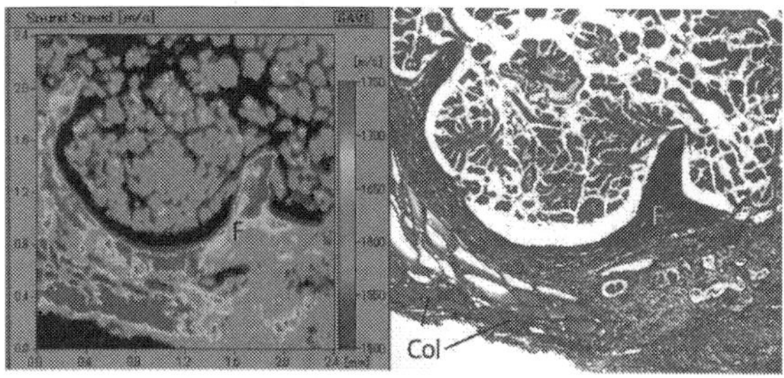

Figure 13. SOS image of papillary carcinoma of the thyroid. A papillary carcinoma comprises thin elongating branching cores with uniform low SOS values compared with the surrounding fibrous stroma (F). Compressed thyroid follicles have high SOS values due to their colloid contents (Col). The corresponding LM image with Masson trichrome staining is shown on the right.

Hashimoto's thyroiditis (or chronic lymphocytic thyroiditis) is characterized by many lymphoid follicles among the thyroid follicles (**Figure 14**). The lymphoid follicles have low SOS values in the thyroid. The SOS values of colloids vary according to their concentration, and a high concentration of colloids is seen in a low functioning thyroid; thus, SOS imaging may be used to predict the thyroid activity.

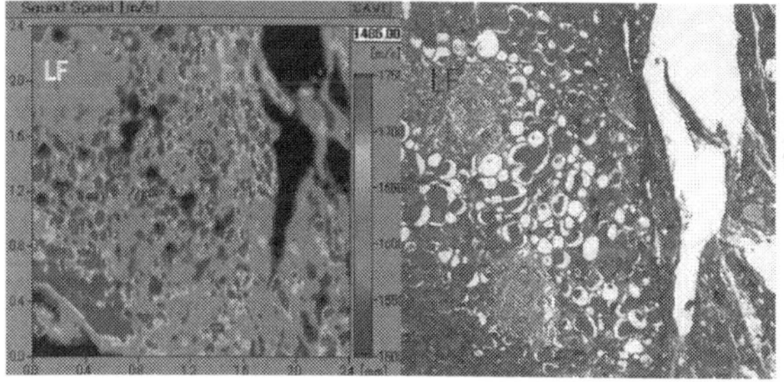

Figure 14. SOS image of Hashimoto's thyroiditis. The concentrated colloids in the thyroid follicles have high SOS values compared with the patchy lymph follicles (LF). The corresponding LM image with Masson trichrome staining is shown on the right.

5.10. Cytology

Differentiating malignant and benign cells in body fluids is possible using SAM [8]. Malignant cells have higher SOS values and irregular shapes compared with benign cells.

In urine cytology, cells with larger sizes and irregular shapes correspond to urothelial carcinoma cells, most of which have high SOS values (**Figure 15A**).

Among brain tumors, glioblastomas are the most common malignant tumor, and the cells have an irregular spindle shape with variable size (**Figure 15B**), where they are connected with each other. These features are characteristic of malignant brain gliomas.

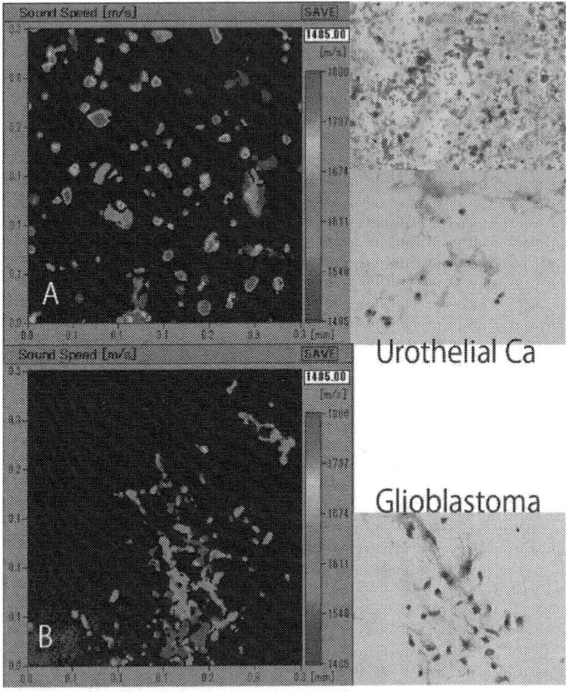

Figure 15. SOS image of urothelial carcinoma and glioblastoma. Cells with various different shapes and sizes are present in the urine. Large cells with irregular shapes correspond to urothelial carcinoma cells, most of which are shown in red in the SOS image (**A**). A touch smear of a brain tumor was obtained by SAM (**B**). Irregular spindle-shaped cells are connected to each other, and the size and shape of the cells vary. LM images with Giemsa or Papanicolaou staining obtained using the same specimen are shown on the right.

The thin preparation method in cytology (http://www.bdj.com/jps/brush) can be employed to assemble cells in a single cell layer at the center of a slide. Each cell or cell cluster is separate, which makes it easy to find cancer cells. SAM is a useful tool for screening cells without the necessity for staining.

6. Differentiation between malignant and benign effusions

When malignant and benign effusions are compared, the former have higher SOS values than the latter, but there is no significant difference in the mean thicknesses of the two cell types.

The original SOS image can be converted into a bipolar image comprising up and down cutoff point values. By adjusting the observation range, the malignant cells can be readily differentiated from benign cells (**Figure 16**).

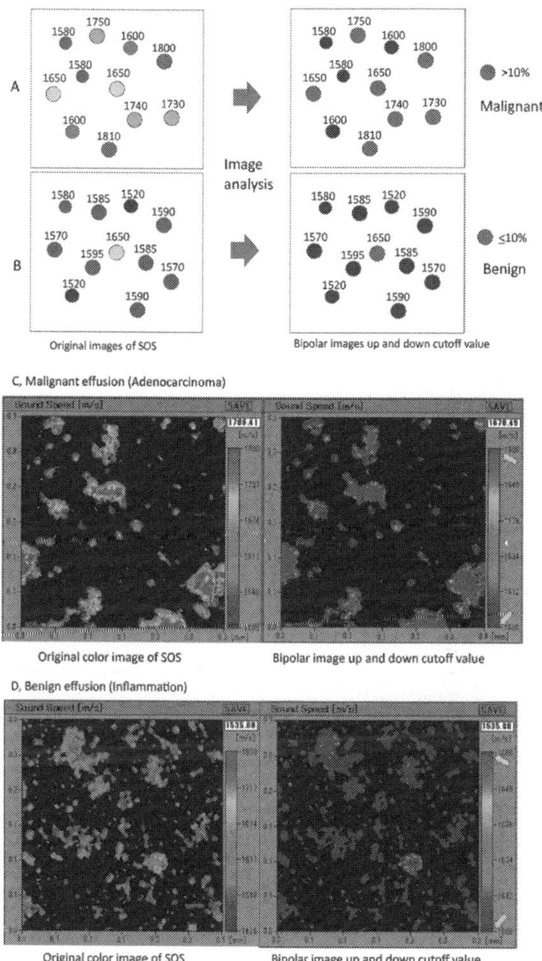

Figure 16. Differentiating benign and malignant effusions. **A** (malignant) and **B** (benign) show schematic color images obtained by SAM, where the SOS values are shown in the cells. These images were converted into bipolar red and blue images using up and down cutoff point values. If more than 10% of the single cells or small clusters had values higher than the cutoff point, they were regarded as a malignant effusion in SAM. Real SOS images of malignant (**C**) and benign (**D**) effusions are shown on the left. The bipolar images on the right were obtained based on the up and down cutoff point values. These images were obtained by limiting the range of the color scale bar on the right-hand side of the screen (yellow arrows).

7. SOS changes by fixation

The two main types of cross-linking and noncross-linking fixatives are formalin and ethanol, respectively. Changes in the SOS values have been reported after formalin fixation [6]. The differences in the SOS values and thickness were also compared using formalin and ethanol fixation. Using tannic acid fixation, the SOS values increase according to the concentration and duration of fixation (**Figure 17**).

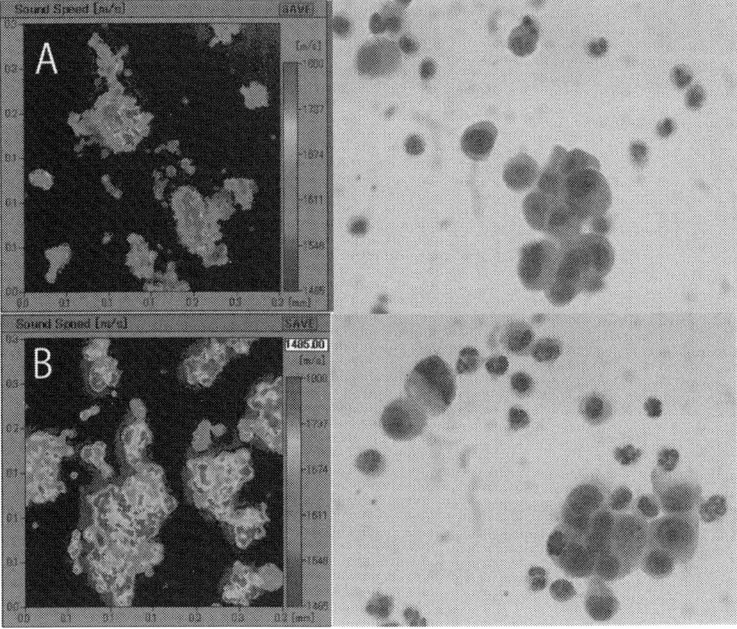

Figure 17. Effects of tannic acid fixation on cytology specimens. A cytological slide with an adenocarcinoma specimen was fixed in 1% tannic acid for 10 min. The SOS images obtained before (**A**) and after (**B**) fixation are compared. After fixation, the SOS values increased in the adenocarcinoma, which indicates that the cells became harder. The outer contour became clearer, and thus it was easier to detect the large cells/clusters as well as the irregular high SOS values. The corresponding LM image with Papanicolaou staining is shown on the right.

8. Effects of PAS reaction on SOS imaging

After the PAS reaction, the SOS values increase according to the strength of optical staining (**Figure 18**), where the more glycosylated regions appear as higher SOS areas. The degree of glycosylation may be evaluated objectively to allow comparisons among lesions and cases.

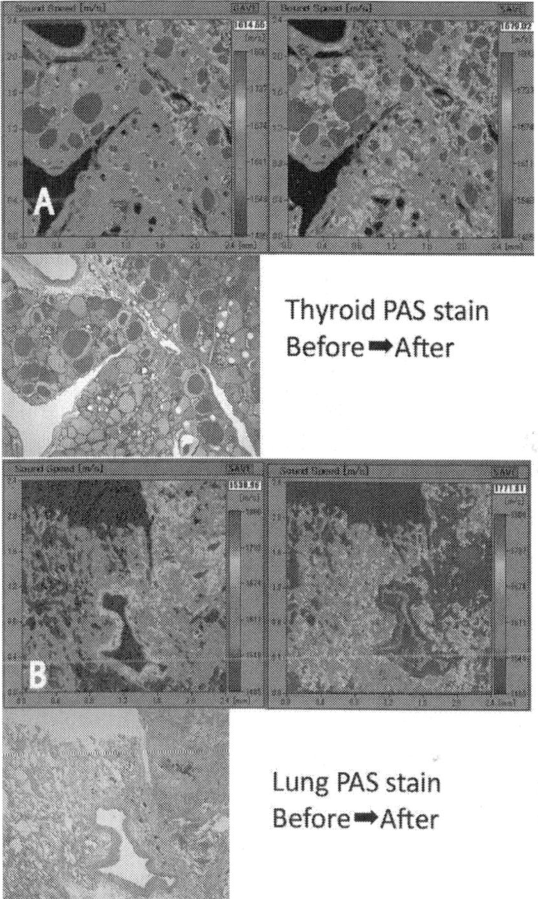

Figure 18. Effects of periodic acid-Schiff staining on SOS images. The SOS values increased according to the intensity of periodic acid-Schiff staining. In a thyroid sample (**A**), the colloids in the thyroid follicles have high SOS values, where the intensity correlates with the concentration of the colloids. A lung sample (**B**) has high SOS values in the fibrosis and vascular walls.

9. Effects of collagenase on SOS values

Fibrosis is a consequence of inflammation, ischemia, cancer invasion, and other pathological events. Myofibroblasts produce collagen fibers that undergo chemical modifications such as glycation and bridging, where the degree of modification varies among tissues or with age. Younger collagen fibers without modification may appear in granulation tissues, and they are vulnerable to protease digestion (**Figure 19A**). By contrast, the fibrous scar tissue seen in old myocardial infarctions (**Figure 19B**) resists protease breakdown because the chemical modifications interfere with the activity of protease.

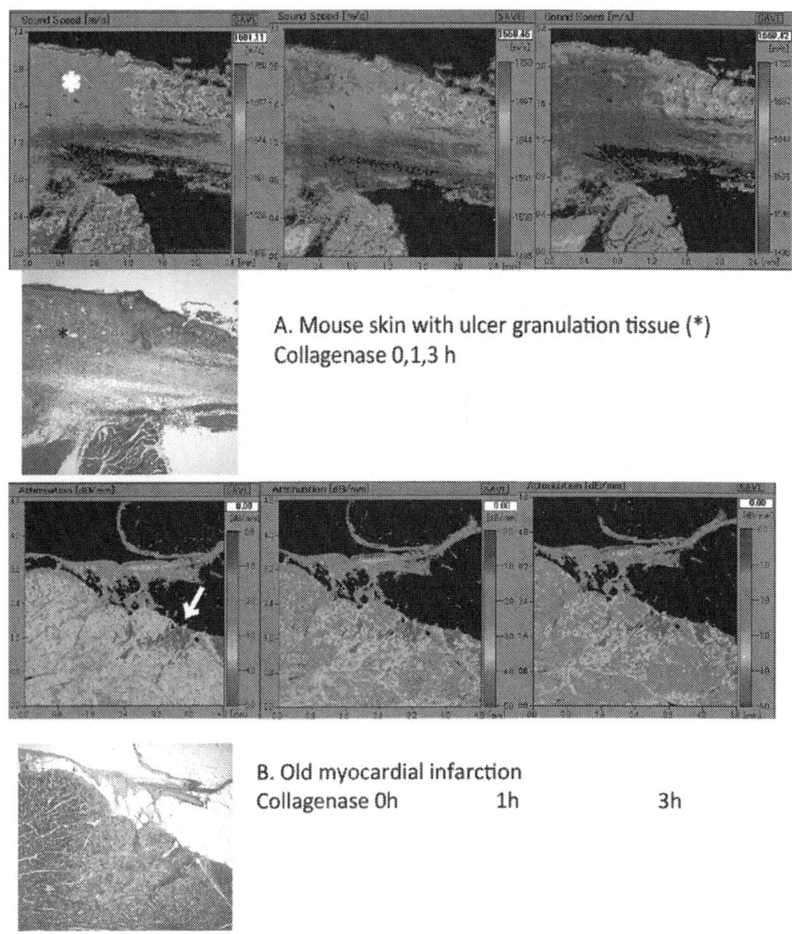

Figure 19. Effects of collagenase on SOS values. Changes in the SOS values in a section after collagenase treatment are shown. A newly formed fibrosis in granulation tissues in a skin ulcer (**A**) is vulnerable to collagenase digestion, whereas the old fibrous scar tissues (arrow) due to myocardial infarction (**B**) is resistant to the enzyme. The SOS values for the background epidermis and smooth muscles in **A** and the cardiac muscles in **B** decrease gradually.

The susceptibility to collagenases is also influenced by aging, where the dermal collagens in juvenile skin (**Figure 20A**) are digested more easily than those in elderly skin (**Figure 20B**).

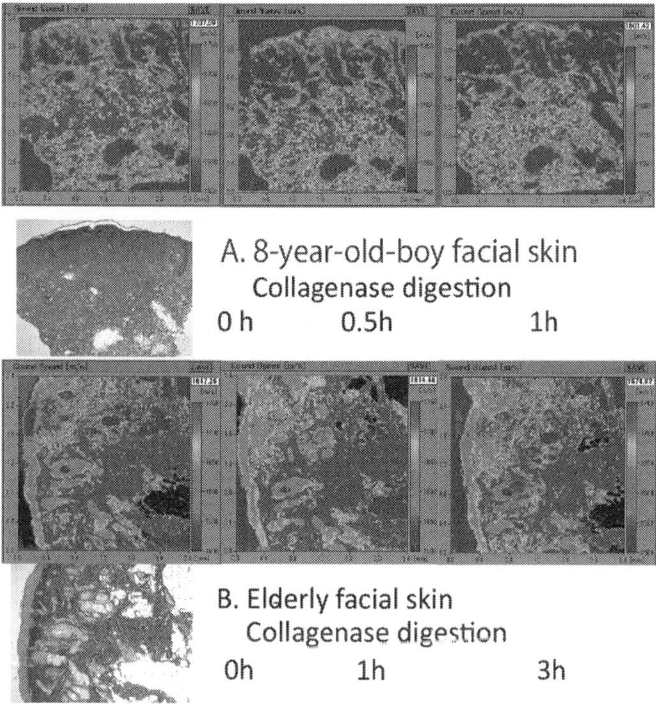

Figure 20. Differences in the susceptibility of juvenile and elderly skin to collagenases. The SOS values after collagenase digestion of the sections are compared in juvenile (**A**) and elderly (**B**) skin. The dermal SOS values of the juvenile skin decreased gradually, whereas those of the elderly skin were stable.

10. Statistical analysis of SOS

The SOS data obtained from each image were expressed as the mean and standard deviation (SD; m/s). Before the statistical tests mentioned below, normal distribution was confirmed using the test for difference of mean. One-way analysis of variance (ANOVA) was used to evaluate the SOS values. Significant differences were evaluated using multiple comparisons with the Tukey–Kramer's test, with $P < 0.05$ considered statistically significant.

The mean and SD SOS values for squamous cell carcinomas, adenocarcinomas, mesotheliomas, lymphomas, and inflammatory cells in fluid are summarized in **Figure 21** and **Table 1**. Each cell type has a characteristic SOS value, and there are statistically significant differences ($P < 0.01$) between malignant and benign cells, as well as between epithelial cells and free cells in the blood. In general, malignant and epithelial cells have higher SOS values than benign cells and free cells in the blood, respectively.

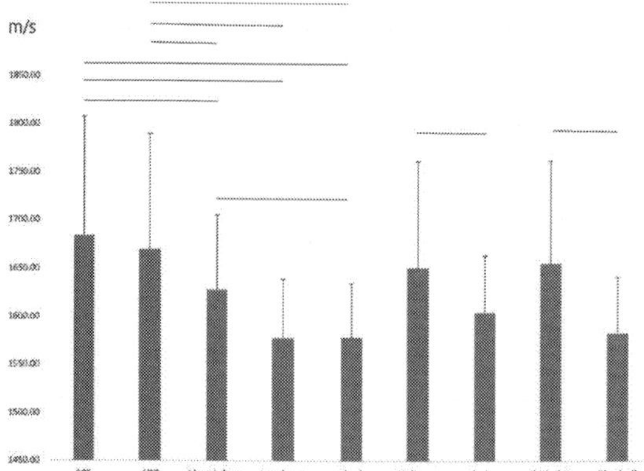

Figure 21. Mean and SD of SOS values among various cells in fluids. There SOS values differ significantly among cell types. The malignant cell group has significantly greater SOS values than the benign cell group ($P < 0.01$), and the epithelial group has higher SOS values than the blood cell group ($P < 0.01$). ADC, adenocarcinoma; Inf cell, inflammatory cells; SCC, squamous cell carcinoma.

Cell types	SOS AVE (m/s)	SD	n
SCC	1684.10	123.46	111
ADC	1669.61	120.10	356
Mesothelioma	1627.94	77.47	132
Lymphoma	1577.56	61.55	85
Inf cells	1578.38	56.23	81
Malignant	1651.10	110.99	711
Benign	1604.95	59.09	211
Epithelial	1655.69	107.31	729
Blood cell	1583.30	57.85	193

Table 1. Acoustic values of speed of sound among different cells in fluid. SCC, squamous cell carcinoma; ADC, adenocarcinoma; Inf cell, inflammatory cells.

In tissue sections, each tissue component has a specific SOS value, as shown in **Figure 22** and **Table 2** for the gastric walls, where the mucosal layers have lower SOS values than the muscularis mucosae or muscularis propria layers. Neoplasms usually have similar SOS values to their original tissues. Carcinomas arising from mucosal epithelia have almost the same SOS values as the mucosal layer, but poorly differentiated carcinomas have higher SOS values due to their desmoplastic reactions. Malignant lymphomas comprising free cells in the blood have low SOS values, as found with fluid cytology.

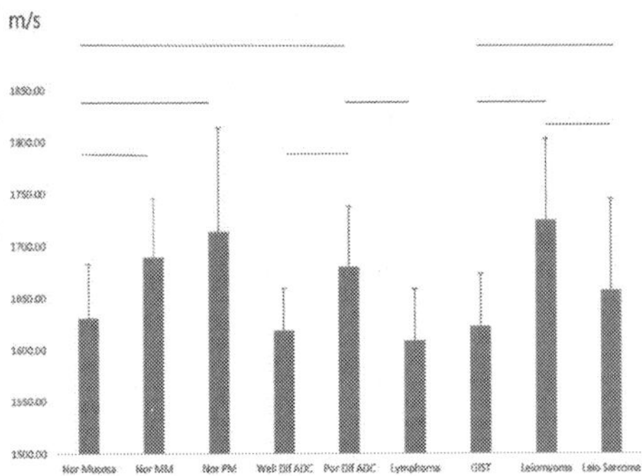

Figure 22. Acoustic values of SOS among gastric tissues. Comparison of the mean ± SD of the SOS values among gastric tissues. There are statistically significant differences among different tissue components ($P < 0.01$). GIST, gastrointestinal stromal tumor; leio sarcoma, leiomyosarcoma; nor MM, normal muscularis mucosae; nor mucosa, normal mucosa; nor PM, normal muscularis propria; Por Dif ADC, poorly differentiated adenocarcinoma; Well Dif ADC, well-differentiated adenocarcinoma.

Gastric tissues	SOS Ave(m/s)	SD	n
Nor mucosa	1630.51	52.14	105
Nor MM	1689.47	56.43	57
Nor PM	1713.34	99.97	90
Well Dif ADC	1618.89	40.32	155
Por Dif ADC	1680.05	57.52	210
Lymphoma	1608.93	49.87	196
GIST	1622.96	50.18	171
Leiomyoma	1724.30	78.38	162
Leio sarcoma	1657.07	87.72	113

Nor Mucosa, normal mucosa, Nor MM, normal muscularis mucosae; Nor PM, normal muscularis propria; Well Dif ADC, well-differentiated adenocarcinoma; Por Dif ADC, poorly differentiated adenocarcinoma; GIST, gastrointestinal stromal tumor; Leio Sarcoma, leiomyosarcoma.

Table 2. Acoustic values of speed of sound among different gastric tissues.

11. Conclusion

SAM has five unique features compared with optical microscopy. First, the measurement procedure is rapid and easy, without any requirement for special staining. The magnification of the images can be changed simply by selecting the area for scanning. Second, the observation ranges of SOS are adjustable, which facilitates the detection of specific cells such as malignant cells. Third, observations are repeatable under different conditions using the same slide so time course observations may be obtained before and after events. Fourth, the microscopic acoustic properties correspond to the echo intensity and texture in clinical echography. Finally,

SAM data can help to assess the biomechanical characteristics of tissues and cells, such as elasticity, to statistically compare their digital values [14].

However, there are some limitations of this technique. First, appropriate sample preparations are necessary such as flat 10-μm sections or single-layered cytology samples. Second, observation areas are limited. In our system, observation range is from 0.3×0.3 mm^2 to 4.8×4.8 mm^2 at the maximum. We usually refer to LM image to determine the observation area of SAM. Third, discrimination between nuclei and cytoplasm of the cell is so difficult that we can obtain only cell contours.

For overcoming the current limitations, new technologies are been developed. The wavelength of ultrasound can be shorter than that of visible light, so the resolution of ultrasound overcomes that of the LM to facilitate the observation of organelles in cells. New transducers with much shorter wave lengths are been developed. For observing living cells, impedance mode of SAM observation without contact with cells is on the way [15]. For making contrast images to specify particular structures, new techniques are necessary. We are developing new acoustic staining or immunostaining with high SOS value materials. Acoustic images can illustrate different properties of tissues and cells compared with that by LM images. Developing new acoustic techniques may help to reveal more precise properties of tissues and cells in the future.

Acknowledgements

The author thanks T. Moriki and Y. Egawa for providing clinical samples and Y. Kawabata, N. Suzuki, and K. Sugaya for preparing the tissue sections. The author also thanks K. Kobayashi for his technical support and advice on SAM observation. This study was supported by the Japan Society for the Promotion of Science KAKENHI, Scientific Research Grant Number (c) 15K08375.

Author details

Katsutoshi Miura

Address all correspondence to: kmiura@hama-med.ac.jp

Division of Health Science, Hamamatsu University School of Medicine, Hamamatsu, Japan

References

[1] Lemons RA, Quate CF. Acoustic microscopy: biomedical applications. Science. 1975;188(4191):905–11.

[2] Briggs G, Kolosov O. Acoustic microscopy. Second edition. New York: Oxford University Press; 2010. 174–81 p.

[3] Miura K, Yamamoto S. Pulmonary imaging with a scanning acoustic microscope discriminates speed-of-sound and shows structural characteristics of disease. Lab Investig. 2012;92(12):1760–5.

[4] Miura K, Yamamoto S. Histological imaging of gastric tumors by scanning acoustic microscope. Br J Appl Sci & Technol. 2014;4(1):1–17.

[5] Miura K, Mineta H. Histological evaluation of thyroid lesions using a scanning acoustic microscope. Pathol Lab Med Int. 2014;6:1–9.

[6] Miura K, Egawa Y, Moriki T, Mineta H, Harada H, Baba S, et al. Microscopic observation of chemical modification in sections using scanning acoustic microscopy. Pathol Int. 2015;65(7):355–66.

[7] Miura K, Nasu H, Yamamoto S. Scanning acoustic microscopy for characterization of neoplastic and inflammatory lesions of lymph nodes. Sci Rep.2013;3:1255.

[8] Miura K, Yamamoto S. A scanning acoustic microscope discriminates cancer cells in fluid. Sci Rep. 2015;5:15243.

[9] Barr RJ, White GM, Jones JP, Shaw LB, Ross PA. Scanning acoustic microscopy of neoplastic and inflammatory cutaneous tissue specimens. J Invest Dermatol. 1991;96(1): 38–42.

[10] Saijo Y. Acoustic microscopy: latest developments and applications. Imaging Med. 2009;1(1):47–63.

[11] Fenske NA, Lober CW. Structural and functional changes of normal aging skin. J Am Acad Dermatol. 1986;15(4 Pt 1):571–85.

[12] Venazquez E, Murphy G. Histology of the skin. In: Elder D, Elenitsa R, Johnson B, Murphy G, Xu X, editors. Histopathology of the skin. Philadelphia, PA, USA: Lippincott Williams & Wilkins; 2009.

[13] Mitchell RN. Blood vessels. In: Kumar V, Abbas AK, Aster JC, editors. Pathologic basis of disease. Philadelphia, PA: Elsevier; 2015. p. 483–522.

[14] Saijo Y. Recent applications of acoustic microscopy for quantitative measurement of acoustic properties of soft tissues. In: Mamou J, Oelza ML, editors. Quantitative ultrasound in soft tissues. Heiderberg, Germany: Springer; 2013. p. 291–313.

[15] Gunawan AI, Hozumi N, Takahashi K, Yoshida S, Saijo Y, Kobayashi K, et al. Numerical analysis of acoustic impedance microscope utilizing acoustic lens transducer to examine cultured cells. Ultrasonics. 2015;63:102–10.

Chapter 16

Applying High-Frequency Ultrasound to Examine Structures and Physical Properties of Cells and Tissues

Frank Winterroth

Additional information is available at the end of the chapter

http://dx.doi.org/10.5772/63732

Abstract

Medical ultrasound is a diagnostic imaging technique used for visualizing subcutaneous body structures. The frequencies used in conventional diagnostic ultrasound are typically 2–10 MHz. For scanning acoustic microscopy (SAM), the frequencies applied to image cells and tissues are >50 MHz. Increasing the frequency increases spatial resolution, but reduces the depth that can be imaged. The advantages of using SAM over conventional light and electron microscopes include imaging specimens without requiring any preparations which may kill or alter them; this provides a more accurate representation of them. SAM's main components are similar to those found on typical light microscopes, but the lens is often replaced by a confocal transducer. The ultrasound signal encountering the specimen generally has three results: scatter, transmission, or reflection; these signals are then merged to form the image as either a B-Scan or C-Scan. The acoustic parameters determining the image quality are absorption and scattering. SAM can objectively quantify the surface characteristics of the specimen being scanned and can also study the elastic properties of cells and tissues to discern differences between healthy and affected conditions. SAM has the potential as a major instrument of detection and analyses in biomedical research and clinical studies.

Keywords: acoustics, B-Scan, C-Scan, RMS data, elastography

1. Introduction

Ultrasound imaging is an extremely broad and elaborate topic covering many facets that are too complex to be covered over the course of a single book chapter. Although this chapter is dedicated to the arena of acoustic microscopy, this topic in and of itself also entails far more

details than can be covered here. For pedagogical purposes, the material has been condensed to introduce the basic concepts and definitions for this unique form of microscopy.

1.1. Ultrasound in imaging: definition, operation, and image types

Medical ultrasound (or ultrasonography) is a common diagnostic imaging technique used to noninvasively visualize subcutaneous body structures including tendons, muscles, joints, vessels, and internal organs for possible pathology or lesions; its most common use is in obstetrics to image the developing fetus. In physics, the term "ultrasound" applies to all sound waves with a frequency above the audible range of human hearing, approximately 20,000 Hz. The frequencies used in conventional diagnostic ultrasound—including for obstetrics—are typically 2–10 MHz [1, 2].

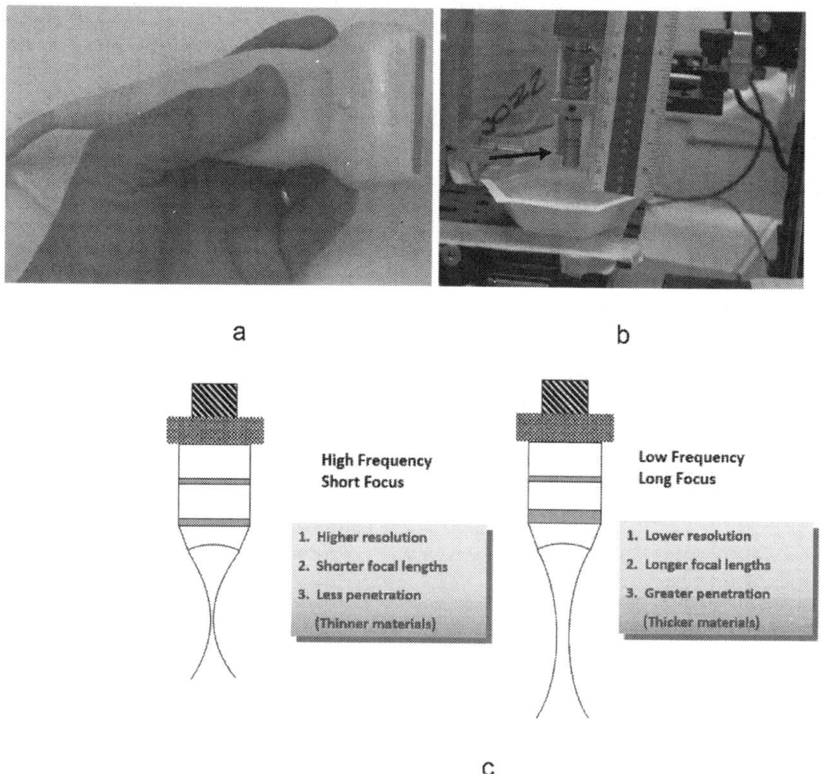

Figure 1. Comparison between a typical multielement, linear array transducer using in standard medical ultrasound, (a) and a single element confocal transducer on the SAM (b). Note the former device is handheld, while the SAM's transducer (arrow) is mounted above the specimen stage and is approximately the size of a standard pen cap. An illustration contrasting the differences between the two transducer types is included (c).

The most common type of 2D image by using conventional ultrasound is a B-Mode (brightness mode) image which is produced from the acoustic impedance (Z) of the object being scanned by the system [1, 3, 4]. The speed of sound varies as it travels through different materials and is dependent on the acoustic impedance of the material. However, the sonographic instrument assumes that the acoustic velocity (C) is constant. An effect of this assumption is that in a real body with nonuniform tissues, the beam becomes somewhat defocused and image resolution is reduced. Typical acoustic velocity in degassed/deionized water is at 25 °C = 1497 m/s.

The transducers used in conventional ultrasound imaging usually consist of a handheld device with multiple elements (**Figure 1a**)—and sometimes a servomotor—allowing simultaneous radiofrequencies which are then merged into a real-time image.

The sound from the transducer is focused either by the shape of the transducer, a lens in front of the transducer, or a complex set of control pulses originating from the ultrasound scanner (beamforming). This focusing produces an arc-shaped sound wave from the face of the transducer. The wave travels into the body and comes into focus at a desired depth.

The return of the sound wave to the transducer results in the same process as sending the sound wave, except in reverse. The returned sound wave vibrates the transducer which converts the vibrations into electrical pulses that travel to the ultrasonic scanner where they are processed and transformed into a digital image.

To produce an image, the ultrasound scanner must determine two factors from each received echo:

1. The difference in the time it takes for the echo to be received from the time when the sound was transmitted.

2. The intensity of the echo.

Once the ultrasonic scanner determines these two factors, it can locate which pixel in the image to light up and to what intensity.

Transforming the received signal into a digital image may be explained by using an analogy involving a long, column-like, multilayered box. Picture placing a transducer at the top of the box and then sending pulses down the length of the box; listen for any return echoes. When an echo is heard, how long it took for the echo to return is noted; the longer the wait, the deeper the layer (1, 2, 3, etc.). The strength of the echo determines the brightness setting for that layer (white for a strong echo, black for a weak echo, and varying shades of gray for everything in between). When all the echoes are recorded for the box, a grayscale image can be produced for it.

To generate an acquired 2D ultrasound image, the ultrasonic beam is swept across the area being examined. A transducer may sweep either mechanically by rotating or swinging it manually; a 1D phased array transducer may also be used to sweep the beam electronically. The received data is processed and used to construct the image. The acquired image is then a 2D representation of the slice into the body.

3D images can be generated by acquiring a series of adjacent 2D images. Commonly, a specialized probe that mechanically scans a conventional 2D image transducer is used. However, since the mechanical scanning is slow, it is difficult to make 3D images of moving tissues. Recently, 2D phased array transducers that can sweep the beam in 3D have been developed. These can image faster and can even be used to make live 3D images of a beating heart [5].

Ultrasound also applies Doppler sonography in order to study physiological processes such as blood flow and other organ functions [1, 2]. The different detected speeds are represented in color for ease of interpretation. To image and differentiate tissues and organs—in particular healthy and affected—the damaged tissue often shows up in a unique color (i.e., flash of different color in a blood vessel leak). Colors can alternatively represent differences in amplitudes between any received echoes. Doppler ultrasound is generally applied in conventional ultrasound, but not in acoustic microscopy; it will not be discussed in detail here.

1.2. Acoustic microscopy: background, specifications, applications, and advantages

The scanning acoustic microscope (SAM) has been employed since the 1970s—mostly in materials sciences and engineering—typically to study failure analyses and nondestructively assess structural properties. In the past two decades, it has been similarly utilized in biology and medicine to examine the structures and functions of cells and tissues [18]. For SAM, the frequencies applied to image materials are usually greater than 50 MHz; this provides a significant dynamic range ratio between the lowest and highest values used in ultrasonic.

In contrast to the aforementioned conventional ultrasound system, the acoustic microscope often employs a smaller transducer—similar in size to a pencil eraser—mounted on the body of the microscope (**Figure 1b**). The SAM's transducer is generally a single-element piezoelectric confocal unit, producing a signal similar to its conventional counterpart.

Given such high frequencies for imaging, there is a trade-off between penetration depth of the signal through an object and spatial resolution. With higher frequency ultrasound, there is an increased attenuation coefficient, resulting in it being more readily absorbed by the material it is scanning. This is analogous with light microscopes: Increasing the magnification enhances a particular area, but reduces the field of view of the imaged specimen; likewise, with SAM, increasing the frequency increases spatial resolution, but reduces the depth that the specimen can be imaged. **Figure 1c** illustrates the differences between high- and low-frequency ultrasound imaging, comparing the benefits and limitations between them.

A typical SAM transducer used in imaging biological specimens is approximately 50–60 MHz. Its parameters are the following:

1. 15 μm scanning step size in both the transverse and horizontal directions.

2. Axial resolution (R_{ax}) equals 24 μm.

3. Lateral resolution (R_{lat}) equals 37 μm.

4. Depth of field (DOF) equals 223 μm.

5. The f–number is 1.5.

6. A routine scan is sampled at 300 mega samples/second.

Note: All of these parameters can be manually adjusted to accommodate a specific scan of the desired object.

The axial resolution depends on the dimensions of the pressure pulse, which is related to the transducer bandwidth (BW) and system electronics:

$$R_{ax} = \frac{c}{2 \times BW} \tag{1}$$

where c is the sound speed in the tissue and BW is 32 MHz. Likewise, the lateral resolution at the focal point can be estimated by:

$$R_{lat} = \lambda \times f \ / \ number \tag{2}$$

where λ is the wavelength (25 µm).

The depth of field (DOF) is calculated by:

$$DOF = 4(f \ / \ number)^2 \lambda \tag{3}$$

Due to a small f/number (which provides a tight beamwidth), the DOF is also limited.

Lateral resolution is the resolution orthogonal to the propagation direction of the ultrasound wave.

Acoustic impedance (Z): The impedance determines the amplitude of the reflected and transmitted waves at the fluid–tissue interface. Complex scattering properties of tissues are due to acoustic impedance interfaces in microstructure of tissues:

$$Z = \rho c \tag{4}$$

$$R = |(Z_2 - Z_1) / (Z_2 + Z_1)| \tag{5}$$

where ρ is the density of the material, and R is the reflection coefficient.

Table 1. Logistics of the SAM operating mechanisms and principles.

R_{ax} is the resolution in the direction of propagation and is determined by the length of the ultrasound pulse propagating in the tissue; R_{lat} is the resolution orthogonal to the propagation direction of the ultrasound wave. The ultrasound signal encountering the specimen has three possibilities: scatter where the signal travels in directions orthogonal to the transducer, transmission where the signal passes through the specimen, or reflection where the signal is received back to the transducer for processing into an image. Essentially, it is this impedance mismatch between the signal propagating through water and encountering the specimen which results in the processed image. **Table 1** provides further comprehensive details of the SAM's operating mechanisms, including axial and lateral resolutions, DOF, and acoustic impedance.

The advantages of using SAM over conventional light and electron microscopes include being able to image materials—including cells and tissues—without performing any preparations which could potentially kill or alter them. This provides a more accurate representation of the materials' natural properties [18]. In biological imaging, it can also provide evidence as to any degree(s) of differentiation which the cells undergo without chemically affecting their properties [6–9]. In some SAM systems, there is also the ability to scan across three axes, allowing for a three-dimensional composite image of the desired specimen to be produced. This is covered in greater detail later in this chapter.

2. Principal components of the scanning acoustic microscope

2.1. Transducer and general design layout

The SAM's main components are similar to those found on typical light microscopes. In place of the magnifying lens, there is often a single-element confocal transducer attached to stepper motors which control the transducer's movement in three dimensions: horizontal, transverse, and vertical. For the stage, there is also a small tank filled with fluid—usually degassed/deionized water—in the specimen being imaged as immersed (**Figure 2**). Just as with any conventional ultrasound, when imaging with the SAM, all materials being imaged must be submerged in a fluid medium or have coupling gel between the transducer and the area being scanned as ultrasound energy scatters when exposed to air.

Figure 2. Close up of the SAM transducer immersed in a tank of degassed and deionized water, above the material to be imaged.

2.2. Principal features of the SAM and its operation

In addition to the stage, other key features of the SAM include the following: a broadband high-frequency pulser where the excitation signal originates; a diode expander which reduces the baseline noise before the signal travels into the transducer where it is converted into ultrasonic waves. Signals received by the transducer go into a diode limiter which blocks some of the high-amplitude excitation signal from the receiving electronics. The received signal is generally amplified by a set of attenuators and 32 dB fixed-gain amplifiers. Finally, the signal is captured by a high-speed 8-bit digitizing board.

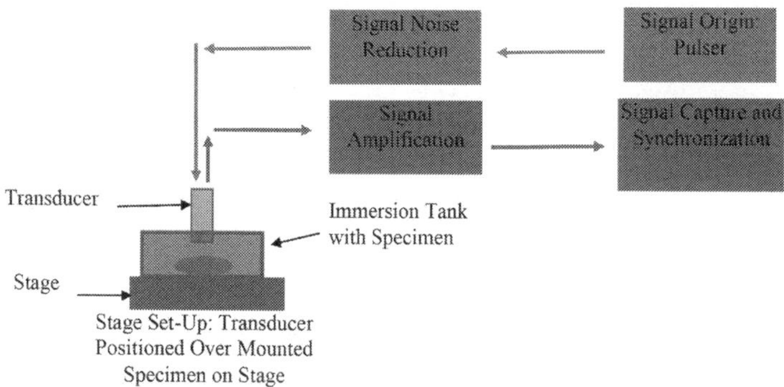

Figure 3. The principal operating units relative to the SAM's overall layout. In addition to the stage, other key features of the SAM include the following: a broadband high-frequency pulser where the excitation signal originates; a diode expander which reduces the baseline noise before the signal travels into the transducer where it is converted into ultrasonic waves. Signals received by the transducer go into a diode limiter which blocks some of the high-amplitude excitation signal from the receiving electronics. The received signal is generally amplified by a set of attenuators and 32-dB fixed-gain amplifiers. Finally, the signal is captured by a high-speed 8-bit digitizing board. The SAM signals are then synchronized to reduce any signal to signal jitter and improve the efficiency of signal averaging.

A custom-built circuit takes the 10 MHz synchronized output from the digitizing board and divides it down to a 9.76 kHz square wave that triggers both the pulser and the digitizing board. The SAM signals are synchronized to reduce any signal to signal jitter and improve the efficiency of signal averaging. A basic illustration of the SAM signal transmission and reception is shown in **Figure 3**: the principal operating units relative to the SAM's overall layout.

3. Basic operations: set-up and acquisition of standard 2D images

3.1. Specimen setup

Because the desired object for imaging is often submerged in an imaging tank (often no larger than 20 cm^3), the object itself should be no greater than 1.0 cm^2 in area (both the object's size and its scanning areas can be increased as long as the operator knows and recalls the exact region being examined is) and no thicker than approximately 5.0 mm, less than the focal length of the transducer. Further, it must not be composed partly or completely of material that is miscible in water after being submerged for approximately 2–3 h. Ideally, the object should have greater density than the water, but with most biological materials, this is not the case. Therefore, a method of mounting the object either with push pins or embedding the base of it deep into mounting clay at the base of the immersion tank would be suitable. The area being imaged must be fully exposed to the transducer. Examples of mounting the specimens are illustrated in **Figure 4** with respect to their positions to the SAM transducer. The tank with the specimen must be placed on the stage below the transducer with enough fluid in the tank so that it is approximately 1–2 mm below the rim.

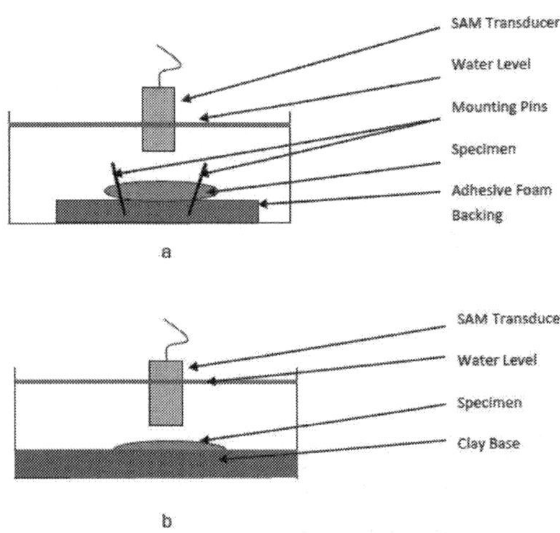

Figure 4. Two examples of mounting specimens in the immersion tank for imaging. The first involves using mounting pins through the object (away from the imaging area) into a softer backing, usually an adhesive sponge (a); the second involves embedding the specimen into mounting clay at the base of the tank, exposing the top surface of the object over the transducer (b).

3.2. Powering the microscope

The SAM requires turning on both the computer and each hardware component responsible for powering the scope. These include the power supplies for both the pulser-receiver and the digitizer. Additionally, the digitizer and pulser-receiver units must be manually powered on.

Because of the single-element configuration of the transducer—typically lacking a servomotor which allows the transducer to fluctuate—its movement is dependent on a stepper motor system. Therefore, initialization of the stepper motor driver card is imperative; this card connects to and controls the stepper motor driver.

After powering up the microscope and its accompanying computer, the operator must check the computer's appropriate site for its devices and interfaces, scroll down, and select the driver card. When the prompt screen appears, "Initialize" is selected to engage its software features.

The operator will then select an icon or tab labeled "Move"; a screen with three-dimensional axes will appear. The x-axis moves the transducer along the longitudinal direction; the y-axis moves the transducer along the transverse direction; the z-axis moves the transducer along the vertical direction. Each axis is programmed to move a specific step size (typically between 1.25 and 2.5 μm for all three axes); therefore, inputting the number of desired steps multiplies the distance of the desired step size by that number. For example: inputting 20 steps and selecting once along an axis set to 2.5 μm will move the transducer 50 μm. When operating the stepper motor, selecting "In" on the Move screen will move the transducer toward the stepper motors for each of the 3 selections, while "Out" moves it away from the motors. **Figure 5** provides a screenshot of the SAM software display for the X, Y, Z stepper motors.

Once the transducer is positioned over the prepared tank, it is slowly lowered in using the "Move" feature to control the stepper motor along the z-axis. It is recommended that once the transducer is touching the fluid surface to descend the transducer at a rate of 10–20 steps per "Move" selection. While the transducer is immersed in the fluid, but close to the surface, its bottom face is inspected under the fluid for the presence air bubbles. All air bubbles must be removed as they could scatter the ultrasound energy. Removal can often be accomplished using a cotton swab or a plastic twist tie that has a slight kink at the bottom to gently remove the bubbles off of the bottom face of the transducer; **no sharp or metallic objects must ever be applied directly to the transducer.**

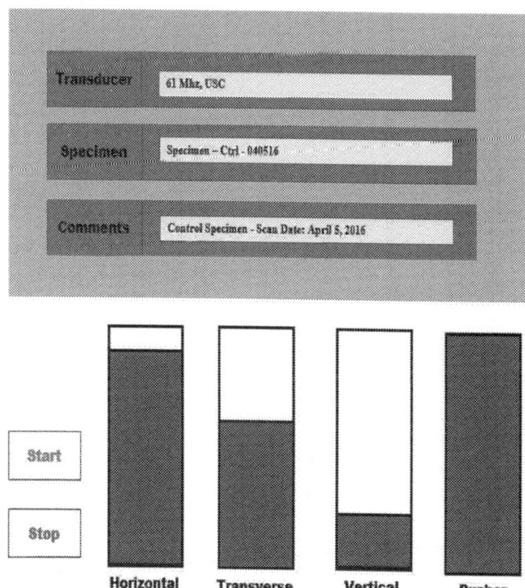

Figure 5. Illustration of a typical screenshot display for the software used to operate the SAM. The gray area displays the logistics of the transducer, the specimen being scanned, and any comments about the experiment. The 3D axes x, y, and z are labeled horizontal, transverse, and vertical, respectively. The blue bars in each of the axes' fields indicate how much of each scan has been completed. The bar labeled "Pusher" refers to the compressor used for elastography.

3.3. Image acquisition

Initiation: Once the transducer is positioned over the desired area for scanning, it is necessary to obtain the RF wave of the ultrasound. This is performed with an oscilloscope and accompanying command screen. On the command screen toolbar, the operator will select the initial/recall settings; a command prompt will appear with the SAM operation commands. After selecting data acquisition commands, it is imperative to choose the acquisition modality, either "Auto" or "Normal" acquisition (one is automated; the other manually controlled): a signal wave will appear on the oscilloscope, with an acquisition rate of approximately 1000–2000/s. A trigger delay on the command window must be selected at approximately 3×10^{-6} to provide a 3.0 μsec delay; this prevents the scan from "freezing" during the signal acquisition off of the specimen. The operator then selects the appropriate channel of the acquisition software which will provide a single band for the signal. This is followed by selecting "Connect Data Points" under the "Tools" command which will enhance the signal coming off of the oscilloscope. At this point, there will appear a single rapid signal moving from right to left on the oscilloscope (**Figure 6a**).

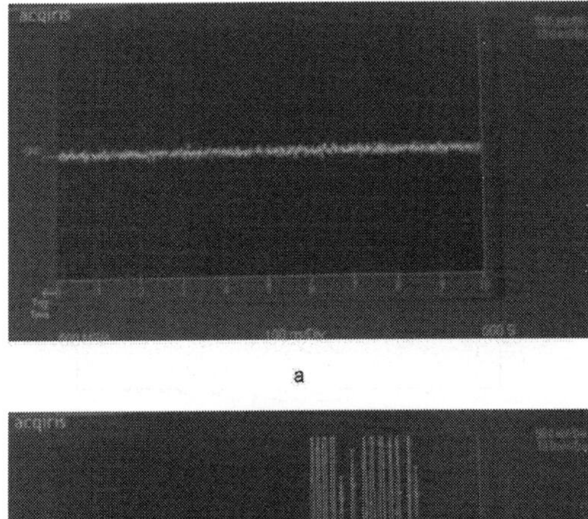

a

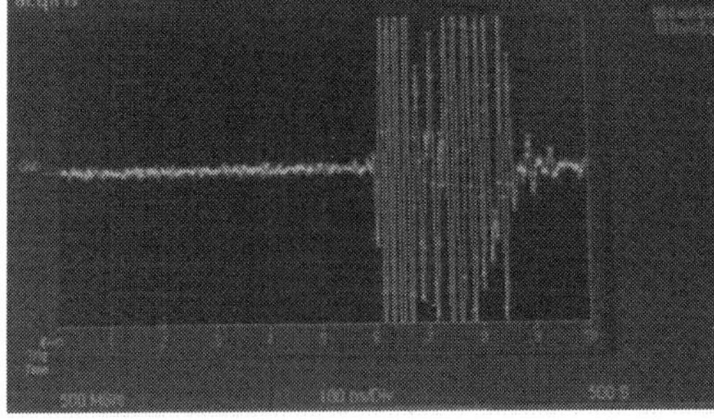

b

Figure 6. Oscilloscope images showing the background signal of the SAM in water (a) and upon encountering the specimen (b). The RF signal is generated by the impedance (Z) mismatch between the ultrasound energy in water and the specimen.

Focusing the transducer: The operator now restores the positioning functions of the transducer, namely the z-axis, moving the transducer in the "Out" direction—away from the stepper motor and toward the specimen—in increments of approximately 10 steps. Upon the ultrasound signal encountering the specimen, an RF signal wave appears on the right side of the oscilloscope (**Figure 6b**). At this point, the transducer's vertical position is adjusted so that the RF signal is in the center of the oscilloscope. It is imperative to pay attention to both the distance that the wave moves and to check the transducer's position relative to the object in the tank. If the signal does not appear very strong, the attenuators can be adjusted, by clicking first on the receiving attenuator, then the transmitting one; the lower the number, the greater the signal should increase. However, the trade-off is that there will be an increase in signal noise as both the low and high pass filters are being reduced. The signal is continuously adjusted until it is centered. Once this point on the oscilloscope is recorded, the signal is adjusted further until it attenuates. Here, it is best to zero out the position of the transducer in the vertical position, followed by reversing the adjustments made (by selecting "In" for the z-axis) until the signal returns back to the right edge of the scope. The operator records the distance travelled under the z-axis which is the approximate thickness of the specimen.

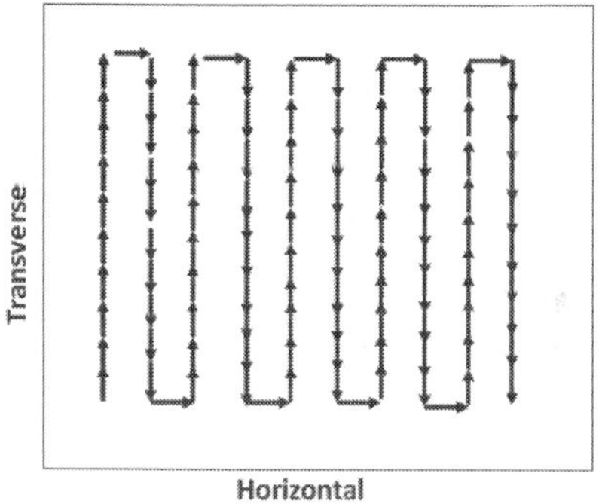

Figure 7. Top-down view illustrating the scanning pattern made by the SAM transducer. Upon scanning in the transverse direction, the transducer shifts over in the horizontal direction and the scanning process repeats itself. After the desired area is scanned, the transducer shifts in the vertical direction in order to scan multiple layers of the specimens as desired.

Once all the configurations are in place, the start key can be selected; the scope will start scanning the desired area of the specimen. A typical scan using the parameters listed above will take approximately 20–30 min. It is recommended that the SAM computer not be used during the scanning session as this often will disrupt the scan and may potentially shut down the system, which will require reconfiguring the SAM set-up. The transducer will acquire the specimen image first along the transverse line, firing 31 scans into the specimen before moving to the next point which will be 1.25 μm × the number of steps which were entered for the scan. After scanning the desired distance along the transverse axis (usually 1.0 mm), the transducer will shift 2.5 μm × the number of selected steps in the horizontal direction and repeat scanning in the transverse direction, followed by a repeated shifting 2.5 μm × the number of selected steps in the horizontal direction and repeating in the transverse direction. It will form a "zigzag" pattern until it has completed the scan over the desired area of the specimen surface. The transducer will then shift 2.5 μm × the number of steps down in the Z direction and the process will repeat itself; **Figure 7** illustrates this scanning pattern.

After the scan is complete, both the dataset and logistics on the scan will be produced (an example of the latter is shown in **Table 2**). These are generally saved under preselected LabVIEW files.

31	Number of averages
256	Number of points per trace
923	Offset from the trigger
6	Number of 2.50 μm pulses between horizontal sampling points (X)
12	Number of 1.25 μm pulses between transverse sampling points (Y)
48	Number of 2.50 μm pulses between vertical sampling points (Z)
10	Number of 1.5875 μm pulses between compressor sampling points
70	Number of sites along horizontal axis (X)
70	Number of sites along transverse axis (Y)
7	Number of sites along vertical axis (Z)
1–5	Number of sites along compressor
0.50	Voltage scale (V)
14	Initial setting for transmission attenuator
14	Initial setting for reception attenuator
0.00	Voltage offset (V)
250	Sampling frequency (MHz)
3.000	Trigger level (V)
External	Trigger source
Positive	Trigger slope
9.76	Avtech repetition rate (kHz)
8	Avtech frequency setting (1–10)
7	Avtech amplitude setting (1–10)
1.555	F–number of the transducer
3.111	Focal length of the transducer (mm)
61.0	Center of frequency of the transducer (MHz)
30.0	Bandwidth of the transducer (MHz)

Table 2. Example of the settings and operating parameters applied to perform a standard SAM scan.

Shut down: Once the scan is complete, the transducer will reposition itself back to its original starting point. After completing all scans, the operator restores the positioning functions and raises the transducer out of the tank. The specimen can then be removed from the tank and handled in the appropriate manner. The scan is saved under the files of the scan program. The operator can go to the data folder (such as in LabVIEW), select the files from the scan, and copy and paste them into a flash drive. To power off the SAM, it is a matter of turning off all the power supplies to the SAM and shutting down the computer.

3.4. Image processing

Upon completing the specimen scan, the RF data is first translated into an acoustic envelope for each scan point. The raw acquired signal initially includes its inverse image; this is removed by eliminating all frequencies below zero frequency. An inverse Fourier transform is applied to convert the data from a frequency domain into a time domain. **Figure 8** is the resulting acoustic envelope of a processed signal; in this case, a threshold is set approximately 20 dB above the background noise so that the acoustic energy encountering the specimen will be displayed as the acoustic envelope.

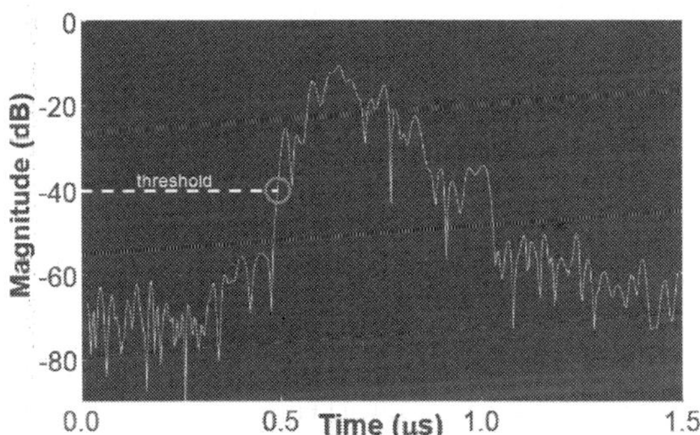

Figure 8. Representation of a typical acoustic envelope developed from the SAM scan of a specimen and derived from the raw RF scan. The raw scan initially includes an inverse image below zero frequency; this is eliminated by removing any frequency below zero. An inverse Fourier transform is performed converting the scan from a frequently domain to one of time. A threshold is set 20–30 dB above the water signal to establish point at which the acoustic signal encounters the specimen. By examining the envelope, one can see at what time the signal attenuates, allowing the calculation of the specimen's thickness.

The signals are then merged to form the image in B-Mode—called a B-Scan—which is an amalgamation of the intensities of the echoes received from the transducer. The two acoustic parameters determining the image quality are absorption which indicates how ultrasonic energy was converted into heat in the specimen, and scattering, specifically backscatter. Scattering means that an ultrasonic wave when striking an obstacle radiates part of its energy orthogonally to the specimen. Quantifying the degree of backscatter from small particles or perturbations in tissues and other materials allows differentiations in the degrees of pixel intensity in the B-Scan images; such scatter properties include the average scatterer (energy scattered) diameter and average scatterer concentration [19].

Template 1 illustrates a program to acquire the image and converge the images using a program (such as in MATLAB) to assist in processing the acquired images from the SAM scans.

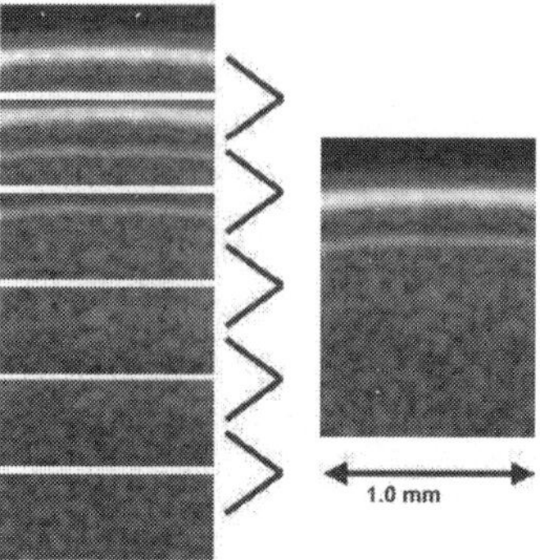

Figure 9. B-Scan merge processing in either the vertical/horizontal plane or the vertical/transverse plane. On the left are unmerged images. On the right is the resulting merged image.

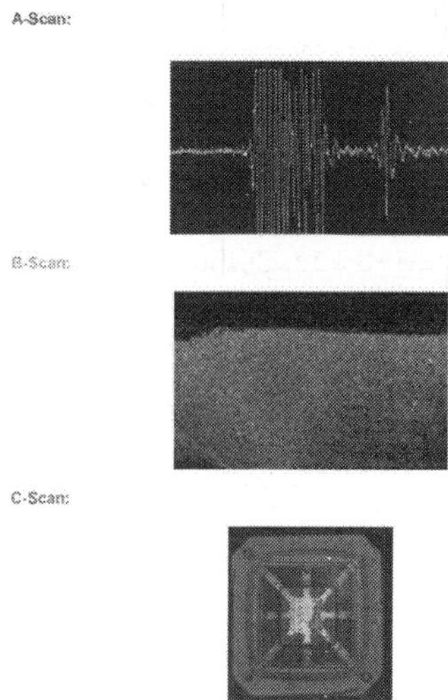

Figure 10. The "ABC's" of acoustics. Defining the raw ultrasonic signals and how they are converged to form images. The A-Scan (or A-Line) is the RF signal from a single point (x, y) in the scan. After they are converted into acoustic envelopes and merged into a vertical cross section, they form a B-Scan. The C-Scan is the collected data from a specified depth over the scanned area (a horizontal cross-section).

Table 2 provides an example of scanning logistics. It includes the number of sampling points per scan, the distances traveled by the transducer in all three dimensions, and the parameters set both manually and automatically regarding the scan. **Figure 9** shows the resultant B-Scan after acquiring and merging the scanned data. Each of the bright ovoid areas shown in the image represents scatter, and each is approximately 20–40 μm in diameter; this is within size range of many structural cells and white blood cells, but not red blood cells which are generally <10 μm in diameter. At 50–60 MHz, SAM can typically image a collective group of cells and tissues, similar to the resolution of histology images seen at 20× magnification. **Figure 10** shows comparative images of histology specimens taken with standard light microscopy (**Figure 10a** and **b**) and their SAM B-Scan counterparts (**Figure 10c** and **d**). The light microscope images were from a standard H & E stain imaged at 20×, showing the differences in surface characteristics between two biological tissues. The SAM B-Scan images were imaged using a 60 MHz transducer of similar biological tissues.

Development of higher-frequency transducers with accommodating digitizing boards have resulted in an acoustic resolutions of subcellular components, in particular noninvasive examination of cells undergoing apoptosis [3, 7].

4. Advanced operations: C-Scans, RMS profiles, and elasticity studies

4.1. Generating C-Scans and RMS data

The C-Scan is the scan along one selected horizontal-transverse plane of the specimen [4]; this is usually the surface of the specimen. It is similar to the B-Scan, but the merged data is orthogonal to those in a typical B-Scan. To acquire the C-Scan, the operator will enter a series of commands into a program such as the MATLAB function shown below after merging the acquired data from the scan (following acquisition of the B-Scan). This command will fit the planar surface and remove any tilt to the image, thus correcting for any irregularities stemming from incorrect mounting of the specimen.

Many of the commands contained in this chapter are custom written MATLAB functions (and some custom subfunctions within those functions). SAM operators will need the custom functions as well as the data files and MATLAB will need to be aware of the folder where the custom functions are. Template 2 is an example of MATLAB commands for generating C-Scan images

Figure 11 illustrates the three basic scanning modes in SAM as the "ABC's" of acoustics.

The C-Scan both profiles and quantifies any surface irregularities of the scanned specimen. This is done by fitting the planar image of the specimen, then subtracting everything above and below that plane. All images at the plane will appear as green; those above it will be orange-red and those below as blue. A typical C-Scan image may appear as **Figure 12**. The resulting image and root-mean-square (RMS) value will be the C-Scan profile. The smaller the number, the less surface variations there are on the specimen.

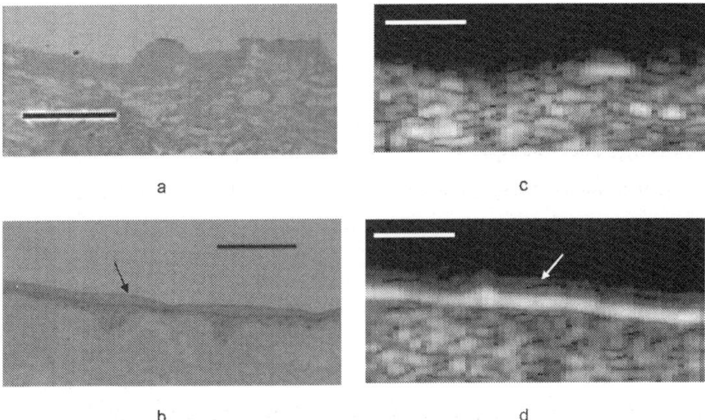

Figure 11. Comparative images of histology specimens taken with standard light microscopy (a and b) and their SAM B-Scan counterparts (c and d). The light microscope images were from a standard H&E stain imaged at 20×, showing the differences in surface characteristics between two biological tissues. The SAM B-Scan images were imaged using a 60 MHz transducer of similar biological tissues. Note the differences in the surface characteristics of light microscope images and compare them with the B-Scans: The keratinized surface shown in (b) (arrow) shows up as a bright band in (d) (arrow). Scale bars equal 100 μm.

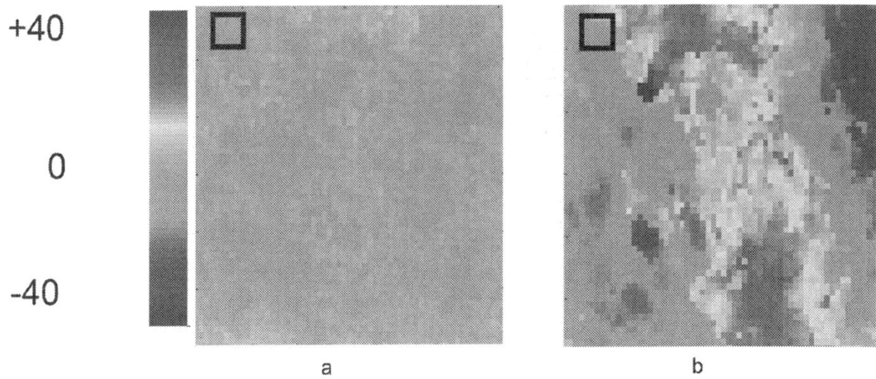

Figure 12. SAM images of C-Scans taken of two tissue specimens contrasting surface characteristics between them. One image shows the area to be essentially void of any irregularities as evidenced by its uniform color pattern (a). The second image displays significant levels of surface irregularities as the high variations in colors demonstrate. The scale boxes represent 10,000 μm². The color scale shows the degrees of variance in the surfaces, relative to their height above or below the set planar surface: The greater the yellow-green in appearance, the closer the specimen is to the planar surface and more uniform in structure. The x- and y-axes represent the number of steps in the scan along the horizontal and transverse directions; approximately 1.0 mm × 1.0 mm.

4.2. Generating 3D images

The 3D images are composites of the 2D B-Scans produced. They are generated by simply compiling and merging the acquisition data multiple times. Provided that the parameters listed under the SAM set-up procedures have been maintained, extra steps can now be taken to produce an image of the complete region that the transducer scanned.

Template 3 is exemplifies what can be generated using MATLAB to assist in processing the acquired images from the SAM scans, following acquisition of the B-Scans, C-Scans, and RMS profiles.

An example of a complete 3D composite produced with SAM (1.0 mm^3 scan) is shown in **Figure 13**.

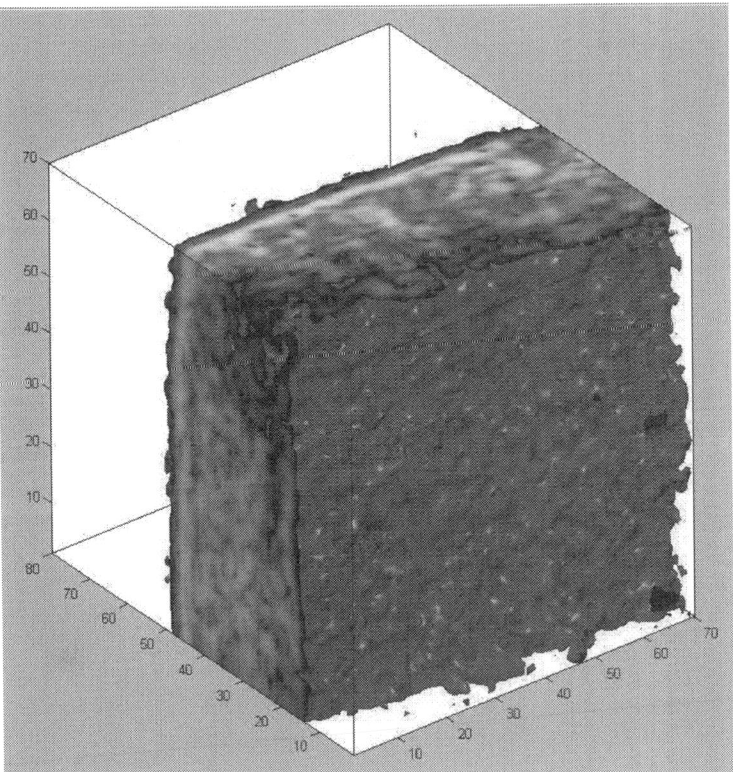

Figure 13. 3D rendering of a 1.0 mm^3 SAM scan of a tissue specimen. The apical surface faces forward and is rendered in blue to signify the threshold detected by the SAM transducer. Any signal detected above the threshold is rendered in grayscale: The greater the backscatter properties of the ultrasound at the point of detection, the brighter the image at that point. Any signal below the threshold is not registered.

4.3. Elastography (ultrasound elasticity imaging)

Ultrasound is also used for elastography, which is a relatively new imaging modality that maps the elastic properties of soft tissue [10, 11, 20]; this modality emerged in the last two decades. Elastography is useful in medical diagnoses as it can discern healthy from unhealthy tissue for specific organs/growths. For example, cancerous tumors will often be harder than the surrounding tissue, and diseased livers are stiffer than healthy ones [10–13].

There are many ultrasound elastography techniques [13]. The most prominent include quasistatic elastography/strain imaging, shear wave elasticity imaging (SWEI), supersonic shear imaging (SSI), acoustic radiation force impulse imaging (ARFI), and transient elastography. The steadily growing clinical use of ultrasound elastography is a result of the implementation of technology in clinical ultrasound machines [20]. Currently, an increase of activities in the field of elastography is observed demonstrating successful application of the technology in various areas of medical diagnostics and treatment monitoring.

Figure 14. The SAM compressor positioned within the imaging tank. The transducer is seen on the upper right as it is being moved over the compressor.

An example of a SAM compressor is shown in **Figures 14** and **15**, by itself and in use with the transducer, respectively. The compressor has a flat aluminum plate with a slit in the center which is approximately 20 mm in length and 2 mm in width. After each scan volume is completed, the plate will move down a designated distance, compressing the specimen as the scan is repeated by the SAM. A full 3D rendering of the scanned specimen at each compression step can then be produced to determine the degree of strain at specific regions of the specimen. Here, the SAM is potentially advantageous over a conventional mechanical compressor in determining what properties in the specimen contribute to the greatest or least changes in overall strain behavior [14, 15]. In this latter case, they can be analyzed through a process called speckle tracking which is not covered in this chapter.

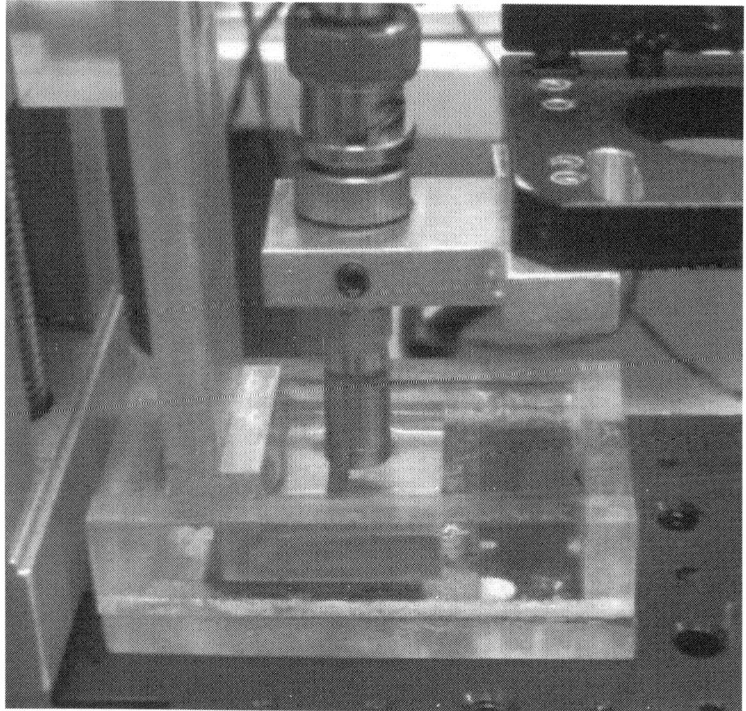

Figure 15. The SAM set up for performing compression studies. The compressor is positioned over the specimen in the tank, while the transducer is located over the slit in the compressor.

4.4. Specimen set-up and positioning

The specimen set-up for the compressor is essentially the same as described earlier with the addition of positioning the compressor plate over the object. After mounting the specimen in the immersion tank, the operator will place the tank on the stage, add enough water to barely cover over the specimen, and slowly lower the compressor toward the tank. When the compressor plate is approximately 1.0 cm from contacting the tank, the tank is raised manually so that the compressor plate fits directly inside and over the specimen. The compressor plate can then be lowered using the "Move" program with one hand while manually lowering the tank down to the stage with the other, moved in step sizes ranging from 20 to 30. Once the tank is back on the stage, the operator continues lowering the plate until it comes in contact with the specimen, without pressing it. Once this is done, the controller axis under the "Move" program is zeroed-out. The operator can then begin moving the transducer into place so that it is directly over the compressor. *Extreme care must be taken so that the transducer is out of the way when positioning the compressor over the specimen and tank. The transducer and compressor must never contact one another.* The operator finally positions the transducer into the tank based on the guidelines described earlier.

4.5. Transducer positioning and signal acquisition

The transducer is moved in the vertical direction toward the compressor plate until a strong reflection signal is seen in the oscilloscope. At this point, the transducer should be positioned over the metallic portion of the plate—preferably to the right of the compression plate's slit. The transducer is moved in the horizontal direction toward the slit until the acquisition signal diminishes, followed by several more steps until the signal completely dissipates. The transducer will now be positioned over the compression plate slit. Care must be taken not to run the transducer into the pusher axis. There is not much room beyond the side of the slit closest to the pusher axis. The axis controller is zeroed out, and the operator slowly (in increment steps of 10–20) continues moving the transducer in the horizontal direction until the reflection signal off the compressor plate returns. Once it returns, the horizontal distance moved should read approximately 2.0 mm (the width of the slit). The transducer is moved in the horizontal direction back into the slit until it is positioned 1.0 mm in the center of the slit (the signal reading on the oscilloscope should be minimal). The transducer is now moved in the vertical direction until the acquisition signal appears again; this will be the signal reflecting off of the specimen. The transducer is moved away from the specimen again until the signal is just visible on the right side of the scope. The operator finally zeroes out the axis controller, minimizes the "Move" program, and closes out the acquisition program. The operator now follows the instructions listed under "Image Acquisition" to scan the specimen. Under "Sampling Sites," the same standards are kept as listed under the basic operations, except for the compressor.

Transverse: 70 (1050 μm); horizontal: 70 (1050 μm); vertical: 7 (960.00 μm); pusher (compressor): 1 (15.875 μm). Note that for the pusher, the number of pusher steps desired for the compression scan is selected. A typical selection ranges from 3 to 10, depending on the thickness of the specimen.

After entering the desired information, the operator starts the scan. Also note, because the compressor is now employed in the scan, the scanning program time will be multiplied by the number of steps entered under the pusher. For instance, entering 3 steps will triple the scan time listed under "Basic Scans."

4.6. Image acquisition and generating 3D movies of the compression data

One process for producing movies resulting from the compression sequences is derived from the MATLAB files in Template 4.

The computer must have Windows Media Player—or its equivalent—in order to access the movies. Open Media Player and select the saved files, and set to continuous run, if preferred (usually, a film involving 6-step compressions runs approximately 0.5 s). The resulting images will combine to form both a movie displaying the result of the compression steps and the degrees of stiffness/compliance—based on the strain maps generated during the compressions —in the properties of the specimen being tested. The SAM can also display which stages within the specimen (if it is heterogeneous) are the greatest/least changes in the compressions. In this latter case, the operator can better correlate what specific properties within the tested specimen contribute to the observed changes.

4.7. Use of apparent integrated backscatter to quantify image surfaces

The use of apparent integrated backscatter (AIB) to quantify image surfaces further allows additional objective methods to assess the conditions of the specimen. AIB is a measure of the frequency-averaged (integrated) backscattered power contained in some portion of a backscattered ultrasonic signal.

Changes in the material's composition will change the impedance mismatch, resulting in different backscatter characteristics. From a materials perspective, the SAM has been shown to analyze changes in material compositions, interstitial fractures, or densities within the surface and/or subsurface regions of the scanned object. From a biological perspective, we can quantify these changes and correlate them to the development of the cellular changes and production of any cellular products which result from the differentiation and evolution of the cells [5, 16].

The MATLAB commands for applying AIB to the image surfaces are exemplified in Template 5.

5. Conclusions and additional applications of ultrasound microscopy

This chapter provides a general overview of acoustic imaging techniques at the microscopic level: its uniqueness as an imaging modality along with its benefits and limitations. It also compares its principal features and differences between conventional ultrasound and light microscopy. Ideally, SAM will evolve to become a standard biomedical tool within laboratory settings—both research and clinical—as with other currently available microscopy systems.

Because SAM requires no preparatory work on the specimens, recent studies have been performed to image cells and tissues under aseptic conditions to effectively monitor their growth and differentiation [9]. Other potential applications include real-time imaging of cells and tissues and even *in vivo* monitoring of tissues, provided there are materials and methods to image the selected tissues within the animal or person.

One possible clinical application has been recently examined: utilizing SAM's elastography abilities to test for physical properties in the cornea. This has a strong potential for improving current screening methods in patients electing to undergo LASIK surgery [16]. It may also be utilized to screen for other vision disorders, including glaucoma and keratoconus.

Another major arena of clinical study for SAM is in dermatology [6]. Because skin is an exposed organ, it is relatively easy to noninvasively examine its structures and functions, both *in vitro* and *in vivo*. SAM's elastography applications can be further applied to study strain maps within different skin specimens, to properly assess how the skin ages, what conditions are affecting it, and detecting healthy skin from pathological.

Differentiating normal versus affected conditions is a major advantage of SAM, especially in examining changes in elastic properties of cells and tissues. A recent study examined stiffness properties in breast tumors and correlating them to degrees of metastasis [17]. This and the aforementioned work on cell growth/development, cornea analyses for LASIK, and skin imaging clearly shows the potential of acoustic microscopy as a major instrument of detection and analyses in both biomedical research and possibly even in clinical patient care.

Appendices

Appendix 1:
List of MATLAB templates used in image processing

1. Producing the wavefield plot

fid = fopen('>filename<', 'r', 'b')

fread 'int32'

256k 9800 double

On the wavefield plot, the horizontal lines mean there is system noise

figure plot rf

Focal zone: set between 100 and 150 and pick area without ultrasound signal denoise ('>filename w/prefix.data<')

(This averages across 1000 signals and subtracts out lower level signals, eliminating much of the data associated with the horizontal lines.)

2. Prefilter "oval filter" used to remove the "side noises"—temporal and spatial frequency filters oval (applies to 2D custom filter on 3D dataset)

prefilter2('filename(n)'); the n at the end stands for denoise

(For this prefilter command, the filter is customized to the selected transducer's specifications. If a different transducer (with different specifications) is used, the new filter customized to the new transducer is needed.)

append 'f' onto the file name

(Which will now be suffixed as filename.nf)

3. Refile2

(Puts the scanned data in order: either xzy and yzx (z-axis depth always in the center)

4. Merge6x (or merge6y)

(Corrects any limit on the signal magnitude. If the signal is bad, then will ignore the correction. Typical use of a correction magnitude is 0.75. After calculating correlations, this will put out a baseband file (labeled 'bb') at the end in a transverse–vertical slice. In a 3D dataset: will compose each of the slices (there are 70, each approx. 15 m thick))

Template 1: Generating standard B-Mode images

```
pos=ithreshold(ab,40,1:300,0);

[xn,yn]=size(pos);

[x1,y1]=meshgrid(1:yn,1:xn);

coeff=planefit(x1,y1,pos);

tilt=coeff(1)*x1+coeff(2)*y1+coeff(3);

pos2=pos-tilt;

figure;imagescale(pos2);colormap jet;axis image;

rms1 = ((sum(sum(pos2.^2)))/(xn*yn)).^0.5

rms2 = 1480*rms1/(2*250)
```

Template 2: Generating C-Scan images

```
for nf = 1:10

bb = merge_load6('>Enter Filename Here<',nf,0,69,0);

ab = 20*log10(abs(bb));

ab1 = ab(:,0:69,0:69);

mask1 = threshold(ab1,70,20:500,0);

p1(nf) = max(max(ithreshold(ab1,70,10:500,0)));

mask2 = threshold(ab1,60,1:700,1);
```

```
ab2 = mask1.*mask2.*ab1;

view3d(ab2,1);

set(gca,'CLim',[90-60,90]);

fr(nf) = getframe(gcf);

clf;

end
```

Template 3: Commands for developing 3D images

```
bb = merge_load6('>Enter Filename Here<'nf,0,44,0);

ab = 20*log10(abs(bb));

ab1 = ab(:,2:42,2:42);

mask1 = threshold(ab1,70,20:500,0);

p1(nf) = max(max(ithreshold(ab1,70,10:500,0)));

mask2 = threshold(ab1,60,1:700,1);

ab2 = mask1.*mask2.*ab1;

view3d(ab2,1);

set(gca,'CLim',[90-60,90]);

fr(nf) = getframe(gcf);

clf;

end

movie2avi(10)

figure;

bb = merge_load6('>Enter Filename Here<',nf,0,69,0);

ab = 20*log10(abs(bb));

ab1 = ab(:,2:68,2:68);

mask1 = threshold(ab1,55,200:325,0);

mask2 = threshold(ab1,65,1:300,1); 30

ab2 = mask1.*ab1;

view3d(ab2,1);

set(gca,'CLim',[75-60,75]);
```

```
fr(nf) = getframe(gcf);

clf;

M(nf) = getframe(gcf);

end

fr2 = flipdim(fr,2);

fr3 = [fr fr2];

movie2avi(fr,' >Enter Filename Here<,'compression','none');

movie2avi(fr3, >Enter Filename Here<','compression','none');

movie(M,10)
```

Template 4: Generating 3D movies in MATLAB

```
/* load merged basebanded signals */

bb = merge_load6('>Enter Filename Here<',1,0,69,0);

/* convert to signal magnitude in dB */

ab = 20*log10(abs(bb));

/* find axial (time) position of surface from thresholding */

pos = ithreshold(ab,30,1:300,0);

/* calculate apparent integrated backscatter */

/* import reference spectrum (sp7) with workspace function */

ib1 = aibs(bb,sp7,pos,76,108:146);

/* subtract amplifier difference between tissue and reference */

ib1 = ib1 - 27;

/* calculate mean aibs over entire surface */

ibm = mean(ib1(:));

/* load merged basebanded signals */

bb = merge_load6('>Enter Filename Here<',1,0,69,0);

/* convert to signal magnitude in dB */

ab = 20*log10(abs(bb));
```

Template 5: Commands for apparent integrated backscatter

Author details

Frank Winterroth

Address all correspondence to: fwinterr@umich.edu

Department of Biomedical Engineering, University of Michigan, Ann Arbor, MI, USA

References

[1]   Martin, K. Introduction to B-Mode imaging. In: Hoskins, P., Martin, K., Thrush, A., eds. *Diagnostic Ultrasound: Physics and Equipment*, 2nd ed. 2010. Cambridge: Cambridge University Press. pp. 11–30.

[2]   Szabo, T. L. *Diagnostic Ultrasound Imaging: Inside Out*. 2nd ed. Boston: Elsevier Science. January 2014. 830 pp.

[3]   Czarnota, G.J., Kolios, M.C., Vaziri, H., Benchimol, S., Ottensmeyer, F.P., Sherar, M.D., Hunt, J.W. Ultrasonic biomicroscopy of viable, dead, and apoptotic cells. *Ultrasound in Medicine and Biology*. 1997. 23:961–965.

[4]   Somekh M.G., Bertoni H.L., Briggs G.A.D., Burton N.J. A two-dimensional imaging theory of surface discontinuities with the scanning acoustic microscope. *Proceedings of the Royal Society of London Series A: Math and Physical Sciences*. 1985. 401: 29–51.

[5]   Matsuyama, T. Valantine, H.A., Gibbons, R., Schnittger, I., Popp, R.L. Serial measurement of integrated ultrasonic backscatter in human cardiac allografts for the recognition of acute rejection. *Circulation* 1990. 81: 829–839.

[6]   Dill-Müller, D., Maschke, J. Ultrasonography in Dermatology. *Journal der Deutschen Dermatologischen Gesellschaft*. 2007. 5:689–707.

[7]   Kolios, M.C. Czarnota, G.J., Lee, M., Hunt, J.W., Sherar, M.D. Ultrasonic spectral parameter characterization of apoptosis. *Ultrasound in Medicine and Biology*. 2002. 28: 589–597.

[8]   Kolios, M.C., Taggart, L., Baddour, R.E., Foster, F.S.., Hunt, J.W., Czarnota, G.J., Sherar, M.D. An investigation of backscatter power spectra from cells, cell pellets, and microspheres. *IEEE Ultrasonics Symposium*. 2003. 752–757.

[9]   Winterroth, F. Kato, H., Kuo, S., Feinberg, S.E., Hollister, S.J., Fowlkes, J.B., Hollman, K.W. High frequency ultrasonic imaging of growth and development in manufactured engineered oral mucosal tissue surfaces. *Ultrasound in Medicine and Biology*. 2014. 40(9): 2244–2251.

[10] Ophir, J., Céspides, I., Ponnekanti, H., Li, X. Elastography: a quantitative method for imaging the elasticity of biological tissues. *Ultrasonic Imaging*. 1991. 13(2): 111–134.

[11] Rognin N., et al. Molecular ultrasound imaging enhancement by volumic acoustic radiation force (VARF): pre-clinical in vivo validation in a murine tumor model. *World Molecular Imaging Congress*. 2013 Savannah, GA, USA.

[12] Sarvazyan, A., Hall, T.J., Urban, M.W., Fatemi, M., Aglyamov, S.R., Garra B.S. Overview of elastography – an emerging branch of medical imaging. *Current Medical Imaging Reviews*, 2011, 7(4):255–282.

[13] Wells, P. N. T. Medical ultrasound: imaging of soft tissue strain and elasticity. *Journal of the Royal Society, Interface 8*. 2011. 64: 1521–1549.

[14] Cohn, N.A., Emelianov, S.Y., Lubinski, M.A., O'Donnell, M. An elasticity microscope. Part I: Methods. *IEEE Transactions in Ultrasonics, Feroelectricity and Frequency Control*. 1997. 44(6): 1304–1319.

[15] Cohn, N.A., Emelianov, S.Y., O'Donnell, M. An elasticity microscope. Part II: Experimental results. *IEEE Transactions in Ultrasonics, Feroelectricity and Frequency Control*. 1997. 44(6): 1320–1331.

[16] Hollman, K.W., Emelianov, S.Y., Neiss, J.H., Jotyan, G., Spooner, G.J.R., Juhasz, T., Kurtz, R., O'Donnell, M. Strain imaging of corneal tissue with an ultrasound elasticity microscope. *Cornea*. 2002. 21(1): 68–73.

[17] Fenner, J., Stacer, A.C., Winterroth, F., Johnson, T.D., Luker, K.E., Luker, G.D. Macroscopic stiffness of breast tumors predicts metastasis. *Nature Scientific Reports*. 2014. 4: 5512. doi:10.1038/srep05512.

[18] Briggs, A., Kolosov, O. *Acoustic Microscopy*. 2nd ed. Oxford: Oxford University Press. November 2009. 384 pp.

[19] O'Donnell, M., Bauwens, D., Mimbs, J.W., Miller, J.G, Broadband integrated backscatter: an approach to spatially localized tissue characterization in vivo. *Proceedings from the IEEE Ultrasonics Symposium (New Orleans)* 1979. 79: 175–178.

[20] Parker, K.J., Doyley, M.M., Rubens, D.J. Imaging the elastic properties of tissue: the 20-year perspective. *Physics in Medicine and Biology*. 2011. 56 (2): 513. doi: 10.1088/0031-9155/57/16/5359.